Moringa oleifera

Moringa oleifera is a small plant found around the globe. It is known as a miracle tree due to its diverse medicinal uses. The various parts of the plant – root, stem, leaf, flower, fruit and seed – have nutritive value. This plant is rich in phenolic content, flavonoids, carotenoids, proteins, vitamins, fatty acids and various micro- and macronutrients.

This book highlights the nutritional, pharmaceutical and medicinal values of *Moringa oleifera* and its utilisation in nutraceuticals and functional foods. It also covers *Moringa oleifera*'s nutritional properties, pharmacological importance and biological activities. Special chapters describe the harvesting, handling and storage practices of *Moringa oleifera*, as well as traditional uses of its various parts.

This book reports on the significance of *Moringa oleifera* in agriculture, horticulture, functional foods, cosmetics, clinical nutrition, pharmaceuticals, bacterial fermentation products, biosafety, industrial applications and toxicology, and also discusses any potential side effects.

The book covers the following topics:

- The phytochemistry of *Moringa oleifera*.
- The active constituents and nutritional effects of *Moringa oleifera*.
- *Moringa oleifera*'s phytochemicals, products and extracts.
- The importance of *Moringa oleifera* in agriculture, horticulture and functional foods.
- The applications of *Moringa oleifera* in cosmetics, clinical nutrition and pharmaceuticals.

Moringa oleifera

Therapeutics, Phytochemistry and Its Nutritional Importance

Edited By
Nagendra Singh Chauhan, Durgesh Nandini Chauhan,
Kamal Shah and Umesh Kumar Patil

CRC Press
Taylor & Francis Group
Boca Raton London New York

CRC Press is an imprint of the
Taylor & Francis Group, an **informa** business

Designed cover image: Shutterstock ID: 2589022297

First edition published 2027
by CRC Press
2385 NW Executive Center Drive, Suite 320, Boca Raton FL 33431

and by CRC Press
4 Park Square, Milton Park, Abingdon, Oxon, OX14 4RN

CRC Press is an imprint of Taylor & Francis Group, LLC

Library of Congress Cataloging-in-Publication Data
Names: Chauhan, Nagendra Singh editor | Chauhan, Durgesh Nandini editor | Shah, Kamal editor | Patil, Umesh editor
Title: Moringa oleifera: therapeutics, phytochemical and its nutritional importance / edited by Nagendra Singh Chauhan, Durgesh Nandini Chauhan, Kamal Shah, and Umesh K. Patil.
Description: First edition. | Boca Raton: CRC Press, 2026. | Includes bibliographical references and index. |
Identifiers: LCCN 2026010505 (print) | LCCN 2026010506 (ebook) | ISBN 9781041021407 hbk | ISBN 9781041021421 pbk | ISBN 9781003617860 ebk
Subjects: LCSH: Moringa oleifera–Therapeutic use | Moringa oleifera–Nutritional aspects | Moringa oleifera–Properties
Classification: LCC RS165.M64 M67 2026 (print) | LCC RS165.M64 (ebook)
LC record available at https://lccn.loc.gov/2026010505
LC ebook record available at https://lccn.loc.gov/2026010506

ISBN: 9781041021407 (hbk)
ISBN: 9781041021421 (pbk)
ISBN: 9781003617860 (ebk)

DOI: 10.1201/9781003617860

Typeset in Times New Roman
by Deanta Global Publishing Services, Chennai, India

Contents

PART 2 Phytochemistry, Technology and Applications

PART 3 Formulation Development

About the Editors

Nagendra Singh Chauhan obtained his MPharm and PhD from the Department of Pharmaceutical Sciences, Dr Hari Singh Gour University, Sagar, in 2006 and 2011, respectively. He is currently working as Senior Scientific Officer Grade II and Government Analyst at the Drugs Testing Laboratory Avam Anusandhan Kendra, Raipur, Chhattisgarh, India. He has professional expertise in natural product isolation and phytopharmacology. He has written more than 75 articles in national and international journals and 32 book chapters, and has edited 5 books. He has more than 3000 citations, with an h-index of 42, an i10 index of 38 (Google Scholar) and a Scopus h-index of 15 (669 citations).

Durgesh Nandini Chauhan, MPharm, has several years of academic (teaching) experience at institutes in India in pharmaceutical sciences and has worked at Columbia College of Pharmacy, Raipur, India. She taught subjects such as pharmaceutics, pharmacognosy, traditional concepts of medicinal plants, drug delivery, phytochemistry, cosmetic technology, pharmaceutical engineering, pharmaceutical packaging, quality assurance, dosage form design and anatomy and physiology. She is a member of the Association of Pharmaceutical Teachers of India, SILAE: Società Italo-Latinoamericana di Etnomedicina (The Scientific Network on Ethnomedicine, Italy) and others. She has attended several workshops, conferences and symposiums, including the AICTE-Sponsored Staff Development Programme on 'Effects of Teaching and Learning Skills in Pharmacy: Tool for Improvement of Young Pharmacy Teachers'. She has written more than 10 articles published in national and international journals and 15 book chapters, and has served as an editor for 10 books. She is also active as a reviewer for several international scientific journals and a participant at national and international conferences, including Bharatiya Vigyan Sammelan and the International Convention of the Society of Pharmacognosy. She joined Ishita Research Organisation, Raipur, India, as a freelance writer and guide for students in pharmacy, Ayurveda and science in their research projects. She is interested in data analysis, proposal writing, questionnaire design and research and policy. She is actively engaged in accurately monitoring, investigating and analysing statistical information, professionally presenting information and reports and investigating and documenting key data and information at the request of senior researchers.

Kamal Shah has more than 18 years of research and teaching experience and is currently working as a Professor at the Institute of Pharmaceutical Research, GLA University, Mathura, India. He completed his BPharm from Rajiv Gandhi Proudyogiki Vishwavidyalaya, Bhopal, in 2003. He was a gold medallist in BPharm and MPharm from the Department of Pharmaceutical Sciences, Sagar University, Sagar, India, and obtained his PhD from APJ Abdul Kalam University, Lucknow, India. He has written more than 90 articles in national and international journals and 20 book chapters, and has edited 5 books. His Scopus h-index is 17, with 905 citations.

Umesh K. Patil completed his BPharm, MPharm and PhD from Dr Hari Singh Gour University, Sagar, and post-doctoral studies from the Institute of Biology, Leiden University, the Netherlands. He is currently Professor of Pharmacognosy at Hari Singh Gour University, Sagar, and serves as General Secretary to the Indian Society of Pharmacognosy (ISP). He has also served as a board member and Vice President of the International Council for Medicinal Plant Research (ICMAP) and Regional Representative of ICMAP for SAARC countries. His area of specialisation is herbal drug technology and natural products. He has more than 22 years of experience in teaching and research. He is the recipient of seven prestigious awards given in the field of HMPs and remains a BOYSCAST fellow of DST, the Government of India. He is a member of many professional bodies, including the International Pharmaceutical Federation (FIP), the International Society of

Ethnopharmacology (ISE), the International Society for Medicinal Plant Research (GA), Germany, the International Pharmaceutical Students Federation, the International Society of Natural Product Development, APTI, IPA, IPGA, ISTE, IASTAM and IAHP. He has published 118 papers in reputable journals and serves on the editorial boards of many journals. Professor Patil has made 28 foreign visits to countries including Spain, Switzerland, the Netherlands, Belgium, Turkey, South Africa, Germany, France, Thailand, Sri Lanka, Malaysia, the USA and Australia for academic and scientific assignments.

Preface

This book highlights the nutritional, pharmaceutical and medicinal aspects of *Moringa oleifera* and their utilisation as nutraceuticals and functional foods. It covers the current importance of *M. oleifera*'s nutritional properties, pharmacology and biological activities. This book integrates ancient knowledge with current research to highlight the importance of *M. oleifera*. It is a precious natural medicine and a potential lead for future therapeutic advances. *M. oleifera* stands as a real wonder of the enduring usefulness of ancient wisdom in addressing worldwide health challenges. Due to its rich nutritional composition, it is regarded as one of the most effective solutions for combating malnutrition. The book comprises 21 chapters, further categorised into four broad categories: nutritional and pharmacological activity; phytochemistry, technology and applications; formulation development; and crop production and agricultural practices. The objective of this book is to explore every aspect of *M. oleifera*.

The first part of the book focuses on the nutritional and pharmacological activity of *M. oleifera*. This category comprises nine chapters. The first chapter emphasises moringa's nutrient profile, encompassing bioactive compounds with their related health benefits and diversified uses. This approach encourages farmers to choose *M. oleifera* for improved economic prospects.

Chapter 2 shows the potential of *M. oleifera* in the treatment of inflammatory illnesses. The inflammatory response involves the release of numerous inflammatory mediators. Moringa's anti-inflammatory action is primarily based on the suppression of pro-inflammatory cytokine production.

Chapter 3 covers the anticancer properties of *M. oleifera*. Chapter 4 explores the immunomodulatory potential of *M. oleifera*, exploring both its phytochemical composition and experimental evidence from in vitro and in vivo studies. Chapter 5 provides an overview of the immunomodulatory effects of *M. oleifera*, its bioactive compounds, mechanisms of action and future therapeutic applications. Chapter 6 focuses on the application of *M. oleifera* as an alternative therapy to synthetic drugs for the management of diabetes. The chapter further explores preclinical and clinical studies on the antidiabetic properties of moringa as well as its possible mechanism of action. Chapter 7 discusses the ancient and modern applications of moringa in traditional medicine, focusing on its abundant phytochemicals and pharmacological actions. Apart from traditional uses of moringa, the plant is also used in oral health care, which is covered in Chapter 8. Chapter 9 emphasises the anticoagulant and thrombolytic properties of this miracle plant.

Another broad category of the book is based on the phytochemistry, technology and applications of *M. oleifera*. This part has seven chapters. These chapters are written based on the plant parts used. The tenth chapter highlights the potential of *M. oleifera* leaf-based products. Chapter 11 underscores the transformative potential of *M. oleifera* fruits in improving global nutrition, health and sustainability. Chapter 12 further explores the cross-disciplinary exploration of this underutilised floral resource. The thirteenth chapter provides a comprehensive overview of *M. oleifera*'s chemistry, extraction processes and applications, emphasising its role as a natural resource for health and environmental solutions. Chapter 14 comprises a comprehensive analytical profiling of *M. oleifera* and its characterisation techniques, such as UV-vis spectrophotometry, high-performance thin-layer chromatography (HPTLC), high-performance liquid chromatography (HPLC), Fourier transform infrared (FTIR) spectroscopy, nuclear magnetic resonance (NMR) and gas chromatography-mass spectrometry (GC-MS). Chapter 15 highlights information regarding extraction methods, physicochemical properties and pharmacological activities of *M. oleifera*'s seed oil. Chapter 16 comprehensively covers the nutritional value, consumer applications, global utilisation and future potential of *M. oleifera* as a functional dietary superfood. The final two chapters of this part discuss the seed of *M. Oleifera*, covering extraction methods, characterisation, physicochemical evaluation and regulatory aspects. The final chapter of this part highlights the role of *M. Oleifera* as a dietary superfood.

The third part of this book discusses the formulation-associated variables of *M. oleifera*, and it has three chapters. These chapters highlight its use as a nutraceutical in drug delivery systems and as a pharmaceutical excipient.

The last part of the book has two chapters; these chapters are focused on agricultural and post-harvesting aspects. The chapters discuss strategies regarding cultivation, collection and post-harvesting aspects.

The editors are thankful to their family members, friends and fellow scientists; without their support, it would not have been possible to complete this book. We also acknowledge the support of the publishers and editorial team, whose guidance and cooperation were instrumental in shaping this volume into its final form.

Finally, we dedicate this book to countless researchers, innovators and clinicians whose continued pursuit of excellence in herbal drugs and phytochemistry improves lives and inspires future generations.

Nagendra Singh Chauhan, Durgesh Nandini Chauhan,
Kamal Shah and Umesh K. Patil

Contributors

J. Abhyudaya
Academy of Scientific and Innovative Research (AcSIR), Food Safety and Analytical Quality Control Laboratory
Central Food Technological Research Institute (CSIR)
Mysore, India

Jonathan Ilemona Achika
Department of Chemistry
Federal University Lokoja
Lokoja, Nigeria

Feyisayo O. Adepoju
Department of Technology for Organic Synthesis
Institute of Chemical Technology
Ural Federal University
Yekaterinburg, Russia

Liakat Ali
Department of Genetics and Plant Breeding, Faculty of Agriculture
Gazipur Agricultural University
Gazipur, Bangladesh

Kamal Krishan Aggarwal
University School of Biotechnology
GGS Indraprastha University
New Delhi, India

Abhishek Ambawatiya
Pharmacology and Toxicology Department
National Institute of Pharmaceutical Education and Research (NIPER)
Hajipur, India

Anurag
Department of Pharmaceutical Technology
MIET
Meerut, India

Muhammad Hammad Anwar
Department of Food Science and Technology
Muhammad Nawaz Shareef University of Agriculture
Multan (60000), Pakistan

Muhammad Junaid Anwar
Department of Food Science and Technology, Faculty of Food Science and Nutrition
Bahauddin Zakariya University
Multan (60800), Punjab, Pakistan

Joy Cecilia Atawodi
Department of Zoology
Federal University Lokoja
Lokoja, Nigeria

Sunday Ene-Ojo Atawodi
Department of Biochemistry
Federal University Lokoja
Lokoja, Nigeria

Adarsha Bachhar
Department of Pharmaceutical Technology
Jadavpur University
Kolkata, India

Kamrun Sanjana Badhon
Department of Horticulture, Faculty of Agriculture
Gazipur Agricultural University
Gazipur, Bangladesh

Marwa Samy Abu Bakr
Department of Pharmacognosy, Faculty of Pharmacy
Al-Azhar University
Cairo, Egypt

Murali Balasubramaniam A.
Department of Oral Pathology and Microbiology
Sathyabama Dental College and Hospital
Chennai, India

Shraddha B. Bhatt
Department of Biochemistry/Biotechnology, Food Testing Laboratory
Junagadh Agricultural University
Junagadh, India

Priya Bisht
Pharmacology and Toxicology Department
National Institute of Pharmaceutical Education and Research (NIPER)
Hajipur, India

Ana R. N. Campos
Federal University of Campina Grande
Academic Unit of Engineering
Campina Grande, Brazil

Nagendra Singh Chauhan
Drugs Testing Laboratory
Ayam Anusandhan Kendra
Raipur, India

Pavan Ramrao Chavan
Pharmacology and Toxicology Department
National Institute of Pharmaceutical Education and Research (NIPER)
Hajipur, India

Shalini Choudhary
Pharmacology and Toxicology Department
National Institute of Pharmaceutical Education and Research (NIPER)
Hajipur, India

Danilo L. Dantas
Department of Chemistry
Federal Rural University of Pernambuco
Recife, Brazil

Reham S. Darwish
Department of Pharmacognosy, Faculty of Pharmacy
Alexandria University
Alexandria, Egypt

Ushasi Das
Department of Pharmaceutical Technology
Jadavpur University
Kolkata, India

Hend M. Dawood
Department of Pharmacognosy, Faculty of Pharmacy
Alexandria University
Alexandria, Egypt

Omayma A. Eldahshan
Pharmacognosy Department, Faculty of Pharmacy
Ain Shams University
Cairo, Egypt

Rawah H. Elkousy
Pharmacognosy Department, Faculty of Pharmacy
Al-Azhar University
Cairo, Egypt

Mohamed A. Farag
Pharmacognosy Department, Faculty of Pharmacy
Cairo University
Cairo, Egypt

Juliano C. R. Freitas
Federal University of Campina Grande
Center for Education and Health
Campina Grande, Brazil

P. Giridhar
Academy of Scientific and Innovative Research (AcSIR), Plant Cell Biotechnology Department
Central Food Technological Research Institute (CSIR)
Mysore, India

Sanjay Kumar Gupta
Department of Pharmacology
Rungta College of Pharmaceutical Sciences and Research, Rungta International Skills University
Bhilai, India

Aneela Hameed
Department of Animal Food Products Technology
Faculty of Food Science and Nutrition, Bahauddin Zakariya University
Multan (60800), Punjab, Pakistan

Saqib Hameed
Pharmacology and Toxicology Department
National Institute of Pharmaceutical Education and Research (NIPER)
Hajipur, India

Jahidul Hassan
Department of Horticulture, Faculty of Agriculture
Gazipur Agricultural University
Gazipur, Bangladesh

Phong T. Huynh
Department of Food Technology, Faculty of Applied Science and Technology
Nguyen Tat Thanh University
Ho Chi Minh City, Vietnam

Asmaa Ayman Saber Ibrahim
IV Mixing Unit,
National Cancer Institute
Cairo University
Cairo, Egypt

Reham S. Ibrahim
Department of Pharmacognosy, Faculty of Pharmacy
Alexandria University
Alexandria, Egypt

Gunawan Indrayanto
School of Medicine
Ciputra University
Surabaya, Indonesia

Muhammad Irfan
Department of Biochemistry and Biotechnology
University of Gujrat
Gujrat, Pakistan

Md. Saiful Islam
Department of Horticulture, Faculty of Agriculture
Gazipur Agricultural University
Gazipur, Bangladesh

Sofia Javed
Department of Biochemistry and Biotechnology
University of Gujrat
Gujrat, Pakistan

Indu Bharkavi S. K.
Department of Oral Pathology and Microbiology
Sathyabama Dental College and Hospital
Chennai, India

K. Kamireddy
Academy of Scientific and Innovative Research (AcSIR), Plant Cell Biotechnology Department
Central Food Technological Research Institute (CSIR)
Mysore, India

P. Karuppasamy
Department of Chemistry
Dayananda Sagar College of Engineering
Bangalore, India

Nabeeha Koya
Pharmacology and Toxicology Department
National Institute of Pharmaceutical Education and Research (NIPER)
Hajipur, India

Sitaraman Krishnan
Department of Chemical and Biomolecular Engineering
Clarkson University
Potsdam, USA

Nitesh Kumar
Pharmacology and Toxicology Department
National Institute of Pharmaceutical Education and Research (NIPER)
Hajipur, India

Pravin Kumar
Pharmacology and Toxicology Department
National Institute of Pharmaceutical Education and Research (NIPER)
Hajipur, India

Ngoc-Tram Le
Department of Chemical and Biomolecular Engineering
Clarkson University
Potsdam, USA

Sanchita Mandal
Department of Pharmaceutical Technology
Jadavpur University
Kolkata, India

Aimen Mazhar
Department of Human Nutrition
Faculty of Food Science and Nutrition, Bahauddin Zakariya University
Multan (60800), Punjab, Pakistan

Nourhan Ibrahim Morad
Department of Pharmacognosy and Natural Products, Faculty of Pharmacy
Menofia University
Menofia, Egypt

Farwa Nasir
Department of Biochemistry and Biotechnology
University of Gujrat
Gujrat, Pakistan

Khaled A. Nematallah
Department of Pharmacognosy, Faculty of Pharmacy
The British University in Egypt
Cairo, Egypt

Emmanuel Chukwudalu Ohaekenyem
Pure and Industrial Chemistry Department
Nnamdi Azikiwe University
Awka, Nigeria

Temitope Orimolade
Department of Chemical and Biomolecular Engineering
Clarkson University
Potsdam, USA

Yukio Ozaki
Laboratory of Horticultural Science, Faculty of Agriculture
Kyushu University
Fukuoka, Japan

Shraddha R. Parmar
Department of Biochemistry/Biotechnology, Food Testing Laboratory
Junagadh Agricultural University
Junagadh, India

Periyakaruppan Pradeepkumar
State Key Laboratory of Molecular Engineering of Polymers, Department of Macromolecular Science
Fudan University
Shanghai, China

P. S. Priyanka
Academy of Scientific and Innovative Research (AcSIR), Plant Cell Biotechnology Department
Central Food Technological Research Institute (CSIR)
Mysore, India

Sahar Qundeel
Department of Food Science and Technology, Faculty of Food Science and Nutrition
Bahauddin Zakariya University
Multan (60800), Punjab, Pakistan

Halida Rahmania
Laboratory of Pharmaceutical Analysis, Department of Pharmaceutical Chemistry, Faculty of Pharmacy
Universitas Gadjah Mada
Yogyakarta, Indonesia

V. Rajapandian
Department of Chemistry
Sri Ramakrishna Mission Vidyalaya College of Arts and Science
Coimbatore, India

Md. Mijanur Rahman Rajib
Department of Horticulture, Faculty of Agriculture
Gazipur Agricultural University
Gazipur, Bangladesh

Chandini Rajkumar
Department of Oral Pathology and Microbiology
Sathyabama Dental College and Hospital
Chennai, India

Saranya Ramsridhar
Department of Oral Pathology and Microbiology
Sathyabama Dental College and Hospital
Chennai, India

Abdul Rohman
Laboratory of Pharmaceutical Analysis, Department of Pharmaceutical Chemistry, Faculty of Pharmacy
Universitas Gadjah Mada
Yogyakarta, Indonesia
and
Indonesia Center of Excellence, Institute of Halal Industry and Systems
Universitas Gadjah Mada
Yogyakarta, Indonesia

Dhiman Roy
Department of Pharmaceutical Technology
Jadavpur University
Kolkata, India
Vibhu Sahani
Department of Pharmacy
Lala Lajpat Rai Memorial Medical College
Meerut, India
Alimuddin Saifi
Department of Pharmaceutical Technology
MIET
Meerut, India

Renato A. C. Santana
Federal University of Campina Grande
Academic Unit of Engineering
Campina Grande, Brazil

Arindam Sarkar
Department of Pharmaceutical Technology
Jadavpur University
Kolkata, India

Sawetaji
University School of Biotechnology
GGS Indraprastha University
New Delhi, India

Jyoti Sehrawat
University School of Biotechnology
GGS Indraprastha University
New Delhi, India

Alok Sharma
Department of Pharmaceutical Technology
MIET
Meerut, India

Neha Sharma
Department of Pharmaceutical Technology
MIET
Meerut, India

Alécia R. A. Silva
Federal Rural University of Pernambuco
Department of Chemistry
Recife, Brazil

Harjeet Singh
Central Ayurveda Research Institute
Patiala, India

Syada Nizer Sultana
College of Agricultural Science
International University of Business Agriculture and Technology (IUBAT)
Dhaka, Bangladesh

Yugo Susanto
Sekolah Tinggi Ilmu Kesehatan (STIKES)
ISFI Banjarmasin
Banjarmasin, Indonesia

Noha Swilam
Department of Pharmacognosy, Faculty of Pharmacy
The British University in Egypt
Cairo, Egypt

Sadia Tasnim
Department of Horticulture, Faculty of Agriculture
Gazipur Agricultural University
Gazipur, Bangladesh

Zainab Yousaf
Department of Biochemistry and Biotechnology
University of Gujrat
Gujrat, Pakistan

Md. Zubayer
Department of Horticulture, Faculty of Agriculture
Gazipur Agricultural University
Gazipur, Bangladesh

Part 1

Nutritional and Pharmacological Activity

1 *Moringa oleifera* as a Potential Source of Food Value Phytoconstituents

Current Trends and Future Prospects

P. S. Priyanka, K. Kamireddy, J. Abhyudaya and P. Giridhar

1.1 INTRODUCTION

Moringa oleifera is often called a miracle tree due to its numerous nutritional benefits and uses in traditional medicine. Due to its rich nutritional composition, it is regarded as one of the most effective solutions for combating malnutrition. This species is widely distributed and used, demonstrating significant potential for producing various products, including food items, medicinal products, nutraceuticals, cosmetics, renewable polymers, fertilisers, fuel wood and water purification systems. Additionally, it can serve as a living fence, be used in alley cropping and provide feed for animals and aquaculture (Gassenschmidt et al., 1995; Ndabigengesere and Narasiah, 1998; Abdulkarim et al., 2005; Sharma et al., 2006; Bhatti et al., 2007; Vissoh et al., 2024; Klimek-Szczykutowicz et al., 2024). Thirteen native species from semi-arid Africa, Madagascar and the Arabian region comprise the Moringaceae family (Namdev and Chauhan, 2023). It is thought to have originated in the southern Indian Himalaya (Makkar and Becker, 1997), and it has since spread to several tropical and sub-tropical nations. This crop is important in India, Ethiopia, the Philippines and Sudan. It is found on the African continent, Tropical Asia, Latin America, the Caribbean, Florida and the Pacific Islands. Genetic diversity exhibited by eight commercially grown Indian cultivars of *M. oleifera* was reported by Saini et al. (2013a).

Moringaceae resembles Capparidaceae (Gassenschmidt et al., 1995) and some Sapindaceae members (Dahlgren, 1989). Moringaceae was reclassified under Violales, near Violaceae, due to similarities in pentamerous zygomorphic flowers and embryological traits (Datta and Mitra, 1947). Based on wood anatomy, it shows a sister relationship with Caricaceae (Olson, 2002). Moringa (Moringaceae) is classified as such, despite its similarities to other plant families. Moringa trees exhibit distinct morphological differences, which allow them to be categorised into four classes: bottle trees, tuberous shrubs, lean trees and sarcorhizal trees. The first three classes show anatomical homogeneity, whereas the last class is markedly different in anatomy despite habitat similarity. Bottle trees, distinguished by massive trunks and swollen roots, are mainly found in Africa and Madagascar. They include *M. drouhardii, M. hildebrandtii, M. stenopetala* and *M. ovalifolia.* Tuberous shrubs, which include *M. borziana, M. longituba, M. pygmaea* and *M. rivae*, are characterised by roots having a diameter greater than the stem. The third class is of slender trees found in India and Arabia, which includes *M. concanensis, M. oleifera* and *M. peregrina.* Sarcorhizal trees, defined by their slender trunks, smooth bark and thick, soft, fleshy and somewhat twisted roots, include *M. arborea* and *M. ruspoliana* (Olson, 2002). *M. oleifera* is the most widely distributed species of the *Moringa* genus, and its significance is reflected in the various names attributed to it across

DOI: 10.1201/9781003617860-2

different countries and regions (Bhuptawat et al., 2007). *M. oleifera* is named 'Shigon' in one Indian traditional medicinal text, 'Shushruta Samhita', which was likely to be composed between the third and fourth century BCE (Muluvi et al., 1999). It is generally known as horseradish tree, drumstick tree, never die tree, West Indian Ben tree, radish tree, kelor, marango, mlonge, mulangay, saijhan, sajna, shagara al rauwaq, moonga, nebeday and ben oil tree (Ramachandran et al., 1980; Anwar and Bhanger, 2003; Abdulkarim et al., 2005; Fahey, 2005; Anwar et al., 2007). In India, various vernacular names have been given to this tree, such as Morunga (Dravidian), Munaga or Saijhan (Hindi), Nuggekayee (Kannada), Shevga (Marathi), Sajana or Sujuna (Oriya), Morigkai (Tamil), Sohanjna (Sanskrit), Mulaga (Telugu), Muringa (Malayalam), Saragvo (Gujarati) and so on. *M. oleifera* and *M. concanensis* are found in India. *M. concanensis* is a small tree that grows widely in Goa, Tamil Nadu, Madhya Pradesh, Maharashtra, Andhra Pradesh, Gujarat and Rajasthan (Singh et al., 2000).

Moringa oleifera thrives in various climates, including hot and humid, semi-arid and wet conditions. Moringa requires a yearly rainfall of around 250–3000 mm. It is relatively resistant to drought for long periods, but its foliage production suffers. It responds well to irrigation. It grows well in temperatures between 25°C and 40°C and high solar radiation. Mature plants can survive at low temperatures for short durations. It grows at altitudes from sea level to 1800 m. *M. oleifera* will adapt to a wide range of pH values, but well-drained or loamy soil with neutral to slightly acidic pH is optimum for its growth and development. It grows well in heavy soil conditions and in the presence of proper irrigation and fertilisation (Radovich, 2009).

Moringa shows considerable morphological variations in growth rate, flowering time, leaflet size, shape, fruit length and other characteristics (Radovich, 2009). Propagation is through seeds and stem cuttings. It is highly cross-pollinating, leading to variability in seeds. Seed propagation begins in the nursery, and after two to three months, seedlings are transferred to the fields. However, propagation through stem cuttings (1–2 m long) is preferred in India to avoid variability due to cross-pollination. Plants propagated through cuttings are often preferred over seeds, which are reportedly slower to bloom and produce fruits of inferior quality (Ramachandran et al., 1980). In northern and eastern India, leaves are shed between December and January, with new leaves appearing by March and May, followed by flowering and fruiting in April and June. In the southern states, fruits and flowers are produced year-round, with peak seasons from June to September and March to April (The Wealth of India, 1962). Trees are often effectively pollarded to enhance fruit and fodder production. This practice helps limit the spread of the crown and promotes healthy new branch growth, contributing to more productive yields (Ramachandran et al., 1980).

To date, many reviews have been published discussing limited dimensions. However, this chapter is a comprehensive review of the plant's nutrient profile, various edible parts and its use as a food plant, its different bioactive compounds, its utility as a nutraceutical and functional food and its prospects encompassing various methods to enhance its nutrient profile.

1.2 BOTANICAL DESCRIPTION

Moringa oleifera is a deciduous perennial tree with a chromosome number of 2n = 28 (Pandey et al., 2011). The typical height reached at maturity generally ranges from 7 to 12 metres, with a diameter of 20–40 cm (Aregheore, 2002; Foidl and Paull, 2008). The plant usually has a single, straight stem, but sometimes the stem can be feebly formed. It has an umbrella-shaped canopy due to disorderly extended drooping branches. Leaves are 20–70 cm long, petiolated with eight to ten pairs of pinnae, each having two sets of opposite, elliptic or obovate leaflets and one at the tip. Two or three alternate leaves mostly grow at the branch tips. Flowers are around 2.5 cm wide, white or cream coloured with yellow dots at the base and are profusely produced in axillary drooping panicles 10–25 cm long. The flower comprises five linear-lanceolate reflexed sepals with five slender petals surrounding five yellow stamens alternating with five staminodes. Fruits 20–60 cm long have a pointed apex with tapering bases and hang down from branches. It has nine ribs and three-lobed pods, each containing 12–35 seeds surrounded by a pith. Seeds are round, ivory-white and brown in colour, with thin deciduous wings. The approximate weight of the seed is 0.3 g, and the kernel-to-hull ratio is 75:25. Every tree produces between 15,000 and 25,000 seeds each year.

1.3 YIELD

Moringa oleifera has a greater yield in warm, dry conditions, with little supplementation of fertilisers and irrigation. *Moringa oleifera* yields can vary based on several factors, including the season, the variety of the plant, the fertilisation methods used and the irrigation practices in place. Home gardeners use beds with 10 cm × 10 cm (4 in × 4 in) spacing for intensive leaf production, but for small fields, 0.75 m × 1 m spacing is possibly the best option. In India, approximately 1.2 million metric tons of immature pods are produced each year across 38,000 hectares (or 94,000 acres) on a commercial scale, with post-harvest handling and processing classified as a large-scale industry (Radovich, 2009).

1.3.1 Pods

In India, each tree yields about 19 kg (42 lb) of pods per year, which corresponds to 31,000 kg/ha per year at 2.5 m × 2.5 m spacing. Each tree bears approximately 230 pods annually, each weighing 80–90 g (2.8–3.2 oz). For optimal pod production, it is suggested to space the plants 2.5 metres apart in both directions, resulting in a spacing of 2.5 m × 2.5 m (Radovich, 2009).

1.3.2 Leaves

Each tree produces leaves (fw) weighing approximately 1–5 kg (2.2–11 lb) annually. This results in a total of 10,000–50,000 kg per hectare each year at a spacing of 1 m × 1 m. The leaf yield of *M. oleifera* sees a notable drop from November to March, reducing to 50–100% of summer levels due to lower sunlight during the winter months. In optimally irrigated, well-drained and fertilised beds with a spacing of 10 cm × 10 cm, moringa leaf production can attain 7–8 kg/m^2 during the initial harvest, with the possibility of up to seven cuts in a year. A recommended spacing is 0.75 m within rows and 1 m between them. In many cases, a spacing of 5 m × 5 m (16.4 ft × 16.4 ft) is also appropriate (Radovich, 2009).

1.3.3 Oil

Each dry pod produces around 3 g (0.1 oz) of flavourful kernels, with a remarkable oil yield of approximately 20% by weight derived from these kernels, and considering the oil's specific gravity of 0.90, it is anticipated that a single hectare of these trees could yield an impressive 250 litres of rich oil. This equates to about 107 quarts per acre, offering a substantial harvest for those cultivating these vibrant trees (Radovich, 2009).

1.4 NUTRIENT PROFILE

A nutritional analysis of *M. oleifera* indicates its richness in macronutrients, micronutrients and disease-preventing nutrients. All parts of the plant are a good source of vitamins like γ and α tocopherols (Gharsallah et al., 2021), β-carotene, proteins, sulphur-containing amino acids (Makkar and Becker, 1997; Saini et al., 2014a), phenolic compounds, carbohydrates, minerals and unsaturated fatty acids, with low concentrations of anti-nutritional factors like glucosinolates, phytates, haemagglutinins, saponins and tannins (Ferreira et al., 2008; Vergara-Jimenez et al., 2017). Moringa, often referred to as a nutritional powerhouse, is believed to boast an extraordinary array of vitamins and minerals. It contains eight times the vitamin C of fresh oranges, providing a potent boost to the immune system. This remarkable plant also offers vitamin A ten times that of carrots, promoting healthy vision and skin. Furthermore, there are claims that it delivers 17 times the calcium found in milk. In terms of protein, moringa provides nine times the amount in creamy yoghurt, making it an excellent choice for muscle repair and growth. Additionally, it is believed to contain 15 times the potassium of bananas, which is essential for maintaining heart and muscle function. Lastly, moringa stands out for its iron content, boasting 25 times the amount found in leafy spinach, making it a valuable source for those seeking to enhance their iron intake (Rockwood et al., 2013). Because of its nutrient richness, it has been considered an important plant to provide food security and improve the nutritional levels of individuals suffering from malnutrition. Some studies report a decrease in anti-nutritional contents during drying with a simultaneous increase in nutrient content (Table 1.1), proving it to be a good ingredient for processed food. Table 1.2 compares the nutrient profile of *M. oleifera* leaves with spinach with reference to relative dietary allowance, proving its utility as a food plant. Proximate analysis studies of different parts of the moringa tree showed that there is a wide range of variation in nutrient profile, which depends on the geographical conditions, soil composition, cultivars and agronomical conditions.

1.4.1 Oils

Oil extracted from *M. oleifera* is used in various ethnobotanical applications (Yang et al., 2007). All edible parts of *M. oleifera* showed the presence of unsaturated fatty acids (Table 1.3), including linoleic acid, linolenic acid and arachidonic acid (C20:4ω6) (Sánchez-Machado et al., 2004a; Saini et al., 2014b). In leaves, polyunsaturated fatty acids (PUFA) were higher than monounsaturated fatty acids (MUFA), containing a very high quantity of linolenic acid (C18:3ω3), followed by palmitic acid (C16:0), together constituting 80% of the total fatty acids (Table 1.3) (Freiberger et al., 1998; Amaglo et al., 2010; Monteiro et al., 2022). Flowers and immature pods contain saturated and unsaturated fatty acids in a 1:2 ratio. Unsaturated fatty acids like palmitic acid, linoleic acid (C18:2ω6), linolenic acid and oleic acid (C18:1ω9) are present in similar amounts, constituting 90% of the total fatty acids. The quantity of rudimentary fat contents of flowers and pods may differ due to variances in farming settings, as observed in different studies (Sánchez-Machado et al., 2010). Immature pods from different locations show variability in PUFA content (Amaglo et al., 2010). *M. oleifera* seed oil has different FA profiles from other commercial seed oils, showing good health effects as it has little SFA and high MUFA, primarily composed of oleic acid (71.57%), palmitic acid (8.22%), stearic acid (5.25%) and behenic acid (4.15%). Crude fat was also found to be highest in seeds (Amaglo et al., 2010; Athikomkulchai et al., 2020).

1.4.2 Essential Oils and Volatile Compounds

Essential oils (EOs) are used to relieve pain, promote healing and act as antimicrobial, antifungal, antioxidant and antiviral agents (Sibanda et al., 2004; Chaieb et al., 2007; Ebrahimi et al., 2008). Along with health benefits, the volatile fraction of oil can also be used as an aromatising agent. A total of 29 components were identified from leaf oil, of which 91.1% were found to be hydrocarbons, and the remaining amount was made up of oxygenated monoterpenes, phenols, oxygenated sesquiterpenes, hexenyl propanoate, phenylethyl alcohol and pseudo phytol. Biological activities, viz., antimicrobial, antifungal and low free radical scavenging activity were reported (Marrufo et al., 2013). In EOs derived from flowers, alkanes constitute the predominant chemical group, making up an impressive 84% of the total composition. Within this category, n-heptacosane emerges as the most significant compound, accounting for 36% of flower EOs, closely followed by n-nonacosane, which comprises 29%.

The chemical profile shifts slightly in leaf EOs, with alkanes (41%) and fatty acids (38%) forming the two leading categories. Noteworthy among these are palmitic acid, which represents a substantial

TABLE 1.1
Effects of Different Drying Methods on the Nutrient and Anti-Nutrient Content of *M. oleifera* Leaves

Nutrient	After sun drying	Increase/ decrease in content	After shade drying	Increase/ decrease in content	After oven drying	Increase/ decrease in content	After cabinet drying	Increase/ decrease in content	Fresh weight
Moisture *Decrease*	6.33	65.59	6.55	65.37	5.12	66.8	5.3	66.62	71.92
Protein *Increase*	21.67	13.22	33.02	24.57	25.56	17.11	32.60	24.15	8.45
Fat *Decrease*	4.345	8.005	5.46	6.89	2.115	10.235	9.85	2.5	12.35
Ash *Increase*	3.05	2.185	1.27	0.405	2.01	1.145	8.70	7.835	0.865
Fibre *Increase*	1.40	0.19	1.53	0.32	1.50	0.29	12.05	10.84	1.21
Carbohydrate *Increase*	63.19	58	52.165	46.97	63.25	58.06	43.55	38.36	5.19
Vitamin B1 (mg/100 g) *Increase*	0.765	0.34	0.87	0.445	0.515	0.09	–	–	0.425
Vitamin A (mg/100 g) *Increase*	2.67	1.145	3.31	1.85	1.975	0.45	–	–	1.525
Iron (mg/100 g) *Decrease*	0.56	1.305	0.39	1.475	0.76	1.105	–	–	1.865
Phosphorus (mg/100 g) *Increase*	2.03	1.33	2.18	1.48	2.15	1.45	–	–	0.7
Calcium (mg/100 g) *Increase*	1.35	0.11	3.075	1.835	2.59	1.35	–	–	1.24

(Continued)

TABLE 1.1 (CONTINUED)
Effects of Different Drying Methods on the Nutrient and Anti-Nutrient Content of *M. oleifera* Leaves

Nutrient	After sun drying	Increase/ decrease in content	After shade drying	Increase/ decrease in content	After oven drying	Increase/ decrease in content	After cabinet drying	Increase/ decrease in content	Fresh weight
Zinc (mg/100 g) *Increase*	0.87	0.31	0.825	0.265	0.63	0.07	–	–	0.56
Tannin (mg/100 g) *Increase*	1.605	1.39	1.61	1.395	1.405	1.19	–	–	0.215
Phytate (mg/100 g) *Decrease*	6.17	3.26	3.9	5.53	3.625	5.805	–	–	9.43
Oxalate (mg/100 g) *Decrease*	3.56	0.135	3.27	0.425	2.795	0.9	–	–	3.695
Saponin (mg/100 g) *Decrease*	0.32	0.555	0.835	0.04	0.49	0.385			0.875

(Joshi and Mehta., 2010; Mbah et al., 2012; Satwase, 2012).

TABLE 1.2
Comparison of the Nutrient Content of *M. oleifera* Leaves with Spinach

Nutrient	Relative dietary allowance	Raw spinach (per 100g) National nutrient database for std ref USDA sr21	*M. oleifera* (per 100g)	Reference
Carbohydrate (g)	682.5 g/d	3.63 g/100 g	13.4 g (fw)	Fuglie, 2001
Fat (g)	30 g/d	0.39 g	1.7 g (fw)	Shindano and Kasase, 2009
Protein (g)	60 g/d	2.86 g	5.7–6.7 g (fw)	Fuglie, 2001
Retinol (ug)	600 ug/d	469 μg	1500 μg (fw)	Yang et al., 2007
Calcium (mg)	600 mg/d	99 mg	346–638 mg (fw)	Aslam, 2005, Yang et al., 2007
Iron (mg)	17 mg/d	2.71 mg	16.9 mg (dw)	Fakankun et al., 2013
Zinc (mg)	12 mg/d	0.53 mg	1.35–24.20 mg (dm)	Foidl and Paull, 2008
Phosphorus (mg)	600mg/d	49 mg	116–136 mg (dm)	Foidl and Paull, 2008
Sodium (mg)	2500 mg/d	79 mg	116–273 mg (dm)	Amaglo et al., 2010
Potassium (mg)	2500 mg/d	558 mg	910–1910 mg (dm)	Adeoye et al., 2005
Manganese (mg)	340 mg	–	4.710–11.390 mg (dm)	Foidl and Paull, 2008
Selenium (ug)	340 ug	–	0.1–2.7 ug (dm)	Amaglo et al., 2010
β-carotene (ug)	4800 ug/d–4,800,000 μg/d	–	7700–19100 μg (fw)	Yang et al., 2007
Thiamine (mg)	1.2 mg/d	0.21 mg	2.64 mg	Leone et al., 2015
Folic acid (mg)	0.2 mg/d–200,000 mg/d	0.194 mg	0.452 mg (dm)	Devi et al., 2008
Niacin (mg)	1.4–18 mg/d	0.8 mg	8.2 mg	Leone et al., 2015
Ascorbic acid (mg)	40 mg/d	28.1 mg	240–250 mg (fw)	Yang et al., 2007
Fibre (g)	40 g/d	2.2 g	1.30–1.99 g (fw)	Yang et al., 2007
Magnesium (mg)		79 mg	390–430 mg (dm)	Amaglo et al., 2010
Copper (mg)		–	4.36–4.40 mg (fw)	Anjorin et al., 2010
γ-tocopherols (mg)		–	0.57–2.780 mg	Sanchez-Machado et al., 2006
α-tocopherols (mg)		2.03 mg	25 mg (fw)	Yang et al., 2007

38% of leaf EOs, and n-nonacosane, at 18%, highlighting the diverse array of compounds present. The chemistry of the seeds reveals a different trend, where fatty acids dominate extensively at 67%, underscoring their importance in seed EOs. The major components include palmitic acid, making up 38% of the total, and oleic acid, which constitutes 29%. This variation in chemical composition across different plant parts signifies the intricate biochemical pathways involved in the production of essential oils, leading to unique profiles across flowers, leaves and seeds (Monteiro et al., 2022). Gas chromatography-mass spectrometry (GC-MS) confirmed the presence of 74 compounds belonging to aliphatic compounds, nitrogen compounds, oxygenated monoterpenes, monoterpene hydrocarbons and oxygenated sesquiterpenes. (E)-nerolidol, α-terpineol and benzyl isothiocyanate were present in a volatile fraction of flowers (Pino, 2013).

TABLE 1.3
Fatty Acid Composition (Percentage of Total Fatty Acids) of the Edible Parts of *M. oleifera*

Fatty acid	Leaves	Immature pods	Flowers
Caprylic acid	0.05	0.24	0.09
Lauric acid	0.12	0.11	0.09
Myristic acid	0.96	0.26	0.59
Myristoleic acid	0.53	0.15	1.96
Palmitic acid	23.28	26.48	23.43
Palmitoleic acid	0.43	nd	nd
Stearic acid	4.08	3.10	4.52
Vaccenic acid	1.15	1.00	2.00
Oleic acid	5.12	17.00	21.55
Linoleic acid	6.11	23.50	18.96
α-linolenic acid	56.87	26.15	23.01
Arachidic acid	0.72	0.39	0.98
Gondoic acid	nd	0.24	0.75
Arachidonic acid	0.21	0.24	0.19
Behenic acid	0.70	1.06	2.06

(Sánchez-Machado et al., 2010)

1.4.3 Minerals

The mineral content of *M. oleifera* and its bioavailability have been important research topics, attracting considerable attention in recent studies (Rajurkar and Damame, 1998; Anjorin et al., 2010). The total K content was very high (as mentioned in Table 1.2), but its availability was very low (Adeoye et al., 2005). The Ca content in the lamina of *M. oleifera* was found to be 3.463 mg g^{-1} (Aslam, 2005), and in mature leaves, it was reported to be around 6.38 mg g^{-1} fw (Yang et al., 2007), which could be a preferable source of Ca to animals or humans. This was found to be much higher than other leaf and seed parts. Seed shells and pods also contain approximately the same Ca content, whereas seed kernel oil had the least Ca content (0.109 mg mL^{-1}) (Anjorin et al., 2010). The Ca content of moringa leaves was found to differ based on agroclimatic conditions (as mentioned in Table 1.1). Mg levels in the leaf lamina and the seed powder have been reported to be the same. They were found to be in the range of 3.463–3.827 mg g^{-1} in leaves and 0.0530–0.185 mg mL^{-1} in seed kernel oil (Oliveira et al., 1999; Anjorin et al., 2010).

The leaf lamina had a Cu content of about 0.0436–0.044 mg g^{-1} (Oliveira et al., 1999). The highest content of Cu, ranging from 0.045–0.1311 mg mL^{-1}, was detected in seed kernel oil, and the smallest content of Cu, ranging from 0.015-0.0342 mg g^{-1}, was reported in seed kernel powder (Oliveira et al., 1999; Anjorin et al., 2010). Some reports showed low Cu content in leaves and pods grown in different locations (Prasad, 1998; Onianwa et al., 2001; Anwar and Bhanger, 2003). Other mineral contents in seed kernel oil, except for Cu, were lower than those found in different parts of the plant.

In the seed shell, the highest quantity of Fe, ranging from (0.204–0.4476 mg g^{-1}, was observed. This was much higher than in other parts of the seed (0.0750–0.1185 mg g^{-1}); the leaf was found to contain five times as much compared with spinach (as mentioned in Table 1.2). This indicated that the seed shell could be a potential source of Fe supplement (Oliveira et al., 1999; Anjorin et al., 2010). Iron from moringa leaves was shown to overcome iron deficiency in Wistar rats and successfully modulated the expression of iron-responsive genes better than conventional iron supplements (Saini et al., 2014e).

A small amount of Zn of 0.0180 mg g^{-1} and 0.0096 mg g^{-1} was reported in the petiole and seed kernel, respectively (Oliveira et al., 1999; Anjorin et al., 2010). The Zn content in leaves is higher than in the petiole and seed kernel (as mentioned in Table 1.2).

1.4.4 Amino Acids and Protein

Moringa oleifera seeds are a very good substitute for legumes, which are scarce in sulphur-containing amino acids. *M. oleifera* seeds have an exceptionally high content of methionine+cysteine (43.6 g kg^{-1} protein) and lysine. From a child's nutrition point of view, methionine, isoleucine and leucine are important (Oliveira et al., 1999). However, in *M. oleifera* leaves, Arg was the amino acid with the maximum level (143.3 mg g^{-1}) of protein, trailed by Asp, Glu, Lys, Leu, Cys and Phe (Benhammouche et al., 2021).

Crude protein content is high in all parts of *M. oleifera*. The protein content for leaves varies from 178–338.2 mg g^{-1}, and for immature pods, it is 206.6 mg g^{-1}. The variation in protein content may be attributed to differences in environmental conditions or variations in soil conditions (Makkar and Becker, 1997; Sena et al., 1998; Gidamis et al., 2003; Reyes Sánchez, 2006; Sánchez-Machado et al., 2010). Mature seeds of *M. oleifera* cultivated in different areas comprise about 300.7–332.6 mg g^{-1} dm of protein (Oliveira et al., 1999; Manzoor et al., 2007).

1.4.5 Antioxidant Contents of Moringa

In a cell, continuous oxidation reactions occur, producing free radicals, which cause cell damage or death. This can be met by antioxidant molecules that help to terminate chain reactions by constraining the development of free radical intermediates, thus helping to avert cancer, coronary heart disease and even altitude sickness, which are widely used in dietary supplements (as mentioned in Table 1.4). Natural antioxidants such as phenols and vitamins A, C and E were reported in *M. oleifera* in extensive quantities compared with vegetables and fruits. Seasonal variation causes a 1.5–3 times change in the content of vitamin A and other antioxidants in moringa leaves; higher vitamin A was obtained in the hot-wet season, while vitamin C was found in the cool-dry season (Yang et al., 2007). *M. olifera* leaves are also reported to be rich in folate, and the bioavailability of folate from dehydrated moringa leaves was estimated to be 81.9% (Saini et al., 2016).

Moringa oleifera tissues are a vital nutritional foundation of antioxidant polyphenolics (Amaglo et al., 2010; Peñalver et al., 2022). The leaves contain the highest levels of flavonoids, as well as the most complex variety of these compounds (as mentioned in Table 1.4), which backs up their high antioxidant activity (Sreelatha and Padma 2009; Verma et al., 2009; Lin et al., 2021). Strong antioxidant properties have been observed for d-(+)-catechin, quercetin and kaempferol (Lakka et al., 2019; Pollini et al., 2020). Additionally, they are naturally occurring antioxidants that have positive health effects and help prevent and manage vascular illnesses, diabetes, obesity and oxidative damage (Jayawardana et al., 2015).

1.4.6 Fibres

The dietary fibre content in immature pods, measuring an impressive 46.78%, was nearly double that found in the leaves and flowers. This remarkable level of fibre aligns closely with the amounts typically reported in seaweed, highlighting the pods' exceptional nutritional value and potential as a significant dietary source (Sánchez-Machado et al., 2004b).

TABLE 1.4
Bioactive Molecules in *M. oleifera* and Their Health Benefits

Plant part	Molecule	Effective against/used as	Reference
Leaves	quercetin 3-O-glucoside quercetin 3-O-(6′′-malo-nylglucoside) quercetin 3-O-rutinoside quercetin 3-O-glucoside quercetin 3-O-(6”-malonylglucoside)	Antidiabetic, anticancerous, antioxidant, antitumour, hepatoprotective	Bennett et al., 2003; Murakami et al., 2008; Amaglo et al., 2010;Rayamajhi et al., 2013; Oldoni et al., 2019
	5-caffeoylquinic acid (5-CQA) 3-caffeoylquinic acid (3-CQA)	Alzheimer's disease, antibacterial, anti-inflammatory, anti-influenza, cardiovascular disease, diabetes type 2	Lindsay et al., 2002; Salazar-Martinez et al., 2004a; Ranheim and Halvorsen, 2005; Almeida et al., 2006; Amaglo et al., 2010;Urushisaki et al., 2011; Braham et al., 2020
	kaempferol 3-O-rutinoside kaempferol 3-O-glucoside kaempferol 3-O-(6”-malonylglucoside)	Antioxidant, antimicrobial, analgesic, antiallergy, anxiolytic-like response, cardioprotective, degenerative disease, neuroprotective, effective against osteoporosis, estrogenic effect	Bennett et al., 2003; Prouillet et al., 2004; Zern et al., 2005; Trivedi et al., 2008; Medeiros et al., 2009; Aguirre-Hernández et al., 2010; Amaglo et al., 2010; Nsonde Ntandou et al., 2010; Hendra et al., 2011; Sreelatha et al., 2011; Bhatia et al., 2021
	lutein	Age-related maculopathy, cardiovascular disease, cataracts, improving visual function, several types of cancer	Granado et al., 2003; Shindano and Kasase, 2009
	2-ethyl-3,6-dimethylpyrazine	Flavouring agent	Shindano and Kasase, 2009; Van Ba et al., 2012
	p-menth-1-en-8-ol	Aroma constituents	Shindano and Kasase, 2009
	2,6,6-trimethylcyclohexa-1,3-dienec arbaldehyde	Saffron aroma compound	Teranishi et al., 1999; Shindano and Kasase, 2009
	α-ionone	Perfumery and flavour agent	Curtis and Williams, 1994; Shindano and Kasase, 2009; Pino and Regalado, 2010
	1,1,6-trimethyl-1,2-dihydronaphthalene	'Petrol' aromas in aged Riesling wines	Shindano and Kasase, 2009; Sacks et al., 2012
	β-damascenone	Impact odorant in red wine	Pineau et al., 2007; Shindano and Kasase, 2009
	β-ionone	Aromas of fruits and flora	Mau et al., 2003; Shindano and Kasase, 2009
	ledene oxide	Antibacterial and antioxidant activities	Shindano and Kasase, 2009; Zhang and Du, 2009;
	(E)-6,10-dimethylundeca-5,9-dien-2-one	Flavour and fragrance agents, food	Shindano and Kasase, 2009

(*Continued*)

TABLE 1.4 (CONTINUED)
Bioactive Molecules in *M. oleifera* and Their Health Benefits

Plant part	Molecule	Effective against/used as	Reference
	2,3,6-trimethyl-naphthalene	Flavour compound	Shindano and Kasase, 2009; Givianrad, 2012
	megastigmatrienone	Spicy aroma	Shindano and Kasase, 2009; Yeast Metabolome Database
	1-[2,3,6-trimethyl-phenyl]-3-buten-2-one	Aroma and antioxidant	Mau et al., 2003
	isolongifolene	Fragnance agents	Shindano and Kasase, 2009; The Good Scents Company
	hexahydrofarnesylactone	Flavour and fragrance agents	Shindano and Kasase, 2009; The Good Scents Company
	farnesylacetone	Flavour and fragrance agents	Shindano and Kasase, 2009; The Good Scents Company
	methyl palmitate	Cosmetic, flavour and fragrance agents	Shindano and Kasase, 2009; The Good Scents Company
	n-hexadecanoic acid	Anti-inflammatory	Shindano and Kasase, 2009;Aparna et al., 2012
	[6E,10E]-7,11,15-trimethyl-methylene-1,6,10,14-hexadeca-tetraene	Antifungal	Chen and Chuang, 2008; Shindano and Kasase, 2009
	(E)-phytol	Antimycobacterial activity	Rajab et al., 1998; Shindano and Kasase, 2009
	docosane	Fragrance agents	Shindano and Kasase, 2009; The Good Scents Company
	1-docosene	Antibacterial	Shindano and Kasase, 2009; The Good Scents Company
	teracosane	Natural substances and extractives	Shindano and Kasase, 2009; The Good Scents Company
	pentacosane	Natural substances and extractives	The Good Scents Company
	hexacosane	Natural substances and extractives	Chuang et al., 2007; The Good Scents Company
	niazinin A niazinin B niazinin A + B	Blood pressure lowering	Faizi et al., 1994; Gilani et al., 1994
	niazimicin	Antitumour	Guevara et al., 1999
	β-sitosterol	Serum cholesterol-lowering	Ghasi et al., 2000
	4-[α-(L-rhamnosyloxy) benzyl]-o-methyl thiocarbamate	Diarrhoea	Gilani et al., 1992
	astragalin	Preventing weight loss and colon shortening	Peng et al., 2020
Petioles	5-caffeoylquinic acid (5-CQA)	Alzheimer's disease, antibacterial, anti-inflammatory, anti-influenza, cardiovascular disease, diabetes type 2	Lindsay et al., 2002; Salazar-Martinez et al., 2004; Santos et al., 2012; Almeida et al., 2006; Amaglo et al., 2010; Urushisaki et al., 2011
	quercetin 3-O-rutinoside quercetin 3-O-glucoside quercetin 3-O-(6"-malonylglucoside)	Antidiabetic, anticancerous, antioxidant, antitumour	Bennett et al., 2003; Murakami et al., 2008; Amaglo et al., 2010; Rayamajhi et al., 2013;

(Continued)

TABLE 1.4 (CONTINUED)
Bioactive Molecules in *M. oleifera* and Their Health Benefits

Plant part	Molecule	Effective against/used as	Reference
	kaempferol 3-O-rutinoside	Anticancerous, antioxidant	Brochado et al., 2003; Amaglo et al., 2010
Flowers	5-Caffeoylquinic acid (5-CQA)	Alzheimer's Disease, Antibacterial, Anti Inflammatory Anti-Influenza, Cardiovascular Disease, Diabetes Type 2	Lindsay et al., 2002, Salazar-Martinez et al., 2004, Ranheim and Halvorsen 2005, Almeida et al., 2006, Amaglo et al., 2010, Urushisaki et al., 2011
	Quercetin 3-O-rutinoside Quercetin 3-O-glucoside Quercetin 3-O-(6"- malonylglucoside)	Anticancerous, Antidiabetic, Antioxidant, Antitumor	Bennett et al., 2003, Murakami et al., 2008, Amaglo et al., 2010, Rayamajhi et al., 2013
	Kaempferol 3-O-rutinoside Kaempferol 3-O-glucoside Kaempferol 3-O-(6"-malonylglucoside)	Analgesic, Anti Allergy, Anticancerous, Antimicrobial, Antioxidant, Anxiolytic Like Response, Cardioprotective, Effective Against Osteoporosis, Estrogenic Effect, Neuroprotective	Brochado et al., 2003, Prouillet et al., 2004, Zern et al., 2005, Trivedi et al., 2008, Medeiros et al., 2009, Sreelatha and Padma, 2009b, Aguirre-Hernández et al., 2010, Amaglo et al., 2010, Hendra et al., 2011
Seeds	4(α-L-rhamnosyloxy)-benzyl isothiocyanate	Anti bacterial, Antitumoral	Galuppo et al.,2013
	niazimicin	Blood pressure lowering	Jideani and Diedericks, 2014
	3-O-(6′-O-oleoyl-β-D-glucopyranosyl)-β-sitosterol β-sitosterol-3-O-β-D-glucopyranoside β-sitosterol	Antidiabetic, Anti-Inflammatory, Antioxidant, Antipyretic, healing of ulcers and alleviating cramps.	Gupta et al., 1980, Gupta et al., 2011, Mishra et al., 2011
	niazirin	Antitumor and Anti-inflammatory	Jideani and Diedericks, 2014
	oil glycosides and thiocarbamate glycosides	Blood pressure lowering	Faizi et al., 1995
Pods	Thiocarbamate isothiocyanate glycosides Methyl p-hydroxybenzoate β-sitosterol	Hypotensive	Faizi et al., 1995
Gum	O-(β-D-glucopyranosyluronic acid) (1→6)-β-D-galactopyranosyl (1→6)-D-galactose	Colon-specific drug delivery	Mishra et al., 2011, Singh et al., 2012
Stem	vanillin	Anticlastogen, Antioxidant, Antitumor, Skin protection	Mishra et al., 2011
	β-sitosterone	Anti-Inflammatory	Mishra et al., 2011, Leu et al., 2012
	octacosonoic acid	Skin protection and lightening	Mishra et al., 2011

(Continued)

TABLE 1.4 (CONTINUED)
Bioactive Molecules in *M. oleifera* and Their Health Benefits

Plant part	Molecule	Effective against/used as	Reference
	Moringine Moringinine	Antiasthamatic, Appetite suppressant, Cardiac stimulant, Decongestant, Rise in blood pressure, Stimulant and to treat hypotension	Paul and Didia, 2012
	Pterygo-spermin	Antibiotic	Das et al., 1957, Ruckmani et al., 1998, Paul and Didia, 2012
	4-(α -L-rhamnopyranosyloxy)-benzylglucosinolate and benzyl-glucinolate	Antibacterial, Antitumoral	Menhnaz, K, 2009, Paul and Didia, 2012, Galuppo et al., 2013
	Spirochin	Antibacterial	Paul and Didia, 2012
Root bark	N- benzyl,S ethyl thioformate	Antibacterial, Antifungal	Nikkon et al., 2003
oil	Toluene, 5-tert-Butyl-1,3-cyclopentadiene, Benzaldehyde, 5-Methyl-2-furaldehyde, Benzeneacetaldehyde, Linaloloxide, 2-Ethyl-3,6-dimethylpyrazine, Undecane, α –Isophoron, Benzylnitrile 2,6,6-Trimethyl-2-cyclohexane-1,4-dione, 2,2,4-Trimethyl-pentadiol, 2,3-Epoxycarane, p-Menth-1-en-8-ol, 2,6,6-Trimethylcyclohexa-1,3-dienecarbaldehyde, Indole, Tridecane, α –Ionone, 1,1,6-Trimethyl-1,2-dihydronaphthalene, α-Ionene, β-Damascenone, β-Ionone, Ledene oxide, 2-tert-Butyl-1,4-dimethoxybenzene, (E)-6,10-dimethylundeca-5,9-dien-2-one, 4,6-Dimethyl-dodecane, 3,3,5,6-Tetramethyl-1-indanone, Dihydro-actiridioide, 2,3,6-Trimethyl-naphthalene, Megastigmatrienone, 1-[2,3,6-Trimethyl-phenyl]-2-butanone, 1-[2,3,6-Trimethyl-phenyl]-3-buten-2-one, Isolongifolene Hexahydrofarnesylactone, Farnesylacetone, Methyl palmitate, n-Hexadecanoic acid.	Antifungal activity	Chuang et al., 2007

1.4.7 Glucosinolates

Moringaceae is very similar to Capparidaceae, which is known to contain glucosinolates in its leaves and seeds. Though it was classified as an anti-nutritional compound, studies have shown its positive effects on human health due to the production of isothiocyanates (ITC) and nitriles generated when cooking (Mithen et al., 2000; Johnson, 2002; Juge et al., 2007). Glucomoringin-derived ITC is more effective than sulforaphane, commonly found in Brassica vegetables, in constraining NF-kB and inducing apoptosis (Brunelli et al., 2010).

All tissues except roots have 4-*O*-(a-L-rhamnopyranosyl-oxy)-benzylglucosinolate (glucomoringin) predominantly, whereas in roots, benzylglucosinolate (glucotropaeolin) was greater (Bennett et al., 2003; Amaglo et al., 2010). Four additional glucosinolates were reported in petioles, leaves and flowers: 4-hydroxybenzylglucosinolate was reported in *M. oleifera*, probably the biosynthetic precursor of the rhamnose-substituted glucosinolates and three mono-acetyl-rhamnose isomers; 4-O-(a-L-rhamnopyranosyloxy)-benzylglucosinolate was reported alone in tender fruits and seeds.

1.5 *MORINGA OLEIFERA* AS FOOD FOR HUMANS AND LIVESTOCK

All parts of moringa are highly nutritious and are used for food, oil and medicine, especially in Asia and Africa. Young pods are eaten as vegetables; young pods resemble string beans, while larger, fibrous pods are added to soups and curries. In Bihar and Orissa, tender pods are cooked with mustard seed paste and served with rice (panibhaat). Scraped drumstick pulp is used to make moringa bharta (Watt, 1893; Anwar et al., 2005).

The nutritious leaves and flowers are used in various dishes, including soups, stews and stir-fries (Radovich, 2009) such as 'Sanjana saag'. Dried leaves can be powdered for off-season use. Moringa is suggested as a food complement for pregnant women, lactating mothers and those with osteoporosis or bone fractures (Pandey et al., 2011).

In South India, pods aged 35 to 50 days are used in dishes like sambhar, avial and fish curry (Pandey et al., 2011). In West Bengal and certain regions of Bangladesh, they are commonly referred to as sojne fool and are typically prepared with green peas and potatoes, especially in spring. In the northern regions of India, the Punjabi, Sindhi and Multani communities cook the flowers by either boiling or frying them with curd (as noted by Ms N. K. Chaudhari, former member of NBPGR, New Delhi) (Arora and Pandey, 1996; Pandey et al., 2011). In Bihar and Orissa, flower buds and tender leaves are deep-fried in gram flour batter, while immature seeds can be eaten raw or cooked (Ramachandran et al., 1980; Pandey et al., 2011). The infusion from flowers is consumed as tea.

Tender seeds are pressed into sweet, non-desiccating oil, commercially known as 'Ben oil', which is of high quality. Mature seeds become bitter and are generally used for oil extraction, as seeds contain 35% oil, which does not turn rancid and also burns without smoke. Roots are used to prepare the sauce, but they should not be excessively consumed. Gum found in the bark can be used to season food.

Several studies proved that using moringa with fodder significantly affects animals (Aregheore, 2002). A diet containing 70% moringa was shown to be the best diet for pigs. Excessive feeding with moringa seems to reduce weight gain in livestock (Foidl and Paull, 2008).

1.6 AS A WATER PURIFIER

In the picturesque rural landscapes of Asia and Africa, the seeds of *M. oleifera* have become a remarkable solution for transforming contaminated water into transparent, drinkable sources. This natural method not only highlights the ingenuity of local communities but also showcases the incredible potential of nature in addressing water purification challenges (Faizi et al., 1995). Of all known 14 species, *M. oleifera* has gained considerable importance as a flocculant (Gassenschmidt et al., 1995), and its coagulation properties have been investigated (Muyibi and Evison, 1995; Gupta et al., 2005; Nadeem et al., 2006; Pandey et al., 2022; Irikannu et al., 2024). *M. oleifera* inhibits microbes with two hours of treatment (Madsen et al., 1988). It was found to inhibit the growth of the parasite *Schistosoma mansoni cercariae* (Olsen, 1987). *M. oleifera* can efficiently be used to substitute alum to dewater chemical sludge (Ghebremichael and Hultman, 2004). The pressed cake obtained after oil extraction from moringa seeds is highly protein-rich. Proteins of about 7–17 kDa are found to be active cationic polyelectrolytes. The cationic polyelectrolytes neutralise the colloids in muddy or dirty water. Moringa seeds can be used as a non-toxic method for water purification. They contain natural polypeptides that aid in sedimenting mineral particles and organic matter (Makkar and Becker, 1997). Seeds and protein fractions obtained from it were found to coagulate cadmium (Cd), iron (Fe) (Gupta et al., 2007, Kituyi et al., 2013), chromium (Cr), nickel (Ni) (Sharma et al., 2007), Zinc (Zn) (Bhatti et al., 2007, Kituyi et al., 2013), arsenic (Ar), mercury (Hg) (Sajidu et al., 2008), silver (Ag) (Araújo et al., 2013), lead ions (Pb^{2+}) (Kumari et al., 2006; Gupta et al., 2007; Meneghel et al., 2014) and fluoride (Agnihotri et al., 2013). Seed cake was also found to help remove humic acid (Santos et al., 2012), benzene, toluene, ethylbenzene, p-xylene and o-xylene from aqueous solution, thus reducing the cost of waste treatment (Almeida et al., 2012). Seed powder of this plant has been used to remove turbidity, alkalinity, total dissolved solids and hardness of water (Muyibi and Evison, 1995; Okuda et al., 2001; Kumari et al., 2006). *M. oleifera* pods were reported to possibly eliminate organics from aqueous media and methyl parathion pesticide (MP) from surface and ground waters through sorption (Akhtar et al., 2007a; Akhtar et al., 2007b). It also removes phenol from sewage water at alkaline pH (Krishnan, 2012). As well as seeds, leaves, bark and pods effectively remove metal ions and other pollutants (Araújo et al., 2013). *M. olifera* leaves were effective as a water purifier in decontaminating faecal coliform bacteria (Abd-Elhalim, 2022).

1.7 AS A BIOFUEL

Diminishing petroleum reserves and increasing environmental and health hazards due to their use have made it compulsory to find an alternative fuel source. Most parts of the plant, except for the roots, contain a compound called 4-O-(α-L-rhamnopyranosyl-oxy)-benzylglucosinolate (glucomoringin). However, the roots have more benzylglucosinolate (glucotropaeolin), according to studies by Bennett et al. (2003) and Amaglo et al. (2010). Researchers found four additional glucosinolates in the petioles, leaves and flowers. One of these is 4-hydroxybenzylglucosinolate, reported for the first time in *M. oleifera* and is likely a precursor for the rhamnose-substituted glucosinolates. The other three are mono-acetyl-rhamnose isomers. In the pods and seeds, only 4-O-(α-L-rhamnopyran osyloxy)-benzylglucosinolate was found. Along with the cetane number, other fuel properties such as cloud point, kinematic viscosity and oxidative stability were also determined, based on which *M. oleifera* oil was an acceptable feedstock for biodiesel (Rashid et al., 2008). Because of its ability to thrive in a variety of conditions, *M. oleifera* has a lot of potential as a biofuel compared with other lignocellulosic biomass. Pyrolysis can be used to turn seeds' extremely valuable chemical products – carbohydrates and lipids – into fuel (Sukarni et al., 2024). *M. oleifera* leaves were used to create a heterogeneous catalyst for the transesterification reaction that produces biodiesel (Aleman-Ramirez et al., 2021).

1.8 NUTRIENT ENHANCEMENT AND PEST MANAGEMENT

Experiments conducted using a moringa foliar spray on different crops like onion, bell pepper, coffee, tea, sorghum, soybean, chilli, safflower and common beans were found to enhance crop yield, plant height, cob diameter, grains per cob, seedling growth and germination rate (Fuglie, 2000; Mvumi et al., 2013; Abbas et al., 2013; Hafeez et al., 2022). A study conducted on late-sown wheat also proved the efficacy of foliar spray during different stages, especially during heading, enhancing sink capacity during the grain filling period, leaf surface area and yield and delayed maturity (Yasmeen et al., 2012). Primary studies also revealed that moringa leaf powder could be effectively used to protect crops in fields and storehouses and eliminate insect pest infestations (Musa, 2013; Sowley et al., 2014; Okwor et al., 2021).

1.9 THE HEALTH BENEFITS OF MAJOR BIOACTIVE CONSTITUENTS

All tissues, except for roots, contain 4-O-(α-L-rhamnopyranosyl-oxy)-benzylglucosinolate (glucomoringin), while roots have a higher concentration of benzylglucosinolate (glucotropaeolin) (Bennett et al., 2003; Amaglo et al., 2010). Four additional glucosinolates were found in petioles, leaves and flowers, including 4-hydroxybenzylglucosinolate, a likely precursor of rhamnose-substituted glucosinolates, along with three mono-acetyl-rhamnose isomers. In pods and seeds, only 4-O-(α-L-rhamnopyranosyloxy)-benzylglucosinolate was detected (Bernhoft, 2010).

Moringa oleifera is rich in compounds called glucosinolates and isothiocyanates (Fahey et al., 2001; Bennett et al., 2003; Wolff et al., 2023), alkaloids, namely moringine and moringinine (Kerharo, 1969; Pathak et al., 2015), vanillin, β-sitosterol, β-sitostenone, 4-hydroxymellin, octacosanoic (Faizi et al., 1994) and some flavonoid pigments such as kaempferol, rhamnetin, isoquercitrin and kaempferitrin (Faizi et al., 1994; Siddhuraju and Becker, 2003) that have potential therapeutic uses.

M. oleifera leaves contain phenolics, which are well known as excellent antioxidants, good scavengers of reactive metabolites, such as OH·, O^{-2} and ROO·, and oxidative DNA-nicking protective activity (Chumark et al., 2008). Leaf extract (LE), fruit extract (FE) and seed extract (SE) contain gallic acid, chlorogenic acid, ellagic acid, ferulic acid, kaempferol, quercetin and vanillin. LE is the richest in phenolics and may be a good source of natural antioxidants (listed in Table 1.4), and can be used as a likely food additive or as a nutraceutical (listed in Table 1.5). Antioxidant components of drumstick extract revealed high thermal stability (Arabshahi-Delouee et al., 2011). All tissues, except roots, predominantly contain 4-O-(α-L-rhamnopyranosyl-oxy)-benzylglucosinolate (glucomoringin), while roots have a higher concentration of benzylglucosinolate (glucotropaeolin) (Bennett et al., 2003; Amaglo et al., 2010). Four additional glucosinolates, including 4-hydroxybenzylglucosinolate – the first identified in *Moringa oleifera* – are found in petioles, leaves and flowers. In pods and seeds, only 4-O-(α-L-rhamnopyranosyloxy)-benzylglucosinolate is present (Saini et al., 2014c).

Moringa leaf juice has a calming effect on blood pressure (The Wealth of India, 1962; Dahot, 1998). Fully acetylated active compounds like nitrile, mustard oil glycosides and thiocarbamate glycosides, extracted from moringa leaves, are responsible for lowering blood pressure (Faizi et al., 1994, 1995; Anwar et al., 2007). Thiocarbamates, namely niazinin A [1], niazinin B [1], niazimicin [4] and niazinin A + B, were first isolated from the leaves of *M. oleifera* as pure compounds, which act as calcium antagonists and showed blood-pressure-lowering and bradycardiac effects in a study carried out on rats (Gilani et al., 1994). The crude extract from moringa leaves has been shown to reduce serum cholesterol levels in rats that are on a high-fat diet, which may be due to the presence of β-sitosterol (Ghasi et al., 2000). The moringa fruit helps lower serum cholesterol and increases the excretion of faecal cholesterol (Mehta et al., 2003). Hypotensive activity was also obtained from extracts of coat, pulp and pod seeds, which, on analysis, were found to contain thiocarbamate, isothiocyanate glycosides, methyl *p*-hydroxybenzoate and β-sitosterol (Faizi et al., 1995). Substances with diuretic properties have been identified in the roots, leaves, flowers, gum and the aqueous infusion of moringa seeds, which also contribute to the overall reduction of blood pressure (Morton, 1991; Caceres et al., 1991).

1.9.1 Antispasmodic Activity

The extracts from the roots and leaves of *M. oleifera* show antispasmodic properties, which support its historical application in treating diarrhoea, likely due to calcium channel inhibition (Caceres et al., 1991; Gilani et al., 1994). The antispasmodic activity of these extracts (Gilani et al., 1992) has been attributed to the presence of the bioactive compound 4-[α-(L-rhamnosyloxy) benzyl]-*O*-methyl thiocarbamate (trans).

TABLE 1.5
Uses and Benefits of Different Parts of the *M. oleifera* Tree

Plant part	Uses	Author
Leaves	Antimicrobial infection, urinary tract infection, Epstein–Barr virus, herpes simplex virus, HIV-AIDS, helminths, trypanosomes, bronchitis, external sores/ulcers, fever, hepatic, cancer therapy/protection, antitumour, prostate cancer, radioprotective, circulatory/endocrine disorders, antianaemic, antihypertensive, diabetes/hypoglycaemia, diuretic, hypocholesterolaemia, thyroid, hepatorenal, digestive disorders, colitis treatment, diarrhoea treatment, dysentery, ulcers/gastritis, inflammation, rheumatism, headache, lactation enhancer, antiseptic, animal feed, food ingredient, growth enhancer of crops, nematocide, pesticide, insecticide, rice crackers, biscuit preparation, functional food powder, skin care, essential oils, saag, soups, stews, teas, fodder, reducing agent in gold nanoparticle synthesis	Fahey, 2005; Radovich, 2009; Farooq et al., 2012; Yasmeen et al., 2012; Abbas et al., 2013; Ali et al., 2013b; Chakraborty et al., 2013; Kar et al., 2013, Manaois et al., 2013; Marrufo et al., 2013; Mvumi et al., 2013; Singh and Prasad, 2013; Sowley et al., 2014; Milla et al., 2021; Jikah et al., 2023
Pods	Diabetes/hypoglycaemia, inflammation, rheumatism, joint pain, Sambar, fish curry, moringa bharta, pani bhaat, vegetable, soups, culinary preparations	Fahey, 2005; Radovich, 2009
Seeds	Warts, schistosomes, skin (dermal), water treatment, antitumour, circulatory/endocrine disorders, digestive disorders, ulcers/gastritis, inflammation, rheumatism, antimicrobial, fever, cancer therapy/protection, inflammation, rheumatism, arthritis, immune-stimulant, antispasmodic, pyoderma, water purification, fertiliser, oil extraction, bioremediation, coagulant, removal of BTEX, humic acid removal	Fahey, 2005; Ted Radovich, 2009; Farooq et al., 2012; Santos et al., 2012; Agnihotri et al. 2013; Araújo et al., 2013; Gautier et al., 2022
Flowers	Antimicrobial infection, parasites – helminths, external sores/ulcers, throat infection, cancer therapy/protection, antitumour, circulatory/endocrine disorders, diuretic, inflammation, rheumatism, hysteria, abortifacient, parasites – helminths, cancer therapy/protection, skin, circulatory/endocrine disorders, antihypertensive, soups, stews, teas	Fahey et al., 2001; Radovich, 2009; Farooq et al., 2012
Bark	Dental caries/toothache, common cold, external sores/ulcers, cancer therapy/protection, antitumour, detoxification, snake bites, scorpion bites, digestive disorders – colitis treatment, digestif, epilepsy, hysteria, headache, abortifacient, aphrodisiac, birth control, paper making	Fahey, 2005; Radovich, 2009; Farooq et al., 2012; Mali et al., 2022
Gum	Dental caries/toothache, syphilis, typhoid, earache, fever, asthma, diuretic, digestive disorders, dysentery, inflammation, rheumatism, headache, abortifacient, rubefacient, seasoning food	Fahey, 2005; Radovich, 2009
Roots	Dental caries/toothache, common cold, trypanosomes, external sores/ulcers, fever, asthma, circulatory/endocrine disorders, cardiotonic, diuretic, hepatorenal, digestive disorders, diarrhoea treatment, flatulence, inflammation, rheumatism, oedema, antispasmodic, epilepsy, hysteria, headache, abortifacient, aphrodisiac, astringent, rubefacient, vesicant, sauce preparation	Fahey, 2005; Radovich, 2009
Wood	Blue dye	Bwana-Simba, 2006
Seed oil	Lubricant, volatile substance in the perfume industry, oxidative stability in butter, cooking purposes, biodiesel	Ramachandran et al., 1980; Rashid et al., 2008; Van Ba et al., 2012; Nadeem et al., 2006; Esmaeili et al., 2019
Whole plant	Biogas production	Radovich, 2009; Tambone et al., 2020

1.9.2 Hepatoprotective Activity

Moringa leaf extracts help protect against liver damage (Khalid et al., 2023; Sahoo et al., 2024). In *M. oleifera* extracts, the presence of quercetin (quercetin-3-*O*-glucoside, quercetin-3-*O*-(6"-malonyl-glucoside), a well-known flavonoid with phenolic hydroxyl groups, was found to be accompanied by hepatoprotective activity (Gilani et al., 1997; Bennett et al., 2003; Siddhuraju and Becker, 2003). They also exhibit many other health benefits, as mentioned in Tables 1.4 and 1.5.

1.9.3 Antioxidant Properties

Moringa leaf extracts exhibit antioxidant activity in vivo and in vitro due to their flavonoids and phenolic content (Khor et al., 2018; Uphadek et al., 2018). Kaempferol (kaempferol-3-*O*-glucoside) and kaempferol-3-*O*-(6"-malonyl-glucoside), flavonoid compounds with phenolic hydroxyl groups, were also found in the leaf extract of *M. oleifera,* and they exhibit high antioxidant properties (Tables 1.4 and 1.5), and thus provide protection to humans against infections and degenerative diseases (Bennett et al., 2003; Siddhuraju and Becker, 2003).

1.9.4 Antibacterial and Antifungal Properties

Extracts from leaves, roots, flowers and the bark of *M. oleifera* possess antibacterial and antifungal activity (Das et al., 1957; Ruckmani et al., 1998). Root extracts have also been found to contain 4-α-L-rhamnosyloxy benzyl isothiocyanate and the chemical N-benzylethylthioformate, an aglycone of deoxyniazimincin which possesses antimicrobial activity (Eilert et al., 1981; Fahey, 2005; Upadhyay et al., 2015). Similarly, the root bark extract was reported to exhibit antibacterial and antifungal activities because of the presence of the active compound aglycone of deoxyniazimicine (N-benzyl, S-ethyl thioformate) (Nikkon, 2003). Extracts from stem and bark were found to be effective against pathogens like *Staphylococcus aureus* (Mehta et al., 2003), *Pseudomonas aeruginosa* and *Pseudomonas aureus* (Caceres et al., 1991). According to Onsare et al. (2015), seed extracts exhibit antibacterial action against Gram-negative bacteria and prevent the growth of bacteriophages (Al-Jadabi et al., 2023). Because of its steroidal ring structure, the fruit of *M. oleifera*, abundant in alkaloids, flavonoids and steroids, has a noticeable inhibitory effect on *Candida albicans* cells. Protein denaturation and the suppression of spore germination pathways are two mechanisms by which this inhibitory impact works (Medeiros et al., 2023).

1.9.5 Anticancer Properties

Numerous phytoconstituents of moringa extracts (leaves, bark and seeds) exhibit anticancer properties in both in vivo and cancer cell line models (Wu et al., 2021). The apoptotic pathway in HeLa cells was initiated by the aqueous extract of *M. oleifera* seeds (Adam et al., 2023). Leaf extracts contained *O*-ethyl-4-(α-L-rhamnosyloxy) benzyl carbamate together with 4(α-L-rhamnosyloxy)-benzyl isothiocyanate, niazimicin and 3-O-[(6′-O-oleoyl)-β-D-glucopyranosyl]-β-sitosterol that showed significant antitumour activity (Makonnen et al., 1997). Niazimicin is a potent chemopreventive agent in chemical carcinogenesis (Guevara et al., 1999); 4-[(4′-*O*-acetyl-α-i-rhamnosyloxy) benzyl] isothiocyanates and niaziminin, a thiocarbamate from the leaves, are found to inhibit Epstein–Barr virus activation by inhibiting tumour promoter (Murakami et al., 2008).

1.10 OTHER THERAPEUTIC APPLICATIONS

Moringa oleifera also exhibits other diverse activities, such as anthelminthic activity (Bhattacharya et al., 1982; Abadu et al., 2021), regulation of the thyroid hormone and treating hyperthyroidism (Pal et al., 1995a, 1995b; Tahiliani and Kar, 2000; Al-Hadidy and Dawood, 2021); reducing glucose levels in rabbits (Makonnen et al., 1997); significant radiation protection (Rao et al., 2001); skin and hair care (Stussi et al., 2002; Garg et al., 2024) and as a prophylactic, therapeutic anti-herpes simplex virus type 1 medicine (Lipipun et al., 2003). Similarly, the antihepatotoxic effect was demonstrated in roots and flower extracts (Pari and Kumar, 2002; Zhu et al., 2007). The aqueous extract of leaves was found to down-regulate nuclear factor kappa β (NF-κB) and thus help to increase the effectiveness of chemotherapy in treating pancreatic cell cancer (Berkovich et al., 2013). It was also shown to be effective against lung cancer cell lines by induction of apoptosis (Bhadresha et al., 2022). *M. oleifera* seed oil showed anticancer activity in oral cancer (Barhoi et al., 2021). *M. oleifera* leaf extract can be used in creams during the winter season as it reduces undesirable skin sebum secretions from sebaceous glands (Ali et al., 2013).

Moringa oleifera is widely used in health formulations in India due to its potential health benefits. Examples include Rumalaya and Septilin (Himalaya Drug Company), Orthoherb (Walter Bushnell Ltd), Kupid Fort (Pharma Products Pvt. Ltd) and Livospin (Herbals APS Pvt. Ltd), which are known medicines for various health disorders (Mehta et al., 2003). Moringa seeds contain specific protein fractions. Purisoft®, derived from moringa seed oil cake, comprises peptides that protect the skin from environmental damage and combat premature ageing. Additionally, *M. oleifera* seed extracts offer dual benefits for hair care by providing antipollution properties and conditioning effects (Table 1.6 lists different patents stating the plant's usefulness).

1.11 COMMERCIAL USES

In India, about 1.2 million metric tonnes (MT) (1.1 million T) of immature pods are produced annually on 38,000 ha (94,000 ac) of commercial land, with post-harvest handling and processing falling under large-scale industry. Andhra Pradesh is a major producer with an area of about 15,661 ha, followed by Karnataka (10,280 ha) and Tamil Nadu (7,408 ha). Moringa products are generally consumed within a very short period, mainly within a day, and the leaves are perishable. To avoid excessive wilting and abscission of leaves, fruits and flowers are packed in plastic bags and stored under cool temperatures with high humidity. Alternatively, fruits may be canned to preserve them for later consumption. Apart from food uses, fruit extract has also been found to be an effective inhibitor of the corrosion of mild steel (Singh et al., 2012). Dried leaves and flowers may be used as a food additive to improve protein content and make tea. Seed kernels are used for culinary and cosmetic purposes. Moringa seed kernel oil has a good amount of antioxidant content, making it slow to go rancid. Production of moringa oil at a small scale using low-tech extraction methods is economically feasible as it can replace imported olive oil in cooking (Sengupta and Gupta, 1970; Radovich, 2009; Van Ba et al., 2012). Because moringa oil has a long shelf life and can retain smells, it is used in cosmetics and as an infusion with other oils. Because of its significant potential in addressing immune disorders, extracts from leaves have been utilised for treating inflammation in both the epidermis and dermis (Choi et al., 2016), while seed extracts have been used for photoprotective (Zhou et al., 2018), hydrating, enlightening and anti-ageing purposes (Athikomkulchai et al., 2020). Enhancing domestic oil production methods will contribute to increased self-sufficiency and profitability. Following oil extraction, seed cakes can be used as fertilisers. They can also be used to remediate wastewater and make feed.

1.12 MARKETS

About 80% of the demand for different *M. oleifera* products is met in India; 11,81,468 tonnes of various *M. oleifera* parts and/or products valued at approximately Rs 46.24 crores (~US$6.25 million) were exported by India in 2017 (Zirmire, 2018). Regarding moringa products, leaves are in high demand on the global market (Venkatesan et al., 2018).

Pods are suited for local markets and are not shipped because of pest concerns, whereas leaves have equal demand in both local and export markets. Due to the increase in knowledge about the

TABLE 1.6
Patents on *Moringa oleifera*

Subject	Part use	Purpose	Patent no.	Patentee
Hydroponic	Seeds	Water purification	US 5,121,708	Nuttle, 1992
Pharmaceutical product	Leaf extract	To treat diabetes	WO 02/094299	Gbodossou, 2002
Herbal biscuit	Leaf powder	Dietary supplements	US 6,517,861 B2	Singh et al., 2003
Herbal formulation	Seeds and leaves	Treatment of cancer	US 6,780,441	Solanki, 2004
Moringa oil	Leaves, seeds and pods	Ultra stable oil	US 6,528,075 B1 US 6,667,047 B2 US 7,691,397 B2	Brown and Kleiman, 2003a Brown and Kleiman, 2003b Brown and Kleiman, 2010
Herbal formulation	Fresh leaves	Treatment of vascular headache, neurological conditions and neurodegenerative disorders	WO 2004/02507 US 7,070,817 B2	Kuppam, 2006 Kuppam, 2006
Moringa seed protein	Seeds	Coagulant proteins	US 2004/0147723 A1	Dorries et al., 2004
Bioenhancer of drug and nutrients	Pods	Niaziridin	US 6,858,588 B2	Khanuja et al., 2005
Coagulants for water purification	Oil extracted seed residues	Coagulant proteins	US 6,890,565 B2 US 7,718,062 B2	Sutherland, 2005; Bauer et al., 2010
Personal care composition	Leaf extract	Activator of glutathione reductase and epidermal hyaluronic acid	WO 2008/104941	Lynn et al., 2008
Phytochemical composition	Leaf extract	Treatment of hyperlipidaemia	WO 2009/06630	Gokaraju et al., 2013
Nutraceutical composition	Leaves, seeds and fruits	Nutritional composition	US 2009/0098230A1	Andrews and Andrews, 2009
Dietary supplementation	Leaf powder	Nutritional supplement	WO 2011/098819	Gomez and Gomez, 2011
Activation of alternative cellular energy pathway	Leaves	Enerceutical	US 2011/0208110 A1	Martin, 2011
Isothiocyanate derivatives as anti-myeloma agents	Mature plant	Use of glucomoringin and of its des-thio-glucoside	US 2011/0053870 A1	Morazzoni et al., 2014
Aggregation of Algae	Seeds	Coagulant proteins	US 2011/0003350	Schafran et al., 2011
Nanoparticles	Leaf	Preventing the growth of cancer cells and treating diabetes by inhibiting α-glucosidase and/or α-amylase activity	US 11617719 B1	Awad et al., 2023

high nutrient content of *M. oleifera*, its markets are health food stores, food additives, juices, smoothies and processed foods. Some moringa products are marketed under different brand names, such as moringa powder as 'Hawaiian Drumstick', cooking oil as 'Organic Hawaiian Cooking Oil', scented massage oil as 'Tropic Passion' and smoothies and drinks as 'Malunggai Energy' (Radovich, 2009).

1.13 ECONOMIC ANALYSIS

Moringa oleifera production cost is comparatively lower than that of other vegetable crops. However, it is somewhat high in the first year but decreases in succeeding years because of its perpetual nature. The estimated cost is about US$5,000/acre, excluding overheads (Radovich, 2009). Using direct seedlings will further reduce the cost by eliminating the necessity for a nursery, containers and media.

The expenses associated with oil production can be quite significant due to the comparatively low output per acre. The approximate cost for producing 250 litres/ha of oil is about US$49/litre (US$46/qt). A report on a commercial project in Malawi estimated a cost of around US$2.60–6.00 per kg (US$1.20–2.70 per lb) to extract oil (Saint Sauveur, 2001). As with all other crops, labour costs disadvantaged Hawaii compared with most other tropical locations. The highest pod-yielding varieties must be used to make oil production viable. To reduce production costs, many things, like seed planting and harvesting, should be mechanised to reduce labour costs, along with optimising the extraction protocol. The expected income per plant will be around US$41 per tree and oil production (Radovich, 2009).

In the near future, the global moringa market is expected to grow considerably. The global moringa products market size was valued at US$8.15 billion in 2023 and is projected to grow at a CAGR of 9.6% from 2024 to 2030 (https://www.grandviewresearch.com/industry-analysis/moringa-products-market; accessed on February 2, 2025). Products made from moringa are taken from various portions of the tree, including oils, capsules, leaf powder and soaps. In contrast, the Asia-Pacific region accounted for the greatest part of the moringa product market in 2018 (35.30%), compared with global contribution (Islam et al., 2021).

1.14 APPROACHES TO ENHANCE NUTRITIONAL VALUE

Plants are a good source of essential nutrients for human and animal health. The composition of nutrients in plants is not uniform; plants rich in one nutrient may be deficient in another. Thus, it is necessary to have a balanced diet to avoid nutrient malnutrition in the population. Due to the increase in knowledge in various disciplines like food and nutritional science, health science, biochemistry, molecular markers, molecular biology, bioinformatics, plant transformation, genomics and biotechnology, tremendous attempts are being made to enhance the nutrient profile of the plant and eliminate anti-nutritional compounds that are harmful to human and animal health.

Greater variation in the nutrient profile was observed among different species of moringa. Differences in agroclimatic conditions were one reason for variation in the nutrient profile of moringa. To overcome nutrient variation and use it as an effective weapon against malnutrition, both non-biotechnological and biotechnological approaches will prove helpful.

1.14.1 Non-Biotechnological Approaches

1.14.1.1 Applications of Fertiliser

Fertilisers are substances that enhance plant growth by providing essential macro and micronutrients to replenish soil nutrients. While applying fertilisers is straightforward and does not require special tools, their effectiveness can be influenced by factors like application method, soil composition, nutrient mobility in plants and accumulation sites (Zhu et al., 2007). Cobalt (Co) deficiency frequently occurs in arid and semi-arid regions where moringa grows. Co and organic fertilisation may enhance moringa productivity (Gad et al., 2019). NPK fertiliser and vermicompost application have increased pods and seed production in *M. olifera* (Atteya et al., 2021).

Biofortification with iodine and selenium fertilisers has been particularly successful because of their mobility in soil and plants (Dai et al., 2004; Hartikainen, 2005). Likewise, Zn is also mobile in the soil. Supplying Zn in the form of $ZnSO_4$ resulted in an increase in crop yield and Zn concentrations in cereals and legumes (White and Broadley, 2005). Alternatively, biofortification of Fe using fertilisers has not been successful because $FeSO_4$ is rapidly converted into Fe(III) by binding to soil particles and thereby reducing the Fe mobility in soil (Grusak and DellaPenna, 1999). Along with an increase in the yield of the crop, fertilisers also showed some deleterious effects on plant growth and other soil organisms, thus decreasing the fertility of the soil (Grusak and Cakmak, 2005).

1.14.1.2 Conventional Breeding

Conventional breeding is the principal approach to crop improvement. It helps to incorporate required traits in plants to develop a new variety (Ames et al., 1990; Gendel and Jenkins, 2006). Introgression of new characteristics into commercially cultivated varieties is generally performed by crossing distantly related parents. The F_1 hybrid is back-crossed with a commercial variety. Progeny with positive traits are carried forward for backcrossing to obtain plants with more positive traits and fewer negative ones. Now, researchers are seeking conventional breeding methods to obtain elite moringa plants, which are expected to be available for commercial cultivation soon. Especially in the cultivation of low-yielding moringa varieties, the lack of organised crop management and significant levels of plant mortality during the nursery phase contribute to lower yields. In this context, researchers continuously monitor the performance of modern high-yielding varieties developed by various breeding programmes under suitable climatic conditions in Africa, Asia and America, among others. The only constraint in this method is that the required traits do not come alone but are embedded within much larger segments of wild chromosomes (so-called linkage drag). Some new genes may produce harmful toxins or allergens, negatively impacting crop quality (Ogawa et al., 2000).

1.14.1.3 Elicitors and Elicitation

Elicitor-mediated triggering of secondary metabolite production, particularly for improving food and pharmaceutically important metabolites, has immense potential in plant biotechnology (Yu et al., 2002; Radman et al., 2003). Despite various reports on elicitor-mediated studies across multiple plant systems (Sudha and Ravishankar, 2003; Rudrappa et al., 2006; Sreedhar et al., 2009), the actual mechanism of elicitor actions is not clearly defined, and many current research activities are directed towards understanding a cascade of events at the cellular level under elicitor treatments. Recently, several attempts have been made towards the empirical use of these elicitors in inducing the synthesis of valuable pharmaceutical products, rarely for nutritionally essential molecules (Zhao et al., 2005; Ruiz-García et al., 2013). The exogenous application of elicitors increases the phenolic content of grapes and, hence, the resulting wine (Ruiz-García et al., 2013). An elicitation study conducted on *M. oleifera* using salicylic acid and methyl jasmonate in a 0.1 mM concentration showed a 35.08% to 54% increase in lutein and 23.75% in carotenoid level within 24 hours when compared with the control. Along with lutein and carotenoids, tocopherol levels were also found to rise to a great extent in both studies (Saini et al., 2014d). When methyl jasmonate and salicylic acid were used in an elicitation investigation, Abed et al. (2023) also observed an increase in secondary metabolites in *M. oleifera* callus cultures. Similar types of studies using different elicitors can be undertaken to obtain an enhancement in nutrient content.

1.14.1.4 Endophytes and Plant Growth-Promoting Rhizobacteria

In plants, defence responses are triggered by the microbes associated with the host plant (Nogueira et al., 2001). Some bacteria and fungi that do not harm the plants enter into it as endophytes and establish mutualistic associations. Plant growth-promoting rhizobacteria mix has been shown to augment Fe, folate, carotenoids and tocopherol content in the foliage of *M. oleifera* (Sonbarse et al., 2017; Sonbarse et al., 2020). Endophytes associated with *M. olifera* were reported to produce an array of secondary metabolites that may assist plant defence and forbearance (Perotti, 1926; Hallmann et al., 1997; Azevedo et al., 2000; Hashem et al., 2023). A vast and diverse array of endophytic organisms is present in plants. Endophytic bacteria have been extracted from a wide range of plant species. No plant species exists without endophytes.

Exploring the role of endophytes native to moringa would help improve soil nutrition and its availability to plants through nitrogen fixation (Sevilla et al., 2001; Hurek et al., 2002; Sessitsch et al., 2002; Iniguez et al., 2004).

1.14.1.5 Plant Tissue Culture

Micropropagating plants in vitro has made it possible to cultivate the crop all year round. It helps produce disease-free plants with less energy and space. To date, many different media compositions have been designed for micropropagating moringa (as mentioned in Table 1.7) to overcome seasonal and other growth barriers.

1.14.2 Biotechnological Approaches

1.14.2.1 Intragenic Approach

The intragenic approach combines the benefits of traditional breeding and genetic engineering and helps avoid several of their issues (Rommens et al., 2004; Rommens, 2004). Marker-free transformation systems are necessary to avoid the entry of foreign DNA. An alternate method is co-transformation with a P-DNA and a second 'life-support' transfer DNA that contains two selectable marker genes (Rommens, 2004). Other marker-free transformation systems can be used as well (Bidney and Scelonge, 2001; de Vetten et al., 2003; Rommens et al., 2006). However, such studies in moringa are lacking and need to be pursued (Rommens et al., 2006).

1.14.2.2 Transgenic Approach

In transgenic approaches, new hereditary information is incorporated into the genome of plants (Brookes and Barfoot, 2005). Though the method helps to increase yield and express foreign protein in the new system, there is concern about the expression of foreign proteins; the issue is of importance from the consumer's point of view. Pea legumin, rich in lysine, was successfully expressed in rice and wheat (Singh Sindhu et al., 1997), and sunflower seed albumin, high in methionine, was expressed in lupin (Molvig et al., 1997). Given this, a similar approach in moringa would help obtain plants with augmented levels of targeted nutritionally essential molecules. An efficient substitute method for raising the bioactive content of plants is the application of RNA interference (RNAi), a genetic engineering technique. According to recent studies, the most sophisticated genome editing technique, CRISPR/Cas9, can improve the manufacturing of beneficial substances and cause a specific mutation (Balasubramani et al., 2023). Similar techniques can be used to enhance *M. oleifera*'s nutritional value. Augmenting the efficacy of uptake and transportation to leaves and fruits is possible through iron uptake in roots via the expression of iron transport proteins (Takahashi et al., 2001; Connolly et al., 2002; White and Broadley, 2005).

1.15 CONCLUSION

Numerous studies on *M. oleifera* have revealed its nutrient content. Supplementing the diet with moringa will help meet the body's requirement of proteins, amino acids, vitamins, minerals, unsaturated fatty acids, antioxidants and bioactive molecules, with good health benefits. It is an affordable and sustainable source of nutrients; however, the bioavailability of its contents is a concern. Though moringa is known to contain rich nutraceuticals and is regarded as a miracle plant, its exploitation to combat malnutrition is hindered by several anti-nutritional factors such as saponins, phytates

TABLE 1.7
In Vitro Propagation of *M. oleifera* on Different Media Composition

Sr. no	Media composition	Explant used	Result	Author
1	MS 2% sucrose + 1 mg/l BAP	Nodal explants (matured field plant)	Multiple shoots (approx. 22)	Kantharajah and Dodd, 1991
2	MS + 1 mg/l BAP	Nodal explants (in vitro grown seedlings)	Multiple shoots (4)	Islam et al., 2005
3	MS +1 mg/l 6-BA + 5 g/l karagum	Seedling stem	Adventitious bud induction	Wang and Wei, 2008
	MS + 0.4 mg/l 6-BA + 0.2 mg/l NAA + 5 g/l karagum	Subculturing		
	1/2 MS + 0.4 mg/l IBA + 0.2 mg/l NAA + 7 g/l karagum	Shoots	Roots	
4	MS + 1.5 mg/l BAP + 1.5 mg 2,4-D/l	Seeds	Profuse callus	Talreja, 2010
5	MS + 2.0 mg/l BA	Stem	10–11 shoots	Riyathong et al., 2010
6	MS + 2.5 mg/l BA	Cotyledons	Callus	Nieves and Aspuria, 2011
7	1/2 MS 2.5% Sucrose + NAA0.1 mg/l + IBA 0.2 mg/l	Shoots	Roots	Weiyin, 2011
8	MS + 10.75 μM NAA	Zygotic embryos	Callus	Devendra et al., 2012
	MS + 13.31 μM BAP	Organogenic callus	17.3 embryos per gram of callus	
9	MS + 4.44 μM BA	Nodal explants	Multiple shoots (9)	Saini et al., 2012
	MS + 2.85 μM IAA + 4.92 μM IBA	Shoots	Roots	
10	MS + 3 mg/l BAP + 3 mg/l IBA	Seeds and nodal segments	Callus	Mathur et al., 2014
	MS + 2 mg/l BAP + 0.2 mg/L KN	Callus	Shoots	
	MS + 3 mg/l IBA + 0.5 mg/l NAA	Shoots	Roots	
	MS + 1 mg/l BAP + 1 mg/l NAA + 0.2 mg/l IAA	Nodal segments	Multiple shoots	
11	MS + 2.5 μM BAP + 3% sucrose	Nodal segment and	Shoot	
12	MS + 4.92 μM IBA	shoot tips	Adventitious root	Jun-jie et al., 2017
	MS + 0.8 mg/l BA+0.2 mg/l KA + 0.05 mg/l NAA	Leaf segments	Multiple shoots (4.40 shoots per explant)	
13	MS + 3 mg/l BA	Nodal segment	Multiple shoots (18)	Gupta et al., 2020
14	MS + 1.0 mg/l BA + 0.05 md/l KT	Cotyledonary nodes Apical buds	Shoot buds (4.40 shoots per explant)	Zheng et al., 2022
	MS + 1.2 mg/l BA + 0.05 md/l KT		Shoot buds (4.30 shoots per explant)	

and phenols. Hence, future studies should be focused on addressing these problems. Various biotechnological approaches can also overcome variations in nutritional content among cultivars due to agroclimatic factors, the application of fertilisers, conventional breeding, intragenic approaches, mutagenesis, transgenics, xenogenics, elicitation studies and endophytes.

REFERENCES

Abadu, N. B., Bernardo, B. F., Salvador, Q. D. C., Mercado, V. J. P., Reyes, J. C. B., Luci, K. Q., Villacorte, E. A., & Angeles, J. M. M. (2021). In vitro anthelminthic activity of aqueous leaf extracts from *Leucaena leucocephala* and *Moringa oleifera* against *Caenorhabditis elegans*. *International Journal of Herbal Medicine*, *9*(4), 92–96.

Abbas, R. N., Tanveer, A., Khaliq, A., Iqbal, A., Ghaffari, A. R., Matloob, A., & Maqsood, Q. (2013). Maize (*Zea mays* L.) germination, growth and yield response to foliar application of *Moringa oleifera* Lam. leaf extracts. *Crop Environment*, *4*(1), 39–45.

Abd-Elhalim, B. T. (2022). Moringa oleifera leaf as a natural water purifier and causes decontamination of fecal-coliform bacteria. *Novel Research in Microbiology Journal*, *6*(6), 1783–1800.

Abdulkarim, S. M., Long, K., Lai, O. M., Muhammad, S. K. S., & Ghazali, H. M. (2005). Some physico-chemical properties of *Moringa oleifera* seed oil extracted using solvent and aqueous enzymatic methods. *Food Chemistry*, *93*(2), 253–263.

Abed, A. S., Ismail, E. N., Majeed, D. M., & Al-Jibouri, A. M. J. (2023). Increasing amounts of secondary metabolites and medicinal compounds in callus culture of *Moringa oleifera* (Lam.) using abiotic elicitors. *Iraqi Journal of Science*, *64*(3), 3902–3913.

Adam, D., Yansen, I. A., Fahria, T. I. B., & Nugraha, A. P. (2023). Potential of *Moringa oleifera* as anti-cancer agents in oral cancer: A review. *Malaysian Journal of Medicine and Health Sciences*, *19*, 140–143.

Adeoye, G. O., Sridhar, M. K. C., Adeoluwa, O. O., & Akinsoji, N. A. (2005). Evaluation of naturally decomposed solid wastes from municipal dump sites for their manurial value in southwest Nigeria. *Journal of Sustainable Agriculture*, *26*(4), 143–152.

Agnihotri, N., Pathak, V. K., Khatoon, N., & Rahman, M. (2013). Removal of fluoride from water by *Moringa oleifera* seed residue after oil extraction. *International Journal of Scientific & Engineering Research*, *4*(10), 106–110.

Aguirre-Hernández, E., González-Trujano, M. E., Martínez, A. L., Moreno, J., Kite, G., Terrazas, T., & Soto-Hernández, M. (2010). HPLC/MS analysis and anxiolytic-like effect of quercetin and kaempferol flavonoids from *Tilia americana* var. *mexicana*. *Journal of Ethnopharmacology*, *127*(1), 91–97.

Akhtar, M., Hasany, S. M., Bhanger, M. I., & Iqbal, S. (2007a). Low-cost sorbents for the removal of methyl parathion pesticide from aqueous solutions. *Chemosphere*, *66*(10), 1829–1838.

Akhtar, M., Hasany, S. M., Bhanger, M. I., & Iqbal, S. (2007b). Sorption potential of *Moringa oleifera* pods for the removal of organic pollutants from aqueous solutions. *Journal of Hazardous Materials, 141*(3), 546–556.

Aleman-Ramirez, J. L., Moreira, J., Torres-Arellano, S., Longoria, A., Okoye, P. U., & Sebastian, P. J. (2021). Preparation of a heterogeneous catalyst from moringa leaves as a sustainable precursor for biodiesel production. *Fuel*, *284*, 118983.

Al-Hadidy, A. A. A., & Dawood, S. T. (2021). The effect of *Moringa oleifera* leaves powder on some hormones to prevent the development of experimental hypothyroidism in rabbits. *Indian Journal of Forensic Medicine & Toxicology*, *15*(3), 1040–1047.

Ali, A., Akhtar, N., Khan, M. S., Khan, M. T., Ullah, A., & Shah, M. I. (2013). Effect of *Moringa oleifera* on undesirable skin sebum secretions of sebaceous glands observed during winter season in humans. *Biomedical Research*, *24*(1), 127–130.

Al-Jadabi, N., Laaouan, M., El Hajjaji, S., Mabrouki, J., Benbouzid, M., & Dhiba, D. (2023). The dual performance of *Moringa oleifera* seeds as eco-friendly natural coagulant and as an antimicrobial for wastewater treatment: A review. *Sustainability*, *15*(5), 4280.

Almeida, A. A. P., Farah, A., Silva, D. A. M., Nunan, E. A., & Glória, M. B. A. (2006). Antibacterial activity of coffee extracts and selected coffee chemical compounds against enterobacteria. *Journal of Agricultural and Food Chemistry*, *54*(23), 8738–8743.

Almeida, I. L. S., Antoniosi Filho, N. R., Alves, M. I. R., Carvalho, B. G., & Coelho, N. M. M. (2012). Removal of BTEX from aqueous solution using *Moringa oleifera* seed cake. *Environmental Technology*, *33*(11), 1299–1305.

Amaglo, N. K., Bennett, R. N., Lo Curto, R. B., Rosa, E. A. S., Lo Turco, V., Curto, A. L., Crea, F., & Timpo, G. M. (2010). Profiling selected phytochemicals and nutrients in different tissues of the multipurpose tree *Moringa oleifera* L., grown in Ghana. *Food Chemistry*, *122*(4), 1047–1054.

Ames, B. N., Profet, M., & Gold, L. S. (1990). Nature's chemicals and synthetic chemicals: Comparative toxicology. *Proceedings of the National Academy of Sciences of the United States of America*, *87*(19), 7782–7786.

Andrews, D. A., & Andrews, K. (2009). Nutraceutical Moringa composition. U.S. Patent Application 12/338,789.

Anjorin, T. S., Ikokoh, P., & Okolo, S. (2010). Mineral composition of *Moringa oleifera* leaves, pods and seeds from two regions in Abuja, Nigeria. *International Journal of Agriculture and Biology, 7*, 431–434.

Anwar, F., & Bhanger, M. I. (2003). Analytical characterization of *Moringa oleifera* seed oil grown in temperate regions of Pakistan. *Journal of Agricultural and Food Chemistry, 51*(22), 6558–6563.

Anwar, F., Ashraf, M., & Bhanger, M. I. (2005). Interprovenance variation in the composition of *Moringa oleifera* oilseeds from Pakistan. *Journal of the American Oil Chemists' Society, 82*(1), 45–51.

Anwar, F., Latif, S., Ashraf, M., & Gilani, A. H. (2007). *Moringa oleifera*: A food plant with multiple medicinal uses. *Phytotherapy Research, 21*(1), 17–25.

Aparna, V., Dileep, K. V., Mandal, P. K., Karthe, P., Sadasivan, C. S., & Haridas, M. (2012). Anti-inflammatory property of n-hexadecanoic acid: Structural evidence and kinetic assessment. *Chemical Biology & Drug Design, 80*(3), 434–439.

Arabshahi-Delouee, S. A. E. D. E. H., Aalami, M., & Urooj, A. (2011). Drumstick (*Moringa oleifera* L.) leaves: A potential source of natural lipid antioxidants. *Journal of Food Process Engineering, 34*(3), 947–959.

Araújo, C. S. T., Carvalho, D. C., Rezende, H. C., Almeida, I. L. S., Coelho, L. M., Coelho, N. M. M., Marques, T. L., & Alves, V. N. (2013). Bioremediation of waters contaminated with heavy metals using *Moringa oleifera* seeds as biosorbent. *Applied Bioremediation – Active and Passive Approaches, 2*(5), 779–786.

Aregheore, E. M. (2002). Intake and digestibility of *Moringa oleifera*–batiki grass mixtures by growing goats. *Small Ruminant Research, 46*(1), 23–28.

Arora, R. K., & Pandey, A. (1996). *Wild edible plants of India: Diversity, conservation and use.* Indian Council of Agricultural Research, National Bureau of Plant Genetic Resources.

Aslam, M. (2005). Mineral composition of *Moringa oleifera* leaves and pods from different regions of Punjab, Pakistan. *Asian Journal of Plant Science, 4*, 417–421.

Athikomkulchai, S., Tunit, P., Tadtong, S., Jantrawut, P., Sommano, S. R., & Chittasupho, C. (2020). *Moringa oleifera* seed oil formulation physical stability and chemical constituents for enhancing skin hydration and antioxidant activity. *Cosmetics, 8*(1), Article 2.

Atteya, A. K. G., Albalawi, A. N., El-Serafy, R. S., Albalawi, K. N., Bayomy, H. M., & Genaidy, E. A. E. (2021). Response of *Moringa oleifera* seeds and fixed oil production to vermicompost and NPK fertilizers under calcareous soil conditions. *Plants, 10*(10), Article 1998.

Awad, M. A. G., Virk, P., Wagealla, M. A. E., Alsai f, S. S. A., Hendi, A. A., Ortashi, K. M. O., & Qindeel, R. (2023, April 4). *Moringa oleifera nanoparticles* (U.S. Patent No. 11,617,719 B1). U.S. Patent and Trademark Office.

Azevedo, J. L., Maccheroni Jr, W., Pereira, J. O., & De Araújo, W. L. (2000). Endophytic microorganisms: A review on insect control and recent advances on tropical plants. *Electronic Journal of Biotechnology, 3*(1), 15–16.

Balasubramani, S., Chen, Q., Zhou, Z., Moola, A. K., Duraisamy, S. M., Prakash, P., Gayathiri, E., Satish, L., & Swamy, M. K. (2023). Transgenic medicinal plants for improved plant metabolites production. In M. K. Swamy & M. S. Akhtar (Eds.), In *Phytochemical genomics: Plant metabolomics and medicinal plant genomics* (pp. 403–415). Springer Nature Singapore.

Barhoi, D., Upadhaya, P., Barbhuiya, S. N., Giri, A., & Giri, S. (2021). Aqueous extract of *Moringa oleifera* exhibit potential anticancer activity and can be used as a possible cancer therapeutic agent: A study involving in vitro and in vivo approach. *Journal of the American College of Nutrition, 40*(1), 70–85.

Bauer, H., Kozianka, F., & Endisch, S. (2010). Wastewater purification plant by means of plants. U.S. Patent No. 7,718,062.

Benhammouche, T., Melo, A., Martins, Z., Faria, M. A., Pinho, S. C. M., Ferreira, I. M. L. P. V. O., & Zaidi, F. (2021). Nutritional quality of protein concentrates from *Moringa oleifera* leaves and in vitro digestibility. *Food Chemistry, 348*, Article 128858.

Bennett, R. N., Mellon, F. A., Foidl, N., Pratt, J. H., Dupont, M. S., Perkins, L., & Kroon, P. A. (2003). Profiling glucosinolates and phenolics in vegetative and reproductive tissues of the multi-purpose trees *Moringa oleifera* L. (horseradish tree) and *Moringa stenopetala* L. *Journal of Agricultural and Food Chemistry, 51*(12), 3546–3553.

Berkovich, L., Earon, G., Ron, I., Rimmon, A., Vexler, A., & Lev-Ari, S. (2013). *Moringa oleifera* aqueous leaf extract down-regulates nuclear factor-kappaB and increases cytotoxic effect of chemotherapy in pancreatic cancer cells. *BMC Complementary and Alternative Medicine, 13*, Article 1.

Bernhoft, A. (2010). A brief review on bioactive compounds in plants. In A. Bernhoft (Ed.), *Bioactive compounds in plants: Benefits and risks for man and animals* (Vol. 50, pp. 11–17). Novus Forlag.

Bhadresha, K., Thakore, V., Brahmbhatt, J., Upadhyay, V., Jain, N., & Rawal, R. (2022). Anticancer effect of *Moringa oleifera* leaves extract against lung cancer cell line via induction of apoptosis. *Advances in Cancer Biology-Metastasis*, *6*, Article 100072.
Bhatia, P., Sharma, A., George, A. J., Anvitha, D., Kumar, P., Dwivedi, V. P., & Chandra, N. S. (2021). Antibacterial activity of medicinal plants against ESKAPE: An update. *Heliyon*, *7*(2), e06310.
Bhattacharya, S. B., Das, A. K., & Banerji, N. (1982). Chemical investigations on the gum exudate from sajna (*Moringa oleifera*). *Carbohydrate Research*, *102*(1), 253–262.
Bhatti, H. N., Mumtaz, B., Hanif, M. A., & Nadeem, R. (2007). Removal of Zn (II) ions from aqueous solution using *Moringa oleifera* Lam. (horseradish tree) biomass. *Process Biochemistry*, *42*(4), 547–553.
Bhuptawat, H., Folkard, G. K., & Chaudhari, S. (2007). Innovative physico-chemical treatment of wastewater incorporating *Moringa oleifera* seed coagulant. *Journal of Hazardous Materials*, *142*(1–2), 477–482.
Bidney, D. L., &-Scelonge, C. J. (2001). Method of plant transformation. U.S. Patent No. 6,265,638.
Braham, F., Carvalho, D. O., Almeida, C. M. R., Zaidi, F., Magalhães, J. M. C. S., Guido, L. F., & Gonçalves, M. P. (2020). Online HPLC-DPPH screening method for evaluation of radical scavenging phenols extracted from *Moringa oleifera* leaves. *South African Journal of Botany*, *129*, 146–154.
Brochado, C. de O., Almeida, A. P. de, Barreto, B. P., Costa, L. P., Ribeiro, L. S., Pereira, R. L. da C., Koatz, V. L. G., & Costa, S. S. (2003). Flavonolrobinobiosides and rutinosides from *Alternanthera brasiliana* (Amaranthaceae) and their effects on lymphocyte proliferation in vitro. *Journal of the Brazilian Chemical Society*, *14*, 449–451.
Brookes, G., & Barfoot, P. (2005). *GM crops: The global economic and environmental impact: The first nine years 1996–2004*. PG Economics Ltd.
Brown, J. H., & Kleiman, R. (2003a). Ultra-stable composition comprising moringa oil and its derivatives and uses thereof. U.S. Patent No. 6,528,075 B1.
Brown, J. H., & Kleiman, R. (2003b). Enhanced slip and/or break strength; massage oils, cosmetics. U.S. Patent No. 6,667,047 B2.
Brown, J. H., & Kleiman, R. (2010). Ultra-stable composition comprising *Moringa* oil and its derivatives and uses thereof. U.S. Patent No. 7,691,397 B2.
Brunelli, D., Tavecchio, M., Falcioni, C., Frapolli, R., Erba, E., Iori, R., Rollin, P., et al. (2010). The isothiocyanate produced from glucomoringin inhibits NF-κB and reduces myeloma growth in nude mice in vivo. *Biochemical Pharmacology*, *79*(8), 1141–1148.
Bwana-Simba, E. (2006). *The various uses of Moringa oleifera*. http://www.runetwork.de/html/en/index.html?article_id=3862
Caceres, A., Cabrera, O., Morales, O., Mollinedo, P., & Mendia, P. (1991). Pharmacological properties of *Moringa oleifera* I: Preliminary screening for antimicrobial activity. *Journal of Ethnopharmacology*, *33*(3), 213–216.
Chaieb, K., Hajlaoui, H., Zmantar, T., Ben Kahla-Nakbi, A., Rouabhia, M., Mahdouani, K., & Bakhrouf, A. (2007). The chemical composition and biological activity of clove essential oil, *Eugenia caryophyllata* (Syzigium aromaticum L., Myrtaceae): A short review. *Phytotherapy Research*, *21*(6), 501–506.
Chakraborty, A., Das, D. K., Sinha, M., Dey, S., & Bhattacharjee, S. (2013). *Moringa oleifera* leaf-extract mediated green synthesis of stabilized gold nanoparticles. *Journal of Bionanoscience*, *7*(4), 415–419.
Chen, H. M., & Chuang, P. H. (2008). Moringa crude extracts and their derived fractions with antifungal activities. U.S. Patent No. 7,404,975 B2.
Choi, E.-J., Debnath, T., Tang, Y., Ryu, Y.-B., Moon, S.-H., & Kim, E.-K. (2016). Topical application of *Moringa oleifera* leaf extract ameliorates experimentally induced atopic dermatitis by the regulation of Th1/Th2/Th17 balance. *Biomedicine & Pharmacotherapy*, *84*, 870–877.
Chuang, P.-H., Lee, C.-W., Chou, J.-Y., Murugan, M., Shieh, B.-J., & Chen, H.-M. (2007). Anti-fungal activity of crude extracts and essential oil of *Moringa oleifera* Lam. *Bioresource Technology*, *98*(1), 232–236.
Chumark, P., Khunawat, P., Sanvarinda, Y., Phornchirasilp, S., Morales, N. P., Phivthong-Ngam, L., Ratanachamnong, P., Srisawat, S., & Klai-Upsorn, S. (2008). The in vitro and ex vivo antioxidant properties, hypolipidaemic and antiatherosclerotic activities of water extract of *Moringa oleifera* Lam. leaves. *Journal of Ethnopharmacology*, *116*(3), 439–446.
Connolly, E. L., Fett, J. P., & Guerinot, M. L. (2002). Expression of the IRT1 metal transporter is controlled by metals at the levels of transcript and protein accumulation. *The Plant Cell*, *14*(6), 1347–1353.
Curtis, T., & Williams, D. G. (1994). *Introduction to perfumery*. Micelle Press.
Dahlgren, G. (1989). The last Dahlgrenogram. System of classification of the dicotyledons. In K. Tan (Ed.), *Plant taxonomy, phytogeography and related subjects: The Davis & Hedge festschrift* (pp. 249–260). Edinburgh University Press.

Dahot, M. U. (1998). Antimicrobial activity of small protein of *Moringa oleifera* leaves. *Journal of the Islamic Academy of Sciences*, *11*(1), 6.

Dai, J.-L., Zhu, Y.-G., Zhang, M., & Huang, Y.-Z. (2004). Selecting iodine-enriched vegetables and the residual effect of iodate application to soil. *Biological Trace Element Research*, *101*, 265–276.

Das, B. R., Kurup, P. A., & Narasimha Rao, P. L. (1957). Antibiotic principle from *Moringa pterygosperma*. VII. Antibacterial activity and chemical structure of compounds related to pterygospermin. *The Indian Journal of Medical Research*, *45*(2), 191–196.

Datta, R. M., & Mitra, J. N. (1947). The systematic position of the family Moringaceae based on the study of *Moringa pterygosperma* Gaertn. (M. oleifera Lamk.). *Journal of the Bombay Natural History Society*, *47*, 335–358

de Vetten, N., Wolters, A.-M., Raemakers, K., van der Meer, I., ter Stege, R., Heeres, E., Heeres, P., & Visser, R. (2003). A transformation method for obtaining marker-free plants of a cross-pollinating and vegetatively propagated crop. *Nature Biotechnology*, *21*(4), 439–442.

Devendra, B. N., Talluri, V. S. S. L. P., & Srinivas, N. (2012). Callus induction and somatic embryogenesis of *Moringa oleifera* Lam.—an anti-radiation plant. *Journal of Agricultural Technology*, *8*(6), 1953–1963.

Devi, R., Arcot, J., Sotheeswaran, S., & Ali, S. (2008). Folate contents of some selected Fijian foods using tri-enzyme extraction method. *Food Chemistry*, *106*(3), 1100–1104.

Dorries, S., Marison, I., Mermod, N., Meyer, E., & Suarez, M. (2004). Coagulation agents for water treatment and/or antibiotic agents; genetic engineering. US Patent Application 2004/0147723 A1.

Ebrahimi, S. N., Hadian, J., Mirjalili, M. H., Sonboli, A., & Yousefzadi, M. (2008). Essential oil composition and antibacterial activity of *Thymus caramanicus* at different phenological stages. *Food Chemistry*, *110*(4), 927–931.

Eilert, U., Wolters, B., & Nahrstedt, A. (1981). The antibiotic principle of seeds of *Moringa oleifera* and *Moringa stenopetala*. *Planta Medica*, *42*(5), 55–61.

Esmaeili, H., Yeganeh, G., & Esmaeilzadeh, F. (2019). Optimization of biodiesel production from *Moringa oleifera* seed oil in the presence of nano-MgO using Taguchi method. *International Nano Letters*, *9*(3), 257–263.

Fahey, J. W. (2005). *Moringa oleifera*: A review of the medical evidence for its nutritional, therapeutic, and prophylactic properties. Part 1. *Trees for Life Journal*, *1*(5), 1–15.

Fahey, J. W., Zalcmann, A. T., & Talalay, P. (2001). The chemical diversity and distribution of glucosinolates and isothiocyanates among plants. *Phytochemistry*, *56*(1), 5–51.

Faizi, S., Siddiqui, B. S., Saleem, R., Siddiqui, S., Aftab, K., & Gilani, A.-H. (1994). Isolation and structure elucidation of new nitrile and mustard oil glycosides from *Moringa oleifera* and their effect on blood pressure. *Journal of Natural Products*, *57*(9), 1256–1261.

Faizi, S., Siddiqui, B. S., Saleem, R., Siddiqui, S., & Aftab, K. (1995). Fully acetylated carbamate and hypotensive thiocarbamate glycosides from *Moringa oleifera*. *Phytochemistry*, *38*(4), 957–963.

Fakankun, O. A., Babayemi, J. O., & Utiaruk, J. J. (2013). Variations in the mineral composition and heavy metals content of *Moringa oleifera*. *African Journal of Environmental Science and Technology*, *7*(6), 371–379.

Farooq, F., Rai, M., Tiwari, A., Khan, A. A., & Farooq, S. (2012). Medicinal properties of *Moringa oleifera*: An overview of promising healer. *Journal of Medicinal Plants Research*, *6*(27), 4368–4374.

Ferreira, P. M. P., Farias, D. F., Oliveira, J. T. d. A., & Carvalho, A. F. U. (2008). *Moringa oleifera*: Bioactive compounds and nutritional potential. *Revista de Nutrição*, *21*, 431–437.

Foidl, N., & Paull, R. (2008). *Moringa oleifera*. In In J. Janick & R. E. Paull (Eds.), *Encyclopedia of fruit and nuts* (pp. 509-512). CABI.

Freiberger, C. E., Vanderjagt, D. J., Pastuszyn, A., Glew, R. S., Mounkaila, G., Millson, M., & Glew, R. H. (1998). Nutrient content of the edible leaves of seven wild plants from Niger. *Plant Foods for Human Nutrition*, *53*, 57–69.

Fuglie, L. J. (2000). *New uses of Moringa studied in Nicaragua*. ECHO Development Notes, 68. ECHO.

Fuglie, L. J. (2001). Combating malnutrition with *Moringa*. In L. J. Fuglie (Ed.), In *Miracle tree: Multiple attributes of Moringa* (pp. 117–136). CTA Publications.

Gad, N., Sekara, A., & Abdelhamid, M. T. (2019). The potential role of cobalt and/or organic fertilizers in improving the growth, yield, and nutritional composition of *Moringa oleifera*. *Agronomy*, *9*(12), Article 862.

Galuppo, M., De Nicola, G. R., Iori, R., Dell'Utri, P., Bramanti, P., & Mazzon, E. (2013). Antibacterial activity of glucomoringin bioactivated with myrosinase against two important pathogens affecting the health of long-term patients in hospitals. *Molecules*, *18*(11), 14340–14348.

Garg, P., Pundir, S., Ali, A., Panja, S., Chellappan, D. K., Dua, K., Kulshrestha, S., & Negi, P. (2024). Exploring the potential of *Moringa oleifera* Lam in skin disorders and cosmetics: Nutritional analysis, phytochemistry, geographical distribution, ethnomedicinal uses, dermatological studies and cosmetic formulations. *Naunyn-Schmiedeberg's Archives of Pharmacology, 397*(6), 3635–3662.

Gassenschmidt, U., Jany, K. D., Tauscher, B., & Niebergall, H. (1995). Isolation and characterization of a flocculating protein from *Moringa oleifera* Lam. *Biochimica et Biophysica Acta (BBA) – General Subjects, 1243*(3), 477–481.

Gautier, A., Duarte, C. M., & Sousa, I. (2022). *Moringa oleifera* seeds characterization and potential uses as food. *Foods, 11*(11), Article 1629.

Gbodossou, E. V. A. (2002). Medicinal plant extracts used in the treatment of diabetic diseases (WO 02/094299) [Patent].

Gendel, S. M., & Jenkins, J. A. (2006). Allergen sequence databases. *Molecular Nutrition & Food Research, 50*(7), 633–637.

Gharsallah, K., Rezig, L., Msaada, K., Chalh, A., & Soltani, T. (2021). Chemical composition and profile characterization of *Moringa oleifera* seed oil. *South African Journal of Botany, 137*, 475–482.

Ghasi, S., Nwobodo, E., & Ofili, J. O. (2000). Hypocholesterolemic effects of crude extract of *Moringa oleifera* leaf in high-fat diet fed Wistar rats. *Journal of Ethnopharmacology, 69*(1), 21–25.

Ghebremichael, K. A., & Hultman, B. (2004). Alum sludge dewatering using *Moringa oleifera* as a conditioner. *Water, Air, and Soil Pollution, 158*, 153–167.

Gidamis, A. B., Panga, J. T., Sarwatt, S. V., Chove, B. E., & Shayo, N. B. (2003). Nutrient and antinutrient contents in raw and cooked young leaves and immature pods of *Moringa oleifera* Lam. *Ecology of Food and Nutrition, 42*(6), 399–411.

Gilani, A. H., Aftab, K., Shaheen, F., Siddiqui, B. S., Siddiqui, S., Saleem, R., & Faizi, S. (1992). Antispasmodic activity of active principle from *Moringa oleifera*. In CapassoF. & N. Mascolo (Eds.), *Natural drugs and the digestive tract* (pp. 60–63). EMSI.

Gilani, A. H., Aftab, K., Suria, A., Siddiqui, S., Saleem, R., Siddiqui, B. S., & Faizi, S. (1994). Pharmacological studies on hypotensive and spasmolytic activities of pure compounds from *Moringa oleifera*. *Phytotherapy Research, 8*(2), 87–91.

Gilani, A. H., Janbaz, K. H., & Shah, B. H. (1997). Quercetin exhibits hepatoprotective activity in rats. *Biochemical Society Transactions, 25*(4), S619–S619.

Givianrad, M. H. (2012). Characterization and assessment of flavor compounds and some allergens in three Iranian rice cultivars during gelatinization process by HS-SPME/GC-MS. *Journal of Chemistry, 9*(2), 716–728.

Gokaraju, G. R., Gokaraju, R. R., Golakoti, T., Chirravuri, V. R., Somepalli, V., & Bhupathiraju, K. (2013). Synergistic phytochemical composition for the treatment of obesity. U.S. Patent No. 8,541,383.

Gomez, R., & Gomez, R. (2011). Dietary supplement composition (WO 2011/098819) [Patent].

Granado, F., Olmedilla, B., & Blanco, I. (2003). Nutritional and clinical relevance of lutein in human health. *British Journal of Nutrition, 90*(3), 487–502.

Grusak, M. A., & Cakmak, I. (2005). Methods to improve the crop-delivery of minerals to humans and livestock. In M. R. Broadley & P. J. White (Eds.), *Plant nutritional genomics* (pp. 265–286). Routledge.

Grusak, M. A., & DellaPenna, D. (1999). Improving the nutrient composition of plants to enhance human nutrition and health. *Annual Review of Plant Biology, 50*, 133–161.

Guevara, A. P., Vargas, C., Sakurai, H., Fujiwara, Y., Hashimoto, K., Maoka, T., Kozuka, M., Ito, Y., Tokuda, H., & Nishino, H. (1999). An antitumor promoter from *Moringa oleifera* Lam. *Mutation Research/Genetic Toxicology and Environmental Mutagenesis, 440*(2), 181–188.

Gupta, M. B., Nath, R., Srivastava, N., Shanker, K., Kishor, K., & Bhargava, K. P. (1980). Anti-inflammatory and antipyretic activities of β-sitosterol. *Planta Medica, 39*(6), 157–163.

Gupta, R., Sharma, A. K., Dobhal, M. P., Sharma, M. C., & Gupta, R. S. (2011). Antidiabetic and antioxidant potential of β-sitosterol in streptozotocin-induced experimental hyperglycemia. *Journal of Diabetes, 3*(1), 29–37.

Gupta, R., Dubey, D. K., Kannan, G. M., & Flora, S. J. S. (2007). Remediation of arsenic-induced oxidative stress in mice via concomitant administration of *Moringa oleifera* seed powder. *Cell Biology International, 31*(1), 44–56.

Gupta, R., Kannan, G. M., Sharma, M., & Flora, S. J. S. (2005). Therapeutic effects of *Moringa oleifera* on arsenic-induced toxicity in rats. *Environmental Toxicology and Pharmacology, 20*(3), 456–464.

Gupta, S., Kachhwaha, S., Kothari, S. L., & Jain, R. (2020). Synergistic effect of cytokinins and auxins enables mass clonal multiplication of drumstick tree (*Moringa oleifera* Lam.): A wonder. *In Vitro Cellular & Developmental Biology–Plant, 56*, 458–469.

Hafeez, A., Tipu, M. I., Saleem, M. H., Al-Ashkar, I., Saneoka, H., & El Sabagh, A. (2022). Foliar application of moringa leaf extract enhanced antioxidant system, growth, and biomass-related attributes in safflower plants. *South African Journal of Botany*, *150*, 1087–1095.

Hallmann, J., Quadt-Hallmann, A., Mahaffee, W. F., & Kloepper, J. W. (1997). Bacterial endophytes in agricultural crops. *Canadian Journal of Microbiology*, *43*(10), 895–914.

Hartikainen, H. (2005). Biogeochemistry of selenium and its impact on food chain quality and human health. *Journal of Trace Elements in Medicine and Biology*, *18*(4), 309–318.

Hashem, A. H., Al-Askar, A. A., Elgawad, H. A., & Abdelaziz, A. M. (2023). Bacterial endophytes from *Moringa oleifera* leaves as a promising source for bioactive compounds. *Separations*, *10*(7), Article 395.

Hendra, R., Ahmad, S., Sukari, A., Shukor, M. Y., & Oskoueian, E. (2011). Flavonoid analyses and antimicrobial activity of various parts of *Phaleria macrocarpa* fruit. *International Journal of Molecular Sciences*, *12*(6), 3422–3431.

Hurek, T., Handley, L. L., Reinhold-Hurek, B., & Piché, Y. (2002). *Azoarcus* grass endophytes contribute fixed nitrogen to the plant in an unculturable state. *Molecular Plant–Microbe Interactions*, *15*(3), 233–242.

Iniguez, A. L., Dong, Y., & Triplett, E. W. (2004). Nitrogen fixation in wheat provided by *Klebsiella pneumoniae* 342. *Molecular Plant–Microbe Interactions*, *17*(10), 1078–1085.

Irikannu, K. C., Umeh, A. C., Okenu, J. K., Nwagbo, C. C., Nzeukwu, C. I., Chukwubude, C. B., Nwadike, C. C., Aniefuna, C. O., Obiefule, I. E., & Nnatuanya, I. O. (2024). Preliminary assessment of *Moringa oleifera* seed as a flocculation agent for purification of drinking water. *Nigerian Journal of Parasitology*, *45*(1), 21–27.

Islam, S., Jahan, M. A. A., & Khatun, R. (2005). In vitro regeneration and multiplication of year-round fruit-bearing *Moringa oleifera* L. *Journal of Biological Sciences*, *5*, 145–148.

Islam, Z., Islam, S. R., Hossen, F., Mahtab-ul-Islam, K., Hasan, M. R., & Karim, R. (2021). *Moringa oleifera* is a prominent source of nutrients with potential health benefits. *International Journal of Food Science*, *2021*, Article 6627265.

Jayawardana, B. C., Liyanage, R., Lalantha, N., Iddamalgoda, S., & Weththasinghe, P. (2015). Antioxidant and antimicrobial activity of drumstick (*Moringa oleifera*) leaves in herbal chicken sausages. *LWT – Food Science and Technology*, *64*(2), 1204–12108.

Jideani, V. A., & Diedericks, C. F. (2014). Nutritional, therapeutic, and prophylactic properties of *Vigna subterranea* and *Moringa oleifera*. In O. Oguntibeju (Ed.). *Antioxidant-antidiabetic agents and human health* (Vol. 9, pp. 187–201). IntechOpen.

Jikah, A. N., & Edo, G. I. (2023). *Moringa oleifera*: A valuable insight into recent advances in medicinal uses and pharmacological activities. *Journal of the Science of Food and Agriculture*, *103*(15), 7343–7361.

Johnson, I. T. (2002). Glucosinolates: Bioavailability and importance to health. *International Journal for Vitamin and Nutrition Research*, *72*(1), 26–31.

Joshi, P., & Mehta, D. (2010). Effect of dehydration on the nutritive value of drumstick leaves. *Journal of Metabolomics and Systems Biology*, *1*(1), Article 5.

Juge, N., Mithen, R. F., & Traka, M. (2007). Molecular basis for chemoprevention by sulforaphane: A comprehensive review. *Cellular and Molecular Life Sciences*, *64*, 1105–1127.

Jun-jie, Z., Yue-sheng, Y., Meng-fei, L., Shu-qi, L., Yi, T., Han-bin, C., & Xiao-yang, C. (2017). An efficient micropropagation protocol for direct organogenesis from leaf explants of an economically valuable plant, drumstick (*Moringa oleifera* Lam.). *Industrial Crops and Products*, *103*, 59–63.

Kantharajah, A. S., & Dodd, W. A. (1991). Rapid clonal propagation of *Moringa oleifera* Lam., using tissue culture. *South Indian Horticulture*, *39*, 224–228.

Kar, S., Mukherjee, A., Ghosh, M., & Bhattacharyya, D. K. (2013). Utilization of *Moringa* leaves as a valuable food ingredient in biscuit preparation. *International Journal of Applied Sciences and Engineering*, *1*(1), 29–37.

Kerharo, P. J. (1969). Un remèdepopulairesénégalais: Le "Nebreday" (Moringa oleifera Lann.) emploisthérapeutiquesen milieu africain: chimie et pharmacologie. *Plantes Médicinales et Phytothérapie*, *3*, 14–219.

Khalid, S., Arshad, M., Raza, K., Mahmood, S., Siddique, F., Aziz, N., Khan, S., Khalid, W., Al-Farga, A., & Aqlan, F. (2023). Assessment of hepatoprotective, nephroprotective efficacy, and antioxidative potential of Moringa oleifera leaf powder and ethanolic extract against PCOS-induced female albino mice (Mus musculus). *Food Science & Nutrition*, *11*(11), 7206–7217.

Khanuja, S. P. S., Arya, J. S., Tiruppadiripuliyur, R. S. K., Saikia, D., Kaur, H., Singh, M., Gupta, S. C., et al. (2005). Nitrile glycoside useful as a bioenhancer of drugs and nutrients, process of its isolation from Moringa oleifera. U.S. Patent No. 6,858,588.

Khor, K. Z., Lim, V., Moses, E. J., & Samad, N. A. (2018). *The in vitro and in vivo anticancer properties of Moringa oleifera. Evidence-Based Complementary and Alternative Medicine*, *2018*, Article 1071243.

Kituyi, J. L., Foulkes, M., Worsfold, P., Ongulu, R. A., Kiplagat, A., & Gachanja, A. (2013). *Efficiency of pre-treated Moringa oleifera for the removal of Cd^{2+} and Zn^{2+} ions from wastewaters. Ecohydrology & Hydrobiology*, *13*(4), 267–271.

Klimek-Szczykutowicz, M., Gaweł-Bęben, K., Rutka, A., Blicharska, E., Tatarczak Michalewska, M., Kulik-Siarek, K., ... Szopa, A. (2024). Moringa oleifera (drumstick tree)—nutraceutical, cosmetological and medicinal importance: A review. *Frontiers in Pharmacology*, *15*, 1288382.

Krishnan, T. (2012). *Investigation of Moringa oleifera as potential green coagulant for removal of phenol from aqueous solution* [Unpublished undergraduate thesis]. UniversitiTeknologi PETRONAS.

Kumari, P., Sharma, P., Srivastava, S., & Srivastava, M. M. (2006). Biosorption studies on shelled Moringa oleifera Lamarck seed powder: Removal and recovery of arsenic from aqueous system. *International Journal of Mineral Processing*, *78*(3), 131–139.

Kuppam, C. R. (2006). Herbal composition for treating or alleviating vascular headaches, neurological conditions and neurodegenerative diseases. U.S. Patent No. 7,070,817.

Lakka, A., Grigorakis, S., Kaltsa, O., Karageorgou, I., Batra, G., Bozinou, E., Lalas, S., & Makris, D. P. (2019). The effect of ultrasonication pretreatment on the production of polyphenol-enriched extracts from Moringa oleifera L. (drumstick tree) using a novel bio-based deep eutectic solvent. *Applied Sciences*, *10*(1), 220.

Leone, A., Spada, A., Battezzati, A., Schiraldi, A., Aristil, J., & Bertoli, S. (2015). Cultivation, genetic, ethnopharmacology, phytochemistry and pharmacology of Moringa oleifera leaves: An overview. *International Journal of Molecular Sciences*, *16*(6), 12791–12835.

Leu, Y. -L., Hwang, T. -L., Kuo, P. -C., Liou, K. -P., Huang, B.-S., & Chen, G.-F. (2012). Constituents from Vigna vexillata and their anti-inflammatory activity. *International Journal of Molecular Sciences*, *13*(8), 9754–9768.

Lin, X., Wu, L., Wang, X., Yao, L., & Wang, L. (2021). Ultrasonic-assisted extraction for flavonoid compounds content and antioxidant activities of India Moringa oleifera L. leaves: Simultaneous optimization, HPLC characterization and comparison with other methods. *Journal of Applied Research on Medicinal and Aromatic Plants*, *20*, 100284.

Lindsay, J., Laurin, D., Verreault, R., Hébert, R., Helliwell, B., Hill, G. B., & McDowell, I. (2002). *Risk factors for Alzheimer's disease: A prospective analysis from the Canadian study of health and aging. American Journal of Epidemiology*, *156*(5), 445–453.

Lipipun, V., Kurokawa, M., Suttisri, R., Taweechotipatr, P., Pramyothin, P., Hattori, M., & Shiraki, K. (2003). Efficacy of Thai medicinal plant extracts against herpes simplex virus type 1 infection in vitro and in vivo. *Antiviral Research*, *60*(3), 175–180.

Lynn, B. D., Jackson, G. L., Lynn, M. C., Rosemarie, O., & Richard, R. L. (2008). Personal care composition comprising botanical extract of ficus benghalensis. (WO 2008/104941).

Madsen, M., Schlundt, J., & Omer, E. F. E. (1988). Effect of water coagulation by seeds of Moringa oleifera on bacterial concentrations. *Journal of Ethnopharmacology*, *22*(3), 329–331.

Makkar, H. P. S., & Becker, K. (1997). Nutrients and antiquality factors in different morphological parts of the Moringa oleifera tree. *The Journal of Agricultural Science*, *128*(3), 311–322.

Makonnen, E., Hunde, A., & Damecha, G. (1997). Hypoglycaemic effect of Moringa stenopetala aqueous extract in rabbits. *Phytotherapy Research*, *11*(2), 147–148.

Mali, S., Bendre, S., & Patil, S. (2022). Overview of pharmacognostical and pharmacological properties of Moringa oleifera. *Asian Journal of Pharmacy and Technology*, *12*(1), 77–83.

Manaois, R. V., Morales, A. V., & Abilgos-Ramos, R. G. (2013). Acceptability, shelf life and nutritional quality of moringa-supplemented rice crackers. *Philippine Journal of Crop Science*, *38*(2), 1–8.

Manzoor, M., Anwar, F., Iqbal, T., & Bhanger, M. I. (2007). Physico-chemical characterization of Moringa concanensis seeds and seed oil. *Journal of the American Oil Chemists' Society*, *84*, 413–419.

Marrufo, T., Nazzaro, F., Mancini, E., Fratianni, F., Coppola, R., De Martino, L., Agostinho, A. B., & De Feo, V. (2013). Chemical composition and biological activity of the essential oil from leaves of Moringa oleifera Lam. cultivated in Mozambique. *Molecules*, *18*(9), 10989–11000.

Martin, W. J. (2011, filed August 25). Moringa oil mediated activation of the alternative cellular energy pathway in the therapy of diseases (U.S. Patent Application 12/712,110).

Mathur, M., Yadav, S., Katariya, P. K., & Kamal, R. (2014). In vitro propagation and biosynthesis of steroidal sapogenins from various morphogenetic stages of *Moringa oleifera* Lam., and their antioxidant potential. *Acta Physiologiae Plantarum*, *36*, 1749–1762.

Mau, J.-L., Ko, P.-T., & Chyau, C.-C. (2003). Aroma characterization and antioxidant activity of supercritical carbon dioxide extracts from *Terminalia catappa* leaves. *Food Research International*, *36*(1), 97–104.

Mbah, B. O., Eme, P. E., & Ogbusu, O. F. (2012). Effect of cooking methods (boiling and roasting) on nutrients and anti-nutrients content of *Moringa oleifera* seeds. *Pakistan Journal of Nutrition*, *11*(3), 211.

Medeiros, K. C. de P., Faustino, L., Borduchi, E., Nascimento, R. J. B., Silva, T. M. S., Gomes, E., Piuvezam, M. R., & Russo, M. (2009). Preventive and curative glycoside kaempferol treatments attenuate the TH2-driven allergic airway disease. *International Immunopharmacology*, *9*(13–14), 1540–1548.

Medeiros, M. L. S., Alves, R. R. V., Napoleão, T. H., Paiva, P. M. G., Coelho, L. C. B. B., Bezerra, A. C. D. S., & Silva, M. D. C. (2023). Anthelmintic effect of a water-soluble *Moringa oleifera* lectin in rodents experimentally infected with *Haemonchuscontortus*. *Parasitology International*, *92*, 102656.

Mehta, K., Balaraman, R., Amin, A. H., Bafna, P. A., & Gulati, O. (2003). Effect of fruits of *Moringa oleifera* on the lipid profile of normal and hypercholesterolemic rabbits. *Journal of Ethnopharmacology*, *86*(2–3), 191–195.

Meneghel, A. P., Gonçalves Jr, A. C., Tarley, C. R. T., Stangarlin, J. R., Rubio, F., & Nacke, H. (2014). Studies of Pb^{2+} adsorption by *Moringa oleifera* seeds from an aqueous medium in a batch system. *Water Science and Technology*, *69*(1), 163–169.

Milla, P. G., Peñalver, R., & Nieto, G. (2021). Health benefits of uses and applications of *Moringa oleifera* in bakery products. *Plants*, *10*(2), 318.

Millikin, C. L., Goodman, L. J., Bissett, D. L., Robinson, L. R., & Osborne, R. (2016). Personal care composition comprising botanical extract (U.S. Patent No. 9,358,263).

Mishra, G., Singh, P., Verma, R., Kumar, S., Srivastav, S., Jha, K. K., & Khosa, R. L. (2011). Traditional uses, phytochemistry and pharmacological properties of *Moringa oleifera* plant: An overview. *Der Pharmacia Lettre*, *3*(2), 141–164.

Mithen, R. F., Dekker, M., Verkerk, R., Rabot, S., & Johnson, I. T. (2000). The nutritional significance, biosynthesis and bioavailability of glucosinolates in human foods. *Journal of the Science of Food and Agriculture*, *80*(7), 967–984.

Molvig, L., Tabe, L. M., Eggum, B. O., Moore, A. E., Craig, S., Spencer, D., & Higgins, T. J. V. (1997). Enhanced methionine levels and increased nutritive value of seeds of transgenic lupins expressing a sunflower seed albumin gene. *Proceedings of the National Academy of Sciences*, *94*(16), 8393–8398.

Monteiro, J., Scotti-Campos, P., Pais, I., Figueiredo, A. C., Viegas, D., & Reboredo, F. (2022). Elemental composition, total fatty acids, soluble sugar content and essential oils of flowers and leaves of *Moringa oleifera* cultivated in Southern Portugal. *Heliyon*, *8*(12), e12647.

Morazzoni, P., Manzotti, C., Fontana, G., Riva, A., & Iori, R. (2014). Use of isothiocyanate derivatives as anti-myeloma agents (U.S. Patent No. 8,772,251).

Morton, J. F. (1991). The horseradish tree, *Moringa pterygosperma* (Moringaceae)—a boon to arid lands? *Economic Botany*, *45*, 318–333.

Muluvi, G. M., Sprent, J. I., Soranzo, N., Provan, J., Odee, D., Folkard, G., McNicol, J. W., & Powell, W. (1999). AFLP analysis of genetic variation in *Moringa oleifera* Lam. *Molecular Ecology*, *8*(3), 463–470.

Murakami, A., Ashida, H., & Terao, J. (2008). Multitargeted cancer prevention by quercetin. *Cancer Letters*, *269*(2), 315–325.

Musa, A. K. (2013). Influence of plant powders on infestation by adults and larvae of Khapra Beetle, *Trogoderma granarium* Everts (Coleoptera: Dermestidae) in stored groundnut. *Australian Journal of Basic and Applied Sciences*, *7*(6), 427–432.

Muyibi, S. A., & Evison, L. M. (1995). *Moringa oleifera* seeds for softening hard water. *Water Research*, *29*(4), 1099–1105.

Mvumi, C., Tagwira, F., & Chiteka, A. Z. (2013). Effect of moringa extract on growth and yield of maize and common beans. *Greener Journal of Agricultural Sciences, 3*(1), 55–62.

Nadeem, M., Mahmood, A., Shahid, S. A., Shah, S. S., Khalid, A. M., & McKay, G. (2006). Sorption of lead from aqueous solution by chemically modified carbon adsorbents. *Journal of Hazardous Materials*, *138*(3), 604–609.

Nadeem, M., Abdullah, M., & Hussain, I. (2014). Improvement of the oxidative stability of butter oil by blending with *Moringa oleifera* oil. *Journal of Food Processing and Preservation*, *38*(4), 1491–1500.

Namdev, S., & Chauhan, C. (2023). Moringa species: The potential candidate for traditional and future applications: A review. In *AIP conference proceedings* (Vol. 2535, No. 1). AIP Publishing.

Ndabigengesere, A., & Subba Narasiah, K. (1998). Use of *Moringa oleifera* seeds as a primary coagulant in wastewater treatment. *Environmental Technology*, *19*(8), 789–798.

Nikkon, F. (2003). In vitro antimicrobial activity of the compound isolated from chloroform extract of *Moringa oleifera* Lam. *Pakistan Journal of Biological Sciences*, *6*(22), 1888–1890.

Nogueira, E. de M., Vinagre, F., Masuda, H. P., Vargas, C., Muniz de Pádua, V. L., da Silva, F. R., dos Santos, R. V., Baldani, J. I., Ferreira, P. C. G., & Hemerly, A. S. (2001). Expression of sugarcane genes induced by inoculation with *Gluconacetobacterdiazotrophicus* and *Herbaspirillumrubrisubalbicans. Genetics and Molecular Biology, 24*, 199–206.

Ntandou, G. F. N., Banzouzi, J. T., Mbatchi, B., Elion-Itou, R. D. G., Etou-Ossibi, A. W., Ramos, S., Benoit-Vical, F., Abena, A. A., & Ouamba, J. M. (2010). Analgesic and anti-inflammatory effects of *Cassia siamea* Lam. stem bark extracts. *Journal of Ethnopharmacology, 127*(1), 108–111.

Ogawa, T., Samoto, M., & Takahashi, K. (2000). Soybean allergens and hypoallergenic soybean products. *Journal of Nutritional Science and Vitaminology, 46*(6), 271–279.

Okuda, T., Baes, A. U., Nishijima, W., & Okada, M. (2001). Coagulation mechanism of salt solution-extracted active component in *Moringa oleifera* seeds. *Water Research, 35*(3), 830–834.

Okwor, J. I.,Onah, I. E., Oboho, D. E., Haruna, S. A., Okafor, F. C., & Eyo, J. E. (2021). Biopesticidal potential of *Moringa oleifera* on storage insect pests. *Research on Crops, 22*(3), 666.

Oldoni, T. L. C., Merlin, N., Karling, M., Carpes, S. T., Alencar, S. M., Morales, R. G. F., Silva, E. A. d., & Pilau, E. J. (2019). Bioguided extraction of phenolic compounds and UHPLC-ESI-Q-TOF-MS/MS characterization of extracts of *Moringa oleifera* leaves collected in Brazil. *Food Research International, 125*, 108647.

Oliveira, J. T. A., Silveira, S. B., Vasconcelos, I. M., Cavada, B. S., & Moreira, R. A. (1999). Compositional and nutritional attributes of seeds from the multiple purpose tree Moringa oleifera Lamarck. *Journal of the Science of Food and Agriculture, 79*(6), 815–820.

Olsen, A. (1987). Low technology water purification by bentonite clay and Moringa oleifera seed flocculation as performed in Sudanese villages: effects on Schistosoma mansoni cercariae. *Water Research, 21*(5), 517–522.

Olson, M. E. (2002). Intergeneric relationships within the Caricaceae-Moringaceae clade (Brassicales) and potential morphological synapomorphies of the clade and its families. *International Journal of Plant Sciences, 163*(1), 51–65.

Onianwa, P. C., Adeyemo, A. O., Idowu, O. E., & Ogabiela, E. E. (2001). Copper and zinc contents of Nigerian foods and estimates of the adult dietary intakes. *Food Chemistry, 72*(1), 89–95.

Onsare, J. G., & Arora, D. S. (2015). Antibiofilm potential of flavonoids extracted from *Moringa oleifera* seed coat. *Journal of Applied Microbiology, 118*(2), 313–329.

Pal, S. K., Mukherjee, P. K., & Saha, B. P. (1995a). Studies on the antiulcer activity of *Moringa oleifera* leaf extract on gastric ulcer models in rats. *Phytotherapy Research, 9*(6), 463–465.

Pal, S. K., Mukherjee, P. K., Saha, K., Pal, M., & Saha, B. P. (1995b). Antimicrobial action of the leaf extract of *Moringa oleifera* Lam. *Ancient Science of Life, 14*(3), 197–199.

Pandey, A., Pradheep, K., Gupta, R., Nayar, E. R., & Bhandari, D. C. (2011). 'Drumstick tree' (*Moringa oleifera* Lam.): A multipurpose potential species in India. *Genetic Resources and Crop Evolution, 58*, 453–460.

Pandey, P., & Khan, F. (2022). *Moringa oleifera* plant as potent alternative to chemical coagulant in water purification. *Brazilian Journal of Pharmaceutical Sciences, 58*, e201158.

Pari, L., & Ashok Kumar, N. (2002). Hepatoprotective activity of *Moringa oleifera* on antitubercular drug-induced liver damage in rats. *Journal of Medicinal Food, 5*(3), 171–177.

Pathak, J., Kanwar, N., & Bhargava, S. (2015). Phytoconstituents and pharmacological activities of *Moringa oleifera*: A review. In M. K. Swamy & M. S. Akhtar (Eds.), *Moringa oleifera: Nutritional and pharmacological importance* (pp. 1–24). Springer Nature Singapore.

Paul, C. W., & Didia, B. C. (2012). The effect of methanolic extract of *Moringa oleifera* Lam. roots on the histology of kidney and liver of guinea pigs. *Asian Journal of Medical Sciences, 4*(1), 55–60.

Peñalver, R., Martínez-Zamora, L., Lorenzo, J. M., Ros, G., & Nieto, G. (2022). Nutritional and antioxidant properties of *Moringa oleifera* leaves in functional foods. *Foods, 11*(8), 1107.

Peng, L., Gao, X., Nie, L., Xie, J., Dai, T., Shi, C., ... Sheng, J. (2020). Astragalin attenuates dextran sulfate sodium (DSS)induced acute experimental colitis by alleviating gut microbiota dysbiosis and inhibiting NF-κB activation in mice. *Frontiers in Immunology, 11*, 2058.

Perotti, R. (1926). On the limits of biological enquiry in soil science. *Proceedings of the International Society of Soil Science, 2*, 146–158.

Pineau, B., Barbe, J.-C., Van Leeuwen, C., & Dubourdieu, D. (2007). Which impact for β-damascenone on red wines aroma? *Journal of Agricultural and Food Chemistry, 55*(10), 4103.

Pino, J. A. (2013). Floral scent composition of *Moringa oleifera* Lam. *Journal of Essential Oil Bearing Plants, 16*(3), 315–317.

Pino, J. A., & Regalado, E. L. (2010). Volatile constituents of *Thalassiatestudinum* Banks ex König leaves. *Journal of Essential Oil Research*, *22*(5), 421–423.

Pollini, L., Tringaniello, C., Ianni, F., Blasi, F., Manes, J., & Cossignani, L. (2020). Impact of ultrasound extraction parameters on the antioxidant properties of *Moringa oleifera* leaves. *Antioxidants*, *9*(4), 277.

Prasad, A. S. (1998). Zinc in human health: An update. *The Journal of Trace Elements in Experimental Medicine*, *11*(2–3), 63–87.

Prouillet, C., Mazière, J.-C., Mazière, C., Wattel, A., Brazier, M., & Kamel, S. (2004). Stimulatory effect of naturally occurring flavonols quercetin and kaempferol on alkaline phosphatase activity in MG-63 human osteoblasts through ERK and estrogen receptor pathway. *Biochemical Pharmacology*, *67*, 1307–1313.

Radman, R., Saez, T., Bucke, C., & Keshavarz, T. (2003). Elicitation of plants and microbial cell systems. *Biotechnology and Applied Biochemistry*, *37*(1), 91–102.

Radovich, T. (2009). *Farm and forestry production and marketing profile for Moringa oleifera*. Permanent Agriculture Resources.

Rajab, M. S., Cantrell, C. L., Franzblau, S. G., & Fischer, N. H. (1998). Antimycobacterial activity of (E)-phytol and derivatives: A preliminary structure–activity study. *Planta Medica*, *64*(1), 2–4.

Rajurkar, N. S., & Damame, M. M. (1998). Mineral content of medicinal plants used in the treatment of diseases resulting from urinary tract disorders. *Applied Radiation and Isotopes*, *49*(7), 773–776.

Ramachandran, C., Peter, K. V., & Gopalakrishnan, P. K. (1980). Drumstick (*Moringa oleifera*): A multipurpose Indian vegetable. *Economic Botany*, *34*, 276–283.

Ranheim, T., & Halvorsen, B. (2005). Coffee consumption and human health – beneficial or detrimental? Mechanisms for effects of coffee consumption on different risk factors for cardiovascular disease and type 2 diabetes mellitus. *Molecular Nutrition & Food Research*, *49*(3), 274–284.

Rao, A. V., Devi, P. U., & Kamath, R. (2001). In vivo radioprotective effect of *Moringa oleifera* leaves. *Indian Journal of Experimental Biology*, *39*(9), 858–863.

Rashid, U., Anwar, F., Moser, B. R., & Knothe, G. (2008). *Moringa oleifera* oil: A possible source of biodiesel. *Bioresource Technology*, *99*(17), 8175–8179.

Rayamajhi, N., Kim, S.-K., Go, H., Joe, Y., Callaway, Z., Kang, J.-G., Ryter, S. W., & Chung, H. T. (2013). Quercetin induces mitochondrial biogenesis through activation of HO-1 in HepG2 cells. *Oxidative Medicine and Cellular Longevity*, *2013*, 154279.

Reyes Sánchez, N. (2006). *Moringa oleifera and Cratylia argentea: Potential fodder species for ruminants in Nicaragua* (Doctoral dissertation No. 2001:1). SLU Acta Universitatis Agriculturae Sueciae.

Riyathong, T., Dheeranupattana, S., Palee, J., & Shank, L. (2010). *Shoot multiplication and plant regeneration from in vitro cultures of drumstick tree* (*Moringa oleifera* Lam.). Proceedings of the 8th International Symposium on Biocontrol and Biotechnology. King Mongkut's Institute of Technology Ladkrabang and Khon Kaen University, Nongkhai Campus, pp. 99–104.

Rockwood, J. L., Anderson, B. G., & Casamatta, D. A. (2013). Potential uses of *Moringa oleifera* and an examination of antibiotic efficacy conferred by *M. oleifera* seed and leaf extracts using crude extraction techniques available to underserved indigenous populations. *International Journal of Phytotherapy Research*, *3*(2), 61–70.

Rommens, C. M. (2004). All-native DNA transformation: A new approach to plant genetic engineering. *Trends in Plant Science*, *9*(9), 457–464.

Rommens, C. M., Humara, J. M., Ye, J., Yan, H., Richael, C., Zhang, L., ... & Swords, K. (2004). Crop improvement through modification of the plant's own genome. *Plant Physiology*, *135*(1), 421–431.

Rommens, C. M., Ye, J., Richael, C., & Swords, K. (2006). Improving potato storage and processing characteristics through all-native DNA transformation. *Journal of Agricultural and Food Chemistry*, *54*(26), 9882–9887.

Ruckmani, K., Kavimani, S., Anandan, R., & Jaykar, B. (1998). Effect of *Moringa oleifera* Lam. on paracetamol-induced hepatotoxicity. *Indian Journal of Pharmaceutical Science*, *60*, 33–35.

Rudrappa, T., Neelwarne, B., Lakshmanan, V., Venkataramareddy, S. R., & Aswathanarayana, R. G. (2006). Elicitation of peroxidase activity in genetically transformed root cultures of *Beta vulgaris* L. *Electronic Journal of Biotechnology*, *9*(5). https://www.ejbiotechnology.info/index.php/ejbiotechnology/article/view/v9n5-11

Ruiz-García, Y., Gil-Muñoz, R., López-Roca, J. M., Martínez-Cutillas, A., Romero-Cascales, I., & Gómez-Plaza, E. (2013). Increasing the phenolic compound content of grapes by preharvest application of abscisic acid and a combination of methyl jasmonate and benzothiadiazole. *Journal of Agricultural and Food Chemistry*, *61*(16), 3978–3983.

Sacks, G. L., Gates, M. J., Ferry, F. X., Lavin, E. H., Kurtz, A. J., & Acree, T. E. (2012). Sensory threshold of 1,1,6-trimethyl-1,2-dihydronaphthalene (TDN) and concentrations in young Riesling and non-Riesling wines. *Journal of Agricultural and Food Chemistry, 60*(12), 2998–3005.

Sahoo, N., Bishi, D. K., Srivastava, V., Pattanayak, C., & Sarkar, S. (2024). Hypolipidemic and hepatoprotective effect of *Linum usitatissimum* and *Moringa oleifera* seeds in diabetic-induced damage in rats. *Journal of Pharmacology and Pharmacotherapeutics*. https://doi.org/10.1177/0976500X241303668.

Saini, R. K., Saad, K. R., Ravishankar, G. A., Giridhar, P., & Shetty, N. P. (2013a). Genetic diversity of commercially grown *Moringa oleifera* Lam. cultivars from India by RAPD, ISSR and cytochrome P450-based markers. *Plant Systematics and Evolution, 299*, 1205–1213.

Saini, R. K., Harish Prashanth, K. V., Shetty, N. P., & Giridhar, P. (2014d). Elicitors, SA and MJ enhance carotenoids and tocopherol biosynthesis and expression of antioxidant related genes in *Moringa oleifera* Lam. leaves. *Acta Physiologiae Plantarum, 36*, 2695–2704.

Saini, R. K., Shetty, N. P., & Giridhar, P. (2014b). GC-FID/MS analysis of fatty acids in Indian cultivars of *Moringa oleifera*: Potential sources of PUFA. *Journal of the American Oil Chemists' Society, 91*(6), 1029–1034.

Saini, R. K., Shetty, N. P., Prakash, M., & Giridhar, P. (2014a). Effect of dehydration methods on retention of carotenoids, tocopherols, ascorbic acid and antioxidant activity in *Moringa oleifera* leaves and preparation of a RTE product. *Journal of Food Science and Technology, 51*, 2176–2182.

Saini, R. K., Shetty, N. P., Giridhar, P., & Ravishankar, G. A. (2012). Rapid in vitro regeneration method for *Moringa oleifera* and performance evaluation of field grown nutritionally enriched tissue cultured plants. *3 Biotech, 2*, 187–192.

Saini, R. K., Manoj, P., Shetty, N. P., Srinivasan, K., & Giridhar, P. (2014e). Dietary iron supplements and *Moringa oleifera* leaves influence the liver hepcidin messenger RNA expression and biochemical indices of iron status in rats. *Nutrition Research, 34*(7), 630–638.

Saini, R. K., Manoj, P., Shetty, N. P., Srinivasan, K., & Giridhar, P. (2016). Relative bioavailability of folate from the traditional food plant *Moringa oleifera* L. as evaluated in a rat model. *Journal of Food Science and Technology, 53*, 511–520.

Saini, R. K., Shetty, N. P., & Giridhar, P. (2014c). Carotenoid content in vegetative and reproductive parts of commercially grown *Moringa oleifera* Lam. cultivars from India by LC–APCI–MS. *European Food Research and Technology, 238*, 971–978.

Saint Sauveur, A. de. (2001). Moringa exploitation in the world: State of knowledge and challenges. In *Development potential for Moringa products* (Vol. 29). International Workshop.

Salazar-Martinez, E., Willett, W. C., Ascherio, A., Manson, J. E., Leitzmann, M. F., Stampfer, M. J., & Hu, F. B. (2004). Coffee consumption and risk for type 2 diabetes mellitus. *Annals of Internal Medicine, 140*(1), 1–8.

Sanchez-Machado, D. I., Lopez-Cervantes, J., & Rios Vazquez, N. J. (2006). High-performance liquid chromatography method to measure α-and γ-tocopherol in leaves, flowers and fresh beans from Moringa oleifera. *Journal of Chromatography A, 1105,* 111–114.

Sánchez-Machado, D. I., López-Cervantes, J., Lopez-Hernandez, J., & Paseiro-Losada, P. (2004a). Fatty acids, total lipid, protein and ash contents of processed edible seaweeds. *Food chemistry, 85*(3), 439–444.

Sanchez-Machado, D. I., Lopez-Cervantes, J., Lopez-Hernandez, J., Paseiro-Losada, P., & Simal-Lozano, J. (2004b). Determination of the uronic acid composition of seaweed dietary fibre by HPLC. *Biomedical Chromatography, 18*(2), 90–97.

Sánchez-Machado, D. I., Núñez-Gastélum, J. A., Reyes-Moreno, C., Ramírez-Wong, B., & López-Cervantes, J. (2010). Nutritional quality of edible parts of Moringa oleifera. *Food analytical methods, 3*(3), 175–180.

Santos, A. F., Paiva, P. M., Teixeira, J. A., Brito, A. G., Coelho, L. C., & Nogueira, R. (2012). Coagulant properties of Moringa oleifera protein preparations: application to humic acid removal. *Environmental Technology, 33*(1), 69–75.

Sajidu, S. M. I., Persson, I., Masamba, W. R. L., & Henry, E. M. T. (2008). Mechanisms for biosorption of chromium (III), copper (II) and mercury (II) using water extracts of *Moringa oleifera* seed powder. *African Journal of Biotechnology, 7*(6), 842–849.

Schafran, G.C., Kolb, J.F., Stubbins, A., and Schoenbach, K.H. (2011). System and method for high-voltage pulse assisted aggregation of algae.(US 2011/0003350)

Sena, L. P., Vanderjagt, D. J., Rivera, C., Tsin, A. T., Muhamadu, I., Mahamadou, O., ... Glew, R. H. (1998). Analysis of nutritional components of eight famine foods of the Republic of Niger. *Plant Foods for Human Nutrition, 52*(1), 17–30.

Sengupta, A., & Gupta, M. P. (1970). Studies on the seed fat composition of Moringaceae family. *Fette, Seifen, Anstrichmittel, 72*(1), 6–10.

Sessitsch, A., Howieson, J. G., Perret, X., Antoun, H., & Martínez-Romero, E. (2002). Advances in *Rhizobium* research. *Critical Reviews in Plant Sciences*, *21*(4), 323–378.

Sevilla, M., Burris, R. H., Gunapala, N., & Kennedy, C. (2001). Comparison of benefit to sugarcane plant growth and $^{15}N_2$ incorporation following inoculation of sterile plants with *Acetobacter diazotrophicus* wild-type and nif mutant strains. *Molecular Plant–Microbe Interactions*, *14*(3), 358–366.

Sharma, P., Kumari, P., Srivastava, M. M., & Srivastava, S. (2006). Removal of cadmium from aqueous system by shelled *Moringa oleifera* Lam. seed powder. *Bioresource Technology*, *97*(2), 299–305.

Sharma, P., Kumari, P., Srivastava, M. M., & Srivastava, S. (2007). Ternary biosorption studies of Cd(II), Cr(III) and Ni(II) on shelled *Moringa oleifera* seeds. *Bioresource Technology*, *98*(2), 474–477.

Shindano, J., & Kasase, C. (2009). *Moringa (Moringa oleifera): A source of food and nutrition, medicine and industrial products*. ACS Publications.

Sibanda, S., Chigwada, G., Poole, M., Gwebu, E. T., Noletto, J. A., Schmidt, J. M., Rea, A. I., & Setzer, W. N. (2004). Composition and bioactivity of the leaf essential oil of *Heteropyxisdehniae* from Zimbabwe. *Journal of Ethnopharmacology*, *92*(1), 107–111.

Siddhuraju, P., & Becker, K. (2003). Antioxidant properties of various solvent extracts of total phenolic constituents from three different agroclimatic origins of drumstick tree (*Moringa oleifera* Lam.) leaves. *Journal of Agricultural and Food Chemistry*, *51*(8), 2144–2155.

Sindhu, A. S., Zheng, Z., & Murai, N. (1997). The pea seed storage protein legumin was synthesized, processed, and accumulated stably in transgenic rice endosperm. *Plant Science*, *130*(2), 189–196.

Singh, A., Ahamad, I., Yadav, D. K., Singh, V. K., & Quraishi, M. A. (2012). The effect of environmentally benign fruit extract of Shahjan (*Moringa oleifera*) on the corrosion of mild steel in hydrochloric acid solution. *Chemical Engineering Communications*, *199*(1), 63–77.

Singh, A. K., Bindra, R. L., Singh, J., & Kumar, S. (2003). Composition of herbal biscuits for lactating mothers acting as dietary supplement and process for preparation thereof (U.S. Patent No. 6,517,861).

Singh, N. P., Vohra, J. N., Hazra, P. K., & Singh, D. K. (2000). Ziziphus. In N. P. Singh, J. N. Vohra, P. K. Hazra, & D. K. Singh (Eds.), *Flora of India* (Vol. 5, pp. 221–245). Botanical Survey of India.

Singh, Y., & Prasad, K. (2013). *Moringa oleifera* leaf as functional food powder. *Characterization & Uses*, *4*, 317–324.

Solanki, R. (2004). Composition of eleven herbals for treating cancer (U.S. Patent No. 6,780,441).

Sonbarse, P. P., Kiran, K., Sharma, P., & Parvatam, G. (2020). Biochemical and molecular insights of PGPR application for augmentation of carotenoids, tocopherols, and folate in foliage of *Moringa oleifera*. *Phytochemistry*, *179*, 112506.

Sonbarse, P. P., Sharma, P., & Parvatam, G. (2017). PGPR mix treatment to *Moringa* improved plant growth and iron content in foliage as substantiated by biochemical and molecular methods. *Journal of Plant Interactions*, *12*(1), 526–532.

Sowley, E. N. K., Kankam, F., & Adomako, J. (2014). Management of root-knot nematode (*Meloidogyne* spp.) on sweet pepper (*Capsicum annuum* L.) with *Moringa oleifera* Lam. leaf powder. *Archives of Phytopathology and Plant Protection*, *47*(13), 1531–1538.

Sreedhar, R. V., Roohie, K., Maya, P., Venkatachalam, L., & Bhagyalakshmi, N. (2009). Biotic elicitors enhance flavour compounds during accelerated curing of vanilla beans. *Food Chemistry*, *112*(2), 461–468.

Sreelatha, S., Jeyachitra, A., & Padma, P. R. (2011). Antiproliferation and induction of apoptosis by *Moringa oleifera* leaf extract on human cancer cells. *Food and Chemical Toxicology*, *49*(6), 1270–1275.

Sreelatha, S., & Padma, P. R. (2009). Antioxidant activity and total phenolic content of *Moringa oleifera* leaves in two stages of maturity. *Plant Foods for Human Nutrition*, *64*, 303–311.

Stussi, I. A., Freis, O., Moser, P., & Pauly, G. (2002). *Anti-pollution hair care claims: Moringa proteins enable formulators to make a new claim for hair care products*. Laboratoires Sérobiologiques, Pulnoy, France. Retrieved from https://www.thefreelibrary.com/Anti-pollution+hair+care+claims%3a+Moringa+proteins+enable+formulators...-a095552927

Sudha, G., & Ravishankar, G. A. (2003). Elicitation of anthocyanin production in callus cultures of *Daucus carota* and the involvement of methyl jasmonate and salicylic acid. *Acta Physiologiae Plantarum*, *25*, 249–256.

Sukarni, S., Anis, S., Aminullah, A. Y., Assidiq, M. A., Mufti, N., Tuan Abdullah, T. A., & Johari, A. (2024). *Moringa oleifera* seeds potential as biofuel via thermal conversion method based on morphological and chemical content evaluation. In *E3S Web of conferences* (Vol. 473, p. 01016). EDP Sciences.

Sutherland, J. P. (2005). Extracting coagulant protein derivatives from seeds of trees within the family Moringaceae, especially those of *Moringa oleifera* Lam (syn. *Moringa pterygosperma*Gaertn.) (U.S. Patent No. 6,890,565 B2).

Tahiliani, P., & Kar, A. (2000). Role of *Moringa oleifera* leaf extract in the regulation of thyroid hormone status in adult male and female rats. *Pharmacological Research*, *41*(3), 319–323.
Takahashi, M., Nakanishi, H., Kawasaki, S., Nishizawa, N. K., & Mori, S. (2001). Enhanced tolerance of rice to low iron availability in alkaline soils using barley nicotianamine aminotransferase genes. *Nature Biotechnology*, *19*(5), 466–470.
Talreja, T. (2010). Effect of phenylalanine on quantitative production of flavonoids in unorganized tissue of Moringa oleifera. *Current Biotica*, 373–379.
Tambone, F., Pradella, M., Bedussi, F., & Adani, F. (2020). *Moringa oleifera* Lam. as an energy crop for biogas production in developing countries. *Biomass Conversion and Biorefinery*, *10*, 1083–1093.
The Good Scents Company. http://www.thegoodscentscompany.com/index.html#toflibottom
The Wealth of India (1962). *A dictionary of indian raw materials and industrial products*. CSIR.
Teranishi, R., Wick, E. L., & Hornstein, I. (1999). Flavor Chemistry. 3Island Press, *146*, 1941–1951.
Trivedi, R., Kumar, S., Kumar, A., Siddiqui, J. A., Swarnkar, G., Gupta, V., Kendurker, A., Dwivedi, A. K., Romero, J. R., & Chattopadhyay, N. (2008). Kaempferol has osteogenic effect in ovariectomized adult Sprague–Dawley rats. *Molecular and Cellular Endocrinology*, *289*(1–2), 85–93.
Upadhyay, P., Yadav, M. K., Mishra, S., Sharma, P., & Purohit, S. (2015). *Moringa oleifera*: A review of the medical evidence for its nutritional and pharmacological properties. *International Journal of Research in Pharmacy & Science*, *5*(2), 12–16.
Uphadek, B., Shinkar, D. M., Patil, P. B., & Saudagar, R. B. (2018). *Moringa oleifera* as a pharmaceutical excipient. *International Journal of Current Pharmaceutical Research*, *10*(2), 13–16.
Urushisaki, T., Takemura, T., Tazawa, S., Fukuoka, M., Hosokawa-Muto, J., Araki, Y., & Kuwata, K. (2011). Caffeoylquinic acids are major constituents with potent anti-influenza effects in Brazilian green propolis water extract. *Evidence-Based Complementary and Alternative Medicine*, *2011*, 254914.
Van Ba, H., Hwang, I., Jeong, D., &-Touseef, A. (2012). Principle of meat aroma flavors and future prospect. *Latest Research into Quality Control*, *2*, 145–176.
Venkatesan, N., Sekhar, C., Murugananthi, M., & Vidhyavathi, A. (2018). Marketing and price spread analysis of moringa in Tamil Nadu, India. *Horticulture International Journal*, *2*(4), 212–221.
Vergara-Jimenez, M., Almatrafi, M. M., & Fernandez, M. L. (2017). Bioactive components in *Moringa oleifera* leaves protect against chronic disease. *Antioxidants*, *6*(4), 91.
Verma, A. R., Vijayakumar, M., Mathela, C. S., & Rao, C. V. (2009). In vitro and in vivo antioxidant properties of different fractions of *Moringa oleifera* leaves. *Food and Chemical Toxicology*, *47*(9), 2196–2201.
Vissoh, A. C. S., Klotoé, J. R., Fah, L., Agbodjento, E., Koudokpon, H., Togbe, E., Saïdou, S., & Dougnon, V. (2024). Knowledge and practices of traditional management of child malnutrition and associated pathologies in Benin. *Journal of Ethnobiology and Ethnomedicine*, *20*(1), 47.
Wang, H., Wang, H., & Wei, W. (2008). Establishment of regeneration system in vitro for *Moringa oleifera* with stem. *Journal of Zhejiang Forestry Science and Technology*,40–43.
Watt, G. (1893). *A dictionary of the economic products of India* (Vol. 6). Superintendent of Government Printing.
Weiyin, Z. (2011). Tissue culture and rapid propagation of *Moringa oleifera* Lam. *Anhui Agricultural Sciences Bulletin*, *7*, 25.
White, P. J., & Broadley, M. R. (2005). Biofortifying crops with essential mineral elements. *Trends in Plant Science*, *10*(12), 586–593.
Wolff, K., Moroianu, S., & Cho, E. (2023). Chemical composition of *Moringa oleifera* and cruciferous plants and their anti-inflammatory properties. *Zbornik Matice Srpske za PrirodneNauke*, *145*, 23–37.
Wu, Y.-Y., Xu, Y.-M., & Lau, A. T. Y. (2021). Anti-cancer and medicinal potentials of moringa isothiocyanate. *Molecules*, *26*(24), 7512.

Yang, R.-Y., Chang, L.-C., Hsu, J.-C., Weng, B. B. C., Palada, M. C., Chadha, M. L., & Levasseur, V. (2007). Nutritional and functional properties of moringa leaves—from germplasm, to plant, to food, to health. In L. J. Fuglie, E. M. Makkar, & K. Becker (Eds.), *Moringa leaves: Strategies, standards and markets for a better impact on Nutrition in Africa*. Moringanews, Moringa and Plant Resources Network.
Yasmeen, A., Basra, S. M. A., Ahmad, R., & Wahid, A. (2012). Performance of late-sown wheat in response to foliar application of *Moringa oleifera* Lam. leaf extract. *Chilean Journal of Agricultural Research*, *72*(1), 92–97.
Yeast Metabolome Database. http://www.ymdb.ca/compounds/YMDB01739.
Yu, K.-W., Gao, W., Hahn, E.-J., & Paek, K.-Y. (2002). Jasmonic acid improves ginsenoside accumulation in adventitious root culture of *Panax ginseng* CA Meyer. *Biochemical Engineering Journal*, *11*(2–3), 211–215.

Zern, T. L., Wood, R. J., Greene, C., West, K. L., Liu, Y., Aggarwal, D., Shachter, N. S., & Fernandez, M. L. (2005). Grape polyphenols exert a cardioprotective effect in pre- and postmenopausal women by lowering plasma lipids and reducing oxidative stress. *The Journal of Nutrition*, *135*(8), 1911–1917.

Zhang, X., & Du, P. (2009). GC-MS analysis of chemical constituents of volatile oil from *Lingderathomsonii* Allen fruits. *Food Science*, *16*, 56.

Zhao, J., Davis, L. C., & Verpoorte, R. (2005). Elicitor signal transduction leading to production of plant secondary metabolites. *Biotechnology Advances*, *23*(4), 283–333.

Zheng, M., Yang, H., Yang, E., Zou, X., Chen, X., & Zhang, J. (2022). Efficient in vitro shoot bud proliferation from cotyledonary nodes and apical buds of *Moringa oleifera* Lam. *Industrial Crops and Products*, *187*, 115394.

Zhou, Y., Yang, W., Li, Z., Luo, D., Li, W., Zhang, Y., Wang, X., Fang, M., Chen, Q., & Jin, X. (2018). *Moringa oleifera* stem extract protects skin keratinocytes against oxidative stress injury by enhancement of antioxidant defense systems and activation of PPARα. *Biomedicine & Pharmacotherapy*, *107*, 44–53.

Zhu, C., Naqvi, S., Gomez-Galera, S., Pelacho, A. M., Capell, T., & Christou, P. (2007). Transgenic strategies for the nutritional enhancement of plants. *Trends in Plant Science*, *12*(12), 548–555.

Zirmire, J. (2018). A study on economic viability of *Moringa* in dryland regions of Maharashtra: With special reference to value addition and export aspects. *Journal of Pharmacognosy and Phytochemistry*, *7*(2), 374–376.

2 *Moringa oleifera* as an Anti-Inflammatory Drug

Rawah H. Elkousy, Marwa Samy Abu Bakr,
Nourhan Ibrahim Morad,
Asmaa Ayman Saber Ibrahim and Omayma A. Eldahshan

2.1 INTRODUCTION

2.1.1 General Overview of *Moringa oleifera*

The 'miracle tree', *Moringa oleifera* (MO), belongs to the Moringaceae family and is well-known for its significant medicinal and nutritional benefits (Boopathi and Abubakar 2021). It is often referred to as the 'drumstick tree' or 'horseradish tree' (Mishra et al. 2011; Fahey 2005; Liu et al. 2022). MO originated in India and has been recognised for its health benefits. It is cultivated all over the world, particularly in tropical and subtropical nations. A multitude of vitamins, minerals and medicinally active compounds are abundant in every part of the moringa tree (Boopathi and Abubakar 2021). For instance, essential biologically active compounds, including flavonoids, phenolic acids, alkaloids, carotenoids, glucosinolates and saponins, along with tannins, are found in MO leaves that have gained much of the attention (Hodas, Zorzenon and Milani 2021). Moringa is an excellent source of fatty acids, including omega-3, omega-6, linoleic and linolenic acid. Besides its therapeutic content of phytochemicals, MO represents a wealthy nutritional source due to its content of amino acids and minerals such as calcium, potassium, iron and magnesium, in addition to vitamins such as vitamin A, vitamin C, vitamin E and vitamin B. The nutritional value of moringa exceeds many known rich food resources (Abo-Elsoud, Ahmed Mohamed Abdelaziz, Attia Abd Eldaim and Hazzaa 2022; Peñalver et al. 2022).

2.1.2 Background on Inflammation and Its Health Complications

The development of numerous chronic illnesses, including diabetes, cancer, arthritis, cardiovascular disorders and ulcerative colitis, is significantly influenced by inflammation (Sayed, Omar, Emam and Farag 2022; Yeung, Aziz, Guerrero-Castilla and Arguelles 2018). Inflammation encompasses both systemic and local manifestations. Local manifestations include redness, swelling, pain, heat and dysfunction. The systemic inflammatory response includes the release of inflammatory cytokines such as interleukin-1 (IL-1), interleukin-6 (IL-6) and tumour necrosis factor (TNF) (Xu and Chenli 2024). The inflammatory response is also mediated by several key transcription factors, including nuclear factor kappa B (NF-κB), which essentially regulates the production of IL-1β and IL-6 cytokines. Nitric oxide (NO) produced by the inducible nitric oxide synthase (iNOS) enzyme and PGE2 produced by the COX-2 enzyme are two crucial inflammatory mediators, and multiple inflammatory diseases can be brought on by their overproduction (Sayed, Omar, Emam and Farag 2022; Arulselvan et al. 2016). The activation of inflammatory cytokine cascades by triggering results in various events, including vasodilation, an increase in vascular permeability, infiltration and recruitment of immune cells (Ray, Wolf and Mowa 2015).

DOI: 10.1201/9781003617860-3

2.1.3 Objective and Scope

In this chapter, we outline how MO contributes to inflammation and inflammatory diseases by reviewing several studies on this plant using different parts and extracting solvents, highlighting crucial phytochemicals isolated from moringa and their anti-inflammatory role, in addition to demonstrating molecular pathways of inflammation modulated by moringa.

2.2 ETHNOMEDICINAL PERSPECTIVES

2.2.1 Traditional Uses of *Moringa oleifera* in Medicine

Moringa has historically been used as a diuretic, expectorant, stimulant and antispasmodic (Mishra et al. 2011; Dixit, Tripathi and Kumar 2016). Different parts of the MO tree have been utilised for the treatment of skin infections and sores, helminthiasis, bladder and prostate problems, scurvy and paralysis, as well as abdominal malignancies (Singh et al. 2009; Liu et al. 2022). Furthermore, over 80 countries have used MO leaves, flowers, pods, seeds, bark and gum to alleviate vitamin and mineral deficiencies, manage anaemia, enhance cardiovascular health and regulate blood sugar levels, in addition to inflammation and oxidative stress relief (Dixit, Tripathi and Kumar 2016). Almost every part of this tree was used to treat a wide range of ailments. Roots, for example, were used as circulatory stimulants and cardiotonics, as anti-inflammatories in rheumatism, and to relieve renal pain and constipation. Likewise, leaves were used as antidiabetic, antipyretic and anticancer treatments, as well as in bronchitis. Flowers were consumed for their anti-inflammatory, anti-hyperlipidaemic, anti-hypercholesterolaemic and anti-tumour properties. Seeds of MO were also known for their antipyretic, antioxidant and antimicrobial activities (Anwar, Latif, Ashraf and Gilani 2007).

2.3 PHYTOCHEMICAL PROFILE OF *MORINGA OLEIFERA*

2.3.1 Key Active Compounds in MO

Based on previous research, MO has approximately 110 bioactive compounds, all of which possess medicinal or protective benefits. The composition of phytonutrients differs among the various parts of the tree. Moreover, it is a highly nutrient-rich and adaptable tree. The moringa tree contains the main types of bioactive phytochemicals, including carotenoids, terpenes, alkaloids, sulphur compounds, polyphenols and sterols (Figure 2.1) (Camilleri and Blundell 2024). They are distributed in all plant parts. The polyphenolic content of MO is represented by phenolic acids and flavonoids, particularly tannins, with the highest concentration of polyphenols in leaves. The most prevalent flavonoids in different parts of MO, aside from the seeds and roots, are kaempferol glycosides, quercetin, apigenin, myricetin and isorhamnetin (Saini, Sivanesan and Keum 2016; Chiş et al. 2023). while caffeic acid, coumaric acid, gentisic acid, chlorogenic acid, gallic acid and ellagic acid are the most common phenolic acids found in MO (Pareek et al., 2023). Seeds have the highest concentrations of gallic acid and ellagic acid (Camilleri and Blundell 2024). In addition to this, moringa leaves contain a significant amount of tannins (Chiş et al. 2023).

Glucosinolates and their enzymatic catalysis product, isothiocyanates, are the most significant sulphur-containing compounds found in moringa. Moreover, the leaves, flowers, pods, seeds and stems of the moringa tree contain glucosinolate (glucomoringin). Other glucosinolates are abundant in the plant roots, including benzyl glucosinolate and (4-(α-L-rhamnopyranosyloxy)-benzyl glucosinolate (Chiş et al. 2023; Saini, Sivanesan and Keum 2016). Isothiocyanates chiefly identified in MO are niazidin, niazicin and niazinin (Cheenpracha et al. 2010).

Moringa leaves contain several alkaloids, including the indole alkaloid N,α-L-rhamnopyranosylvincosamide, as well as two alkaloid glycosides, marumosides A and marumosides B. In addition, trigonelline and moringinine alkaloids were also isolated from MO. Moringa

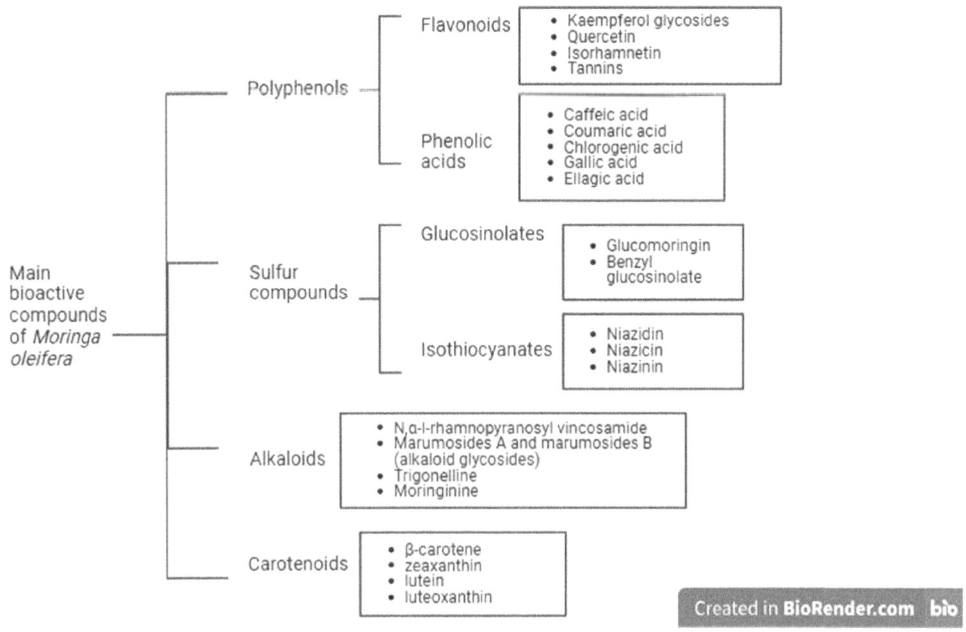

FIGURE 2.1 Structures of Some Bioactive Phytochemicals Found in *Moringa oleifera*.

leaves were identified as a very rich source of carotenoids (β-carotene) that are even higher than those of carrots. Other carotenoids were found in moringa leaves, flowers and fruits, such as zeaxanthin, lutein and luteoxanthin (Chiş et al. 2023). Moreover, β-sitosterol, stigmasterol and campesterol are the most prevalent phytosterols in MO seeds (Pareek et al. 2023) (Figure 2.2).

2.4 ANTI-INFLAMMATORY MECHANISMS OF *MORINGA OLEIFERA* AND ITS BIOACTIVE PHYTOCHEMICALS

2.4.1 Anti-Inflammatory Pathways and Antioxidant Activity Targeted by MO Phytochemicals

One of the key mechanisms that governs the anti-inflammatory activity of MO is the inhibition or regulation of inflammatory cytokines. The NF-κB transcription factor is normally coupled with its inhibitory protein IKBα; phosphorylation and degradation of this protein release free NF-κB, triggering an inflammatory response via regulating gene expression of various inflammatory cytokines such as TNF-α, IL-1β and IL-6, as well as iNOS and COX-2 enzymes (Figure 2.3) (Zheng et al. 2023; Shah and Oza 2022; Su et al. 2023). This pathway was found to be targeted by MO, which suppresses the degradation of IKBα and consequently inhibits IL-6 and TNF-α, as well as COX-2 and iNOS enzymes responsible for PGE-2 and NO production (Anzano et al. 2021; Arulselvan et al. 2016). Isothiocyanates from MO leaves, seeds and flowers were reported by different in vitro studies to inhibit iNOS and COX-2 enzymes, eliciting an anti-inflammatory activity. It has been proven that isothiocyanates have an anti-inflammatory effect by downregulating pro-inflammatory TNF-α and IL-1β while upregulating anti-inflammatory cytokines such as IL-10 (Chiş et al. 2023). This role is significant and has been investigated for the treatment of diseases such as rheumatoid arthritis, which is characterised by joint pain, oedema and inflammation. Serum levels of inflammatory cytokines, including TNFα, IL-1 and IL-6, as well as rheumatoid factors, decreased in arthritis-affected rats given the ethanolic extract of MO seeds. Quercetin and kaempferol flavonoids abundant in

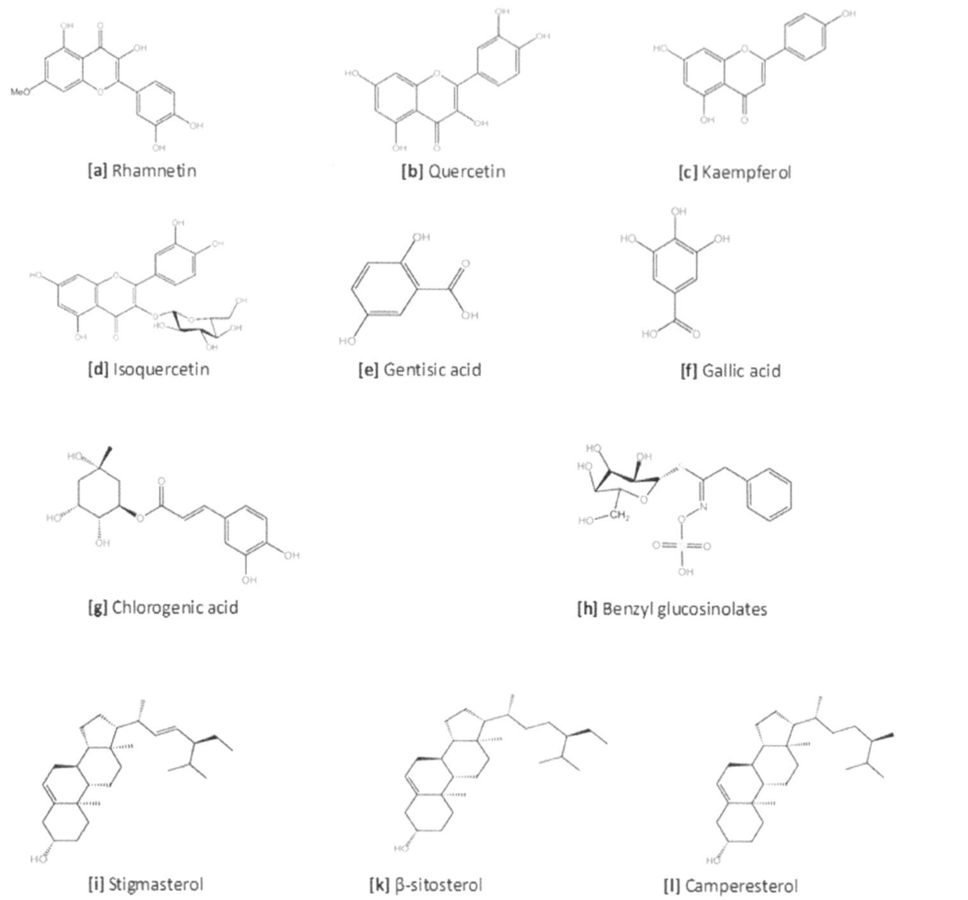

FIGURE 2.2 Inhibitory Pathway of *Moringa oleifera* Against Pro Inflammatory Enzymes.

MO have been demonstrated to inhibit the activity of pro-inflammatory enzymes such as cyclooxygenase (COX) and lipoxygenase (LOX), the inflammatory enzymes that are important in the synthesis of inflammatory mediators such as prostaglandins and leukotrienes (Lopez-Rodriguez, Gaytán-Martínez, de la Luz Reyes-Vega and Loarca-Piña 2020; Chiş et al. 2023). COX-2 and iNOS enzymes were also mitigated by phenolic glycosides in moringa (Wilujeng et al. 2021).

Proteins isolated from moringa, called lectins, namely coagulant lectin (cMoL) and water-soluble lectin (WSMoL), are carbohydrate-binding proteins that can reduce water turbidity because of their coagulant activity, and exhibit anti-inflammatory effects by decreasing TNF-α and IL-1β cytokines (Araújo et al. 2013). Similarly, glucomoringin and vitexin, in addition to a recently isolated compound named Moringa A from seeds of MO, had an inhibitory effect on IL-6, IL-1β and TNF-α levels (Xiong, Riaz Rajoka, Zhang and He 2022). Furthermore, apigenin (one of the flavonoids in moringa) was demonstrated to suppress the expression of intercellular adhesion molecule-1 (ICAM-1) and vascular cell adhesion molecule-1 (VCAM-1) by lowering the steady-state mRNA levels brought on by TNF-α. Additionally, cells pretreated with apigenin were shown to exhibit inhibitions of prostaglandin E2, IL-1β and IL-6 produced by TNF-α (Read 1995).

The p38 MAPK protein is thought to exert a regulatory effect on the activation of NF-κB and subsequent NO and PGE2 production. This protein was a target for MO leaf in a study in which its inhibition resulted in alleviating inflammation (Wilujeng et al. 2021). The inflammatory cytokines TNF-α, IL-1β and IL-6, along with COX-2 enzymes, are all stimulated by VEGF overexpression. MO showed an inhibitory effect on VEGF, which reduced inflammation and aided in wound healing (Shah and Oza 2022).

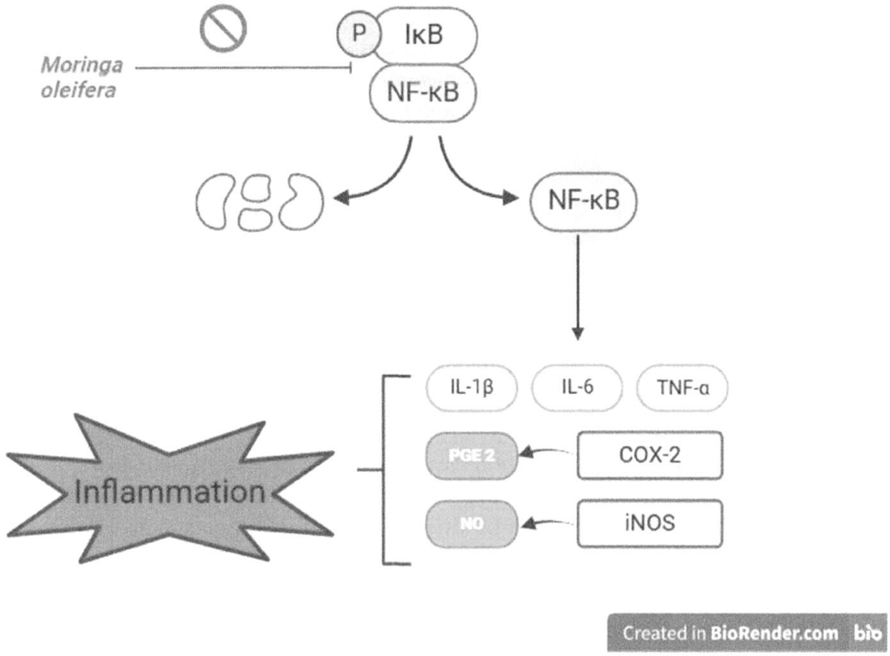

FIGURE 2.3 Main Bioactive Compounds Found in *Moringa oleifera.*

Numerous transcription factors and pathways of inflammation could be induced by oxidative stress, eventually causing chronic inflammatory diseases (Hussain et al. 2016). Polyphenols and flavonoids of MO have demonstrated antioxidant activity that is achieved by stimulation of essential antioxidant enzymes, including glutathione S-transferase, superoxide dismutase and catalase, blocking NO synthase, besides attenuating lipid peroxidation, resulting in oxidative stress relief; this mechanism was exclusively observed with moringa leaves (Divya et al. 2024). Flowers of MO have shown antioxidant in vitro activity as well as anti-inflammatory activity (Alhakmani, Kumar and Khan 2013).

2.4.2 Immunomodulator Effects of *Moringa oleifera*

After receiving MO leaf extract, it was found that it increased red blood cell count, haemoglobin and white blood cells, which are important in enhancing immunity against infections in mice (O'Connell et al. 2015). MO has shown promise in strengthening the immune system by increasing the production of antibodies and white blood cells. In addition to inhibiting the release of histamine, it can also lessen the severity of allergic and anti-inflammatory reactions (Azlan et al. 2023). Extracts from MO seeds raised the content of IgM, while the root extract enhanced TNF-α concentration. A well-regulated immune response helps the body manage inflammation more effectively by preventing excessive immune reactions that could lead to chronic inflammation. The extract's stimulatory impact on lymphocytes and bone marrow haematopoietic cells was enhanced when the weight of the thymus and spleen increased along with their cell counts; their stimulation by MO may contribute to controlling inflammatory processes (Van Pham et al. 2021).

2.5 PRECLINICAL AND CLINICAL RESEARCH EVIDENCE

2.5.1 In Vitro Assessment of Anti-Inflammatory Activity

Almost all parts of the moringa tree were investigated for their anti-inflammatory potential using different types of extracts and different experimental designs, which were compared with various potent anti-inflammatory drugs. Using a bovine serum anti-inflammatory assay, an in vitro study evaluated the anti-inflammatory effectiveness of aqueous leaf extract in comparison with ethanolic extract. The study found that both extracts had concentration-dependent anti-inflammatory activity, with the highest activity occurring at 50 µg/mL and the lowest activity occurring at 10 µg/mL. The aqueous extract was superior to the alcoholic extract and a sodium diclofenac standard at all concentrations, and it was equal to the standard at 50 µL. The alcoholic extract had a slightly decreased anti-inflammatory efficacy compared with the standard and water extract. The anti-inflammatory effect of the extract has been linked to quercetin (Varadarajan and Balaji 2022).

Another study also assessed the protein hydrolysate of MO leaves that was produced via in vitro gastrointestinal digestion. Different extract concentrations, a vehicle as a control and indomethacin (30 ug/mL) as a reference medication were applied to the RAW 264.7 murine macrophage cell line. Following that, macrophages were incubated with the bacterial endotoxin lipopolysaccharide (LPS), which caused an inflammatory response linked to the creation of NO and a subsequent rise in vascular permeability, which is a hallmark of inflammation. NO levels were recorded. Significant dose-dependent suppression of NO generation was seen at protein hydrolysate doses of 50 and 100 µg/ml in comparison with the control. NO suppression was attributed to the presence of arginine in the MO protein hydrolysate, an inhibitor of nitric oxide synthase. Leucine and glycine were also responsible for the anti-inflammatory effect (Avilés-Gaxiola et al. 2021).

The anti-inflammatory effect of leaves was compared with that of seeds in a study using ethanolic extract of both parts of MO grown in Kenya by assessing the production of NO by RAW 264.7 macrophages. Cells were cultured and observed for their nitrite production; they were then treated with 11.11, 33.33 and 100 µg/mL serial dilutions of leaf and seed extracts. MO leaves and seed extract significantly reduced NO generation; however, leaves with a higher flavonoid content showed a stronger anti-inflammatory effect than seeds (Xu, Chen and Guo 2019).

The anti-inflammatory activities of cMoL and WSMoL were also investigated in vitro, along with an aqueous extract of MO seeds, diluted seed extract and extracted proteins (lectins). Mice peritoneal macrophages were stimulated by LPS-triggered NO production due to the induction of inflammation. LPS-stimulated macrophages received a non-cytotoxic dosage of diluted seed extract, aqueous seed extract and isolated cMoL or WSMoL. After a 24-hour incubation period, the production of NO and inflammatory cytokines was evaluated. The aqueous seed extract reduced NO production at a dose of 6.25 µg/mL, even though it was not significantly affected by diluted seed extract at 50 µg/mL. Moreover, cMoL and WSMoL (6.25 µg/mL) were able to successfully lower NO production in comparison with LPS-only control cells, indicating that their in vitro anti-inflammatory properties are a contributing factor in their ability to regulate NO formation. Furthermore, the levels of TNF-α and, subsequently, IL-1β, produced by LPS-stimulated macrophages were significantly decreased by the aqueous seeds, diluted seed extracts and cMoL. WSMoL decreased the TNF-α production and increased IL-1β levels. None of the evaluated MO samples affected the IL-6 level (Araújo et al. 2013). Moringa A, glucomoringin and vitexin are three compounds isolated from moringa seeds that exhibited anti-inflammatory activity in a study on RAW264.7 cells infected with the H1N1 virus. They have shown a significant reduction in levels of inflammatory cytokines IL-6, IL-1β and TNF-α, indicating their anti-inflammatory potential (Xiong, Riaz Rajoka, Zhang and He 2022).

In vitro, the anti-inflammatory potential of MO flowers cultivated in Oman was also demonstrated by a study that used the protein denaturation method and involved different doses of ethanolic flower extract (100, 200 and 500 µg/mL) compared with a standard of diclofenac sodium (100 and 200 µg/mL), which were each incubated with egg albumin followed by thermal denaturation. After measuring the absorbance of each mixture, the percentage inhibition of protein denaturation was calculated. A dose-dependent, substantial reduction of egg albumin denaturation was observed with MO flower extract, which was comparable to the reference drug diclofenac sodium. MO flowers' high polyphenol content is responsible for their anti-inflammatory properties (Alhakmani, Kumar and Khan 2013). These findings were supported by a novel study of MO flowers grown in Egypt, where moringa flower hexane extract at a non-cytotoxic dose of 1 μg/mL demonstrated remarkable anti-inflammatory activity by inhibiting NO, IL-6 and TNF-α in LPS-induced RAW 264.7 macrophages (Fahmy et al. 2024).

2.5.2 Laboratory Research on Animal Models of Inflammation

The anti-inflammatory efficacy of ethanolic and hexane extracts of MO leaves was investigated in vivo in animal groups of six rats. They were subjected to a carrageenan test in which paw oedema was induced via SC injection of carrageenan; before this, experimental rats were administered ethanolic and hexane extracts in varying doses (100 mg/kg and 300 mg/kg), along with 10 mg/kg of naproxen as a reference medication. All treatment groups were compared with the control group, which was given only the vehicle. Inflammation was assessed at various time intervals through paw volume measurement. After four and five hours of inducing inflammatory oedema, ethanol extract showed a remarkable anti-inflammatory response that was independent of its dose, whereas hexane extract did not affect carrageenan-induced inflammation in rats. Earlier in the test, naproxen significantly reduced inflammation in rats two hours after treatment (Martínez-González et al. 2017). Aqueous MO leaf extract was investigated in a different study on rat models (five groups, six rats each) administered aqueous leaf extract of MO in 10, 30 and 100 mg/kg doses, ibuprofen 100 mg/kg as a reference drug or normal saline as a control, before treatment with carrageenan. Paw volume was measured at different time intervals. A significant dose-dependent anti-inflammatory effect was reported with all doses of the extract. Even though the extract had considerable anti-inflammatory activity, it was not as effective as the reference drug ibuprofen. Flavonoids were suggested to be responsible for the anti-inflammatory activity of the extract (Sulaiman et al. 2008).

The anti-inflammatory activity of ethanolic leaf extract was demonstrated in a study using an animal model of liver fibrosis, given 150, 300 and 600 mg/kg of the extract compared with a control. The extract's anti-inflammatory properties in liver fibrosis were confirmed by the considerable inhibition of TNF-α and p38 MAPK levels at all doses (Wilujeng et al. 2021). Furthermore, a number of in vivo investigations evaluated the topical MO leaf extract's capacity to enhance wound healing through anti-inflammatory and antioxidant pathways (Mohammad Shafie et al. 2022). An in vivo investigation was performed to determine the effect of MO alcoholic leaf extract on inflammation and ulceration of the stomach. Bisphenol A was used to induce gastric ulcers in rat models, and the treatment group received 200 mg/kg of the extract. MO leaf extract prevented the increase in TNF-α, IL-6 and NF κB elevated by bisphenol A, besides exhibiting an antioxidant effect (Abo-Elsoud, Ahmed Mohamed Abdelaziz, Attia Abd Eldaim and Hazzaa 2022).

MO seeds extracted using petroleum ether, chloroform, ethanol and water were investigated in an in vivo study using rat models. Rats were randomly assigned to six groups (n=6): the control group; the standard group that received ibuprofen 100 mg/kg; and four treatment groups that received 300 mg/kg doses of petroleum ether, chloroform, ethanol and aqueous MO seed extracts. Carrageenan was administered 30 minutes after treatment. After three hours of oedema induction, the ethanol extract demonstrated the most effective inhibition of paw oedema; this effect was similar to the anti-inflammatory effect of ibuprofen. Nevertheless, all other extracts exhibited less significant anti-inflammatory activity (Usman and Barhate 2012).

The carrageenan-induced pleurisy model was used to evaluate the in vivo anti-inflammatory properties of the seed extract mimicking inflammatory reactions in asthma patients. Different dosages of moringa seed extract (125 mg/kg, 250 mg/kg and 500 mg/kg) were administered to rats before pleurisy induction and in relation to dexamethasone. All doses showed dose-dependent reduction of leukocyte migration. A dose of 500 mg/kg produced the strongest inhibitory effect. The extract considerably reduced NO production, which is consistent with the findings of an in vitro study of the same extract (Araújo et al. 2013).

To evaluate the in vivo anti-inflammatory activity of MO root, an aqueous extract was examined in three groups of rats pretreated before the induction of rat paw oedema via carrageenin: the control group was given distilled water; the treatment group was administered aqueous MO root extract at doses of 750 mg/kg or 1000 mg/kg; and the third group was treated with 10 mg/kg indomethacin as a reference drug. MO root extract at a 750 mg/kg dose exhibited a considerable inhibitory effect on the development of paw oedema comparable to that of indomethacin. No significant increase in the oedema-inhibitory effect of the extract was noted when the dose was increased to 1000 mg/kg (Ndiaye et al. 2002). Furthermore, a review of the effects of moringa seed and leaf extracts on gastroenteritis found that MO had anti-inflammatory activities due to its quercetin content and could be suggested as a substitute treatment for individuals with bacterial gastroenteritis (Adji et al. 2022).

2.5.3 Evidence for Clinical Trials in Inflammatory Diseases

There are quite a few clinical trials conducted on MO regarding its effect on inflammatory diseases. The impact of MO aqueous leaf extract on patients with rheumatoid arthritis (RA) was evaluated in a single randomised controlled experiment that involved two groups of 30 patients. The first group was the control group that received a placebo, whereas the treatment group was given 40.50 mg/kg daily MO for a month. IL-6 levels were evaluated before and after treatment. IL-6 levels decreased in the treatment group, so the study concluded that the anti-inflammatory effect was due to isothiocyanates in moringa leaf extract (Prabowo et al. 2023). Similarly, 15 RA patients were subjected to a study where they received 20 g of moringa leaf dry powder and were compared with ten control patients. The MO-treated group reported improvement in their RA symptoms, indicating the anti-inflammatory potential of MO (Precious, Eboigbe and Zazdravnov 2020).

The effect of MO on inflammatory markers of prediabetic patients was evaluated in a clinical study, but no significant changes were noted, suggesting the need for further trials (Díaz-Prieto et al. 2022).

2.6 *MORINGA OLEIFERA* SIDE EFFECTS AND CONTRAINDICATIONS

No side effects were documented with conventional oral doses of MO leaves in clinical trials. However, moringa was suggested to have drug interactions with antidiabetic and thyroid medications. Due to the moringa content of alkaloids, it may be toxic at high doses (Meireles et al. 2020). A systematic review of a number of case study reports concluded that MO caused side effects either by itself or combined with synthetic drugs involving respiratory, muscular and skin symptoms, which evolved based on dose, duration and method of administration (Wijaya et al. 2022).

2.7 THERAPEUTIC POTENTIALS OF *MORINGA OLEIFERA* IN MODERN MEDICINE

Moringa oleifera's rich nutrient profile and diverse bioactive phytochemicals in its composition allow for a multi-target approach, making it a promising candidate for various therapeutic applications in modern medicine (Pareek et al. 2023). The following are some of the key therapeutic potentials of *M. oleifera*:

Antioxidant potential: the development of antioxidant supplements or functional foods because moringa leaves, seeds and pods are rich in antioxidants, such as quercetin, chlorogenic acid and vitamin C, which help combat oxidative stress and reduce the risk of chronic diseases (Varadarajan and Balaji 2022).

Anti-inflammatory potential: the creation of anti-inflammatory medications or nutraceuticals. Isothiocyanates and flavonoids in moringa leaves and seeds have demonstrated significant anti-inflammatory properties in preclinical trials, which are effective in the treatment of arthritis and asthma (Fahmy et al. 2024).

Antimicrobial potential: the development of natural antimicrobial and antifungal agents for infections and wound healing (Pareek et al. 2023).

Antidiabetic potential: the formulation of herbal supplements for diabetes management. Clinical studies suggest that it may enhance insulin sensitivity and reduce postprandial blood sugar spikes due to the presence of isothiocyanate and polyphenol (Precious, Eboigbe and Zazdravnov 2020).

Cancer prevention and treatment: cancer prevention strategies and complementary therapies. Moringa leaves and bark extracts possess a significant potential for breast and colorectal cancer (Anwar, Latif, Ashraf and Gilani 2007).

Nutritional supplements: the development of nutritional supplements because moringa is rich in vital vitamins, minerals and amino acids. It can help malnourished individuals (Varadarajan and Balaji 2022).

Moringa oleifera's diverse pharmacological properties make it a valuable resource for modern medicine.

2.8 CONCLUSION AND FUTURE PROSPECTS

Ultimately, through a variety of routes investigated in several in vivo and in vitro studies, *M. oleifera* exhibits potential for the treatment of inflammatory diseases. Nevertheless, additional clinical investigations should be conducted to validate this biological activity and determine whether using MO by itself or in conjunction with synthetic medications may positively impact the inflammatory phase of various chronic inflammatory diseases. To fully comprehend the moringa strategy and optimise its applicability, more research is recommended.

REFERENCES

Abo-Elsoud, Reda Abo Elfath Ahmed, Seham Ahmed Mohamed Abdelaziz, Mabrouk Attia Abd Eldaim, and Suzan Moustafa Hazzaa. 2022. Moringa oleifera alcoholic extract protected stomach from bisphenol A–induced gastric ulcer in rats via its anti-oxidant and anti-inflammatory activities. *Environmental Science and Pollution Research* 29:68830–68841.

Adji, Arga Setyo, Nabila Atika, Yemima Billyana Kusbijantoro, Atiyatum Billah, Astrid Putri, and Fitri Handajani. 2022. A review of leaves and seeds moringa oleifera extract: the potential moringa oleifera as antibacterial, anti-inflammatory, antidiarrhoeal, and antiulcer approaches to bacterial gastroenteritis. *Open Access Macedonian Journal of Medical Sciences* 10:305–313.

Alhakmani, Fatma, Sokindra Kumar, and Shah Alam Khan. 2013. Estimation of total phenolic content, in–vitro antioxidant and anti–inflammatory activity of flowers of Moringa oleifera. *Asian Pacific Journal of Tropical Biomedicine* 3:623–627.

Anwar, Farooq, Sajid Latif, Muhammad Ashraf, and Anwarul Hassan Gilani. 2007. Moringa oleifera: a food plant with multiple medicinal uses. *Phytotherapy Research: An International Journal Devoted to Pharmacological and Toxicological Evaluation of Natural Product Derivatives* 21:17–25.

Anzano, Attilio, Mohammad Ammar, Marina Papaianni, Laura Grauso, Mohammed Sabbah, Rosanna Capparelli, and Virginia Lanzotti. 2021. Moringa oleifera lam: a phytochemical and pharmacological overview. *Horticulturae* 7:409.

Araújo, Larissa Cardoso Corrêa, Jaciana Santos Aguiar, Thiago Henrique Napoleão, Fernanda Virgínia Barreto Mota, André Luiz Souza Barros, Maiara Celine Moura, Marília Cavalcanti Coriolano, Luana Cassandra Breitenbach Barroso Coelho, Teresinha Gonçalves Silva, and Patrícia Maria Guedes Paiva. 2013. Evaluation of cytotoxic and anti-inflammatory activities of extracts and lectins from Moringa oleifera seeds. *PLoS One* 8:e81973.

Arulselvan, Palanisamy, Woan Sean Tan, Sivapragasam Gothai, Katyakyini Muniandy, Sharida Fakurazi, Norhaizan Mohd Esa, Abdullah A. Alarfaj, and S. Suresh Kumar. 2016. Anti-inflammatory potential of ethyl acetate fraction of Moringa oleifera in downregulating the NF-κB signaling pathway in lipopolysaccharide-stimulated macrophages. *Molecules* 21:1452.

Avilés-Gaxiola, Sara, Josefina León-Félix, Yazmín B. Jiménez-Nevárez, Miguel A. Angulo-Escalante, Rosalio Ramos-Payán, Juventino Colado-Velázquez III, and J. Basilio Heredia. 2021. Antioxidant and anti-inflammatory properties of novel peptides from Moringa oleifera Lam. leaves. *South African Journal of Botany* 141:466–473.

Azlan, Ummi Kalthum, Nur Aisyah Khairul Annuar, Ahmed Mediani, Wan Mohd Aizat, Hanafi Ahmad Damanhuri, Xiaohui Tong, Daijiro Yanagisawa, Ikuo Tooyama, Wan Zurinah Wan Ngah, and Ibrahim Jantan. 2023. An insight into the neuroprotective and anti-neuroinflammatory effects and mechanisms of Moringa oleifera. *Frontiers in Pharmacology* 13:1035220.

Boopathi, N. Manikanda, and Bashir Yusuf Abubakar. 2021. Botanical descriptions of Moringa spp. In *The Moringa Genome*, edited by Manikanda BoopathiN. , M. Raveendran and Chittaranjan Kole, 11–20. Cham: Springer International Publishing.

Camilleri, Emma, and Renald Blundell. 2024. A comprehensive review of the phytochemicals, health benefits, pharmacological safety and medicinal prospects of Moringa oleifera. *Heliyon* 10:e27807.

Cheenpracha, Sarot, Eun-Jung Park, Wesley Y. Yoshida, Chaz Barit, Marisa Wall, John M. Pezzuto, and Leng Chee Chang. 2010. Potential anti-inflammatory phenolic glycosides from the medicinal plant Moringa oleifera fruits. *Bioorganic & medicinal chemistry* 18:6598–6602.

Chiş, Adina, Paul Aimé Noubissi, Oana-Lelia Pop, Carmen Ioana Mureşan, Michel Archange Fokam Tagne, René Kamgang, Adriana Fodor, Adela-Viviana Sitar-Tăut, Angela Cozma, and Olga Hilda Orăşan. 2023. Bioactive compounds in Moringa oleifera: mechanisms of action, focus on their anti-inflammatory properties. *Plants* 13:20.

Díaz-Prieto, Ligia E., Sonia Gómez-Martínez, Iván Vicente-Castro, Carlos Heredia, Elena A. González-Romero, María del Carmen Martín-Ridaura, Mercedes Ceinos, María J. Picón, Ascensión Marcos, and Esther Nova. 2022. Effects of Moringa oleifera Lam. supplementation on inflammatory and cardiometabolic markers in subjects with prediabetes. *Nutrients* 14:1937.

Divya, Surisetti, Vinay Kumar Pandey, Ritik Dixit, Sarvesh Rustagi, Tejas Suthar, David Atuahene, Vivien Nagy, Diána Ungai, Abdelhakam Esmaeil Mohamed Ahmed, and Béla Kovács. 2024. Exploring the phytochemical, pharmacological and nutritional properties of moringa oleifera: a comprehensive review. *Nutrients* 16:3423.

Dixit, Saurabh, Astha Tripathi, and Prabhat Kumar. 2016. Medicinal properties of Moringa oleifera: a review. *International Journal of Education and Science Research Review* 3:173–185.

Fahey, Jed W. 2005. Moringa oleifera: a review of the medical evidence for its nutritional, therapeutic, and prophylactic properties. Part 1. *Trees for life Journal* 1:1–15.

Fahmy, Nouran M., Shaimaa Fayez, Radwa Wahid Mohamed, Ahmed M. Elissawy, Omayma A. Eldahshan, Gokhan Zengin, and Abdel Nasser B. Singab. 2024. Moringa oleifera flowers: insights into their aroma chemistry, anti-inflammatory, antioxidant, and enzyme inhibitory properties. *BMC Complementary Medicine and Therapies* 24:286.

Hodas, Fabiane, Maria Rosa T. Zorzenon, and Paula G. Milani. 2021. Moringa oleifera potential as a functional food and a natural food additive: a biochemical approach. *Anais da Academia Brasileira de Ciencias* 93:e20210571.

Hussain, Tarique, Bie Tan, Yulong Yin, Francois Blachier, Myrlene C.B. Tossou, and Najma Rahu. 2016. Oxidative stress and inflammation: what polyphenols can do for us? *Oxidative Medicine and Cellular Longevity* 2016:7432797.

Liu, Rong, Jing Liu, Qi Huang, Shao Liu, and Yueping Jiang. 2022. Moringa oleifera: a systematic review of its botany, traditional uses, phytochemistry, pharmacology and toxicity. *Journal of Pharmacy and Pharmacology* 74:296–320.

Lopez-Rodriguez, Norma A, Marcela Gaytán-Martínez, María de la Luz Reyes-Vega, and Guadalupe Loarca-Piña. 2020. Glucosinolates and isothiocyanates from Moringa oleifera: chemical and biological approaches. *Plant Foods for Human Nutrition* 75:447–457.

Martínez-González, Claudia Lizbeth, Laura Martínez, Efraín J. Martínez-Ortiz, María Eva González-Trujano, Myrna Déciga-Campos, Rosa Ventura-Martínez, and Irene Díaz-Reval. 2017. Moringa oleifera, a species with potential analgesic and anti-inflammatory activities. *Biomedicine & Pharmacotherapy* 87:482–488.

Meireles, Diana, João Gomes, Lara Lopes, Mariana Hinzmann, and Jorge Machado. 2020. A review of properties, nutritional and pharmaceutical applications of Moringa oleifera: integrative approach on conventional and traditional Asian medicine. *Advances in Traditional Medicine* 20:495–515.

Mishra, Garima, Pradeep Singh, Ramesh Verma, Sunil Kumar, Saurabh Srivastav, K.K. Jha, and R.L. Khosa. 2011. Traditional uses, phytochemistry and pharmacological properties of Moringa oleifera plant: an overview. *Der Pharmacia Lettre* 3:141–164.

Mohammad Shafie, Norulhuda, Raja Nurul Izazi Raja Shahriman Shah, Premanand Krishnan, Nurul Abdul Haleem, and Thuan Yee Cheryl Tan. 2022. "Scoping Review: Evaluation of Moringa oleifera (Lam.) for Potential Wound Healing in In Vivo Studies." Molecules 27 (17): 5541. https://doi.org/10.3390/molecules27175541.

Ndiaye, M., A.M. Dieye, F. Mariko, A. Tall, and B. Faye. 2002. Contribution to the study of the anti-inflammatory activity of Moringa oleifera (Moringaceae). *Dakar Medical* 47:210–212.

O'Connell, Karyn E., Amy M. Mikkola, Aaron M. Stepanek, Andyna Vernet, Christopher D. Hall, Chia C. Sun, Eda Yildirim, John F. Staropoli, Jeannie T. Lee, and Diane E. Brown. 2015. Practical murine hematopathology: a comparative review and implications for research. *Comparative Medicine* 65:96–113.

Pareek, Ashutosh, Malvika Pant, Madan Mohan Gupta, Pushpa Kashania, Yashumati Ratan, Vivek Jain, Aaushi Pareek, and Anil A. Chuturgoon. 2023. Moringa oleifera: an updated comprehensive review of its pharmacological activities, ethnomedicinal, phytopharmaceutical formulation, clinical, phytochemical, and toxicological aspects. *International Journal of Molecular Sciences* 24:2098.

Peñalver, Rocío, Lorena Martínez-Zamora, José Manuel Lorenzo, Gaspar Ros, and Gema Nieto. 2022. Nutritional and antioxidant properties of Moringa oleifera leaves in functional foods. *Foods* 11:1107.

Prabowo, Nurhasan Agung, Arief Nurudhin, Yulyani Werdiningsih, Dikha Dwi Putra, Desy Puspa Putri, and Retno Widyastuti. 2023. Moringa Oleifera extract decreases interleukin 6 levels and disease activity in rheumatoid arthritis patients. *Bangladesh Journal of Medical Science* 22:416–421.

Precious, Itohan Ighorodje, Victory Eseose Eboigbe, and A. Zazdravnov. 2020. The clinical relevance of moringa oleifera extracts in the treatment of hyperglycaemia in Nigerian patients with rheumatoid arthritis. PhD diss., XHMY.

Ray, Savannah Jordan, T.J. Wolf, and C.N. Mowa. 2015. Moringa oleifera and inflammation: a mini-review of its effects and mechanisms. *I International Symposium on Moringa* 1158, 317-330.

Read, Margaret A. 1995. Flavonoids: naturally occurring anti-inflammatory agents. *The American Journal of Pathology* 147:235.

Saini, Ramesh Kumar, Iyyakkannu Sivanesan, and Young-Soo Keum. 2016. Phytochemicals of Moringa oleifera: a review of their nutritional, therapeutic and industrial significance. *3 Biotech* 6:1–14.

Sayed, Abeer M. El, Farghaly A. Omar, Marwa Mostafa Abdel-Aziz Emam, and Mohamed A. Farag. 2022. UPLC-MS/MS and GC-MS based metabolites profiling of Moringa oleifera seed with its anti-Helicobacter pylori and anti-inflammatory activities. *Natural Product Research* 36:6433–6438.

Shah, Kaushal H., and Manisha J. Oza. 2022. Comprehensive review of bioactive and molecular aspects of Moringa Oleifera lam. *Food Reviews International* 38:1427–1460.

Singh, Brahma N., B.R. Singh, R.L. Singh, D. Prakash, R. Dhakarey, G. Upadhyay, and H.B. Singh. 2009. Oxidative DNA damage protective activity, antioxidant and anti-quorum sensing potentials of Moringa oleifera. *Food and Chemical Toxicology* 47:1109–1116.

Su, Xinyue, Guanzheng Lu, Liang Ye, Ruyu Shi, Maomao Zhu, Xinming Yu, Zhiyong Li, Xiaobin Jia, and Liang Feng. 2023. Moringa oleifera Lam: a comprehensive review on active components, health benefits and application. *RSC Advances* 13:24353–24384.

Sulaiman, Mohd Roslan, Z.A. Zakaria, A.S. Bujarimin, M.N. Somchit, D.A. Israf, and S. Moin. 2008. Evaluation of Moringa oleifera aqueous extract for antinociceptive and anti-inflammatory activities in animal models. *Pharmaceutical Biology* 46:838–845.

Usman, Mohammed Rageeb Mohammed, and S. Barhate. 2012. Phytochemical investigation and study of antiinflammatory activity of Moringa oleifera Lam. *International Journal of Pharmaceutical Research and Development* 3:114–119.

Van Pham, Anh Thi, Minh Hien Luong, Hang Thi Thu Dinh, Thanh Phuong Mai, Quang Vinh Trinh, and Long Hoang Luong. 2021. Immunostimulatory effect of moringa oleifera extracts on cyclophosphamide-induced immunosuppressed mice. *Journal of Herbs, Spices & Medicinal Plants* 27:377–385.

Varadarajan, Saranya, and Thodur Madapusi Balaji. 2022. Assessing the in vitro antioxidant and anti-inflammatory activity of Moringa oleifera crude extract. *The Journal of Contemporary Dental Practice* 23:438.

Wijaya, Deborah Felinda Hari, Alfian Hendra Krisnawan, Oeke Yunita, Budipramana Krisyanti, Karina Citra Rani, and Nikmatul Ikhrom Eka Jayani. 2022. Systematic literature review: Side effects of *Moringa oleifera* Lam. https://doi.org/10.2991/978-94-6463-322-1_2

Wilujeng, Lestari Kanti, Febryana Nur Safitri, Supriono Supriono, Handono Kalim, and Sri Poeranto. 2021. The effect of Moringa oleifera (Lam) leaves ethanol extracts as anti-inflammatory and anti-fibrotic through TNF-α and p38-MAPK expression: in Vivo model of liver fibrosis approach. *AIP Conference Proceedings* 2353:030046. https://doi.org/10.1063/5.0052531

Xiong, Yongai, Muhammad Shahid Riaz Rajoka, MengXun Zhang, and Zhendan He. 2022. Isolation and identification of two new compounds from the seeds of Moringa oleifera and their antiviral and anti-inflammatory activities. *Natural Product Research* 36:974–983.

Xu, Yong-Bing, Gui-Lin Chen, and Ming-Quan Guo. 2019. Antioxidant and anti-inflammatory activities of the crude extracts of Moringa oleifera from Kenya and their correlations with flavonoids. *Antioxidants* 8:296.

Xu, Zhang, and Zhang Chenli. 2024. Inflammation. In *Textbook of Pathologic Anatomy: For Medical Students*, pp. 75–105. Singapore: Springer Nature Singapore.

Yeung, Yiu T, Faisal Aziz, Angelica Guerrero-Castilla, and Sandro Arguelles. 2018. Signaling pathways in inflammation and anti-inflammatory therapies. *Current pharmaceutical design* 24:1449–1484.

Zheng, Long, Xuli Lu, Shengtao Yang, Ying Zou, Fanke Zeng, Shaohao Xiong, Yupo Cao, and Wei Zhou. 2023. The anti-inflammatory activity of GABA-enriched Moringa oleifera leaves produced by fermentation with Lactobacillus plantarum LK-1. *Frontiers in Nutrition* 10:1093036.

3 *Moringa oleifera* as an Anticancer Drug

Neha Sharma, Alok Sharma, Anurag, Alimuddin Saifi, Harjeet Singh and Vibhu Sahani

3.1 INTRODUCTION

It is known that cancer is the leading cause of death worldwide, affecting millions. For patients experiencing severe side effects with conventional chemotherapy, other treatments and drugs are currently under investigation. Because current cancer therapies are costly and not universally available, the unconventional development of medications and therapeutic approaches is necessary (Fahey J. W. 2005). Natural products such as plants and trees are widely used in Ayurveda and traditional medicine as part of regimens to cure a wide range of diseases and disorders, with significant success. One among them is *Moringa oleifera* (MO); nearly every part has been proven to be effective against a number of diseases, including cancer. Novel drugs against cancer are required to overcome drug resistance, side effects and costs. Anticancer drugs have been designed based on the lead from natural agents. Included in the group is the terpene paclitaxel, isolated from *Taxus baccata*, and the *Vinca* alkaloids from the biological activity of *Catharanthus*. *Moringa oleifera* is a medicinal tree with high potential for use in complementary and alternative therapy (Al-Asmari et al. 2015). It is a Himalayan native, extensively cultivated in all tropical and some subtropical regions of the world, including Saudi Arabia. The plant is widely used in herbal medicine for the treatment of several conditions, including skin diseases, respiratory disorders, ear and oral infections, dental abscesses, hypertension, diabetes, anaemia and cancer. MO has been traditionally used to manage cancer, bacterial and viral infections, inflammation and hyperglycaemia. The activity of MO can mainly be attributed to its high content of bioactive compounds and antioxidants. We describe the cancer chemopreventive activity of several components of moringa in this chapter and the function of different bioactive compounds as anticancer agents and their link to structure and activity. Moringa has significant action against various cancers like breast, liver, lung and colon cancer. The mechanism involved in anticancer action has also been described in the chapter. In this chapter, we condense the mechanistic action of moringa extracts as anticancer agents, allowing researchers to design better medicines to cure cancer (Ferreira et al. 2014). MO has the potential to be further developed into an anticancer agent based on its pharmacological actions and chemical constituents.

3.2 CHEMICAL CONSTITUENTS AND SAR STUDIES

Moringa oleifera phytochemicals' structure–activity relationship (SAR) examines how changes to their molecular structures affect biological activity, including anticancer effects. The primary components of moringa are glucosinolates, flavonoids, phenolic compounds and isothiocyanates (Figure 3.1). A significant family of bioactive chemicals found in moringa are isothiocyanates, of which moringin (4-(α-L-rhamnosyloxy)benzyl isothiocyanate) is a well-known example. The potency and selectivity of ITCs can be influenced by the specific side chains attached to the isothiocyanate group, which can affect the compound's ability to penetrate cell membranes, bind to target proteins and induce apoptosis. It can also inhibit cancer cell proliferation and migration, potentially

DOI: 10.1201/9781003617860-4

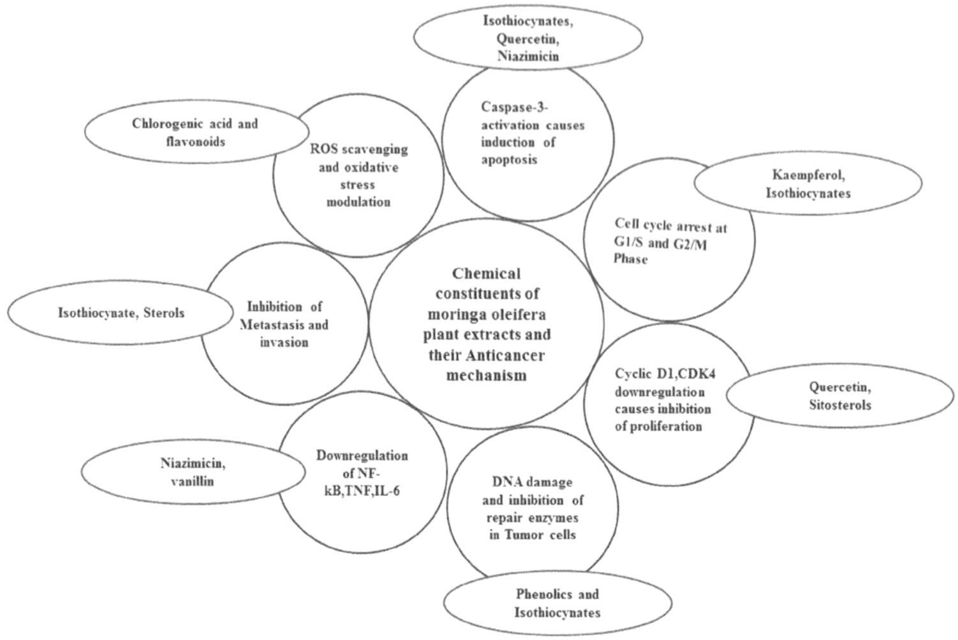

FIGURE 3.1 Anticancer Mechanisms of Various Chemical Constituents of *Moringa oleifera* Extract.

impeding metastasis. It targets multiple pathways to induce apoptosis (programmed cell death) in cancer cells (Wu et al. 2021).

Moringa leaves have been found to contain active pharmaceutical ingredients, including caffeic acid, chlorogenic acid and quercetin, with antioxidant and anti-inflammatory activities. These compounds may have general anticancer activity if they scavenge free radicals and reduce oxidative stress associated with tumour development (Mumtaz et al. 2021).

The flavonoids quercetin and kaempferol, present in moringa, have shown potential to induce apoptosis and inhibit cancer cell growth. They can also act as antioxidants, protecting cells from free radical damage. Moringa also contains glucosinolates, the precursors of ITCs. ITCs are produced when enzymes degrade glucosinolates and have anticancer activities. Other bioactive compounds in moringa, like terpenes, tannins, alkaloids, saponins and steroids, may also play a role in its anticancer effects (Wu et al. 2021). The non-planar, three-dimensional shape of the molecule also affects interaction with biological targets. How different moringa components interact with proteins and enzymes associated with cancer can be inferred from molecular docking experiments (drug-binding) assays (Sultan et al. 2023). Numerous bioactive ingredients, including isothiocyanates, have been suggested to contribute to moringa's anticancer activity. They are candidates for cancer prevention and treatment due to their ability to induce apoptosis and inhibit multiple signalling pathways. Subsequent studies on structure-activity correlations of these compounds may lead to the development of more potent, selective anticancer drugs (Moremane et al. 2023).

3.3 MECHANISMS OF THE CHEMICAL CONSTITUENTS OF MO IN CANCER

A recent study observed that moringa extract can inhibit cell populations, migration and proliferation and potentially express changes in different signalling pathways and revealed that *M. oleifera* has anticancer properties. It was shown that MO extract has a cytotoxic effect on melanoma cells in vitro, mainly through mitochondria-dependent caspase enzyme-mediated, as well as caspase enzyme-independent, apoptosis. This is comparable to selenium in cancer prevention in that it can also promote cancer apoptosis through mitochondria. MO can be used as a cancer therapy medication since it has the ability to stop tumour growth without interfering with the body's regular physiology and function (Wu et al. 2021).

The anticancer potential of MO can be mainly attributed to its abundant bioactive compounds that target various cellular pathways. Key components, such as benzyl isothiocyanate (BITC), niazimicin, quercetin and kaempferol, are reported to be responsible for mediating apoptosis in cancerous cells by a consecutive activation of intrinsic (mitochondrial) pathways via caspase-3 and caspase-9 and by increasing the pro-apoptotic (Bax) expression along with the repression of anti-apoptotic (Bcl-2) proteins.

Moreover, extracts of moringa are able to induce cell cycle arrest, including G2/M and G0/G1, in turn, leading to inhibition of cancer cell growth. These effects are at least partly the result of the regulation of the cell cycle proteins cyclins and cyclin-dependent kinases. It also decreases ROS and increases SOD, CAT and GPx levels, which are MODs (modified one dependent) that prevent normal cells from oxidative DNA damage and selectively stress malignant cells. Moreover, moringa exhibits anti-angiogenic activity by reducing the expression of vascular endothelial growth factor and matrix metalloproteinases (MMP-2/9), which are linked to preventing the development of new blood vessels necessary for metastasis and the growth of tumours. Altogether, these multitargeted actions emphasise that MO could be an attractive candidate as an adjuvant cancer therapy (Tiloke et al. 2013).

ITCs are naturally occurring small compounds that improve carcinogen detoxification and decrease carcinogen activation. They are produced from glucosinolate precursors found in cruciferous vegetables. In comparison with other ITCs found in different crops, the ITCs produced through the biotransformation of moringa glucosinolates contain other sugars in their chemical configurations, which enhance the stability of these bioactive compounds (Lopez-Rodriguez et al. 2020). MIC-1 is an ITC that is particularly abundant in *M. oleifera* seeds (Shehzad et al. 2013) and is currently the most abundant phytochemical in MO seed extracts (Table 3.1).

TABLE 3.1
Role of MO Chemical Constituents in Cancer

Plant part	Chemical constituents	Anticancer mechanism	Cancer	Reference
Leaves extract	Quercetin, kaempferol, chlorogenic acid	Apoptosis induction, anti-angiogenesis, ROS modulation	Breast, lung, colon	Tiloke et al. 2013; Leone et al. 2015
	Isothiocyanates	Cell cycle arrest, apoptosis	Pancreatic, colorectal	Sreelatha and Padma 2011
Seeds	β-sitosterol, stigmasterol	Apoptosis, immune modulation	Colon, prostate	Choudhary et al. 2013
	4-(α-L-rhamnosyloxy) benzyl isothiocyanate	Cytotoxicity inhibits metastasis	Liver, breast	Waterman et al., 2014
Bark	Thiocarbamates: niaziminin A and B	Inhibition of tumour growth	General	Anwar et al. 2007
Root	Moringine	DNA intercalation, cytotoxic activity	Leukaemia, cervical	Abdull Razis et al. 2014
Flowers	Flavonoids (e.g. myricetin, kaempferol)	ROS generation, anti-proliferative	Cervical, breast	Singh et al. 2016
	Essential oils (linalool, eugenol)	Induce apoptosis, inhibit angiogenesis	Lung, prostate	Mbikay 2012

3.4 MO IN LIVER CANCER

MO can be utilised as a cancer therapy medication since it has the ability to stop tumour growth without interfering with the body's regular physiology and function. ITCs are naturally occurring small compounds that improve carcinogen detoxification and decrease carcinogen activation. They are derived from the precursors of glucosinolate in cruciferous vegetables. Unlike other ITCs found in other crops, the ITCs generated from *M. oleifera* GLs via biotransformation have additional sugars on their molecular structure, and these additional sugars provide stability to these bioactive molecules. MIC-1 is an ITC present in MO seeds and is the most prevalent phytochemical extracted from them to date. However, the extract of the bark is rich in hydrocarbons, phenolics, long-chain fatty acids and carboxylic acids; the leaf extract is abundant in thiocyanates, hydrocarbons and fatty acids. Seed analyses demonstrated the presence of hydrocarbons, long-chain fatty acids and their derivatives. The standardisations by GC-MS of the leaves, bark and seed powder proved that the leaf extract possesses isopropyl isothiocyanate, D-allose and cetene, and the bark extract contains eugenol, dibutyl phthalate, 2-chloropropionic acid and 5-eicosene (Al-Asmari et al. 2015).

Flavonoids of various pharmacological activities and phenolics are abundant in MO. In another study, the antiproliferative action of moringa fruit extract in HepG2 human liver cancer cells was simulated computationally to observe cell death. HepG2 cells were treated with different concentrations ranging from 25 to 200 µg/ml of MOF extract for 48 h in culture. MTT assay, Annexin V-FITC/propidium iodide double staining, nuclear condensation, ROS generation and the executioner enzyme of apoptosis, caspase-3, were employed to investigate the antiproliferative activity. By enhancing cellular death, MOF extract dramatically decreased cell viability ($p < 0.05$), as demonstrated by the annexin V-FITC/PI staining experiment. Drug-like candidates with no anticipated toxicity were found by toxicity testing and drug-likeness analysis. The study demonstrated the extract's possible anticancer properties, making it a potentially useful source for developing anticancer drugs. It is interesting to note that the fruit of MO has active phytoconstituents that show potential as drugs without any known harm. As MO fruit is relatively less toxic to normal cells, it might be a potential resource for the generation of anticancer drugs for human liver cancer (Siddiqui et al. 2021).

3.5 MO IN BREAST CANCER

The main motive of the study was to determine whether MO has potential as a CDK-2 inhibitor for hormone receptor-positive breast tumours. To this end, the study suggests screening MO compounds based on in silico molecular docking experiments and also analyses their anticancer potential against protein targets of CDK-2. Out of 36 MO compounds mentioned in the literature, ellagic acid, kaempferol, quercetin, niazirin and chlorogenic acid showed good binding affinities to the target. The chemical interactions were revealed using PYMOL software. Results from the webservers SWISS and ProTox II were used to predict in silico toxicity and ADME profiles. The research investigated the anticancer effects of phytochemicals derived from MO using sophisticated in silico techniques targeting CDK2. The study evaluated the interactions between the ligands and their targets through molecular docking, prime MM-GBSA energy calculations and molecular dynamics (MD) simulations. The Michigan Cancer Foundation-7 (MCF-7) breast cancer cell line and methanolic extract of MO were used in laboratory procedures to assess the ability of the stigma discoid to fight cancer. The main objective of the current study was to identify natural products that may serve as CDK-2 inhibitors present in MO to induce the treatment of breast hormone-positive cancers. The MTT assay results showed that no significant inhibition of the proliferation of Mcf-7 cells was observed for 24 h with fraction A with petroleum ether. But at 200 µg/mL of fraction B (ethyl acetate), an antiproliferative outcome was detected, with cell viability at 40%. The anticancer potential of chemical constituents of MO targeting CDK-2 inhibition in oestrogen receptor-positive breast cancer was also studied using in silico and in vitro approaches (Sultan et al. 2023).

Moringa oleifera, one of the most widely cultivated vegetables in the world since ancient times, contains a profile of important phenolic and other types of phytochemicals, which possess several biological properties. The purpose of this study was to compare the phenolic composition and anticancer activity of MO leaf extracts in vitro and in silico. Docking studies identified a binding pose of the phytochemicals in the target protein. The results indicated that each fraction of the solvents exhibited anticancer activity on HeLa cancer cell lines in vitro. At a concentration of 416 μg mL^{-1}, the n-hexane fraction exhibited an IC 50 value of 50% inhibition in the growth of HeLa cancer cells, with a VC value of 50% inhibitory effect. The ligands exhibit strong binding energies in molecular docking studies and are bound within the target protein's active region. According to this study, MO leaves may be able to prevent the growth of cancer cells and enhance human health in addition to creating new food ingredients. Phytochemicals from MO leaves have the potential to be the next big thing in cancer treatment, according to in vitro and in silico studies (Amin et al. 2024).

3.6 MO IN LUNG CANCER

In a dose-dependent way, the extracts from MO leaves dramatically reduced the proliferation of the human cancer cell line A549 (Guevara et al. 1999). According to morphological investigations, the moringa leaf extract induced apoptosis, as evidenced by nuclear fragmentation, chromatin condensation, blebbing and cell shrinkage. Under different treatment circumstances, quantitative RT-PCR studies of the genes Bax and Bcl-2 revealed aberrant expression profiles. In addition to improving human health and creating novel food ingredients, this study shows that MO leaves may inhibit the formation of cancer cells. According to in vitro research, the phytochemicals found in MO leaves may be used as the main drugs to treat cancer. The MTT assay was used to assess the cytotoxicity of MO leaf aqueous extracts over lung cancer A549 cell lines at various concentrations of 100–500 μg/ml. Cell viability of A549 after treatments versus control after an additional 24 hours is shown in the distribution. Cell viability of sensitive A549 cells decreased dose-dependently ($p<0.05$) to the inhibited rates of 30% and 15% (400 and 500 μg/ml), respectively, compared with the control (Bhadresha et al. 2022).

A few reports have evaluated the antitumour properties of MO in lung cancer, especially in non-small lung carcinoma (NSCLC). Notably, Al-Asmari et al. (2015) reported potent antiproliferative effects and apoptosis-inducing activity in A549 lung carcinoma cells via a dose-dependent manner mediated by WP extract compared with the control. Similarly, Tiloke et al. (2013) found that the moringa water extract induced apoptosis through ROS formation, mitochondrial damage, DNA fragmentation and caspase-3 activation. Adebayo et al. (2020) further reinforced this concept by demonstrating, using in silico and in vitro experiments, that BITC, a bioactive component of moringa, exhibits robust binding potential to the epidermal growth factor receptor (EGFR), a major therapeutic target in LC. Further, in an in vivo study by Manoharan et al. (2006) on DMBA-induced lung cancer in rats, it was observed that moringa seed extract decreased tumour incidence, decreased lipid peroxidation and ameliorated antioxidant enzymes SOD and CAT. Taken together, these data indicate that MO has a strong chemopreventive activity as well as being an active agent in the chemotherapy of lung cancer via multiple pathways: induction of oxidative stress, apoptosis, inhibition of EGFR and immune response. However, more clinical trials are necessary to evaluate its therapeutic effects in humans.

3.7 MO IN PROSTATE CANCER

Moringa oleifera contains several phenolic molecules that possess numerous biological activities and can be developed into new medicines to treat various cancers (Ju et al. 2018). The present study was carried out to estimate the phenolic composition and anticancer activity of MO leaf extract in vitro and in silico. The extraction of the solvent fraction from MO leaves was accomplished using n-hexane, chloroform, ethyl acetate, butanol and aqueous solvents. The solvent of the fractions was determined by reversed-phase high-performance chromatography, after which the anticancer activity of the fractions was evaluated against HeLa cancer cell lines. Docking studies were performed with molecular docking to determine the binding conformations of the phytochemicals to the target protein. The results indicated that all the solvent fractions exhibited anticancer effects on the HeLa cancer cell line in vitro. At 416 μg mL^{-1}, the n-hexane fraction demonstrated a 50% decrease in HeLa cancer cell viability in comparison with the control. According to molecular docking experiments, the ligands exhibit good binding energy values and bind within the target protein's active region. According to this study, MO leaves may be able to prevent the growth of cancer cells and enhance human health in addition to creating new food ingredients (Mumtaz et al. 2021).

Recent investigations indicate that (abnormal) activation of the Notch signalling cascade plays a role in the development and progression of a number of carcinomas, including those of the prostate (Farhan et al. 2020). Phytochemicals from various parts of MO are among the most frequently used plants in traditional medical practices for the treatment of diseases, including cancer. One study was tentatively designed to evaluate the anticancer potential of MO methanolic leaf extract against human prostatic cancer DU145 cells, and the underlying mechanism of action was also determined. After treatment with varying concentrations of MO methanolic leaf extract, cell viability was determined by phase-contrast microscopy and MTT assay. Apoptotic cell populations, as well as nuclear morphology, were evaluated through FITC-Annexin V/PI double staining and DAPI staining. qRT-PCR was performed to detect the mRNA levels of Bax, Bcl-2 and Notch signalling-related genes. Concentration-dependent loss of viability with marked morphological and nuclear changes was observed with the methanolic leaf extract of MO. Activation of apoptosis was also associated with an increase in Bax gene expression compared with the Bcl-2 gene expression level. Additionally, by downregulating the expression of Notch-1 and Hes-1, extract administration suppresses the aberrant Notch signalling pathway. These findings unequivocally demonstrated that the extract from MO methanolic leaves inhibits the growth of prostate cancer DU145 by inducing apoptosis and inhibiting the aberrant Notch signalling pathway (Khan et al. 2020).

Moringa extract's regulatory impact on A549 cells is assessed by examining their migration, apoptosis, cell cycle and proliferation to determine MO's potential mode of action. After testing four different types of lung cancer cells and five different types of cancer cells, it was discovered that moringa extract exhibited minimal damage to GES-1 cells (human gastric mucosal epithelial cells) but the strongest growth-inhibitory effect on A549 cells. At the same time, moringa extract caused A549 cells to undergo apoptosis and expressed more caspase-3 and caspase-9, two proteins linked to apoptosis. Additionally, moringa extract caused cell cycle arrest in the S phase by increasing the expression of the protein p21 and decreasing the expression of cyclin D1 and cyclin E. Additionally, moringa extract reduced the production of the migration-related proteins MMP-2 and MMP-9 and limited the capacity of A549 cells to migrate. Additionally, Moringa extract-treated A549 cells showed a decrease in the phosphorylation level of the JAK2 and STAT3 proteins. Additionally, moringa extract and AZD1480 (a JAK inhibitor) both prevented A549 cells from proliferating and migrating and caused cell death; their effects were not cumulative. According to these findings, moringa extract suppresses A549 cell migration and proliferation while causing apoptosis and cell cycle arrest via a mechanism linked to the suppression of JAK2/STAT3 pathway activation. Lung cancer may therefore be prevented and treated using this extract (Xie et al. 2021).

A series of succinct evaluations aimed to evaluate the effectiveness of MO flower extract as an anti-prostate cancer agent. Cell cycle progression studies were examined using flow cytometry, and the extract's cell viability was assessed using MTT at various time points. Several apoptotic indicators were assessed using immunoblotting. At doses of 22.61 μg/mL and 6.25 μg/mL in PC-3 cells at 24 and 48 hours, respectively, MO flower extract produced 50% inhibition of prostate cancer cell growth. Using annexin V labelling, MO floral extract caused apoptosis and G1 phase cell cycle arrest to accumulate. Additionally, immunoblot detection of PARP breakage increases caspase-3 activity, protein expression and Bax signals, indicating apoptotic activation. All findings point to the possibility that MO flower extract inhibits prostate cancer development in PC-3 cells via disrupting the AKT pathway (Gothai et al. 2018).

3.8 NANOTECHNOLOGY IN MORINGA FORMULATIONS

The green production of nanoparticles can use plant extracts such as MO. Metal ions are reduced into phytonanoparticles as a result. Synthetic silver phytonanoparticles (Althomali et al. 2022) and selenium nanoparticles (Abdel Wahab et al. 2022) of moringa have unique structures that are stable and have biological uses. Controlling the size and form of the phytonanoparticle results in high yields. Numerous investigations have demonstrated the application of MO in phytonanoparticle production, as MO was used by Anand et al. (2015) to synthesise gold nanoparticles (AuNPs). Throughout the synthesis process, the MO served as a reducing agent, and it also exhibited catalytic activity. AuNPs showed promise in the breakdown of industrial waste products by reducing nitroaniline and nitrophenol. Additionally, hydroxyapatite nanoplates were synthesised using the MO extract. The synthesis was aided by the polyphenolic hydroxyl groups present in the extract.

MO's leaves and roots were utilised to create nanocomposites that were cytotoxic to HepG2 cells, MCF7 cells and HCT116/Caco2 cells. To evaluate the antiproliferative qualities, the leaves were utilised in a one-pot production of gold phytonanoparticles. With little impact on healthy PBMCs, AuNPs showed anticancer potential against A549 lung cancer cells after effectively synthesising them (Khan et al. 2020). AuNPs caused cancer cells to undergo apoptosis by activating caspase-9's alternative splicing and by increasing the activity and production of apoptotic chemicals. Iron oxide nanoparticles were also synthesised using MO leaves. Green synthesis is simple to use and safe for the environment. When compared with a commercially available adsorbent, it demonstrated promise in the adsorption of fluoride ions (Tiloke et al. 2018).

3.9 LIMITATIONS AND FUTURE ASPECTS

Although MO shows potential anticancer effects, there are obstacles to its clinical use. A first hurdle to overcome is the lack of standardisation of extracts in terms of the extraction procedure, dosage and active compound(s) concentration, making it difficult to compare among different investigations. Furthermore, the majority of available data are based on in vitro and preclinical experience in animals, and only a few human studies have been performed to assess the safety, efficacy and pharmacokinetics of these agents. Furthermore, certain highly active phytochemicals, including benzyl isothiocyanate, are structurally insoluble and have low bioavailability, limiting their clinical potential. In order to break through these barriers, future experimental work has to be aimed towards achieving standardised formulations, including nanoparticle carrier-based delivery systems, that can effectively increase bioavailability and achieve targeted effects. In addition, well-designed clinical trials are also needed to determine the efficacy of Moringa as an adjunctive therapy in oncology. Combining pharmacogenomic strategies and exploiting synergies with classical chemotherapeutics will shed additional therapeutic light on this promising agent.

3.10 CONCLUSION

This chapter demonstrated the anticancer properties of various sections of MO. We discussed the active substances that have been shown to support these actions. The published results of the research, which focused on extracts from different parts of the moringa leaf, indicate that the extract is cytotoxic to a broad range of malignant cells. During in vivo tests on mice and rats, the extract was found to have very low toxicity when given orally, and it appears to have relatively low toxicity against normal cells. Any variations in anticancer efficacy due to the various extraction methods and solvents require further research. Even though the results varied, several studies also showed comparable results, particularly those that examined the mechanisms by which moringa leaf extract induces cancer cell apoptosis. Multiple studies indicate that the extract increases ROS levels in cells, resulting in DNA damage and ultimately leading to apoptosis. Overall, the studies discussed here suggest that *M. oleifera* extracts hold significant promise for development into an anticancer drug. Nonetheless, further research and development are needed to advance this potential.

Ethics approval and consent to participate: NA.

Consent for publication: We give our consent for the publication.

Availability of data and material: Data can be made available if asked.

Competing interests: None.

Funding: None.

Conflict of interest: The authors declare that they have no conflicts of interest.

Acknowledgement: The authors are thankful to the Department of Pharmaceutical Technology, MIET, Meerut, for providing the necessary facilities.

REFERENCES

Abdel Wahab, A. H., El-Baz, H. A., & Soliman, M. M. (2022). Selenium nanoparticles functionalized with *Moringa oleifera* leaf extract and vitamin C mitigate hepatocellular carcinoma in rats. *Cancer Nanotechnology*, *13*, Article 23. https://doi.org/10.1186/s12951-022-01234-2

Abdull Razis, A. F., Ibrahim, M. D., &Kntayya, S. B. (2014). Health benefits of *Moringa oleifera*. *Asian Pacific Journal of Cancer Prevention, 15*(20), 8571–8576. https://doi.org/10.7314/APJCP.2014.15.20.8571

Adebayo, A., Smith, J., Kumar, R., & Chen, L. (2020). The impact of medicinal plants on cancer treatment: A review. *Journal of Ethnopharmacology, 250*(3), 112–120. https://doi.org/10.1016/j.jep.2020.112345

Al-Asmari, A. K., Albalawi, S. M., Athar, M. T., Khan, A. Q., Al-Shahrani, H., & Islam, M. (2015). *Moringa oleifera* as an anti-cancer agent against breast and colorectal cancer cell lines. *PLoS One*, *10*(8), e0135814. https://doi.org/10.1371/journal.pone.0135814

Althomali, A. M., Al-Salmi, F. A., Alqarni, M. H., Khan, M. I., & Alghamdi, S. A. (2022). Biosynthesized silver nanoparticles from *Moringa oleifera* leaf extract exhibit potent anticancer activity in colon cancer cells. *Green Processing and Synthesis, 11*(1), 565–574. https://doi.org/10.1515/gps-2022-0052

Amin, M. F., Ariwibowo, T., Putri, S. A., & Kurnia, D. (2024). *Moringa oleifera*: A review of the pharmacology, chemical constituents, and application for dental health. *Pharmaceuticals, 17*(1), 142. https://doi.org/10.3390/ph17010142

Anand, S. P., & Sati, N. (2015). Bioactive potential of *Moringa oleifera* Lam. against some human pathogenic bacteria. *International Journal of Pharmaceutical Sciences and Research*, *6*(2), 505–510.

Anwar, F., Latif, S., Ashraf, M., & Gilani, A. H. (2007). Moringa oleifera: A food plant with multiple medicinal uses. *Food Chemistry, 105*(3), 1188–1194. https://doi.org/10.1016/j.foodchem.2007.04.014

Bhadresha, K., Thakore, V., Brahmbhatt, J., Upadhyaya, V., Jain, N., & Rawal, R. (2022). Anticancer effect of *Moringa oleifera* leaves extract against lung cancer cell line via induction of apoptosis. *Advances in Cancer Biology - Metastasis*, *6*, 100072. https://doi.org/10.1016/j.acb.2022.100072

Choudhary, M. K., Bodakhe, S. H., & Gupta, S. K. (2013). Assessment of the antidiarrheal activity of *Moringa oleifera* Lam. root extract. *Journal of Pharmacy &Bioallied Sciences, 5*(1), 30–36. https://doi.org/10.4103/0975-7406.106569

Fahey, J. W. (2005). Moringa oleifera: A review of the medical evidence for its nutritional, therapeutic, and prophylactic properties. *Trees for Life Journal*, *1*(5). https://www.tfljournal.org/article.php/20051201124931586

Farhan, F., Pandey, P., Jha, N. K., Jafri, A., & Khan, I. (2020). Antiproliferative effect of *Moringa oleifera* methanolic leaf extract by down-regulation of Notch signaling in DU145 prostate cancer cells. *Gene Reports*, *19*, 100619. https://doi.org/10.1016/j.genrep.2020.100619

Ferreira, P. M. P., de Araújo, É. J. F., da Silva, J. do N., de Freitas, R. M., de Jesus Costa, N. D., de Carvalho Oliveira, S. F., et al. (2014). Safety and efficacy of *Moringa oleifera* Lamarck (1785)—Therapeutic and toxicological properties. In D. S. J. T. Gowder (Ed.), *Pharmacology and Therapeutics*. InTech. https://doi.org/10.5772/57422

Gothai, S., Muniandy, K., Gnanaraj, C., Ibrahim, I. A., Shahzad, N., & Esa, N. M. (2018). Anticancer potential of *Moringa oleifera* flower extract in human prostate cancer PC-3 cells via induction of apoptosis and downregulation of AKT pathway. *Pharmacognosy Magazine*, *14*(58), 477–481. https://doi.org/10.4103/pm.pm_447_17

Guevara, A. P., Vargas, C., Sakurai, H., Fujiwara, Y., Hashimoto, K., Maoka, T., Kozuka, M., Ito, Y., Tokuda, H., & Nishino, H. (1999). An antitumor promoter from *Moringa oleifera* Lam. *Mutation Research/Fundamental and Molecular Mechanisms of Mutagenesis, 440*(2), 181–188. https://doi.org/10.1016/S1383-5718(99)00025-X

Ju, J., Gothai, S., Hasanpourghadi, M., Nasser, A. A., Ibrahim, I. A., Shahzad, N., & Esa, N. M. (2018). Anticancer potential of *Moringa oleifera* flower extract in human prostate cancer PC-3 cells via induction of apoptosis and downregulation of AKT pathway. *Pharmacognosy Magazine, 14*(58), 477–481. https://doi.org/10.4103/pm.pm_447_17

Khan, F., Pandey, P., Jha, N. K., Jafri, A., & Khan, I. (2020). Antiproliferative effect of *Moringa oleifera* methanolic leaf extract by down-regulation of Notch signaling in DU145 prostate cancer cells. *Gene Reports, 19*, 100619. https://doi.org/10.1016/j.genrep.2020.100619

Khan, M. A., Khan, T., Perveen, S., &Siddiquah, A. (2020). Phytogenic fabrication of gold nanoparticles using *Moringa oleifera* leaf extract and their anticancer activity against lung cancer cells. *Arabian Journal of Chemistry, 13*(1), 1738–1746. https://doi.org/10.1016/j.arabjc.2018.12.021.

Leone, A., Spada, A., Battezzati, A., Schiraldi, A., Aristil, J., & Bertoli, S. (2015). *Moringa oleifera* seeds and oil: Characteristics and uses for human health. *International Journal of Molecular Sciences, 16*(6), 12791–12835. https://doi.org/10.3390/ijms160612791

Lopez-Rodriguez, N. A., Gaytán-Martínez, M., Reyes-Vega, M. de la L., & Loarca-Piña, G. (2020). Glucosinolates and isothiocyanates from *Moringa oleifera*: Chemical and biological approaches. *Plant Foods for Human Nutrition, 75*, 447–457. https://doi.org/10.1007/s11130-020-00851-x

Manoharan, S., Silvan, S., Senthil, N., & Renju, G. L. (2006). Modulating effect of *Moringa oleifera* leaf extract on DMBA-induced oral carcinogenesis in hamsters. *Indian Journal of Pharmaceutical Sciences, 68*(2), 235–238.

Mbikay, M. (2012). Therapeutic potential of *Moringa oleifera* leaves in chronic hyperglycemia and dyslipidemia: A review. *Frontiers in Pharmacology, 3*, 24. https://doi.org/10.3389/fphar.2012.00024

Moremane, M. M., Abrahams, B., & Tiloke, C. (2023). *Moringa oleifera*: A review on the antiproliferative potential in breast cancer cells. *Current Issues in Molecular Biology, 45*(8), 6880–6902. https://doi.org/10.3390/cimb45080434

Mumtaz, M. Z., Kausar, F., Hassan, M., & Jabeen, F. (2021). Anticancer activities of phenolic compounds from *Moringa oleifera* leaves: In vitro and in silico mechanistic study. *Beni-Suef University Journal of Basic and Applied Sciences, 10*, Article 12. https://doi.org/10.1186/s43088-021-00101-2

Shehzad, A., & Lee, Y. S. (2013). Molecular mechanisms of curcumin action: Signal transduction. *BioFactors, 39*(1), 27–36. https://doi.org/10.1002/biof.1065

Siddiqui, S., Upadhyay, S., Ahmad, I., Hussain, A., & Ahamed, M. (2021). Cytotoxicity of *Moringa oleifera* fruits on human liver cancer and molecular docking analysis of bioactive constituents against caspase-3 enzyme. *Journal of Food Biochemistry, 45*(5), e13720. https://doi.org/10.1111/jfbc.13720.

Singh, D., Arya, P., Aggarwal, S., & Verma, R. (2016). Evaluation of antioxidant and antimicrobial potential of different extracts of *Moringa oleifera* leaves. *Journal of Applied Pharmaceutical Science, 6*(6), 157–163. https://doi.org/10.7324/JAPS.2016.60623

Sreelatha, S., & Padma, P. R. (2011). Modulatory effects of Moringa oleifera extracts against hydrogen peroxide-induced cytotoxicity and oxidative damage. *Plant Foods for Human Nutrition, 66*(4), 303–308. https://doi.org/10.1007/s11130-011-0259-9

Sultan, R., Ahmed, A., Wei, L., Saeed, H., Islam, M., & Ishaq, M. (2023). The anticancer potential of chemical constituents of *Moringa oleifera* targeting CDK-2 inhibition in estrogen receptor-positive breast cancer using in-silico and in vitro approaches. *BMC Complementary Medicine and Therapies, 23*, Article 396. https://doi.org/10.1186/s12906-023-04198-z

Tiloke, C., Anand, K., Gengan, R. M., & Chuturgoon, A. A. (2018). *Moringa oleifera* and their phytonanoparticles: Potential antiproliferative agents against cancer. *Biomedicine & Pharmacotherapy, 108*, 457–466. https://doi.org/10.1016/j.biopha.2018.09.068

Tiloke, C., Phulukdaree, A., & Chuturgoon, A. A. (2013). The antiproliferative effect of Moringa oleifera crude aqueous leaf extract on cancerous human alveolar epithelial cells. *BMC Complementary and Alternative Medicine, 13*, 226. https://doi.org/10.1186/1472-6882-13-226

Waterman, C., Cheng, D. M., Rojas-Silva, P., Poulev, A., Dreifus, J., Lila, M. A., & Raskin, I. (2014). Stable, water extractable isothiocyanates from Moringa oleifera leaves attenuate inflammation in vitro. *PLoS One, 9*(2), e95492. https://doi.org/10.1371/journal.pone.0095492

Wu, Y. Y., Xu, Y. M., & Lau, A. T. Y. (2021). Anti-cancer and medicinal potentials of *Moringa* isothiocyanate. *Molecules, 26*(24), 7512. https://doi.org/10.3390/molecules26247512

Xie, J., Peng, L.-J., Yang, M.-R., Jiang, W.-W., Mao, J.-Y., Shi, C.-Y., Tian, Y., & Sheng, J. (2021). Alkaloid extract of *Moringa oleifera* Lam. exerts antitumor activity in human non-small-cell lung cancer via modulation of the JAK2/STAT3 signaling pathway. *Evidence-Based Complementary and Alternative Medicine*, *2021*, Article 5591687. https://doi.org/10.1155/2021/5591687

4 *Moringa oleifera* as an Immunomodulator

Noha Swilam, Khaled A. Nematallah and Omayma Eldahshan

4.1 INTRODUCTION

The immune system is classically divided into two complementary branches: innate and adaptive immunity. Innate immunity constitutes the body's immediate and nonspecific defence against invading pathogens, toxins and tumour cells. It is evolutionarily older and operates through physical barriers, phagocytic cells, natural killer (NK) cells and pattern recognition receptors (PRRs) that detect conserved pathogen-associated molecular patterns (PAMPs) (Riera Romo, Perez-Martinez and Castillo Ferrer 2016; Martins, Ribeiro-Gomes and Daniel-Ribeiro 2023). Adaptive immunity, on the other hand, is slower to activate upon first exposure but offers high specificity and long-term memory through T and B lymphocytes that generate antigen-specific responses (Hu et al. 2022; Mujal, Delconte, and Sun 2021). Though once considered functionally separate, it is now clear that the innate and adaptive branches are tightly integrated: innate immune cells not only initiate but also regulate the adaptive immune response (Martins, Ribeiro-Gomes and Daniel-Ribeiro 2023).

Recent studies have challenged the classical view that immunological memory is exclusive to the adaptive system. It has been demonstrated that innate immune cells, including monocytes and NK cells, can develop memory-like properties through a process known as trained innate immunity. This process involves metabolic and epigenetic reprogramming that leads to enhanced responsiveness upon re-exposure to pathogens (Hu et al. 2022). Autophagy – a conserved cellular degradation process – has also emerged as a crucial mechanism linking innate and adaptive immunity, facilitating antigen processing, pathogen clearance and regulation of inflammation (Xu and Eissa 2010). These emerging insights underscore the complexity of immune regulation and suggest that natural agents, such as phytochemicals, may exert multifaceted immunomodulatory effects by interacting with both innate and adaptive pathways.

Natural immunomodulators are compounds derived primarily from plant sources that influence immune system functions either by stimulation or suppression. These bioactive compounds have attracted a lot of interest because of their potential to help correct immune dysregulation, particularly in diseases like cancer, autoimmune disorders and chronic infections (Zebeaman et al., 2023). Unlike synthetic immunosuppressive or immunostimulatory drugs, natural agents offer a safer and multi-targeted approach, making them favourable candidates for long-term use.

Phytochemicals such as flavonoids, alkaloids, terpenoids and polysaccharides have demonstrated significant immunomodulatory activity through various mechanisms, including cytokine regulation, enhancement of phagocytic activity and modulation of T cell and NK cell functions. As highlighted by Zebeaman et al. (2023), compounds like curcumin, quercetin and resveratrol not only suppress the overproduction of pro-inflammatory cytokines like interleukin-6 (IL-6) and tumour necrosis factor-alpha (TNF-α) but also promote antioxidant defences, which are essential for protecting immune cells from oxidative damage. Similarly, Mohamed, Jantan and Haque (2017) reported that plant-derived molecules exert their actions by modulating transcription factors such as NF-κB and STAT3, thereby influencing key signalling pathways in both innate and adaptive immune responses.

DOI: 10.1201/9781003617860-5

Importantly, immunomodulatory plant extracts and compounds such as polysaccharides from *Astragalus membranaceus* or saponins from *Panax ginseng* – as reviewed by Mohamed, Jantan and Haque (2017) – have shown the ability to activate antigen-presenting cells, stimulate macrophage activity and enhance antibody production. These findings underscore the broad-spectrum immune regulatory potential of natural products. The increasing scientific focus on such agents paves the way for exploring underutilised plants like *Moringa oleifera*, whose immunopharmacological potential remains promising and warrants deeper investigation.

Among its many pharmacological attributes, *M. oleifera* has recently drawn scientific attention for its promising immunomodulatory effects. Preliminary studies suggest that bioactive compounds in *M. oleifera*, particularly flavonoids, glucosinolates and phenolic acids, may regulate immune functions by modulating oxidative stress, cytokine expression and immune cell activity (Kumar et al. 2024; Szlachetka, Kut, and Stępień 2020). While much of the research to date has focused on its antioxidant, anti-inflammatory or metabolic effects, its role in influencing both innate and adaptive immunity is an emerging area of investigation (Zebeaman et al. 2023; Mohamed, Jantan and Haque 2017). This chapter critically explores the immunomodulatory potential of *M. oleifera*, examining its active constituents, mechanistic pathways, experimental evidence and relevance in immune-related diseases, while identifying future research directions.

4.2 PHYTOCHEMICALS IN *MORINGA OLEIFERA* WITH IMMUNOMODULATORY POTENTIAL

Moringa oleifera is a reservoir of structurally diverse phytochemicals that may contribute to its immunomodulatory actions. Rather than acting through a single mechanism, *M. oleifera*'s potential arises from the synergistic interaction of multiple classes of compounds. While some studies have shown immune-related activity in moringa extracts, deeper mechanistic insights can be inferred by examining known immunopharmacological roles of its phytoconstituents, especially when these compounds are also bioactive in other medicinal plants (Table 4.1).

4.2.1 Flavonoids

Flavonoids are among the most abundant secondary metabolites in *M. oleifera*, particularly in its leaves (Coppin et al. 2013; Lin, Zhang and Chen 2018). Quercetin is present in high concentrations and is a well-documented immunomodulator (Michalski et al. 2020). It modulates Th1/Th2

TABLE 4.1
Key Immunomodulatory Phytochemicals in *M. oleifera*

Compound class	Example(s)	Reported immunomodulatory effects	References
Flavonoids	Quercetin, kaempferol, apigenin	↓ TNF-α, IL-6; promotes Th1/Th2 balance; enhances Treg populations	Michalski et al. 2020; Li et al. 2016
Phenolic acids	Chlorogenic, ferulic, caffeic	↑ Splenocyte proliferation, NK/CTL activity; ↓ PGE2, TNF-α	Kilani-Jaziri et al. 2017; Liu et al. 2023
Glucosinolates/isothiocyanates	Glucomoringin, BITC	↑ Macrophage activity; modulates NF-κB/NRF2 pathways	Galuppo et al. 2014; Park et al. 2022

balance, suppresses pro-inflammatory cytokines (TNF-α, IL-6) and inhibits mast cell degranulation – mechanisms demonstrated across several studies outside of *M. oleifera*, such as *Ginkgo biloba* (Park et al. 2009; Ravikumar and Kavitha 2020; Li et al. 2020; Lee et al. 2011)

Kaempferol and apigenin, also found in moringa leaves, influence both innate and adaptive immune functions (Coppin et al. 2013; Lin, Zhang and Chen 2018). Kaempferol has been reported to promote anti-inflammatory macrophage phenotypes (M2 polarisation) and suppress NF-κB activation in lipopolysaccharide (LPS)-stimulated macrophages (Ryyti et al. 2022; Zhai et al. 2024; Kim et al. 2015). Apigenin, a less abundant but potent flavone, inhibits dendritic cell maturation and blocks IL-12 and IL-23 expression, favouring regulatory T cell (Treg) expansion (Liu et al. 2017; Li et al. 2016).

4.2.2 Phenolic Acids

The presence of phenolic acids such as chlorogenic, ferulic and caffeic acids in *M. oleifera* contributes significantly to its immunological profile (Hamdy 2024; Leone et al. 2015). Chlorogenic acid inhibits inflammatory mediators and scavenges free radicals, helping reduce tissue injury and inflammation (Singh et al. 2025; Ye et al. 2025). It also alleviates colitis by enriching SCFA-producing gut microbiota and suppressing immune-related signalling pathways (Ye et al. 2025). It enhances NK and Cytotoxic T-Lymphocyte (CTL) activity and promotes splenocyte proliferation, suggesting a role in B cell activation and humoral immunity (Kilani-Jaziri et al. 2017; Karasawa et al. 2011). Ferulic acid enhances NK and CTL cytotoxicity and boosts lysosomal enzyme activity in macrophages, supporting immune cell activation (Kilani-Jaziri et al. 2017), with its effects partly attributed to antioxidant properties (Kilani-Jaziri et al. 2017). Caffeic acid downregulates inflammatory markers such as PGE2, IL-6 and TNF-α (Liu et al. 2023), reverses LPS-induced neuroinflammation and behavioural changes (Mudgal et al. 2020) and promotes splenocyte proliferation, NK and CTL activity and macrophage enzyme function (Kilani-Jaziri et al. 2017).

4.2.3 Glucosinolates and Isothiocyanates

Glucomoringin is a unique glucosinolate in moringa seeds and leaves. Upon enzymatic hydrolysis, it yields benzyl isothiocyanate (BITC) (Galuppo et al. 2014), a compound that has shown immune-stimulatory effects by enhancing lymphocyte count and splenic macrophage activity. BITC modulates redox-sensitive transcription factors like NRF2 and NF-κB, supporting its immune-balancing activity (Hoch et al. 2024; Park et al. 2022).

4.3 *MORINGA OLEIFERA* IMMUNOMODULATION MECHANISMS

4.3.1 Effects on Innate Immunity

Moringa oleifera extract has been shown to have significant effects on innate immunity through various mechanisms, as evidenced by multiple studies. One of the primary immunostimulatory effects of *M. oleifera* extract is its ability to improve total leukocyte counts and increase levels of key cytokines such as interferon-gamma (IFN-γ) and TNF in immunosuppressed models (Milad et al. 2023). This suggests a robust activation of the immune system. Additionally, the extract enhances phagocytic activity, as demonstrated by increased neutrophil adhesion and phagocytic index in various assays (Sudha et al. 2010; Van Pham et al. 2021), indicating an improved ability of the immune system to engulf and destroy pathogens. In aquatic models, supplementation with *M. oleifera* extract has also been shown to improve mucosal immunity by enhancing the activities of enzymes such as phosphatase, peroxidase and lysozyme, and by upregulating immune-related genes (Mansour et al. 2020). These effects collectively suggest that *M. oleifera* extract can strengthen the first line of defence against pathogens.

Beyond its immunostimulatory effects, *M. oleifera* extract also possesses notable anti-inflammatory and antioxidant properties. It reduces the levels of pro-inflammatory cytokines such as interleukin-1 beta (IL-1β), TNF-α and IL-6, and inhibits NF-κB activation, a critical pathway in inflammation regulation (Lee et al. 2013). This anti-inflammatory effect helps manage immune responses and prevents excessive inflammation. In addition, the extract boosts antioxidant defences by increasing the activities of enzymes like superoxide dismutase and glutathione peroxidase, which protect immune cells from oxidative stress (Lim, Hong and Lyu 2024; Zhou et al. 2022; Akbary et al. 2021). This antioxidant activity is important for preserving immune cell function and maintaining immune health.

The immune-enhancing effects of *M. oleifera* extract have also been demonstrated in disease models. For example, it has been shown to improve resistance to infections, particularly in fish models, where it increased survival rates and enhanced immune responses against bacterial infections (Zhou et al. 2022; Elgendy et al. 2021; Abdelkader et al. 2025). Furthermore, *M. oleifera* extract exhibits immunomodulatory activity by increasing the production of immunoglobulins and enhancing the function of immune cells such as macrophages and lymphocytes (Sudha et al. 2010; Wuryandari, Widjajanto and Rifa'i 2019; Khan et al. 2022). This balanced modulation supports effective immune activation while avoiding an overactive response.

In conclusion, *M. oleifera* extract exerts a comprehensive impact on innate immunity. It enhances leukocyte function, strengthens mucosal immunity, reduces inflammation and provides antioxidant protection. These combined actions contribute to a more robust and resilient immune system, capable of effectively responding to infections and supporting overall health. The evidence suggests that *M. oleifera* extract can be a valuable natural supplement for improving innate immune function (Milad et al. 2023).

4.3.2 Effects on Adaptive Immunity

Moringa oleifera extract has been shown to influence adaptive immunity through various mechanisms, primarily by modulating immune cell populations and cytokine production.

Regarding immune cell populations, *M. oleifera* extract has been observed to increase the number of $CD11b^+$ and $B220^+$ cells, which are markers for macrophages and B cells, respectively (Wuryandari, Widjajanto and Rifa'i 2019). This indicates an enhancement in the activation and proliferation of these immune cells, which play key roles in the adaptive immune response. However, in a different study, the extract was found to significantly inhibit the percentage of lymphocytes (CD11b and CD11c), NK cells and intracellular cytokines such as TNF-α, IFN-γ and IL-6 in mice (Ajedi, Widyarti and Rifa'i 2019). This finding suggests that *M. oleifera* extract can also exert an immunosuppressive effect, particularly when used as a protective agent.

Concerning cytokine production, *M. oleifera* extract has demonstrated the ability to modulate both pro-inflammatory and anti-inflammatory cytokines. It can reduce the levels of pro-inflammatory cytokines such as TNF-α and IFN-γ, which are critical for regulating immune responses (Ajedi, Widyarti and Rifa'i 2019; Abdel-Latif et al. 2018). In contrast, it can increase the levels of anti-inflammatory cytokines, including interleukin-4 (IL-4), interleukin-5 (IL-5) and interleukin-9 (IL-9), which are associated with the T helper 2 (Th2) immune response and contribute to a more balanced immunological state (Abdel-Latif et al. 2018).

The extract also affects humoral immunity by modulating immunoglobulin levels. It has been shown to elevate serum concentrations of immunoglobulins such as immunoglobulin M (IgM) and immunoglobulin A (IgA), which are essential for the body's antibody-mediated defence against pathogens (Abdel-Latif et al. 2018; Van Pham et al. 2021). This enhancement of immunoglobulin production underscores the potential of *M. oleifera* extract to support and strengthen the humoral branch of adaptive immunity.

In terms of therapeutic and protective applications, the effects of *M. oleifera* extract can vary depending on how it is used. When employed as a protective agent, the extract exhibits immunosuppressive properties, whereas in therapeutic contexts, it can stimulate the immune system. For instance, it has been shown to ameliorate the immunosuppressive effects of cyclophosphamide in mice, highlighting its dual functionality (Ajedi, Widyarti and Rifa'i 2019; Van Pham et al. 2021). Moreover, the immunomodulatory effects of the extract are dose-dependent. Lower doses are often more effective in stimulating immune responses than higher doses, indicating the importance of dosing in achieving the desired immunological outcomes (Sudha et al. 2010).

In summary, *M. oleifera* extract significantly influences adaptive immunity by regulating immune cell populations, modulating cytokine profiles and enhancing immunoglobulin levels. Its dual immunosuppressive and immunostimulatory actions, depending on dosage and context, make it a versatile natural agent for immune modulation (Sudha et al. 2010; Abdel-Latif et al. 2018; Ajedi, Widyarti, and Rifa'i 2019; Wuryandari, Widjajanto, and Rifa'i 2019; Van Pham et al. 2021).

4.4 EXPERIMENTAL EVIDENCE OF IMMUNOMODULATORY ACTIVITY

4.4.1 *Moringa oleifera* In Vitro Immunomodulatory Assays

The immunomodulatory potential of *M. oleifera* has garnered significant attention for its ability to influence both innate and adaptive immune responses. Several in vitro studies have explored the immunological effects of *M. oleifera* extracts and purified compounds, particularly focusing on macrophage activation, lymphocyte proliferation, cytokine secretion and molecular signalling pathways (Table 4.2).

One of the most comprehensive investigations into the immunostimulatory activity of *M. oleifera* was conducted by Dong et al. (2018), who isolated a novel polysaccharide fraction named MOP-2 from the hot-water extract of *M. oleifera* leaves. This polysaccharide, with an average molecular weight of approximately 155.35 kDa, was found to be composed primarily of arabinose (35.8%), glucose (6.67%) and galactose (57.53%). Using the RAW264.7 murine macrophage cell line, the researchers assessed the effects of MOP-2 on cell viability, pinocytosis and immune mediator production. At concentrations ranging from 25 to 400 μg/mL, MOP-2 significantly stimulated macrophage proliferation and pinocytic activity. Additionally, it enhanced the production of reactive oxygen species (ROS) and nitric oxide (NO), as well as the pro-inflammatory cytokines TNF-α and IL-6. These effects were confirmed at the molecular level by an upregulation of inducible nitric oxide synthase (iNOS), IL-6 and TNF-α gene expression, demonstrating a robust activation of macrophage immune function via transcriptional regulation (Dong et al. 2018).

A structurally distinct immunostimulatory compound from *M. oleifera*, termed MOLP-PE, was characterised by Gu et al. (2024). This high-purity arabinogalactan polysaccharide was derived through dual enzymolysis and resin purification techniques and was shown to have a complex backbone of (→3,4)-α-D-galactopyranosyl-(1→, →3)-β-D-galactopyranosyl-(1→ and →2,4)-β-D-rhamnopyranosyl-(1→) residues, with various side-chain sugars including arabinose and galacturonic acid. In vitro experiments using RAW264.7 macrophages revealed that MOLP-PE promoted cell proliferation and phagocytosis in a dose-dependent manner when administered at concentrations between 100 and 800 μg/mL. The compound also significantly increased the secretion of TNF-α and IL-6. Interestingly, computational docking studies revealed a strong binding affinity of MOLP-PE to Toll-like receptor 4 (TLR4), suggesting that its immunostimulatory effects may be mediated through TLR4 signalling pathways. The structural uniqueness of MOLP-PE, particularly its high content of 1,3-linked galactose and 1,5-linked arabinose residues, appeared to contribute to its potent immune-enhancing effects (Gu et al. 2024).

Another study by Rachmawati and Rifa'i (2014) focused on adaptive immune cells and used splenocytes harvested from *Mus musculus* (Balb/C mice) to evaluate the influence of an aqueous extract of *M. oleifera* leaves. Cells were treated in vitro with three concentrations of the extract (0.1, 1.0 and 10.0 μg/mL) and cultured for four days. Flow cytometry was then used to quantify changes in $CD4^+$ and $CD8^+$ T lymphocyte populations, as well as $B220^+$ B cells. The results demonstrated a

TABLE 4.2
Summary of Experimental Evidence on *M. oleifera*'s Immunomodulatory Activity

Model	Extract/compound	Concentration	Effects	Reference
In vitro models				
RAW264.7 macrophages	MOP-2 polysaccharide	25–400 µg/mL	↑ Proliferation, pinocytosis, ROS, NO, TNF-α, IL-6; ↑ iNOS, TNF-α, IL-6 gene expression	Dong et al. (2018)
RAW264.7 macrophages	MOLP-PE (arabinogalactan)	100–800 µg/mL	↑ Proliferation, phagocytosis, TNF-α, IL-6; binds TLR4	Gu et al. (2024)
Mouse splenocytes	Aqueous leaf extract	0.1, 1.0, 10.0 µg/mL	↑ $CD4^+$/$CD8^+$ T cells at a low dose; ↑ $B220^+$ B cells at a high dose	Rachmawati and Rifa'i (2014)
BV173 cells and PBMCs	Ethanol leaf extract (MOETE)	Various	↓ BV173 viability (IC_{50} = 125 µg/mL); ↑ PBMC proliferation (EC_{50} = 28 µg/mL); apoptosis in cancer cells	Lawal et al. (2022)
Shrimp haemocytes	Aqueous extract	100–250 ppm	↑ Phenoloxidase, phagocytic rate, superoxide anion	Abidin et al. (2021)
In vivo models				
Rats and mice	Methanol leaf extract	250 and 750 mg/kg orally	↑ Neutrophil adhesion, phagocytic index, antibody titres, survival	Sudha et al. (2010)
Mice	Leaf extract	10, 30, 100 mg/kg orally	↑ Phagocytic index, leukocyte count	Dillasamola et al. (2018)
Immunosuppressed mice	Seed and root extracts	2 g/kg orally	↑ DTH, leukocytes, IgM, TNF-α; restored spleen architecture	Van Pham et al. (2021)
Cyclophosphamide-treated rats	Ethanol extract	400 mg/kg orally	↑ SOD, GSH, catalase; ↓ lipid peroxidation; ↑ leukocytes	El Shanawany et al. (2019)
Nile tilapia	Fermented/non-fermented leaf and seed	5% and 10% in the diet	↑ Survival, phagocytosis, lysozyme, cytokines, IgM	Nassar et al. (2024)
Lambs	Ethanol extract	Not specified	↑ Protein, globulin, antibody titres post-vaccine	Al-Majali et al. (2017)
Salmonella-infected rats	Leaf extract	Orally for ten days	↑ Leukocyte and lymphocyte counts	Putu Khrisna Dharma et al. (2023)

dose-dependent response: the lowest dose (0.1 µg/mL) significantly increased $CD4^+$ and $CD8^+$ T cell populations, whereas the highest dose (10 µg/mL) led to a marked increase in $B220^+$ B cells. These findings suggest that the aqueous extract modulates both T and B cell responses in a concentration-dependent manner, with differential targeting of immune subsets at varying dosages (Rachmawati and Rifa'i 2014).

An important dimension of *M. oleifera*'s immunological potential lies in its selective action on healthy versus malignant cells, as shown by Lawal et al. (2022). In their study, *M. oleifera* ethanol leaf extract (MOETE) was applied to two distinct cell types: the BV173 leukemic cell line and peripheral blood mononuclear cells (PBMCs) derived from healthy human donors. The extract exhibited concentration-dependent cytotoxic effects on the BV173 cells, with an IC_{50} of 125 ± 6 µg/mL. In contrast, it stimulated proliferation in PBMCs, with an EC_{50} of 28 ± 3 µg/mL. Furthermore, the extract induced cell cycle arrest and apoptosis in leukemic cells, while enhancing the viability and cell cycle progression of PBMCs, even in the presence of phytohaemagglutinin (PHA), a T cell mitogen. These dual effects – immunostimulation of healthy lymphocytes and inhibition of tumour cells – highlight the therapeutic promise of MOETE in immune support and cancer control (Lawal et al. 2022).

Although most studies focus on mammalian systems, Abidin et al. (2021) explored the immunostimulatory effects of *M. oleifera* in an aquatic model using whiteleg shrimp (*Penaeus vannamei*). In vitro assays were performed on shrimp haemocytes incubated with aqueous *M. oleifera* extract at concentrations of 100 to 250 ppm. The extract significantly increased phenoloxidase (PO) activity, phagocytic rate (PR) and superoxide anion production, all of which are key indicators of innate immune activation in invertebrates. While this study was conducted in a non-mammalian model, the results underscore the broad-spectrum immunostimulant activity of *M. oleifera* across species (Abidin et al. 2021).

Collectively, these studies provide robust evidence for the in vitro immunomodulatory activity of *M. oleifera*, whether through direct stimulation of macrophages, selective proliferation of lymphocytes, modulation of cytokine secretion or enhancement of phagocytic activity. Its effects are dose-dependent and structurally mediated, particularly via interactions with innate immune receptors such as TLR4.

4.4.2 *Moringa oleifera* In Vivo Immunomodulatory Assays

In vivo experiments conducted using *M. oleifera* demonstrated a broad spectrum of immunomodulatory activities across several animal models and dosing regimens.

Sudha et al. (2010) investigated the methanolic leaf extract of *M. oleifera* in rats and mice, assessing both cellular and humoral immune responses. For cellular immunity, they employed the neutrophil adhesion test, the cyclophosphamide-induced neutropenia model and the carbon clearance assay. Treatment with *M. oleifera* extract at doses of 250 and 750 mg/kg orally significantly increased neutrophil adhesion, improved the phagocytic index and ameliorated cyclophosphamide-induced neutropenia. In humoral immune assessments, such as the indirect haemagglutination assay, serum immunoglobulin levels and lethality test using *Pasteurella multocida*, the extract markedly enhanced circulating antibody titres and survival rates, indicating significant immune enhancement. Interestingly, the lower dose (250 mg/kg) showed more consistent efficacy than the higher dose, suggesting a possible bell-shaped dose-response curve.

Dillasamola et al. (2018) evaluated the immunostimulant potential of *M. oleifera* leaf extract in male white mice using the carbon clearance method. Mice were administered oral doses of 10, 30 and 100 mg/kg for six consecutive days. On the seventh day, a carbon ink suspension was injected intravenously, and the phagocytic index was determined. The study found a dose-dependent increase in phagocytic activity, as indicated by the significantly higher phagocytic index (PI > 1) in the treated groups compared with the control. Additionally, total leukocyte counts were elevated, confirming that the extract enhanced innate immune responses through the stimulation of phagocytic cell activity.

Van Pham et al. (2021) explored the immunostimulatory effects of *M. oleifera* seed and root extracts in a cyclophosphamide-induced immunosuppressed mouse model. Extracts were administered orally at a dose of 2 g/kg body weight daily for seven days, while cyclophosphamide was given intraperitoneally on day four to induce immunosuppression. The treatment significantly ameliorated the immunosuppressive effects, as evidenced by improved delayed-type hypersensitivity responses, increased leukocyte counts, normalisation of relative organ weights (particularly spleen and thymus) and elevated IgM and TNF-α levels. Histological analysis supported these findings by showing restored spleen architecture in treated animals. This study provided substantial in vivo evidence that *M. oleifera* could reverse drug-induced immune suppression.

El Shanawany et al. (2019) studied the immunomodulatory effects of *M. oleifera* ethanolic extract in cyclophosphamide-treated rats. The extract was administered at 400 mg/kg orally for 14 days. Significant enhancements were observed in the antioxidant parameters (SOD, GSH and catalase), along with reductions in lipid peroxidation markers. Haematological parameters, such as total leukocyte count and lymphocyte count, were also improved, indicating the restoration of immune competence. The findings suggest that the extract can counter oxidative and immunosuppressive stress induced by chemotherapy agents.

Nassar et al. (2024) conducted an aquatic in vivo study using Nile tilapia (*Oreochromis niloticus*) to assess the effects of fermented and non-fermented *M. oleifera* leaf and seed mixtures as dietary supplements over a 30-day feeding trial. Fish were divided into groups receiving 5% or 10% of either fermented or non-fermented *M. oleifera*, with a control group on a basal diet. Following the trial, fish were challenged with *Aeromonas hydrophila*. The fermented extract groups showed significantly higher survival rates, reduced bacterial counts and enhanced immune parameters, including phagocytic activity, lysozyme levels and upregulation of immune-related cytokines and IgM gene expression. These findings emphasise the added benefit of fermentation in enhancing the bioactivity and immunomodulatory efficacy of *M. oleifera*.

In another in vivo model by Al-Majali et al. (2017), lambs were orally administered *M. oleifera* ethanolic extract to assess its effect on immune responses following vaccination against enterotoxaemia. Lambs that received the extract exhibited significantly higher levels of total protein, globulin and antibody titres compared with controls, reflecting an enhanced humoral response to the vaccine challenge. This indicates that *M. oleifera* can act as a vaccine adjuvant in large animal models.

Finally, Putu Khrisna Dharma et al. (2023) examined *M. oleifera* leaf extract on albino rats infected with *Salmonella typhi*. The extract was administered orally for ten days, leading to a significant increase in the total number of leukocytes and lymphocytes. These results confirmed an immune-stimulatory effect against bacterial infection, further supporting its role in enhancing host immune defence mechanisms in infectious disease models.

4.5 SAFETY, TOXICOLOGY AND REGULATORY CONSIDERATIONS OF *MORINGA OLEIFERA* AS AN IMMUNOSTIMULANT

Moringa oleifera has been widely investigated for its pharmacological properties, particularly its immunomodulatory effects. Toxicological evaluations in animal models, as well as limited human studies, generally support its safety profile. Research indicates that *M. oleifera* is well tolerated, with no significant adverse effects observed when administered in low to moderate doses (Popoola et al. 2020; de Barros et al. 2020).

Regarding organ-specific safety, a study conducted on male white rats revealed that oral administration of *M. oleifera* ethanol leaf extract resulted in no severe hepatic damage. However, minimal to mild liver tissue alterations were noted at higher doses (140 mg/kg body weight) following 14 and 21 days of treatment (Aldi et al. 2023). Acute toxicity assessments in mice identified dose-dependent toxicity at very high concentrations ranging from 5.1 to 10.0 g/kg. Meanwhile, subchronic toxicity studies in rats treated with higher doses (500 and 1000 mg/kg/day) indicated some variations in blood parameters and organ weights. These changes, although measurable, remained within normal physiological ranges and did not induce dose-related pathological lesions in visceral organs (Chivapat et al. 2012).

Genotoxicity and mutagenicity analyses further affirm the safety of *M. oleifera*. Studies evaluating these parameters demonstrated that the plant exhibits low genotoxic and mutagenic potential at commonly tested concentrations (de Barros et al. 2020; de Barros et al. 2022). Nevertheless, the safety of *M. oleifera* remains dose-dependent. While low to moderate doses are generally safe, higher doses may cause mild to moderate histological changes, particularly in liver and kidney tissues (Aldi et al. 2023; de Barros et al. 2022).

Long-term administration poses additional concerns. Chronic consumption of *M. oleifera* leaf powder at elevated doses (500 and 1000 mg/kg) has been associated with potential hepatic and renal damage, underscoring the importance of cautious dosing, especially during prolonged use (de Barros et al. 2022). Moreover, it is worth noting that *M. oleifera* seeds have exhibited immunosuppressive effects in models of immune inflammation, suggesting a complex pharmacological profile that may vary with dosage, plant part and context (Mahajan and Mehta 2010).

From a regulatory perspective, the approval and widespread use of *M. oleifera* as an immunostimulant require stringent standardisation and quality control practices. These include the development of consistent extraction methods, defined preparation protocols and comprehensive toxicological evaluations, particularly chronic toxicity assessments, to ensure consumer safety (de Barros et al. 2022). Additionally, due to variations in bioactivity depending on the plant part used and extraction techniques employed, regulatory authorities must rely on robust, well-structured datasets to substantiate claims and grant approval for therapeutic or supplemental use (de Barros et al. 2020; Liu et al. 2022).

In conclusion, *M. oleifera* presents as a promising natural immunostimulant when used at appropriate doses. Nonetheless, high-dose or long-term administration can lead to organ toxicity, necessitating careful monitoring and risk-benefit evaluation. Regulatory endorsement will depend on the standardisation of preparations and thorough safety documentation. Continued research is essential to fully elucidate the toxicological profile and support its safe application in human health.

4.6 FUTURE DIRECTIONS AND RESEARCH GAPS

Although current findings support the immunomodulatory potential of *M. oleifera*, several gaps remain that warrant future investigation. Most studies to date focus on crude extracts or partially characterised fractions, making it difficult to attribute observed effects to specific compounds. Advanced phytochemical profiling and isolation of pure constituents are needed to identify the exact bioactive molecules responsible for immunomodulation. Furthermore, while in vitro and in vivo animal models have demonstrated promising outcomes, there is a lack of well-designed clinical trials to evaluate safety, efficacy, dosage optimisation and long-term effects in human subjects. Mechanistic studies exploring molecular targets and pathways – particularly involving adaptive immunity and immunological memory – are also limited. Future research should prioritise standardised extract development, the elucidation of synergistic interactions among *M. oleifera* compounds and the integration of systems biology tools to map the immunoregulatory networks involved.

4.7 CONCLUSION

This chapter provides a comprehensive overview of the immunomodulatory potential of *M. oleifera*, with particular emphasis on its phytochemical constituents and their experimentally observed effects on the immune system. The data compiled from in vitro and in vivo studies demonstrate that *M. oleifera* exerts its immunomodulatory actions through the suppression of pro-inflammatory cytokines, such as TNF-α, IL-6 and IL-1β, as well as through the enhancement of phagocytic activity and the regulation of lymphocyte function. Key compounds, including quercetin, kaempferol, chlorogenic acid, caffeic acid, ferulic acid, glucomoringin, benzyl isothiocyanate and lupeol, were highlighted as potential contributors to these effects. These findings support the rationale for further mechanistic and pharmacological studies on isolated compounds from *M. oleifera* to validate their roles in immune modulation and advance the development of plant-based immunotherapeutics.

REFERENCES

Abdel-Latif, M., G. El-Shahawi, S. M. Aboelhadid, and H. Abdel-Tawab. 2018. "Modulation of murine intestinal immunity by Moringa oleifera extract in experimental hymenolepiasis nana." *Journal of Helminthology* 92 (2):142–153. doi: 10.1017/S0022149X17000293.

Abdelkader, G. S., E. S. Y. El-Naenaeey, H. M. Abdallah, E. H. Abu-Zeid, I. F. Rehan, ZigoF. , and G. A. Elmowalid. 2025. "Immune enhancement and disease resistance against Aeromonas hydrophila infection by dietary Lactobacillus plantarum-fermented Moringa oleifera leaves in Oreochromis niloticus." *Frontiers in Veterinary Science* 12. doi: 10.3389/fvets.2025.1557671.

Abidin, Z., H. T. Huang, Z. H. Liao, B. Y. Chen, Y. S. Wu, Y. J. Lin, and F. H. Nan. 2021. "Moringa oleifera leaves' extract enhances nonspecific immune responses, resistance against *Vibrio alginolyticus*, and growth in whiteleg shrimp (Penaeus vannamei)." *Animals (Basel)* 12 (1). doi: 10.3390/ani12010042.

Ajedi, A. S. S., S. Widyarti, and M. Rifa'i. 2019. "The aqueous extract of Moringa oleifera and Marrubium vulgare L. Leaf inhibit inflammatory response in mice infected with Salmonella typhimurium." *Drug Invention Today* 11:2658–2665.

Akbary, P., A. Gholamhosseini, M. Ali, A. Jahanbakhshi, K. R. Tavabe, B. S. Kuchaksaraei, and A. T. Mirghaed. 2021. "Dietary administration of Moringa oleifera extract enhances growth, fatty acid composition, antioxidant activity and resistance of shrimp Litopenaeus vannamei against Photobacterium damselae." *Agriculture and Natural Resources* 55 (2):161–170. doi: 10.34044/j.anres.2021.55.2.01.

Al-Majali, I. S., S. A. Al-Oran, M. R. Hassuneh, H. N. Al-Qaralleh, W. A. Rayyan, O. Y. Al-Thunibat, E. Mallah, A. Abu-Rayyan, and S. Salem. 2017. "Immunomodulatory effect of Moringa peregrina leaves, ex vivo and in vivo study." *Central European Journal of Immunology* 42 (3):231–238. doi: 10.5114/ceji.2017.70964.

Aldi, Y., A. Aisyah, R. Abdillah, and A. A. Alianta. 2023. "Activity of Moringa oleifera Lam. on liver function and histology in white male rats." *Pharmacognosy Journal* 15 (5):913–920. doi: 10.5530/pj.2023.15.174.

Chivapat, S., P. Sincharoenpokai, P. Suppajariyawat, A. Rungsipipat, PhattarapornchaiwatS. , and V. Chantarateptawan. 2012. "Safety evaluations of ethanolic extract of moringa oleifera lam. Seed in experimental animals." *Thai Journal of Veterinary Medicine* 42 (3):343–352.

Coppin, Julia P., Yanping Xu, Hong Chen, Min-Hsiung Pan, Chi-Tang Ho, Rodolfo Juliani, James E. Simon, and Qingli Wu. 2013. "Determination of flavonoids by LC/MS and anti-inflammatory activity in Moringa oleifera." *Journal of Functional Foods* 5 (4):1892–1899. doi: 10.1016/j.jff.2013.09.010.

de Barros, M. C., R. R. de Vasconcelos Alves, S. P. da Silva, J. D. F. da Silva, P. K. Amorim, L. C. B. B. Coelho, T. H. Napoleão, and P. M. G. Paiva. 2020. "Moringa oleifera toxicology: An overview." In D. S. Figueroa (ed.) *Moringa oleifera: Properties, Applications and Health Effects*. Nova Science Publishers, 35–68.

de Barros, M. C., A. G. B. Silva, T. G. D. S. Souza, C. A. Chagas, J. C. B. Machado, M. R. A. Ferreira, L. A. L. Soares, V. L. Xavier, L. C. C. de Araújo, E. F. D. O. Borba, T. G. da Silva, R. R. D. V. Alves, L. C. B. B. Coelho, A. M. de Oliveira, T. H. Napoleão, and P. M. G. Paiva. 2022. "Evaluation of acute toxicity, 28-day repeated dose toxicity, and genotoxicity of Moringa oleifera leaves infusion and powder." *Journal of Ethnopharmacology* 296. doi: 10.1016/j.jep.2022.115504.

Dillasamola, Dwisari, Yufri Aldi, Mutia Fakhri, Skunda Diliarosta, Biomechy Oktomalio P., and Noverial Noverial. 2018. "Immunomodulatory effect test from Moringa leaf extract (Moringa oleifera L.) with carbon clearance method in male white mice." *Asian Journal of Pharmaceutical and Clinical Research* 11 (9). doi: 10.22159/ajpcr.2018.v11i9.26703.

Dong, Zhuping, Chao Li, Qiang Huang, Bin Zhang, Xiong Fu, and Rui Hai Liu. 2018. "Characterization of a novel polysaccharide from the leaves of Moringa oleifera and its immunostimulatory activity." *Journal of Functional Foods* 49:391–400. doi: 10.1016/j.jff.2018.09.002.

El Shanawany, E. E., E. A. Fouad, H. G. Keshta, S. E. Hassan, A. G. Hegazi, and E. H. Abdel-Rahman. 2019. "Immunomodulatory effects of Moringa oleifera leaves aqueous extract in sheep naturally co-infected with Fasciola gigantica and Clostridium novyi." *Journal of Parasitic Diseases* 43 (4):583–591. doi: 10.1007/s12639-019-01130-6.

Elgendy, M. Y., E. S. Awad, D. A. Darwish, T. B. Ibrahim, W. S. E. Soliman, A. M. Kenawy, I. M. K. Abumourad, H. H. Abbas, and W. T. Abbas. 2021. "Investigations on the influence of Moringa oleifera on the growth, haematology, immunity and disease resistance in Oreochromis niloticus with special reference to the analysis of antioxidant activities by PAGE electrophoresis." *Aquaculture Research* 52 (10):4983–4995. doi: 10.1111/are.15370.

Galuppo, M., S. Giacoppo, G. R. De Nicola, R. Iori, M. Navarra, G. E. Lombardo, P. Bramanti, and E. Mazzon. 2014. "Antiinflammatory activity of glucomoringin isothiocyanate in a mouse model of experimental autoimmune encephalomyelitis." *Fitoterapia* 95:160–174. doi: 10.1016/j.fitote.2014.03.018.

Gu, J., X. Yang, L. Lin, and M. Zhao. 2024. "Identification of an arabinogalactan with special structure from Moringa Oleifera leaf and exploration of its immunomodulatory activity." *International Journal of Biological Macromolecules* 279 (Pt 3):134616. doi: 10.1016/j.ijbiomac.2024.134616.

Hamdy, N. M. 2024. "Exploring the anti-inflammatory potential of the polyphenolic compounds in Moringa oleifera leaf: In silico molecular docking." *Egyptian Journal of Basic and Applied Sciences* 11 (1):367–385. doi: 10.1080/2314808X.2024.2347126.

Hoch, C. C., M. Shoykhet, T. Weiser, L. Griesbaum, J. Petry, K. Hachani, G. Multhoff, A. Bashiri Dezfouli, and B. Wollenberg. 2024. "Isothiocyanates in medicine: A comprehensive review on phenylethyl-, allyl-, and benzyl-isothiocyanates." *Pharmacological Research* 201. doi: 10.1016/j.phrs.2024.107107.

Hu, S., D. Xiang, X. Zhang, L. Zhang, S. Wang, K. Jin, L. You, and J. Huang. 2022. "The mechanisms and cross-protection of trained innate immunity." *Virology Journal* 19 (1):210. doi: 10.1186/s12985-022-01937-5.

Karasawa, K., Y. Uzuhashi, M. Hirota, and H. Otani. 2011. "A matured fruit extract of date palm tree (Phoenix dactylifera L.) Stimulates the cellular immune system in mice." *Journal of Agricultural and Food Chemistry* 59 (20):11287–11293. doi: 10.1021/jf2029225.

Khan, A., M. Tahir, I. Alhidary, M. Abdelrahman, A. A. Swelum, and R. U. Khan. 2022. "Role of dietary Moringa oleifera leaf extract on productive parameters, humoral immunity and lipid peroxidation in broiler chicks." *Animal Biotechnology* 33 (6):1353–1358. doi: 10.1080/10495398.2021.1899936.

Kilani-Jaziri, S., I. Mokdad-Bzeouich, M. Krifa, N. Nasr, K. Ghedira, and L. Chekir-Ghedira. 2017. "Immunomodulatory and cellular anti-oxidant activities of caffeic, ferulic, and p-coumaric phenolic acids: a structure–activity relationship study." *Drug and Chemical Toxicology* 40 (4):416–424. doi: 10.1080/01480545.2016.1252919.

Kim, S. H., J. G. Park, J. Lee, W. S. Yang, G. W. Park, H. G. Kim, Y. S. Yi, K. S. Baek, N. Y. Sung, M. J. Hossen, M. N. Lee, J. H. Kim, and J. Y. Cho. 2015. "The dietary flavonoid kaempferol mediates anti-inflammatory responses via the src, syk, IRAK1, and IRAK4 molecular targets." *Mediators of Inflammation* 2015. doi: 10.1155/2015/904142.

Kumar, Shivendra, Yogesh Murti, Swamita Arora, Wasim Akram, Harsh Bhardwaj, Kapil Gupta, Akshaya Sachdev, Jyoti Devi, Santosh Kumar, Bikash Kumar, Vivek Dwivedi, Shahnawaz Sameem, Nitendra, Pawan Kumar, Kuldeep Singh, and Sunam Saha. 2024. "Exploring the therapeutic potential of Moringa oleifera Lam. in Traditional Chinese Medicine: A comprehensive review." *Pharmacological Research - Modern Chinese Medicine* 12. doi: 10.1016/j.prmcm.2024.100473.

Lawal, Hamza, Satar Jabbar Rahi Algraittee, Ramesh Ranggasamy, Vahid Sarmadi, ThambiahSubashini C. , Johnson Stanslas, and Rajesh Ramasamy. 2022. "*Moringa oleifera* leaf ethanol extract immunomodulatory activity discriminates between chronic myeloid leukaemia cell line and normal lymphocytes." *Malaysian Journal of Medicine and Health Sciences* 18:100–109. doi: 10.47836/mjmhs18.s21.16.

Lee, H. J., Y. J. Jeong, T. S. Lee, Y. Y. Park, W. G. Chae, I. K. Chung, H. W. Chang, C. H. Kim, Y. H. Choi, W. J. Kim, S. K. Moon, and Y. C. Chang. 2013. "Moringa fruit inhibits LPS-induced NO/iNOS expression through suppressing the NF-κB activation in RAW264.7 cells." *American Journal of Chinese Medicine* 41 (5):1109–1123. doi: 10.1142/S0192415X13500754.

Lee, Y. W., J. A. Lin, C. C. Chang, Y. H. Chen, P. L. Liu, A. W. Lee, J. C. Tsai, C. Y. Li, C. S. Tsai, T. L. Chen, and F. Y. Lin. 2011. "Ginkgo biloba extract suppresses endotoxin-mediated monocyte activation by inhibiting nitric oxide- and tristetraprolin-mediated toll-like receptor 4 expression." *Journal of Nutritional Biochemistry* 22 (4):351–359. doi: 10.1016/j.jnutbio.2010.03.002.

Leone, A., G. Fiorillo, F. Criscuoli, S. Ravasenghi, L. Santagostini, G. Fico, A. Spadafranca, A. Battezzati, A. Schiraldi, F. Pozzi, S. Di Lello, FilippiniS. , and S. Bertoli. 2015. "Nutritional characterization and phenolic profiling of moringa oleifera leaves grown in chad, sahrawi refugee camps, and Haiti." *International Journal of Molecular Sciences* 16 (8):18923–18937. doi: 10.3390/ijms160818923.

Li, C., K. Liu, S. Liu, Q. Aerqin, and X. Wu. 2020. "Role of ginkgolides in the inflammatory immune response of neurological diseases: A review of current literatures." *Frontiers in Systems Neuroscience* 14. doi: 10.3389/fnsys.2020.00045.

Li, X., Y. Han, Q. Zhou, H. Jie, Y. He, J. Han, J. He, Y. Jiang, and E. Sun. 2016. "Apigenin, a potent suppressor of dendritic cell maturation and migration, protects against collagen-induced arthritis." *Journal of Cellular and Molecular Medicine* 20 (1):170–180. doi: 10.1111/jcmm.12717.

Lim, W. T., C. E. Hong, and S. Y. Lyu. 2024. "The antioxidant, anti-aging, immunomodulatory, and whitening effects of Moringa oleifera." *Journal of the Korean Society of Food Science and Nutrition* 53 (2):138–148. doi: 10.3746/jkfn.2024.53.2.138.

Lin, Mengfei, Junjie Zhang, and Xiaoyang Chen. 2018. "Bioactive flavonoids in Moringa oleifera and their health-promoting properties." *Journal of Functional Foods* 47:469–479. doi: 10.1016/j.jff.2018.06.011.

Liu, J., M. M. Poojary, L. Zhu, A. R. Williams, and M. N. Lund. 2023. "Phenolic acid-amino acid adducts exert distinct immunomodulatory effects in macrophages compared to parent phenolic acids." *Journal of Agricultural and Food Chemistry* 71 (5):2344–2355. doi: 10.1021/acs.jafc.2c06658.

Liu, R., J. Liu, Q. Huang, S. Liu, and Y. Jiang. 2022. "Moringa oleifera: A systematic review of its botany, traditional uses, phytochemistry, pharmacology and toxicity." *Journal of Pharmacy and Pharmacology* 74 (3):296–320. doi: 10.1093/jpp/rgab131.

Liu, Y. F., X. X. Xue, Z. Y. Li, J. P. Wang, and Y. J. Zhang. 2017. "Effect of apigenin on dendritic cells maturation and function in murine splenocytes." *Yaoxue Xuebao* 52 (3):397–402. doi: 10.16438/j.0513-4870.2016-0949.

Mahajan, S. G., and A. A. Mehta. 2010. "Immunosuppressive activity of ethanolic extract of seeds of Moringa oleifera Lam. in experimental immune inflammation." *Journal of Ethnopharmacology* 130 (1):183–186. doi: 10.1016/j.jep.2010.04.024.

Mansour, A. T., C. Espinosa, J. M. García-Beltrán, L. Miao, D. C. Ceballos Francisco, A. S. Alsaqufi, and M. Á Esteban. 2020. "Dietary supplementation of drumstick tree, Moringa oleifera, improves mucosal immune response in skin and gills of seabream, Sparus aurata, and attenuates the effect of hydrogen peroxide exposure." *Fish Physiology and Biochemistry* 46 (3):981–996. doi: 10.1007/s10695-020-00763-2.

Martins, Y. C., F. L. Ribeiro-Gomes, and C. T. Daniel-Ribeiro. 2023. "A short history of innate immunity." *Memórias do Instituto Oswaldo Cruz* 118:e230023. doi: 10.1590/0074-02760230023.

Michalski, J., A. Deinzer, L. Stich, E. Zinser, SteinkassererA. , and I. Knippertz. 2020. "Quercetin induces an immunoregulatory phenotype in maturing human dendritic cells." *Immunobiology* 225 (4):151929. doi: 10.1016/j.imbio.2020.151929.

Milad, S. S., S. E. Ali, M. Z. Attia, M. S. Khattab, E. S. El-Ashaal, H. A. Elshoky, and A. M. Azouz. 2023. "Enhanced immune responses in dexamethasone immunosuppressed male rats supplemented with herbal extracts, chitosan nanoparticles, and their conjugates." *International Journal of Biological Macromolecules* 250. doi: 10.1016/j.ijbiomac.2023.126170.

Mohamed, S. I. A., I. Jantan, and M. A. Haque. 2017. "Naturally occurring immunomodulators with antitumor activity: An insight on their mechanisms of action." *International Immunopharmacology* 50:291–304. doi: 10.1016/j.intimp.2017.07.010.

Mudgal, J., S. Basu Mallik, M. Nampoothiri, M. Kinra, S. Hall, G. D. Grant, S. Anoopkumar-Dukie, A. K. Davey, C. M. Rao, and D. Arora. 2020. "Effect of coffee constituents, caffeine and caffeic acid on anxiety and lipopolysaccharide-induced sickness behavior in mice." *Journal of Functional Foods* 64. doi: 10.1016/j.jff.2019.103638.

Mujal, A. M., R. B. Delconte, and J. C. Sun. 2021. "Natural killer cells: From innate to adaptive features." *Annu Rev Immunol* 39:417–447. doi: 10.1146/annurev-immunol-101819-074948.

Nassar, Aama, A. A. E. Gharib, S. Y. Abdelgalil, H. M. AbdAllah, and G. A. Elmowalid. 2024. "Immunomodulatory, antioxidant, and growth-promoting activities of dietary fermented Moringa oleifera in Nile tilapia (Oreochromus niloticus) with in-vivo protection against Aeromonas hydrophila." *BMC Veterinary Research* 20 (1):231. doi: 10.1186/s12917-024-04070-3.

Park, H. j, C. M. Lee, I. D. Jung, J. S. Lee, Y. i Jeong, J. H. Chang, S. H. Chun, M. J. Kim, I. W. Choi, S. C. Ahn, Y. K. Shin, S. R. Yeom, and Y. M. Park. 2009. "Quercetin regulates Th1/Th2 balance in a murine model of asthma." *International Immunopharmacology* 9 (3):261–267. doi: 10.1016/j.intimp.2008.10.021.

Park, W. S., J. Lee, G. Na, S. Park, S. K. Seo, J. S. Choi, W. K. Jung, and I. W. Choi. 2022. "Benzyl isothiocyanate attenuates inflammasome activation in *Pseudomonas aeruginosa* LPS-stimulated THP-1 cells and exerts regulation through the MAPKs/NF-κB Pathway." *International Journal of Molecular Sciences* 23 (3). doi: 10.3390/ijms23031228.

Popoola, J. O., O. S. Aworunse, O. L. Oyesola, O. O. Akinnola, and O. O. Obembe. 2020. "A systematic review of pharmacological activities and safety of Moringa oleifera." *Journal of HerbMed Pharmacology* 9 (3):174–190. doi: 10.34172/jhp.2020.24.

Putu Khrisna Dharma, Jaya, Apsari Putu Indah Budi, Dewi Pande Made Alitta Cantika Putri Nadya, Laksemi Dewa Ayu Agus Sri, and I. Ketut Cahyadi Adi Winata Sutarta. 2023. "Effects of Moringa oleifera extract as an immunomodulator of lymphocyte cells and macrophages in BALB/c mice infected with *Plasmodium berghei*." *Folia Medica Indonesiana* 59 (3):214–221. doi: 10.20473/fmi.v59i3.45237.

Rachmawati, Indriya, and Muhaimin Rifa'i. 2014. "In Vitro vitro immunomodulatory activity of aqueous extract of *Moringa oleifera* Lam. Leaf to the CD4 +, CD8+ and B220+ Cells in *Mus musculus*." *Journal of Experimental Life Science* 4 (1):15–20.

Ravikumar, N., and C. N. Kavitha. 2020. "Immunomodulatory effect of Quercetin on dysregulated Th1/Th2 cytokine balance in mice with both type 1 diabetes and allergic asthma." *Journal of Applied Pharmaceutical Science* 10 (3):80–87. doi: 10.7324/JAPS.2020.103010.

Riera Romo, M., D. Perez-Martinez, and C. Castillo Ferrer. 2016. "Innate immunity in vertebrates: An overview." *Immunology* 148 (2):125–39. doi: 10.1111/imm.12597.

Ryyti, R., M. Hämäläinen, T. Leppänen, PeltolaR. , and E. Moilanen. 2022. "Phenolic compounds known to be present in lingonberry (Vaccinium vitis-idaea L.) enhance macrophage polarization towards the anti-inflammatory M2 phenotype." *Biomedicines* 10 (12). doi: 10.3390/biomedicines10123045.

Singh, S., R. P. Singh, S. Bhoopendra, and B. C. Semwal. 2025. "Anti-inflammatory activity of chlorogenic acid and call for further." *Recent Advances in Inflammation and Allergy Drug Discovery*. doi: 10.2174/0127722708351437241121084449.

Sudha, P., S. M. B. Asdaq, S. S. Dhamingi, and G. K. Chandrakala. 2010. "Immunomodulatory activity of methanolic leaf extract of Moringa oleifera in animals." *Indian Journal of Physiology and Pharmacology* 54 (2):133–140.

Szlachetka, Kinga, Paulina Kut, and Agnieszka Stępień. 2020. "Cytotoxic and anticancer activity of Moringa oleifera." *European Journal of Clinical and Experimental Medicine* 18 (3):214–220. doi: 10.15584/ejcem.2020.3.9.

Van Pham, A. T., M. H. Luong, H. T. T. Dinh, T. P. Mai, Q. V. Trinh, and L. H. Luong. 2021. "Immunostimulatory effect of moringa oleifera extracts on cyclophosphamide-induced immunosuppressed mice." *Journal of Herbs, Spices and Medicinal Plants* 27 (4):377–385. doi: 10.1080/10496475.2021.1934620.

Wuryandari, M. M. R. E., E. Widjajanto, and M. Rifa'i. 2019. "The effect of Moringa oleifera Lam leaf extract fermented by Lactobacillus plantarum on the expression of B220+ and CD11b+ cells in mice infected with Salmonella typhi." IOP Conference Series: Earth and Environmental Science.

Xu, Y., and N. T. Eissa. 2010. "Autophagy in innate and adaptive immunity." *Proceedings of the American Thoracic Society* 7 (1):22–28. doi: 10.1513/pats.200909-103JS.

Ye, X., X. An, T. Zhang, Y. Kong, S. Jia, and J. Wu. 2025. "CGA protects against experimental colitis by modulating host purine metabolism through the gut microbiota." *International Immunopharmacology* 153. doi: 10.1016/j.intimp.2025.114547.

Zebeaman, M., M. G. Tadesse, R. K. Bachheti, A. Bachheti, GebeyhuR. , and K. K. Chaubey. 2023. "Plants and plant-derived molecules as natural immunomodulators." *BioMed Research International* 2023:7711297. doi: 10.1155/2023/7711297.

Zhai, H., D. Wang, Y. Wang, H. Gu, J. Jv, L. Yuan, C. Wang, and L. Chen. 2024. "Kaempferol alleviates adipose tissue inflammation and insulin resistance in db/db mice by inhibiting the STING/NLRP3 signaling pathway." *Endocrine Connections* 13 (5). doi: 10.1530/EC-23-0379.

Zhou, X., S. P. Forrester, J. Fan, B. Liu, Q. Zhou, L. Miao, P. Shao, and X. Li. 2022. "Effects of M. oleifera leaf extract on the growth, physiological response and related immune gene expression of crucian carp fingerlings under Aeromonas hydrophila infection." *Fish and Shellfish Immunology* 131:358–367. doi: 10.1016/j.fsi.2022.09.060.

5 *Moringa oleifera* as a Natural Immunomodulator

Phytochemistry, Pharmacological Properties and Therapeutic Potential

Sanjay Kumar Gupta

5.1 INTRODUCTION

Moringa oleifera (MO), commonly known as the 'miracle tree' or 'tree of life', is a fast-growing, drought-resistant tree native to the Indian subcontinent and widely cultivated across tropical and subtropical regions (Pareek et al., 2023). The plant is very important for its interesting nutritional profile and numerous medicinal uses, as almost all of its parts, such as the leaves, pods, seeds, bark, roots and flowers, can be processed to make food and used in traditional medicine (Cao et al., 2023). The leaves are of major interest and contain significant amounts of protein, vitamins, minerals and many bioactive compounds, including flavonoids, phenolic acids, alkaloids and terpenes, which must contribute to their significant antioxidant, anti-inflammatory, antimicrobial and cardioprotective effects (Milla et al., 2021). For centuries, MO has been used to manage a range of conditions, including wounds, pain, ulcers, liver and heart diseases, cancer and inflammatory conditions (Choudhary et al., 2013). The seeds are well known for their ability to purify water and as a natural treatment for digestive and urinary tract disorders.

Pharmacological research has corroborated several of these traditional uses, attracting researchers to assess the plant's properties and the impact moringa may have with respect to reversing chronic diseases such as diabetes, cardiovascular diseases, certain cancers and diseases of the immune system (Milla et al., 2021; Pareek et al., 2023; Zaffar et al., 2025). Research continues to indicate that due to its safety profile, nutritional potential and significant therapeutic applicability, MO remains under investigation and is frequently included in functional foods, dietary supplements and phytopharmaceutical derivations (Gopalakrishnan et al., 2016).

MO has a long history of use among the disciplines of historical civilisations, such as among Indian, Greek, Egyptian and Roman emissaries, dating as far back as 1500 BC (Senthilkumar et al., 2018). The use of MO was first documented in Vedic literature almost 5000 years ago and was a significant part of Ayurvedic medicine for the treatment of an extensive array of conditions (Pareek et al., 2023; Senthilkumar et al., 2018). The plant, other than ben oil, which was made from the seeds and sought after in ancient Egypt, Greece and Rome for its skin protection and use in perfumery, contributed to the tree's value. Almost every part of the tree has been used in traditional medicine: leaves for wound healing and antioxidant potential; seeds for abdominal discomfort and diabetes; the bark for fever and pain; roots for hypertension and digestion; and flowers for their aphrodisiac and digestive potential (Senthilkumar et al., 2018; R. Liu et al., 2022). From an ethnomedicinal perspective, MO has been used for the treatment of wounds and ulcers, liver and heart problems, cancer, inflammatory conditions and malnutrition, particularly in infants and nursing mothers. In

 DOI: 10.1201/9781003617860-6

addition to these uses, MO has been utilised for the treatment of fever, headaches, constipation, muscle pain and skin ailments such as rash, freckles and scabies. The diverse therapeutic relevance and nutritional value make MO an important plant resource within traditional health care systems in Africa and Asia, as well as globally (Choudhary et al., 2013; Pareek et al., 2023).

Immunomodulation is the conscious attempt to change the activity of the immune system, either by stimulating or inhibiting its activity, with the aim of improving the defence against disease conditions. The demand for and the concept of immunomodulation have expanded rapidly and are now recognised as an important aspect of maintaining health and treating various diseases. An immune response that is too strong or even too weak can lead to chronic inflammation, autoimmunity and allergies, as well as increased susceptibility to infections (Rajgopal et al., 2025; Strzelec et al., 2023). Therapeutic immunomodulators can be used to restore immune balance in states where the immune system is overactive (e.g. in autoimmune diseases and chronic inflammation) or underactive (e.g. in immunodeficiency and certain infections). Immunomodulators can be administered with the understanding that the clinician may select a component or pathway of the immune system to achieve the objective of preventing disease progression, reducing tissue or organ damage, or improving the patient's overall health (Shinde et al., 2015; Shukla et al., 2022). Immunomodulatory agents are commonly used in the treatment of autoimmune disease, cancer, allergies and infectious disease, and are being investigated as adjuncts or alternatives to established therapies (Shukla et al., 2022; Strzelec et al., 2023). Given the complexity of immune regulation, natural products such as MO are of particular interest as safe and effective modulators of immune responses.

The goal and purpose of this chapter is to provide a comprehensive and critical assessment of the immunomodulatory capabilities of MO, along the continuum of traditional ethnomedicinal perspectives with modern scientific endeavours. The chapter aims to summarise the phytochemical components relevant to the immune system, collate and analyse experimental, clinical and quasi-experimental evidence of the immunomodulatory effects of MO and provide insights into the immunomechanisms of action, including the regulation of cytokine production and the functional activity of immune cells. Furthermore, the chapter aims to highlight the potential therapeutic effects and safety of using MO as an immunonutrient, while also identifying current knowledge gaps in the literature and proposing future directions. The chapter will have a historical and ethnomedicinal context, a detailed phytochemistry section, evaluation of in vitro, in vivo and human studies and considerations of the implications for human health and nutrition, as well as including the potential limitations and safety and regulatory considerations when promoting MO in contemporary healthcare and public health efforts and in the development of functional foods.

5.2 BOTANICAL PROFILE, NUTRITIONAL VALUE AND PHYTOCHEMISTRY OF MO

Moringa oleifera or munga is one of the most important plants widely cultivated in India, belonging to the Moringaceae family (Figure 5.1) (van den Berg & Kuipers, 2022). MO is a fast-growing, deciduous or evergreen tree that typically reaches heights of 10–12 metres, with a trunk diameter of up to 46 cm. The tree is characterised by its open, umbrella-shaped crown, fragile and drooping branches and feathery foliage composed of tripinnate leaves up to 45 cm long (Choudhary et al., 2013; Dhongade et al., 2017). The bark is whitish-grey and corky, while young shoots are purplish or greenish-white and hairy, and the leaves are compound, with small, oval leaflets that are green and almost hairless on the upper surface. The flowers are fragrant, bisexual and yellowish-white, arranged in spreading or drooping axillary panicles 10–25 cm long, and each flower is about 0.7–1.5 cm long and 2 cm broad, with five unequal, thinly veined petals. The fruit is a long, pendulous, three-sided capsule or pod, which can reach 20–120 cm in length and contains about 26 seeds per pod. The seeds are round, about 1 cm in diameter, with brownish semi-permeable hulls and three papery wings, facilitating wind and water dispersal.

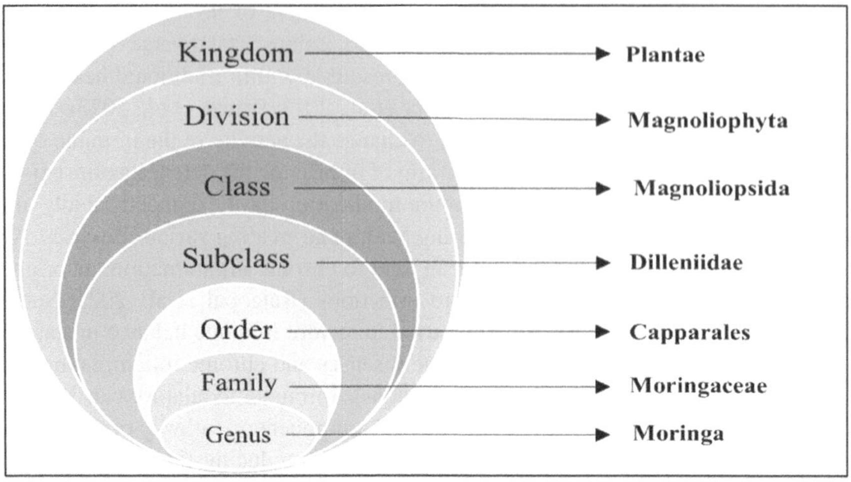

FIGURE 5.1 Taxonomy of *M. oleifera* Plant.

MO is native to the sub-Himalayan tracts of northern India but is now widely cultivated throughout tropical and subtropical regions of Asia, Africa and the Americas (Cao et al., 2023; van den Berg & Kuipers, 2022). Its adaptability to drought and a variety of soil types has contributed to its global distribution and popularity as a multipurpose tree, valued for its nutritional, medicinal and agricultural uses. The global distribution of MO is represented in Figure 5.2.

MO is known for its exceptional nutritional aspects and phytochemical diversity and is considered a functional food and traditional medicine (Núñez-Gastélum et al., 2023). All parts of the plant

FIGURE 5.2 Global Distribution of MO.

are edible and contribute to dietary nutrition, but are especially valuable for diets in areas with malnutrition problems (Domenico et al., 2019). The mature leaves have the highest nutrient density, containing high protein (up to 29% dry weight), many essential amino acids and vitamins, particularly vitamin A (in the form of beta-carotene), vitamin C, vitamin E and B vitamins (Sultana, 2020). Calcium, iron, potassium, magnesium and zinc are also abundantly available minerals, which make MO an important source of micronutrients (Islam et al., 2021). The leaves also have a high fibre concentration compared with fat content, and the seeds contain healthy fats, which are monounsaturated and polyunsaturated fatty acids such as omega-3 and omega-6 fatty acids (Peñalver et al., 2022).

Beyond its nutritional value, MO offers a rich phytochemical profile that contributes to its health-promoting and therapeutic effects (Wilcaso et al., 2025). MO is also a source of bioactive components like flavonoids (quercetin, kaempferol and rutin), phenolic acids (chlorogenic, gallic and caffeic acids), tannins, alkaloids, saponins, glucosinolates and isothiocyanates (Figure 5.3). These components have strong antioxidant, anti-inflammatory, antimicrobial and immunomodulatory properties (Kumar et al., 2024a). Carotenoids such as lutein and beta-carotene alleviate oxidative stress and improve visual health (Duan et al., 2025). Also of interest to researchers are glucosinolates and their derivatives, most notably isothiocyanates, for their possible effects on cancer and immune-modulating potential (Wang & Bao, 2021). When combined, the nutrient density and multitude of phytochemicals found in MO build a robust scientific evidence base for the long-standing use of the plant in traditional medicine, and increasingly, consideration in modern healthcare as a natural immunonutrient and therapeutic agent (Abdel-Latif et al., 2022).

MO is well known for its rich and diverse phytochemical composition, as the evidence presented thus far demonstrates its wide spectrum of pharmacological and nutritional activities. MO contains a multitude of bioactive compounds, including, but not limited to, flavonoids, alkaloids, saponins, tannins, steroids, phenolic acids, terpenoids and glucosinolates. These phytochemicals can be found

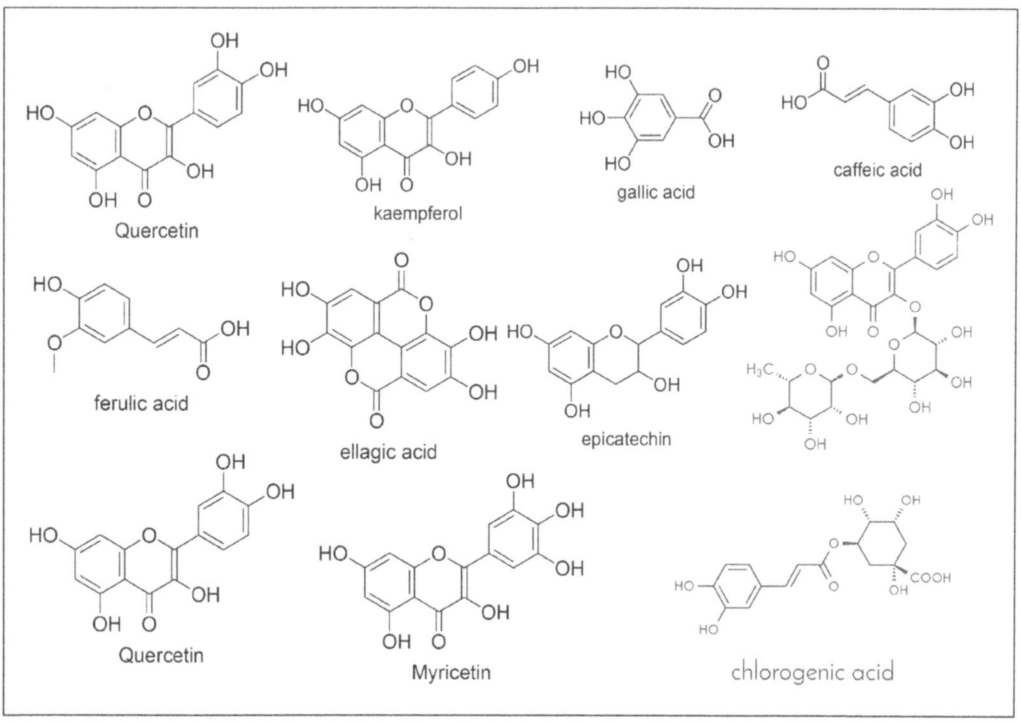

FIGURE 5.3 Structures of the Chief Bioactive Compounds of MO with Immunomodulatory Potential.

throughout the leaves, seeds, pods, flowers and roots, but are usually the most abundant in the leaves (Abd Rani et al., 2018; Farswan et al., 2024; Feng et al., 2025).

Flavonoids are one of the most plentiful and important phytochemicals in MO, especially in the leaves (Owusu-Ansah et al., 2023). The main flavonoids are quercetin, kaempferol, rutin, rhamnetin, apigenin and myricetin (Hassan et al., 2021; Sinha et al., 2025). Flavonoids occur mainly in glycoside forms and contribute to the antioxidant activity of MO, which reduces oxidative damage to immune cells and modulates immune function (Abidin et al., 2022; El-Deep et al., 2019). MO is notable for its content of glucosinolates, especially glucomoringin. When plant tissues are disrupted, glucosinolates are enzymatically converted into isothiocyanates, such as 4-(α-L-rhamnosyloxy) benzyl isothiocyanate (Q. Wu et al., 2024). These substances are thought to contribute to immune regulation by affecting cytokine production and immune cell activity. They have also shown anti-inflammatory, antibacterial and anticancer qualities.

5.3 IMMUNOMODULATION: BASIC CONCEPTS

Immunomodulation refers to the gradual modulation of the immune system response, either by enhancement (immunostimulant) or suppression (immunosuppressant), to achieve an optimum level of homeostasis or host protection (Akter et al., 2023). The immune system is a complex system composed of a diverse array of interrelated cells, tissues and molecules that work together to identify and eliminate pathogens, damaged cells and non-self substances that are structurally harmful to the body's systems. Three categories of immunity are frequently used to describe or categorise it: constitutive, innate and adaptive. The first line of defence is the constitutive immune system, which includes antimicrobial peptides and enzymes, as well as all chemical and physical barriers, including the skin and mucous membranes (Sheta et al., 2024). Innate immunity is triggered upon pathogen exposure through mechanisms involving cells such as neutrophils, macrophages and natural killer (NK) cells, as well as pathways including complement activation. It recognises pathogen-associated molecular patterns (PAMPs) through pattern recognition receptors (PRRs), such as toll-like receptors (TLRs), thereby initiating inflammation and the recruitment of immune cells (Strzelec et al., 2023). Whereas adaptive immunity is characterised by specificity and memory, adaptive immunity involves lymphocytes (T and B cells). T cells mediate cellular immunity (including cytotoxicity and helper activity), while B cells produce antibodies for humoral defence. The adaptive response develops over time, with memory cells providing long-term protection against previously encountered antigens (Phillip West & McGuire, 2025).

Immunomodulators are substances or interventions that modify immune responses to restore or maintain balance, either by stimulation or suppression. Enhancing the immune response increases resistance against infections, malignancy and immune deficiencies. Examples include vaccines, cytokines like interferons and certain plant-derived compounds that activate macrophages, stimulate antibody production or elevate cytokine secretion (Nader, 2017). Reducing or dampening immune activity prevents tissue damage from excessive or misdirected responses, such as those seen in autoimmune diseases and allergies. This is achieved through medications such as corticosteroids, biologics (e.g. TNF inhibitors) or natural agents that inhibit inflammatory cytokines and the proliferation of immune cells. Modifying the balance between immune system arms (innate versus adaptive, pro-inflammatory versus anti-inflammatory) without complete activation or suppression is essential (Komre et al., 2025). This involves the modulation of cellular signalling pathways (such as nuclear factor kappa B (NF-κB) and mitogen-activated protein kinase (MAPK)), cytokine profiles and regulatory cell populations (e.g. regulatory T cells) (A. Gupta & Gupta, 2025; Strzelec et al., 2023).

Natural agents, such as bioactive compounds from plants like MO, offer a promising avenue for safe and effective immunomodulation (Padayachee & Baijnath, 2020). These agents often exhibit both immunostimulatory and immunosuppressive effects depending on the immune context, facilitating immune balance and reducing the risk of adverse effects compared with conventional drugs (S. K. Gupta et al., 2025).

5.4 IMMUNOMODULATORY PROPERTIES OF MO

The immunomodulatory properties of MO have been extensively investigated through both in vitro and in vivo studies, as detailed in the accompanying tables. These studies collectively demonstrate that various preparations of MO, including leaf extracts, seed proteins and polysaccharides, can modulate immune responses by either enhancing or suppressing different immune parameters depending on the cellular context.

5.4.1 In Vitro Studies of MO for Its Immunomodulatory Action

In vitro assays show that MO extracts inhibit pro-inflammatory cytokines such as interleukin-1 beta (IL-1β), interleukin-6 (IL-6) and tumour necrosis factor-alpha (TNF-α), likely through mechanisms involving the downregulation of NF-κB signalling pathways. Additionally, certain polysaccharides stimulate macrophage proliferation and nitric oxide production (Table 5.1).

5.4.2 In Vivo Studies of MO for Its Immunomodulatory Action

In vivo evidence from animal models supports these findings, illustrating that MO administration leads to increased white blood cell counts, enhanced phagocytic activity of neutrophils, promoted proliferation of T lymphocytes and modulation of cytokine profiles towards anti-inflammatory responses (Table 5.2). These collective immunomodulatory effects suggest that MO holds significant potential as a natural immune modulator, which could be harnessed for therapeutic applications in immune-related disorders and inflammation management.

5.5 BIOACTIVE COMPOUNDS RESPONSIBLE FOR IMMUNOMODULATION

A diverse range of bioactive compounds in MO are responsible for its potent immunomodulatory properties, making it an important resource for the development of novel therapeutics targeting immune disorders. These molecules work synergistically to modulate both innate and adaptive immune functions by regulating cytokine production, enhancing or suppressing immune cell proliferation and exerting antioxidant and anti-inflammatory effects through pathways such as NF-κB inhibition. Recent developments in nanotechnology have made it possible to include these bioactive compounds into novel delivery systems such as nanoparticles (Tiwari et al., 2025), liposomes, transferosomes and phytosomes, enhancing their stability, bioavailability and targeted immunomodulatory action, which holds promise for the effective management of immune-related disorders (D. Kumar et al., 2024; Ali et al., 2023; Saroj et al., 2024) (Table 5.3).

5.6 MECHANISMS OF IMMUNOMODULATORY ACTION

Moringa oleifera exerts its immunomodulatory effects through multiple interconnected mechanisms, including modulation of cytokine production, effects on various immune cells and regulation of antioxidant and anti-inflammatory pathways.

5.6.1 Modulation of Cytokine Production

One of the central mechanisms by which MO modulates immunity is through the regulation of cytokines, which are key signalling molecules that orchestrate immune responses. Studies have shown that moringa extracts reduce pro-inflammatory cytokines such as IL-6, TNF-α and IL-1β, which are

TABLE 5.1
Summary of In Vitro Assays on MO, Highlighting Dosage, Effects and Mechanisms of Immunomodulation

In vitro study/assay	MO preparation	Dose/ concentration	Immunomodulatory findings	Immunomodulatory mechanism	References
Murine splenocytes (T/B cell proliferation by flow cytometry)	Aqueous leaf extract	Dose-dependent (low to high)	Increased CD4+, CD8+ and B220+ cells; elevated IL-2 secretion	T and B cell proliferation; stimulation of adaptive immunity	Bhattacharya et al., 2014; Rachmawati & Rifa'i, 2014
Human macrophages (LPS-stimulated)	Hydroethanolic extract	10–100 μg/mL	Suppressed TNF-α, IL-6, IL-8, NF-κB mRNA and protein levels	Inhibition of pro-inflammatory cytokines and the NF-κB pathway	Luetragoon et al., 2020
Mouse macrophage cell line (NO production)	Methanolic leaf extract (polysaccharides)	50–200 μg/mL	Increased nitric oxide production; enhanced macrophage activation	Activation of innate immune functions via NO synthesis	
Murine macrophages and lymphocytes	Leaf extract	50–100 μg/mL	Enhanced phagocytic activity and lymphocyte proliferation	Stimulation of innate and adaptive immune cells	
Neutrophil phagocytic activity assay	Methanolic leaf extract	Need for research	Increased neutrophil adhesion and phagocytosis	Enhancement of cellular innate immunity	Lawal et al., 2022
PBMCs, splenocytes, macrophages	Various extracts and pure compounds	Variable	Upregulation of anti-inflammatory IL-10; inhibition of IL-1β, IL-6 and TNF-α	Cytokine modulation: balancing inflammatory responses	Talib et al., 2025

TABLE 5.2
In Vivo Evidence of MO Effect on Various Pathways for Immunomodulatory Action

MO preparation/ form	Animal model (species, strain, sex, weight)	Dose/route	Duration	Immunomodulatory findings	Mechanism of action	References
Polysaccharide extract	BALB/c mice, male, 20 to 25 g	Not specified/po	Multiple weeks	Modulated gut microbiota; ↓ IL-6, ↑ IL-10	Gut microbiota modulation, anti-inflammatory cytokines	Husien et al., 2024
Methanolic leaf extract	Albino Wistar rats, male, 150 to 180 g	1000 mg/kg/po	Not specified	↑ WBC, neutrophil count; enhanced T-lymphocyte proliferation	Immunostimulation of leukocytes, cytokine modulation	Nfambi et al., 2015
Leaf extract	Swiss albino mice, male, 25 to 30 g	Not specified/po	Not specified	Enhanced humoral and cellular immunity; antioxidant effects	Antioxidant-mediated immune enhancement	Abd-Elhakim et al., 2018
Seed and leaf extract	Sprague Dawley rats, male, 180 to 200 g	Not specified/po	Various	↓ Pro-inflammatory cytokines TNF-α, IL-1β; ↑ anti-inflammatory	NF-κβ inhibition, macrophage activation	Abdel-Daim et al., 2020
Leaf powder (dietary)	Rainbow trout (*Oncorhynchus mykiss*), avg. 20 to 25 g	Diet supplemented/ oral (feed)	Weeks	↑ WBC, immune enzyme activities; enhanced lymphocyte activity	Nutritional and direct immunostimulant action	Sanchooli et al., 2023
Seed extract	Wistar rats, male, 150 to 180 g	200 mg/day/po	3 weeks	Reduced bronchial asthma symptoms; improved respiratory immunity	Reduction in cytokines including TNF-α, IL-4 and IL-6	Mahajan et al., 2007
Hydroethanolic flower extract	BALB/c mice, female, 18 to 22 g	200 mg/kg/po	14 days	↓ Pro-inflammatory cytokines; reduced NO and PGE2, inhibition of inflammation	NF-κβ pathway inhibition	Panova et al., 2025
Aqueous seed extract	New Zealand white rabbits, female, 2.5 to 3.0 kg	500 mg/kg/po	21 days	↑ Leukocyte count; improved antioxidant status	Antioxidant and leukocyte stimulation	Mutwedu et al., 2022
Methanol leaf extract	Wistar rats, male, 160 to 180 g	750 mg/kg/po	28 days	↓ Joint inflammation; reduced TNF-α and IL-6	Anti-inflammatory, cytokine downregulation	Henrrietta Ogadimma Asuzu-Samuel, 2023

TABLE 5.3
MO as an Immunomodulator: Pharmacological Properties, Morphology and Delivery Systems

Pharmacological property	Description	Key bioactive compounds	Mechanism of action	Morphology (part used)	Available drug delivery system	References
Immunostimulatory	Enhances innate and adaptive immunity	Polysaccharides, flavonoids	↑ T cell proliferation, ↑ macrophage activity, ↑ IL-2, and IFN-γ	Leaves, seeds, bark	Oral capsules, nanoemulsion, polysaccharide NPs	Murphy et al., 2023
Anti-inflammatory	Suppresses inflammatory enzymes and cytokines	Quercetin, kaempferol, isothiocyanates	↓ TNF-α, IL-6, COX-2, inhibits NF-κB signalling	Leaves, bark	Topical gel, liposomes, transdermal patches	Aggarwal et al., 2025
Antioxidant	Neutralises ROS and protects immune cells from oxidative damage	Vitamin C, β-carotene, polyphenols	↑ SOD, CAT, GSH; ↓ ROS	Leaves, seeds	Nano-drinks, oral tablets, antioxidant sprays	Xavier et al., 2024
Antimicrobial	Destroys bacteria, fungi and some viruses	Moringin, niazimicin, alkaloids	Cell wall disruption, ROS-mediated cell death	Seeds, leaves	AgNPs (silver NPs), ZnO-NPs, oral rinse, topical ointments	Perumalsamy et al., 2024
Anticancer/ Immunotherapy	Triggers apoptosis and enhances immune recognition of tumours	Benzyl isothiocyanate, niazimicin	Caspase-3 activation, ↓ AKT/mTOR, ↑ NK cell cytotoxicity	Leaves, seeds	Gold NPs, targeted liposomes, SLNs	Y.-Y. Wu et al., 2021
Hepatoprotective	Reduces liver inflammation and protects against immunotoxicity	Flavonoids, saponins	↓ ALT/AST, ↑ liver detox enzymes	Leaves, seeds, bark	Oral extracts, nanoencapsulated oils, phytosomes	Sowunmi & Gonzo, 2023
Antiviral	Blocks viral entry and boosts interferon response	Terpenoids, tannins	ACE-2 inhibition, ↑ IFN-γ response	Leaves	Nasal sprays, nano-liposomes, pulmonary nanoemulsions	Giugliano et al., 2024
Anti-allergic/ Anti-asthmatic	Regulates hypersensitivity and allergic inflammation	Saponins, omega-3s	↓ Histamine, ↓ IgE, IL-4/IL-5 suppression	Leaves, seeds	Inhalers, nanoemulsions, nebuliser-loaded NPs	Abd Rani et al., 2019
Wound healing	Accelerates immune cell recruitment and tissue regeneration	Vitamin E, phenolic acids	↑ fibroblast migration, ↑ VEGF, TGF-β	Leaves, oil	Nano-hydrogel, wound healing films, topical ointments	Al-Ghanayem et al., 2022

Gut immunity enhancer	Strengthens gut mucosa and supports immune microbiota balance	Fibre, polyphenols, prebiotics	↑ IgA secretion, ↓ intestinal inflammation	Leaves, pods	Oral probiotics, microcapsules, prebiotic fibre blends	Hong et al., 2022
Autoimmune modulator	Suppresses overactive immune response in diseases like Rhaumatoid Arthritis (RA) and Inflammatory Bowel Disease (IBD)	Omega-3s, isothiocyanates	↓ Th17 cells, ↓ IL-17, TNF-α; ↑ T-reg cells	Seeds, leaves	Controlled-release capsules, nano-pills	Gandhi et al., 2023
Neuroimmune protector	Protects neurons by regulating neuroinflammation	Phenolics, GABA, vitamin C	↓ Microglial activation, ↓ proinflammatory cytokines	Leaves, seeds	CNS-targeted nanoformulations, transdermal CNS delivery	Azlan et al., 2023

implicated in the amplification of inflammation and tissue damage. Simultaneously, moringa stimulates the production of anti-inflammatory cytokines such as interleukin-10 (IL-10). This balanced cytokine modulation helps suppress excessive inflammation while promoting immune homeostasis and effective pathogen clearance (Sanchooli et al., 2023).

5.6.2 Effects on Immune Cells

MO influences both innate and adaptive immune cells, contributing to an enhanced immune defence:

- *Macrophages*: moringa extracts activate macrophages, increasing their phagocytic activity and nitric oxide production, which are critical for pathogen clearance. Polysaccharides isolated from the plant stimulate macrophage activation and cytokine release, promoting innate immune surveillance.
- *T lymphocytes*: moringa promotes proliferation and activation of T cells, including helper (CD4+) and cytotoxic (CD8+) subsets. These effects enhance cell-mediated immunity, crucial for identifying and destroying infected or malignant cells.
- *B lymphocytes*: the plant also supports B cell proliferation and antibody production, strengthening humoral immunity and improving pathogen neutralisation and clearance.

The immunomodulatory effects on these cells often show dose- and context-dependent responses; for example, low doses may stimulate resting immune cells, while higher doses can dampen overactive inflammatory responses, thus maintaining immune balance

5.6.3 Antioxidant and Anti-Inflammatory Pathways

Inflammation and oxidative stress are closely linked in the pathophysiology of many diseases. By scavenging reactive oxygen species (ROS) and enhancing endogenous antioxidant defences, such as superoxide dismutase (SOD) and catalase activity, MO exerts a significant antioxidant effect. Reducing oxidative stress prevents immune cell damage and dysfunction. Moreover, MO inhibits key inflammatory signalling pathways, notably the NF-κB pathway, which controls the expression of many pro-inflammatory genes, including cytokines, enzymes (e.g. COX-2) and adhesion molecules. Inhibition of NF-κB signalling by MO extracts reduces inflammation and tissue injury, contributing to its anti-inflammatory and immunomodulatory benefits.

5.7 THERAPEUTIC APPLICATIONS

MO's numerous biological uses have been demonstrated, and in recent years, nanotechnology-based formulations have been developed to target specific diseases, including immunological disorders (Table 5.4).

5.7.1 Infectious Diseases and Immune Support

Moringa oleifera exhibits significant therapeutic potential in combating infectious diseases by enhancing the immune system and directly inhibiting various pathogens, including bacteria, viruses, fungi and parasites (Goswami et al., 2024). Its broad-spectrum antimicrobial activity is attributed to bioactive compounds, such as flavonoids, phenolics and glucosinolates, which help eliminate pathogens and modulate immune responses (A. Gupta et al., 2025). Studies indicate that extracts of MO can reduce inflammatory markers, promote the phagocytic activity of macrophages and stimulate adaptive immunity, thereby enhancing pathogen clearance while minimising tissue damage. This dual action supports its traditional use in wound healing and the prevention of infections, especially

TABLE 5.4
MO-Based Novel Formulations as Immunomodulators and Their Biomedical Applications

Nanotechnology	Targeted diseases	Pathway	Mechanism of action	In vitro	In vivo	Application	Result	References
Moringa-loaded silver nanoparticles (AgNPs)	Bacterial, fungal infections	NF-κB, TLR signalling	Reduces pro-inflammatory cytokines (TNF-α, IL-6), ROS scavenging and antimicrobial action	RAW 264.7 macrophage, THP-1 cells	Mice infected with *Staphylococcus aureus*, *Candida albicans*	Antibacterial, wound healing	Reduced inflammatory markers, accelerated wound closure	Dwivedi et al., 2024
Gold nanoparticles (GNPs) synthesised from moringa leaf extract	Breast, colon cancer	PI3K/Akt/mTOR	Induces apoptosis, downregulates survival pathways in tumour cells	MCF-7, HT-29 cell lines	BALB/c mice with breast tumours	Targeted cancer immunotherapy	Tumour reduction, increased immune cell activation	S. Kumar et al., 2024b
Moringa oil nanoemulsion	Autoimmune diseases (rheumatoid arthritis, IBD)	JAK/STAT, COX-2	Suppresses cytokine storms, enhances T-regulatory responses	PBMCs stimulated with LPS	Wistar rats with TNBS-induced colitis or arthritis	Anti-inflammatory oral therapy	Lower cytokine levels, histological improvement	Gandhi et al., 2023; Zou et al., 2015
Moringa-derived ZnO nanoparticles	Immunodeficiency, ageing-related decline	IL-2/IL-10, MAPK	Enhances T cell activation, promotes macrophage function	ROS and WBC proliferation assays	Immunosuppressed mice (cyclophosphamide model)	Nutraceutical immunostimulant	Restoration of immune balance, increased WBCs	
Moringa polysaccharide-stabilised selenium NPs	Viral infections, COVID-19, hepatitis	Nrf2, IFN-γ, ACE-2	Enhances antiviral immunity, antioxidant effect, inhibits viral binding	ACE-2 inhibition assays, HepG2 cells	Preclinical models of viral inflammation	Antiviral adjuvant or supplement	Decreased viral load in vitro, potential as adjunct therapy	Tao et al., 2025

(Continued)

TABLE 5.4 (CONTINUED)
MO-Based Novel Formulations as Immunomodulators and Their Biomedical Applications

Nanotechnology	Targeted diseases	Pathway	Mechanism of action	In vitro	In vivo	Application	Result	References
Moringa-based liposomes	Psoriasis	IL-17, TNF-α	Blocks keratinocyte hyperproliferation, reduces immune cell infiltration	HaCaT cells, ELISA cytokine profiling	Imiquimod-induced psoriatic mice	Topical nano-gel	Reduced scaling, normalised skin architecture	Raina et al., 2023
Moringa alkaloid nano-formulations	Diabetes-induced immune dysfunction	NLRP3 inflammasome, AMPK	Suppresses oxidative stress, regulates inflammasome activation	INS-1, MIN6 β-cell lines	STZ-induced diabetic rat model	Immunoregulatory adjunct in diabetes	Improved pancreatic histology, reduced immune infiltration	Bai et al., 2021
Moringa nano-hydrogel (chitosan-based)	Wound healing in immunocompromised patients	VEGF, TGF-β, TNF-α	Enhances angiogenesis and fibroblast activity, suppresses excessive inflammation	Human dermal fibroblasts, scratch assay	Diabetic wound healing model	Wound dressing, immune-tissue interface	Faster wound closure, reduced pro-inflammatory cytokines	X. Liu et al., 2024
Moringa-synthesised Fe_3O_4 magnetic nanoparticles	Neurodegenerative diseases (Alzheimer's, MS)	NF-κB, Nrf2	Anti-inflammatory and antioxidant activity in microglial cells	BV-2 microglial cell line, ROS assay	Experimental autoimmune encephalomyelitis (EAE) mice	CNS-targeted immunomodulation	Lowered microglial activation, improved motor coordination	Carrotta et al., 2025
Moringa encapsulated solid lipid nanoparticles (SLNs)	Asthma, allergic responses	Th2 cytokines, IL-4, IL-13	Suppresses allergic immune responses, modulates Th1/Th2 balance	ELISA for IgE and IL-4 on splenocyte cultures	Ovalbumin-induced allergic asthma in mice	Pulmonary delivery (nasal spray, inhaler)	Reduced IgE, eosinophil infiltration, airway inflammation	Chatzikleanthous et al., 2021

in resource-limited settings where access to conventional medicines may be limited (Abdel-Daim et al., 2020).

5.7.2 Chronic Inflammatory and Autoimmune Conditions

Chronic inflammation underlies many debilitating diseases, including rheumatoid arthritis, asthma and inflammatory bowel disease. MO has demonstrated potent anti-inflammatory and immunoregulatory properties, capable of downregulating pro-inflammatory cytokines such as TNF-α, IL-6 and IL-1β, while upregulating anti-inflammatory cytokines, including IL-10. Experimental models of rheumatoid arthritis indicate that MO extracts reduce markers of joint inflammation and oxidative stress, likely by inhibiting key signalling pathways, such as NF-κB. Additionally, the immunomodulatory balance achieved by MO helps to mitigate aberrant autoimmune responses, making it a promising natural agent for managing autoimmune disorders with fewer side effects compared with conventional immunosuppressants (Villegas-Vazquez et al., 2025).

5.7.3 Cancer Immunomodulation

Emerging evidence supports the role of MO in cancer therapy through both direct cytotoxic effects on tumour cells and modulation of antitumour immunity. Its bioactive compounds induce apoptosis, inhibit tumour cell proliferation and suppress inflammatory microenvironments that support tumour growth. Moreover, MO extracts enhance immune surveillance by promoting the activity of natural killer cells, T lymphocytes and macrophages against malignant cells. Combinational formulations incorporating MO with other natural products have shown synergistic immunomodulatory and anticancer effects in preclinical studies. This positions MO as a potential adjunct in cancer immunotherapy and chemoprevention strategies (Talib et al., 2025).

5.7.4 Use in Nutritional Supplementation

Due to its rich nutritional profile, high in essential amino acids, vitamins (A, C, E and B complex), minerals and antioxidants, MO serves as a valuable nutritional supplement that supports immune health and metabolic functions. Regular dietary inclusion of MO leaf powder or extracts has been shown to improve growth, enhance immune parameters and bolster resistance to infections in both humans and animals. Its antioxidant and anti-inflammatory phytochemicals also protect against oxidative stress-related damage, further supporting immune resilience. Nutritional supplementation with MO thus offers a dual benefit of meeting micronutrient needs and modulating immune function, which is particularly beneficial in regions affected by malnutrition and immune-compromising conditions (Sanchooli et al., 2023).

5.8 RECENT ADVANCES AND FUTURE PERSPECTIVES

Recent research has significantly expanded our understanding of MO and its immunomodulatory properties, highlighting its potential in both preventive and therapeutic contexts. Significant strides have been made in the discovery of bioactive substances, their mechanisms of action and their potential to modulate immune responses. It has been demonstrated that MO leaves and seeds are rich in polyphenols, flavonoids (such as quercetin and kaempferol), saponins, alkaloids and unique polysaccharides, including arabinogalactans. These compounds can modulate immune responses by suppressing pro-inflammatory cytokines and activating immune effector cells (Kashyap et al., 2022).

Modern investigations have elucidated the underlying mechanisms by which MO exerts its effects. For example, extracts can regulate immunological markers, including the suppression of NF-κB signalling pathways and the reduced expression of pro-inflammatory cytokines such as

IL-6, IL-1 and TNF-α, demonstrating both anti-inflammatory and immunostimulatory activities (Pareek et al., 2023; Xiao et al., 2020). Experimental studies in rodent models have demonstrated that MO leaf extracts significantly enhance phagocytic activity, increase the total leukocyte count and increase the proportion of immune cells, including T lymphocytes and macrophages. These effects are dose-dependent and have been replicated in both standard immune assays and disease models, including models that mimic infection and inflammation (Dillasamola et al., 2018). As an ingredient in functional foods and feed additives, MO is used to promote immune health not only in humans but also in livestock. Its incorporation has been shown to improve illness resistance and immunological function in animals, further demonstrating its utility in a sustainable diet and as a natural health supplement (Abd El-Hack et al., 2018).

Despite these promising findings, several areas require further exploration to translate MO's immunomodulatory potential into mainstream clinical practice. The diversity of MO's bioactive compounds varies widely based on geographic origin, plant part and extraction technique. There is increasing awareness that, in certain cases, lower doses may be more effective than higher doses (Azmi et al., 2025). For this reason, dose-response studies and standardisation procedures should be conducted before making clinical recommendations. While the anti-inflammatory and immunostimulatory effects are well described (El-Sherbiny et al., 2024), there should be further evaluation of the partitioning of constituents across various MOs in terms of potential interactions. Furthermore, adequate toxicological studies should be conducted to address the potential risks associated with chronic and high-dose supplementation. In the meantime, data on potential side effects and contraindications are limited.

5.9 CONCLUSION

Moringa, often known as a 'miracle tree', has the potential to become a powerful natural immunomodulator with significant applications to human health. A review of the peer-reviewed literature and research cited in the chapter reveals that MO provides a wealth of nutrients and health-promoting phytochemicals, as well as numerous immunomodulatory effects. These actions occur via the regulation of cytokine production, stimulation and regulation of innate and adaptive immune cell fates and potent antioxidant and anti-inflammatory action, especially due to inhibition of the NF-κB pathway. The therapeutic purposes of MO range from infectious diseases, chronic inflammatory and autoimmune diseases and cancer immunomodulation to nutritional supplementation. Its bioactive compounds, including flavonoids, phenolics, saponins, alkaloids, glucosinolates and polysaccharides, work in concert to fine-tune immune responses, restore immune homeostasis and protect against a spectrum of diseases. In vitro and in vivo evidence overwhelmingly supports the use of MO extracts and formulations to promote immune balance, enhance resistance to pathogens and mitigate the destructive consequences of chronic inflammation and oxidative stress.

While preclinical and some clinical data are promising, the translation of these findings into routine clinical practice will depend on continued research focused on standardising MO preparations, elucidating optimal dosing regimens and conducting rigorous, large-scale human trials to confirm efficacy and safety. Interdisciplinary collaborations between researchers, clinicians and traditional practitioners will further expand the understanding and therapeutic reach of MO as an immunomodulatory agent. In conclusion, *MO* holds great potential as a functional food and phytomedicine for immunomodulation, offering a safe, accessible and sustainable intervention for both disease prevention and adjunctive therapy. Continued exploration of its mechanisms, clinical applications and integration into public health strategies may make it an invaluable asset in the global pursuit of immune health and well-being.

Acknowledgement: We express our sincere thanks to the management of Rungta International Skills University, Bhilai, for providing the resources and facilities required for writing this chapter.

REFERENCES

Abd El-Hack, M. E., Alagawany, M., Elrys, A. S., Desoky, E.-S. M., Tolba, H. M. N., Elnahal, A. S. M., Elnesr, S. S., & Swelum, A. A. (2018). Effect of forage Moringa oleifera L. (moringa) on animal health and nutrition and its beneficial applications in soil, plants, and water purification. *Agriculture*, *8*(9), 145. https://doi.org/10.3390/agriculture8090145

Abd Rani, N. Z., Husain, K., & Kumolosasi, E. (2018). Moringa genus: A review of phytochemistry and pharmacology. *Frontiers in Pharmacology*, *9*. https://doi.org/10.3389/fphar.2018.00108

Abd Rani, N. Z., Kumolosasi, E., Jasamai, M., Jamal, J. A., Lam, K. W., & Husain, K. (2019). In vitro anti-allergic activity of Moringa oleifera Lam. extracts and their isolated compounds. *BMC Complementary and Alternative Medicine*, *19*(1), 361. https://doi.org/10.1186/s12906-019-2776-1

Abd-Elhakim, Y. M., El Bohi, K. M., Hassan, S. K., El Sayed, S., & Abd-Elmotal, S. M. (2018). Palliative effects of Moringa olifera ethanolic extract on hemato-immunologic impacts of melamine in rats. *Food and Chemical Toxicology*, *114*, 1–10. https://doi.org/10.1016/j.fct.2018.02.020

Abdel-Daim, M. M., Khalil, S. R., Awad, A., Abu Zeid, E. H., El-Aziz, R. A., & El-Serehy, H. A. (2020). Ethanolic extract of moringa oleifera leaves influences NF-κB signaling pathway to restore kidney tissue from cobalt-mediated oxidative injury and inflammation in rats. *Nutrients*, *12*(4), 1031. https://doi.org/10.3390/nu12041031

Abdel-Latif, H. M. R., Abdel-Daim, M. M., Shukry, M., Nowosad, J., & Kucharczyk, D. (2022). Benefits and applications of Moringa oleifera as a plant protein source in Aquafeed: A review. *Aquaculture*, *547*, 737369. https://doi.org/10.1016/j.aquaculture.2021.737369

Abidin, Z., Huang, H.-T., Hu, Y.-F., Chang, J.-J., Huang, C.-Y., Wu, Y.-S., & Nan, F.-H. (2022). Effect of dietary supplementation with Moringa oleifera leaf extract and Lactobacillus acidophilus on growth performance, intestinal microbiota, immune response, and disease resistance in whiteleg shrimp (Penaeus vannamei). *Fish & Shellfish Immunology*, *127*, 876–890. https://doi.org/10.1016/j.fsi.2022.07.007

Aggarwal, D., Chaudhary, M., Mandotra, S. K., Tuli, H. S., Chauhan, R., Joshi, N. C., Kaur, D., Dufossé, L., & Chauhan, A. (2025). Anti-inflammatory potential of quercetin: From chemistry and mechanistic insight to nanoformulations. *Current Research in Pharmacology and Drug Discovery*, *8*, 100217. https://doi.org/10.1016/j.crphar.2025.100217

Akter, R., Rahman, Md. R., Ahmed, Z. S., & Afrose, A. (2023). Plausibility of natural immunomodulators in the treatment of COVID-19–A comprehensive analysis and future recommendations. *Heliyon*, *9*(6), e17478. https://doi.org/10.1016/j.heliyon.2023.e17478

Al-Ghanayem, A. A., Alhussaini, M. S., Asad, M., & Joseph, B. (2022). Moringa oleifera leaf extract promotes healing of infected wounds in diabetic rats: Evidence of antimicrobial, antioxidant and proliferative properties. *Pharmaceuticals*, *15*(5), 528. https://doi.org/10.3390/ph15050528

Ali, A., Saroj, S., Saha, S., Gupta, S. K., Rakshit, T., & Pal, S. (2023). Glucose-responsive chitosan nanoparticle/poly(vinyl alcohol) hydrogels for sustained insulin release *In Vivo*. *ACS Applied Materials & Interfaces*, *15*(27), 32240–32250. https://doi.org/10.1021/acsami.3c05031

Azlan, U. K., Khairul Annuar, N. A., Mediani, A., Aizat, W. M., Damanhuri, H. A., Tong, X., Yanagisawa, D., Tooyama, I., Wan Ngah, W. Z., Jantan, I., & Hamezah, H. S. (2023). An insight into the neuroprotective and anti-neuroinflammatory effects and mechanisms of Moringa oleifera. *Frontiers in Pharmacology*, *13*. https://doi.org/10.3389/fphar.2022.1035220

Azmi, N., Akanno, A., Jacobsen, C., & Fossum, J. O. (2025). Moringa Oleifera as a multifunctional bioresource: Nutritional, medicinal and its pickering emulsifying potential. *Food Hydrocolloids for Health*, *8*, 100231. https://doi.org/10.1016/j.fhfh.2025.100231

Bai, Y., Mu, Q., Bao, X., Zuo, J., Fang, X., Hua, J., Zhang, D., Jiang, G., Li, P., Gao, S., & Zhao, D. (2021). Targeting NLRP3 inflammasome in the treatment of diabetes and diabetic complications: Role of natural compounds from herbal medicine. *Aging and Disease*, *12*(7), 1587. https://doi.org/10.14336/AD.2021.0318

Bhattacharya, A., Behera, R., Agrawal, D., Sahu, P., Kumar, S., & Mishra, S. (2014). Antipyretic effect of ethanolic extract of Moringa oleifera leaves on albino rats. *Tanta Medical Journal*, *42*(2), 74. https://doi.org/10.4103/1110-1415.137810

Cao, J., Shi, T., Wang, H., Zhu, F., Wang, J., Wang, Y., Cao, F., & Su, E. (2023). Moringa oleifera leaf protein: Extraction, characteristics and applications. *Journal of Food Composition and Analysis*, *119*, 105234. https://doi.org/10.1016/j.jfca.2023.105234

Carrotta, R., Vilasi, S., Costa, M. A., Librizzi, F., Martorana, V., Passantino, R., Buzzanca, C., di Stefano, V., Ortore, M. G., Piccirillo, S., Preziuso, A., Magi, S., & Mangione, M. R. (2025). <scp> *Moringa oleifera* </scp> Extracts as strategic phyto-therapy for Alzheimer's disease. *Food Science & Nutrition*, *13*(4). https://doi.org/10.1002/fsn3.70007

Chatzikleanthous, D., O'Hagan, D. T., & Adamo, R. (2021). Lipid-based nanoparticles for delivery of vaccine adjuvants and antigens: Toward multicomponent vaccines. *Molecular Pharmaceutics*, *18*(8), 2867–2888. https://doi.org/10.1021/acs.molpharmaceut.1c00447

Choudhary, M. K., Bodakhe, S. H., & Gupta, S. K. (2013). Assessment of the antiulcer potential of moringa oleifera root-bark extract in rats. *Journal of Acupuncture and Meridian Studies*, *6*(4), 214–220. https://doi.org/10.1016/j.jams.2013.07.003

Dhongade, H. kumar J., Paikra, B. K., & Gidwani, B. (2017). Phytochemistry and pharmacology of Moringa oleifera lam. *Journal of Pharmacopuncture*, *20*(3), 194–200. https://doi.org/10.3831/KPI.2017.20.022

Dillasamola, D., Aldi, Y., Fakhri, M., Diliarosta, S., P, B. O., & Noverial, N. (2018). immunomodulatory effect test from moringa leaf extract (Moringa Oleifera L.) with carbon clearance method in male white mice. *Asian Journal of Pharmaceutical and Clinical Research*, *11*(9), 241. https://doi.org/10.22159/ajpcr.2018.v11i9.26703

Domenico, M., Lina, C., & Francesca, B. (2019). Sustainable crops for food security: Moringa (Moringa oleifera Lam.). In *Encyclopedia of food security and sustainability* (pp. 409–415). Elsevier. https://doi.org/10.1016/B978-0-08-100596-5.22574-2

Duan, H., Liu, G., Liu, J., Li, Y., Bao, S., Song, W., & Yan, W. (2025). Focus on eye health: Analysis of the potential application of carotenoids from Goji berries sources. *Journal of Future Foods*. https://doi.org/10.1016/j.jfutfo.2025.04.004

Dwivedi, S. D., Bhoi, A., Pradhan, M., Sahu, K. K., Singh, D., & Singh, M. R. (2024). Role and uptake of metal-based nanoconstructs as targeted therapeutic carriers for rheumatoid arthritis. *3 Biotech*, *14*(5), 142. https://doi.org/10.1007/s13205-024-03990-z

El-Deep, M. H., Dawood, M. A. O., Assar, M. H., Ijiri, D., & Ohtsuka, A. (2019). Dietary Moringa oleifera improves growth performance, oxidative status, and immune related gene expression in broilers under normal and high temperature conditions. *Journal of Thermal Biology*, *82*, 157–163. https://doi.org/10.1016/j.jtherbio.2019.04.016

El-Sherbiny, G. M., Alluqmani, A. J., Elsehemy, I. A., & Kalaba, M. H. (2024). Antibacterial, antioxidant, cytotoxicity, and phytochemical screening of Moringa oleifera leaves. *Scientific Reports*, *14*(1), 30485. https://doi.org/10.1038/s41598-024-80700-y

Farswan, T. S., Kumari, B., Pandey, M. M., & Rastogi, S. (2024). Influence of seasonality on the targeted phytoconstituents of Moringa oleifera Lam. bark, a potent source of phytonutrients, collected from upper Gangetic plains, India. *Food Chemistry Advances*, *4*, 100706. https://doi.org/10.1016/j.focha.2024.100706

Feng, Y., Huang, J., Gong, J., Peng, W., Yu, C., Huai, Y., Chen, C., Bo, R., Liu, M., & Li, J. (2025). Composition and anti-colitis efficacy analysis of ethanol and aqueous extraction of Moringa oleifera leaves. *Fitoterapia*, 106720. https://doi.org/10.1016/j.fitote.2025.106720

Gandhi, G. R., Mohana, T., Athesh, K., Hillary, V. E., Vasconcelos, A. B. S., Farias de Franca, M. N., Montalvão, M. M., Ceasar, S. A., Jothi, G., Sridharan, G., Gurgel, R. Q., & Xu, B. (2023). Anti-inflammatory natural products modulate interleukins and their related signaling markers in inflammatory bowel disease: A systematic review. *Journal of Pharmaceutical Analysis*, *13*(12), 1408–1428. https://doi.org/10.1016/j.jpha.2023.09.012

Giugliano, R., Ferraro, V., Chianese, A., Della Marca, R., Zannella, C., Galdiero, F., Fasciana, T. M. A., Giammanco, A., Salerno, A., Cannillo, J., Rotondo, N. P., Lentini, G., Cavalluzzi, M. M., De Filippis, A., & Galdiero, M. (2024). Antiviral properties of moringa oleifera leaf extracts against respiratory viruses. *Viruses*, *16*(8), 1199. https://doi.org/10.3390/v16081199

Gopalakrishnan, L., Doriya, K., & Kumar, D. S. (2016). Moringa oleifera: A review on nutritive importance and its medicinal application. *Food Science and Human Wellness*, *5*(2), 49–56. https://doi.org/10.1016/j.fshw.2016.04.001

Goswami, R., Arya, D., Siddiqui, R., Kumar, S., Sarma, O., & Arora, N. (2024). The multifaceted benefits and applications of Moringa oleifera: A comprehensive review. *European Journal of Nutrition & Food Safety*, *16*(8), 1–13. https://doi.org/10.9734/ejnfs/2024/v16i81489

Gupta, A., & Gupta, S. K. (2025). Interplay of novel formulation-based treatment strategies for psoriasis: Advancing therapeutic outcomes. *Next Research*, *2*(3), 100731. https://doi.org/10.1016/j.nexres.2025.100731

Gupta, A., Gupta, S. K., Mahapatra, D. K., Sahu, S., Sinha, S., & Sharma, M. K. (2025). Exploring flavonoids as therapeutic agents for diabetic nephropathy: Insights and innovations. *Current Drug Research Reviews, 18*. https://doi.org/10.2174/0125899775369714251106115731

Gupta, S. K., Verma, A., Verma, N., Shah, K., Chauhan, D. N., & Chauhan, N. S. (2025). Acorus calamus in neurodegenerative disorders. In *Natural scaffolds for prevention and treatment of neurodegenerative disorders* (pp. 320–331). CRC Press. https://doi.org/10.1201/9781003403241-19

Hassan, M. A., Xu, T., Tian, Y., Zhong, Y., Ali, F. A. Z., Yang, X., & Lu, B. (2021). Health benefits and phenolic compounds of Moringa oleifera leaves: A comprehensive review. *Phytomedicine, 93*, 153771. https://doi.org/10.1016/j.phymed.2021.153771

Henrrietta Ogadimma Asuzu-Samuel. (2023). Effect of methanol Moringa oleifera leaves extract on the hematological parameters of cadmium chloride induced hypertensive Wistar rats. *World Journal of Advanced Research and Reviews, 20*(3), 1887–1889. https://doi.org/10.30574/wjarr.2023.20.3.2740

Hong, Z.-S., Xie, J., Wang, X.-F., Dai, J.-J., Mao, J.-Y., Bai, Y.-Y., Sheng, J., & Tian, Y. (2022). Moringa oleifera Lam. Peptide remodels intestinal mucosal barrier by inhibiting JAK-STAT activation and modulating gut microbiota in colitis. *Frontiers in Immunology, 13*. https://doi.org/10.3389/fimmu.2022.924178

Husien, H. M., Rehman, S. U., Duan, Z., & Wang, M. (2024). Effect of Moringa oleifera leaf polysaccharide on the composition of intestinal microbiota in mice with dextran sulfate sodium-induced ulcerative colitis. *Frontiers in Nutrition, 11*. https://doi.org/10.3389/fnut.2024.1409026

Islam, Z., Islam, S. M. R., Hossen, F., Mahtab-ul-Islam, K., Hasan, Md. R., & Karim, R. (2021). Moringa oleifera is a prominent source of nutrients with potential health benefits. *International Journal of Food Science, 2021*, 1–11. https://doi.org/10.1155/2021/6627265

Kashyap, P., Kumar, S., Riar, C. S., Jindal, N., Baniwal, P., Guiné, R. P. F., Correia, P. M. R., Mehra, R., & Kumar, H. (2022). Recent advances in drumstick (Moringa oleifera) leaves bioactive Compounds: composition, health benefits, bioaccessibility, and dietary applications. *Antioxidants, 11*(2), 402. https://doi.org/10.3390/antiox11020402

Komre, P., Gupta, A., Gupta, S. K., Joseph, T. M., & Mahapatra, D. K. (2025). Eosinophilic granulomatosis with polyangiitis: Overcoming diagnostic obstacles and exploring pharmacotherapeutic approaches. *Exploration of Immunology, 5*. https://doi.org/10.37349/ei.2025.1003219

Kumar, D., Gupta, S. K., Verma, A., & Ajazuddin. (2024). Transferosome as nanocarrier for drug penetration enhancement across skin: A comparison with liposome and ethosome. *Current Nanomedicine, 15*. https://doi.org/10.2174/0124681873333341241030065955

kumar, S., Murti, Y., Arora, S., Akram, W., Bhardwaj, H., Gupta, K., Sachdev, A., Devi, J., Kumar, S., Kumar, B., Dwivedi, V., Sameem, S., Nitendra, Kumar, P., Singh, K., & Saha, S. (2024a). Exploring the therapeutic potential of Moringa oleifera Lam. in Traditional Chinese medicine: A comprehensive review. *Pharmacological Research - Modern Chinese Medicine, 12*, 100473. https://doi.org/10.1016/j.prmcm.2024.100473

Kumar, S., Shukla, A., Singh, S. P., Singh, R. K., Patel, A. K., Verma, P. K., Kumar, S., Kumar, N., Singh, V., Wasnik, K., & Acharya, A. (2024b). Synthesized gold nanoparticles with Moringa Oleifera leaf extract induce mitotic arrest (G2/M phase) and apoptosis in Dalton's Lymphoma cells. *Cell Biochemistry and Biophysics, 82*(2), 1043–1059. https://doi.org/10.1007/s12013-024-01257-1

Lawal, H., Rahi Algraittee, S. J., Ranggasamy, R., Sarmadi, V., C. Thambiah, S., Stanslas, J., & Ramasamy, R. (2022). Moringa oleifera ethanol extract immunomodulatory activity discriminates between chronic myeloid leukaemia cell line and normal lymphocytes. *Malaysian Journal of Medicine and Health Sciences, 18*(s21), 101–110. https://doi.org/10.47836/mjmhs.18.s21.16

Liu, R., Liu, J., Huang, Q., Liu, S., & Jiang, Y. (2022). *Moringa oleifera* : A systematic review of its botany, traditional uses, phytochemistry, pharmacology and toxicity. *Journal of Pharmacy and Pharmacology, 74*(3), 296–320. https://doi.org/10.1093/jpp/rgab131

Liu, X., Peng, S., Pei, Y., Huo, Y., Zong, Y., Ren, J., & Zhao, J. (2024). Facile fabrication of chitosan/hyaluronic acid hydrogel-based wound closure material Co-loaded with gold nanoparticles and fibroblast growth factor to improve anti-microbial and healing efficiency in diabetic wound healing and nursing care. *Regenerative Therapy, 26*, 1018–1029. https://doi.org/10.1016/j.reth.2024.10.003

Luetragoon, T., Pankla Sranujit, R., Noysang, C., Thongsri, Y., Potup, P., Suphrom, N., Nuengchamnong, N., & Usuwanthim, K. (2020). Bioactive compounds in Moringa oleifera Lam. Leaves inhibit the pro-inflammatory mediators in lipopolysaccharide-induced human monocyte-derived macrophages. *Molecules, 25*(1), 191. https://doi.org/10.3390/molecules25010191

Mahajan, S. G., Mali, R. G., & Mehta, A. A. (2007). Effect of *Moringa oleifera* Lam. Seed extract on toluene diisocyanate-induced immune-mediated inflammatory responses in rats. *Journal of Immunotoxicology, 4*(2), 85–96. https://doi.org/10.1080/15476910701337472

Milla, P. G., Peñalver, R., & Nieto, G. (2021). Health benefits of uses and applications of Moringa oleifera in bakery products. *Plants*, *10*(2), 318. https://doi.org/10.3390/plants10020318

Murphy, E. J., Fehrenbach, G. W., Abidin, I. Z., Buckley, C., Montgomery, T., Pogue, R., Murray, P., Major, I., & Rezoagli, E. (2023). Polysaccharides—Naturally occurring immune modulators. *Polymers*, *15*(10), 2373. https://doi.org/10.3390/polym15102373

Mutwedu, V. B., Nyongesa, A. W., Kitaa, J. M., Ayagirwe, R. B. B., Baharanyi, C., & Mbaria, J. M. (2022). Effects of Moringa oleifera aqueous seed extracts on reproductive traits of heat-stressed New Zealand white female rabbits. *Frontiers in Veterinary Science*, *9*. https://doi.org/10.3389/fvets.2022.883976

Nader, N. D. (2017). Immunomodulation mechanisms in disease and in the surgical patient. *Immunological Investigations*, *46*(8), 765–768. https://doi.org/10.1080/08820139.2017.1373906

Nfambi, J., Bbosa, G. S., Sembajwe, L. F., Gakunga, J., & Kasolo, J. N. (2015). Immunomodulatory activity of methanolic leaf extract of *Moringa oleifera* in Wistar albino rats. *Journal of Basic and Clinical Physiology and Pharmacology*, *26*(6), 603–611. https://doi.org/10.1515/jbcpp-2014-0104

Núñez-Gastélum, J. A., Arguijo-Sustaita, A. A., López-Díaz, J. A., Díaz-Sánchez, Á. G., Hernández-Peña, C. C., & Cota-Ruiz, K. (2023). Seed germination and sprouts production of Moringa oleifera: A potential functional food? *Journal of the Saudi Society of Agricultural Sciences*, *22*(4), 223–230. https://doi.org/10.1016/j.jssas.2022.12.002

Owusu-Ansah, P., Alhassan, A. R., Ayamgama, A. A., Adzaworlu, E. G., Afoakwah, N. A., Mahunu, G. K., & Amagloh, F. K. (2023). Phytochemical analysis, enumeration, isolation, and antimicrobial activity of lemongrass and moringa leaves extracts. *Journal of Agriculture and Food Research*, *12*, 100579. https://doi.org/10.1016/j.jafr.2023.100579

Padayachee, B., & Baijnath, H. (2020). An updated comprehensive review of the medicinal, phytochemical and pharmacological properties of Moringa oleifera. *South African Journal of Botany*, *129*, 304–316. https://doi.org/10.1016/j.sajb.2019.08.021

Panova, N., Gerasimova, A., Gentscheva, G., Nikolova, S., Makedonski, L., Velikova, M., Beraich, A., Talhaoui, A., Petkova, N., Batovska, D., & Nikolova, K. (2025). Moringa oleifera Lam: A nutritional powerhouse with multifaceted pharmacological and functional applications. *Life*, *15*(6), 881. https://doi.org/10.3390/life15060881

Pareek, A., Pant, M., Gupta, M. M., Kashania, P., Ratan, Y., Jain, V., Pareek, A., & Chuturgoon, A. A. (2023). Moringa oleifera: An updated comprehensive review of its pharmacological activities, ethnomedicinal, phytopharmaceutical formulation, clinical, phytochemical, and toxicological aspects. *International Journal of Molecular Sciences*, *24*(3), 2098. https://doi.org/10.3390/ijms24032098

Peñalver, R., Martínez-Zamora, L., Lorenzo, J. M., Ros, G., & Nieto, G. (2022). Nutritional and antioxidant properties of Moringa oleifera leaves in functional foods. *Foods*, *11*(8), 1107. https://doi.org/10.3390/foods11081107

Perumalsamy, H., Balusamy, S. R., Sukweenadhi, J., Nag, S., MubarakAli, D., El-Agamy Farh, M., Vijay, H., & Rahimi, S. (2024). A comprehensive review on Moringa oleifera nanoparticles: importance of polyphenols in nanoparticle synthesis, nanoparticle efficacy and their applications. *Journal of Nanobiotechnology*, *22*(1), 71. https://doi.org/10.1186/s12951-024-02332-8

Phillip West, A., & McGuire, P. J. (2025). Tipping the balance: innate and adaptive immunity in mitochondrial disease. *Current Opinion in Immunology*, *95*, 102566. https://doi.org/10.1016/j.coi.2025.102566

Rachmawati, I., & Rifa'i, M. (2014). In Vitro immunomodulatory activity of aqueous extract of Moringa oleifera Lam. Leaf to the CD4 +, CD8+ and B220+ Cells in Mus musculus. *The Journal of Experimental Life Sciences*, *4*(1), 15–20. https://doi.org/10.21776/ub.jels.2014.004.01.03

Raina, N., Rani, R., Thakur, V. K., & Gupta, M. (2023). New insights in topical drug delivery for skin disorders: From a nanotechnological perspective. *ACS Omega*, *8*(22), 19145–19167. https://doi.org/10.1021/acsomega.2c08016

Rajgopal, B., Gupta, S. K., Deshmukh, R., Gupta, A., Patel, A., Sakure, K., & Jaiswal, M. (2025). Emerging trends in hydrogels for the treatment of vaginal candidiasis: A comprehensive review. *Recent Advances in Anti-Infective Drug Discovery*, *20*. https://doi.org/10.2174/0127724344348928250220063431

Sanchooli, N., Paknejad, H., Sudagar, M., Shekarabi, S. P. H., & Rahdari, A. (2023). Growth promoting and immunomodulatory effects of dietary Moringa oleifera leaf powder in rainbow trout (Oncorhynchus mykiss). *Aquaculture Reports*, *30*, 101555. https://doi.org/10.1016/j.aqrep.2023.101555

Saroj, S., Saha, S., Ali, A., Gupta, S. K., Bharadwaj, A., Agrawal, T., Pal, S., & Rakshit, T. (2024). Plant extracellular nanovesicle-loaded hydrogel for topical antibacterial wound healing *In Vivo*. *ACS Applied Bio Materials*. https://doi.org/10.1021/acsabm.4c00992

Senthilkumar, A., Karuvantevida, N., Rastrelli, L., Kurup, S. S., & Cheruth, A. J. (2018). Traditional uses, pharmacological efficacy, and phytochemistry of Moringa peregrina (Forssk.) Fiori. —A review. *Frontiers in Pharmacology*, *9*. https://doi.org/10.3389/fphar.2018.00465

Sheta, B., Waheed, O., Ayad, E., Habbak, L., & Hyder, A. (2024). Constitutive immunity is influenced by avian influenza virus-induced modification of gut microbiota in Eurasian teal (Anas crecca). *Comparative Biochemistry and Physiology Part C: Toxicology & Pharmacology*, *278*, 109867. https://doi.org/10.1016/j.cbpc.2024.109867

Shinde, N., Chauhan, A. S., Gupta, S. K., Bodakhe, S. H., & Pandey, D. P. (2015). Antifertility studies of curcumin and andrographolide combination in female rats. *Asian Pacific Journal of Reproduction*, *4*(3), 188–194. https://doi.org/10.1016/j.apjr.2015.06.012

Shukla, M. K., Singh, S. K., Pandey, S., Gupta, P. K., Choudhary, A., Jindal, D. K., Dua, K., & Kumar, D. (2022). Potential immunomodulatory activities of plant products. *South African Journal of Botany*, *149*, 937–943. https://doi.org/10.1016/j.sajb.2022.04.055

Sinha, S., Gupta, S. K., Sahu, P., Gupta, A., & Ajazuddin. (2025). Capsaicin: Unveiling its therapeutic potential and pharmacological actions. *Current Bioactive Compounds*, *21*(6). https://doi.org/10.2174/0115734072338291241129095347

Sowunmi, B. O., & Gonzo, M. (2023). The effect of Moringa oleifera crude extract on liver cell line, HepG2. *BMC Complementary Medicine and Therapies*, *23*(1), 380. https://doi.org/10.1186/s12906-023-04181-8

Strzelec, M., Detka, J., Mieszczak, P., Sobocińska, M. K., & Majka, M. (2023). Immunomodulation—a general review of the current state-of-the-art and new therapeutic strategies for targeting the immune system. *Frontiers in Immunology*, *14*. https://doi.org/10.3389/fimmu.2023.1127704

Sultana, S. (2020). Nutritional and functional properties of Moringa oleifera. *Metabolism Open*, *8*, 100061. https://doi.org/10.1016/j.metop.2020.100061

Talib, W. H., Al Junaidi, H. S., Alshaeri, H. K., Alasmari, M. M., Hadi, R. W., Alsayed, A. R., & Law, D. (2025). Immunomodulatory and anticancer effects of moringa polyherbal infusions: Potentials for preventive and therapeutic use. *Frontiers in Immunology*, *16*. https://doi.org/10.3389/fimmu.2025.1597602

Tao, L., Guan, C., Wang, Z., Wang, Y., Gesang, Q., Sheng, J., Dai, J., & Tian, Y. (2025). Selenium nanoparticles derived from Moringa oleifera Lam. Polysaccharides: Construction, stability, and in vitro antioxidant activity. *Foods*, *14*(6), 918. https://doi.org/10.3390/foods14060918

Tiwari, P., Prajapati, S. K., Dixit, V., Raza, M. A., Ajazuddin, Gupta, S. K., & Sharma, M. (2025). Recent advancements in nanoparticle drug delivery systems to cure neurological disorders via the nasal route. *Neurología Argentina*. https://doi.org/10.1016/j.neuarg.2025.05.003

van den Berg, J., & Kuipers, S. (2022). The antibacterial action of Moringa oleifera: A systematic review. *South African Journal of Botany*, *151*, 224–233. https://doi.org/10.1016/j.sajb.2022.09.034

Villegas-Vazquez, E. Y., Gómez-Cansino, R., Marcelino-Pérez, G., Jiménez-López, D., & Quintas-Granados, L. I. (2025). Unveiling the miracle tree: Therapeutic potential of Moringa oleifera in chronic disease management and beyond. *Biomedicines*, *13*(3), 634. https://doi.org/10.3390/biomedicines13030634

Wang, Q., & Bao, Y. (2021). Nanodelivery of natural isothiocyanates as a cancer therapeutic. *Free Radical Biology and Medicine*, *167*, 125–140. https://doi.org/10.1016/j.freeradbiomed.2021.02.044

Wilcaso, P., Landázuri, A. C., Castelló, M. L., & Ortolá, M. D. (2025). Comprehensive characterization of nutritional, phytochemical, optical, and electrical properties of Moringa oleifera Lam. parts from Ecuador. *LWT*, *228*, 118108. https://doi.org/10.1016/j.lwt.2025.118108

Wu, Q., Zhou, H.-J., Sheng, J., Su, L.-Y., & Tian, Y. (2024). Extraction, structural properties, and bioactivities of Moringa (Moringa oleifera Lam.) isothiocyanates: A review. *Food Bioscience*, *57*, 103447. https://doi.org/10.1016/j.fbio.2023.103447

Wu, Y.-Y., Xu, Y.-M., & Lau, A. T. Y. (2021). Anti-cancer and medicinal potentials of Moringa Isothiocyanate. *Molecules*, *26*(24), 7512. https://doi.org/10.3390/molecules26247512

Xavier, L. E. M. dos S., Reis, T. C. G., Martins, A. S. da P., Santos, J. C. de F., Bueno, N. B., Goulart, M. O. F., & Moura, F. A. (2024). Antioxidant therapy in inflammatory bowel diseases: How far have we come and how close are we? *Antioxidants*, *13*(11), 1369. https://doi.org/10.3390/antiox13111369

Xiao, X., Wang, J., Meng, C., Liang, W., Wang, T., Zhou, B., Wang, Y., Luo, X., Gao, L., & Zhang, L. (2020). Moringa oleifera lam and its therapeutic effects in immune disorders. *Frontiers in Pharmacology*, *11*. https://doi.org/10.3389/fphar.2020.566783

Zaffar, S., Saha, S., Agrawal, T., Gupta, S. K., Gulati, P., Paul, D., Verma, A., Pal, S., & Rakshit, T. (2025). Improved diabetic wound healing via flower extracellular vesicles and carbon-dot-infused alginate-polyethylenimine antibacterial hydrogel. *ACS Biomaterials Science & Engineering*. https://doi.org/10.1021/acsbiomaterials.5c00443

Zou, Y., Li, W.-Y., Wan, Z., Zhao, B., He, Z.-W., Wu, Z.-G., Huang, G.-L., Wang, J., Li, B.-B., Lu, Y.-J., Ding, C.-C., Chi, H.-G., & Zheng, X.-B. (2015). Huangqin-tang ameliorates TNBS-induced colitis by regulating effector and regulatory CD4 + T cells. *BioMed Research International*, *2015*, 1–13. https://doi.org/10.1155/2015/102021

6 *Moringa oleifera* as an Antidiabetic Agent

Feyisayo O. Adepoju

6.1 INTRODUCTION

Diabetes (diabetes mellitus) is a complex chronic disorder that may result in severe health complications, with type 2 diabetes constituting over 90% of all diabetes cases globally. Currently, one in ten adults lives with diabetes, with the incidence of diabetes among adults aged 20–79 predicted to reach 643 million in 2030 (~20% increase), establishing it as a global health challenge (International Diabetes Federation 2021). High blood glucose levels resulting from aberrant insulin secretion or insulin action are the hallmarks of diabetes. Meanwhile, chronic hyperglycaemia leads to secondary comorbidities, including renal failure, cardiovascular diseases, blindness, atherosclerosis and early death; therefore, managing hyperglycaemia is crucial to mitigating the severe sequelae of diabetes. Currently, conventional diabetes treatment typically involves the use of insulin or non-insulin-based agents, such as α-glucosidase inhibitors, biguanide metformin, sulfonylureas and GLP-1 analogues; however, the prolonged usage of these medications has been linked to adverse side effects, including weight gain, gastrointestinal disorders, anaemia and hypoglycaemia (Aljazzaf et al. 2023). There is a continuous search for alternative therapeutics with improved and safer properties derived from natural products to control diabetes.

Medicinal plants, as sources of natural products, bioactive compounds and natural antioxidants, are an excellent choice for natural medicines relative to synthetic drugs. Studies have shown that alternative medicines are emerging as effective remedies for the prevention and treatment of many health disorders due to their naturalness, availability, multitarget mechanism of actions and low toxicity compared with existing drugs (World Health Organization (WHO) 2013). *Moringa oleifera* Lam., widely referred to as the 'tree of life or horseradish tree', is a flowering plant that belongs to the family Moringaceae, consisting of 13 known species, with *M. oleifera* being the most studied and valuable within the family (FAO 2020). Moringa is a cruciferous herbal plant cultivated across Africa, Central America and Asia and can thrive in even the harshest climatic conditions, which has led to its widespread distribution across different geographical locations (Horn et al. 2022; Adepoju and Selezneva 2024). The moringa plant possesses both nutritional and medicinal benefits due to its rich phytonutrients (Yong-Bing, Gui-Lin and Ming-Quan 2019). The whole plant, including leaves, stems, flowers, roots and pods, has been studied for its biological activities. Glucosinolates, isothiocyanates, flavonoids (rutin, kaempferol, vitexin, isoquercitrin, luteolin, naringin, catechin, epicatechin, quercetin), terpenoids (lutein, β-carotene, zeaxanthin, oleanolic acid), sterols and fatty acids (stigmasterol, stearic acid, linoleic acid, palmitic acid), phenolic glucoside (niazirin) and polyphenols (caffeic acid, chlorogenic acid, ellagic acid) are some of the bioactive compounds found in moringa responsible for its different health benefits (Su et al. 2023; Paikra et al. 2017; Adepoju, Selezneva and Okpala 2024). Tuorkey (2016) reported that the oral administration of leaf extract of moringa to diabetic mice lowered blood glucose levels and increased antioxidant enzymes. Meanwhile, other studies have shown that moringa resolves hyperglycaemia by inhibiting α-amylase and α-glucosidase enzymes, hepatic gluconeogenesis and intestinal glucose absorption, while increasing glucose uptake, insulin secretion and sensitivity (Kamrul-Hasan et al. 2023; Khan

DOI: 10.1201/9781003617860-7

et al. 2017). This chapter summarises the therapeutic role of the moringa plant as an antidiabetic agent in both animal and human studies.

6.2 MORINGA AS AN INHIBITOR OF A-GLUCOSIDASE AND A-AMYLASE

α-Amylases and α-glucosidases are some of the key carbohydrate-metabolising enzymes involved in the hydrolysis of polysaccharides, oligosaccharides and/or disaccharides into monosaccharides. α-Amylase breaks down starch molecules and similar polysaccharides into disaccharides by cleaving the α-1,4-glycosidic bond, whereas α-glucosidase catalyses these disaccharides into simpler sugars, which can then be absorbed by glucose transport proteins and enterocytes in the small intestine (Shori 2015). A therapeutic plan for diabetes control would entail managing postprandial hyperglycaemia by blocking gut enzymes in the gastrointestinal tract. Some commercially available α-glucosidase inhibitors include acarbose, miglitol and volglibose, with acarbose the most widely studied. Natural products that effectively inhibit α-glucosidase and α-amylase with fewer side effects may provide an attractive therapeutic strategy for managing/preventing postprandial hyperglycaemia. As shown in Table 6.1, studies have found that the moringa plant can inhibit α-amylase and α-glucosidase enzymes. Magaji and co-authors studied the inhibitory effect of extracts obtained from moringa parts, finding that the methanolic leaf extract demonstrated the strongest α-amylase inhibition (IC_{50} = 8.217 mg/ml), while the hexane extract of the roots demonstrated the strongest α-glucosidase inhibition (IC_{50} = 0.382 mg/ml) (Magaji, Sacan and Yanardag 2020). A recent study reported that an insulin-like protein extract from moringa leaves strongly inhibited α-amylase, with percentage inhibition values of 75.48 and 82.93% and IC_{50} values of 0.6 and 17.6 μM for supernatant and pellet, respectively (Zaib et al. 2025). Another study found that its ethanolic leaf extract exhibited inhibitory activity against α-amylase (IC_{50} = 27.54 μg/mL), with acarbose serving as a positive control (IC_{50} = 19.45 μg/mL). Furthermore, using molecular docking studies, the authors demonstrated that phenolic compounds, such as rutin and nicotiflorin found in moringa, exhibited the most potent binding affinity against human pancreatic α-amylase with docking scores of −9.40 and −9.10 kcal/mol, respectively (Chigurupati et al. 2022). These studies support a clinical trial that examined the effect of the consumption of moringa-leaf-supplemented meals on both healthy and diabetic Sahawi people, finding that it reduced postprandial hyperglycaemia in diabetic subjects. However, no changes were observed in the significant healthy subjects, as reported by Leone et al.

TABLE 6.1
Moringa as an α-Amylase and/or α-Glucosidase Inhibitor

Plant part	Type of extract	Concentration	α-Amylase inhibitory activity	α-Glucosidase inhibitory activity	Reference
Leaf	Methanol	2.5–10.0 mg/ml	IC_{50} 8.22 mg/ml	IC_{50} 44.96 mg/ml	Magaji, Sacan and Yanardag (2020)
Leaf	Ethanol	0.002–0.01 mg/ml	IC_{50} 0.1802 mg/ml	–	Ogundipe et al. (2022)
Leaf	Ethanol	0.01–1 mg/ml	IC_{50} 0.02754 mg/mL	–	Chigurupati et al. (2022)
Leaf	Aqueous	10–500 μg/mL	IC_{50} 52.5 μg/mL	IC_{50} 33.4 μg/mL	Khan et al. (2017)
Leaf	Aqueous	0–15 μg/mL	IC_{50} 6.49 μg/mL	IC_{50} 4.73 μg/ml	Jimoh (2018)
Seed	Methanol	5–15 %	IC_{50} 15.38%	–	Natsir et al. (2024)
Flower	Ethanol	1–5 mg/mL	IC_{50} 37.33 mg/mL	IC_{50} 38.30 mg/ml	Olasehinde et al. (2021)

(2018). All of these findings support moringa as an α-amylase/α-glucosidase inhibitor, as slowing down glucose absorption may be a potential target for the management of diabetes.

6.3 PRECLINICAL STUDIES ON THE ANTIDIABETIC PROPERTIES OF *MORINGA OLEIFERA*

Moringa oleifera, also referred to as the miracle tree, has been reported to cure over 300 diseases (Gopalakrishnan, Doriya and Kumar 2016). The plant contains bioactive compounds, whose biological activities have been explored in recent years to elucidate its mechanisms of action in order to provide a scientific foundation for its application. Olayaki and co-investigators reported that the oral administration of methanol leaf extracts of moringa to rats with alloxan-induced diabetes improved glucose tolerance, serum insulin levels, lipid metabolism, homeostasis and glycogen synthesis when compared with the untreated group (Olayaki et al. 2015). Other studies have examined its aerial parts, roots, seeds and extracts for their antidiabetic properties, with multiple studies reporting their possible glucose-lowering effects (Muzumbukilwa, Nlooto and Owira 2019; Bamagous et al. 2018; Joung et al. 2017). Moreover, studies that corroborated these findings reported that the treatment of diabetic animals with moringa plant extracts significantly improved weight loss, insulin sensitivity, hyperglycaemia, polydipsia, glucose tolerance, lipid metabolism and liver and kidney function (Table 6.2).

One of the proposed mechanisms behind the glucose-lowering effect of moringa is the activation of the Akt insulin-dependent pathway, responsible for glucose uptake, via the GLUT4 transporter (Attakpa et al. 2017). Insulin binds to the insulin receptor (IR), which phosphorylates insulin receptor substrate-1 (IRS-1), resulting in the subsequent phosphorylation of PI3K and ultimately phosphoinositide-dependent kinase-1 (PDK1). Following this, active PDK1 phosphorylates Akt, which promotes the translocation of GLUT4, thereby increasing the insulin-dependent glucose transport into the cell (Figure 6.1). Akt is a key signalling molecule that controls insulin, glucose and lipid metabolism (Świderska et al. 2020). MIC-1, an isothiocyanate-rich moringa extract, was shown to improve insulin sensitivity in VHFD-fed male C57BL/6J mice by increasing the expression of protein kinases, PI3K, IRSs and GLUT4 expression in liver and muscle tissues (Waterman et al. 2015). This is consistent with a recent study that predicted the PI3K-Akt signalling pathway as the possible signalling pathway through which moringa exerts its hypoglycaemic effect (Hong et al. 2023). Other researchers have shown that extracts of moringa seeds and leaves can exert renoprotective effects in HFD/streptozotocin-induced diabetes (Wen et al. 2022; Ahmad et al. 2023). According to Wen and co-investigators, moringa extracts delayed the progression of diabetic kidney disease by exerting anti-renal fibrotic effects via the decreased expression of p-GSK-3β and antioxidant effects through the activation of the Nrf2/HO-1 pathway (Wen et al. 2022). In another study, inducing diabetes in rats with streptozotocin led to increased lipid peroxidation, decreased GSH, SOD and catalase activity and increased production of the pro-inflammatory cytokine, IL-6, all of which improved with treatment with seed extracts of moringa (Al-Malki and El Rabey 2015). Treatment of obese C57BL/6J male mice with isothiocyanate-rich extracts of moringa improved glucose tolerance and insulin sensitivity, decreased adiposity and body weight, improved antioxidant genes and lowered inflammatory genes (Waterman et al. 2015; Jaja-Chimedza et al. 2018). Aljazzaf et al. (2023) have also shown that a therapy combining extracts of leaves and seeds of moringa (2:1 ratio) at a dosage of 500 mg/kg bw increased antioxidant biomarkers and reduced fasting blood glucose, while decreasing liver enzymes, triglycerides, cholesterol, creatinine and oxidant markers.

Islet β-cell mass and function have been observed to progressively decline in diabetic patients, partly due to glucotoxicity, resulting from excessive accumulation of glucose by pancreatic β-cells. In turn, this promotes glycation processes and mitochondrial stress that lead to the production of free radicals at levels exceeding the cells' antioxidant capacity. Consequently, oxidative stress inhibits insulin biosynthesis and secretion, leading to β-cell cytotoxicity and apoptosis (Kaneto et al.

TABLE 6.2
Preclinical Studies on the Antidiabetic Properties of *M. oleifera*

Active component/ dose	Animal model/ duration	Key findings	Reference
Ethanolic leaf extract 200, 400 or 600 mg/kg bw	HFD, male B6/J mice (eight weeks)	↓Plasma glucose, ↑insulin, ↓BG in OGTT, ↓SREBP-1, ↑hepatic PPARα, activation of insulin-dependent Akt pathway (↑pAkt and ↑GLUT4)	Attakpa et al. (2017)
Ethyl acetate leaf extract fraction 200 mg/kg bw	STZ, Sprague-Dawley rats (30 days)	↓HbA1c, ↓blood glucose, ↑insulin, ↓creatinine, ↑albumin, globulin and total protein. ↑Weight gain, ↓food and water intake. Improved lipid profile. ↓Liver enzymes, ↓lipid peroxidation, ↑pancreatic GST, SOD, Vit-E, CAT, GPx, GSH and Vit-C. ↓Inflammatory markers (TNF-α, IL-6 and IL-1β). Restored pancreatic histoarchitectural structure	Bamagous et al. (2018)
Ethanolic leaf extract 150 mg/kg bw	db/db mice (five weeks)	↑Insulin, ↓blood glucose, ↓TG and LDL-C. ↓Renal IL-6, TNFα, iNOS, COX-2 and. improved kidney histopathology	Tang et al. (2017)
Seed powder 50 and 100 mg/kg bw	STZ, albino rats (four weeks)	↑Antioxidant enzymes (CAT, GSH and SOD), ↓MDA, IgA and IgG, ↓IL-6, ↓ HbA1c, α-amylase, FBG. ↓Kidney markers: BUN, uric acid and creatinine. ↓Sodium and ↓potassium. ↑Body weight, ↓water consumption. Ameliorated the histoarchitectural damage of the pancreas and kidney	Al-Malki and El Rabey (2015)
Ethanolic leaf extract 100, 200 and 400 mg/kg bw	HFD-STZ, male Wistar rats (three weeks)	↓Blood glucose ↓AUC glucose, ↓insulin ↓HOMA-IR, ↓IL-1β, IL-6 and TNFα ↑PPARγ	Anwer et al. (2021)
Aqueous leaf extract 100 mg/kg bw	Alloxan, albino mice (two weeks)	↓Hyperglycaemia, ↓HOMA-IR, ↓creatinine, ↓urea, ↑plasma insulin, ↑total antioxidant capacity, ↑CD69, ↑IFN-γ	Tuorkey (2016)
Niazirin 10 and 20 mg/kg/day	db/db mice (four weeks)	↓Hyperglycaemia, ↓BG in OGTT, ↓insulin resistance. ↓TNF-α, ↑anti-inflammatory cytokines, IL-10. Improved lipid profile. ↑Hepatic glycogen, pyruvate kinase and hexokinase. Improved liver histopathology. ↓SREBP-1 and FAS. ↑SirT1 and ↓PGC-1α, HNF-4α and FOXO1. Activated the AMPK pathway to regulate carbohydrate metabolism by ↑p-AMPK α and ↑PFKFB3 expression.	Bao et al. (2020)

(Continued)

TABLE 6.2 (CONTINUED)
Preclinical Studies on the Antidiabetic Properties of *M. oleifera*

Active component/ dose	Animal model/ duration	Key findings	Reference
Methanolic leaf extract 250 mg/kg bw	STZ, male Wistar rats (six weeks)	↓Kidney weight, ↓plasma glucose, ↓inflammatory markers, ↓creatinine, ↑total protein, albumin and globulin. ↓Lipid peroxidation and GPx. ↑CAT, SOD and GSH. Attenuated kidney histopathological conditions in diabetic rats	Omodanisi et al. (2017)
Ethanolic flower extract 100, 200 and 300 mg/kg bw	STZ, male Albino rats (three weeks)	↓FBG, ↑CAT, ↑SOD, ↓MDA. Improved lipid metabolism by ↓LDL-C, TG, TC, HDL-C and atherogenic index. ↓Liver damage biomarkers	Arise et al. (2016)
Aqueous leaf extract 250 mg/kg bw	Alloxan, female Wistar rats (18 days)	↓Serum glucose, ↓lipid peroxidation, ↓serum TG, cholesterol and ↑mRNA expression of FAS, ↓Pyruvate carboxylase expression. Restored the histoarchitectural changes of the liver and pancreas	Abd El Latif et al. (2014)
Fermented leaf extract 250 mg/kg bw	HFD, male B6/J mice (ten weeks)	↓AUC glucose, ↓hepatic lipid accumulation. Improved glucose tolerance, ↓expression of lipogenic genes: SREBP1c, FAS and ACC. ↑Expression of lipid oxidation mediators: HSL, ACOX1, CD36 and CPT1. ↑Hepatic protein expression of p-AMPK. ↓Pro-inflammatory cytokines: IL-12, TNFα and IL-6	Joung et al. (2017)
Methanolic leaf extract 300 and 600 mg/kg bw	Alloxan, male Wistar rats (six weeks)	↓Hyperglycaemia and AUC glucose, ↑insulin secretion, improved glucose tolerance, ↑glycogen synthase, ↑glycogen, ↓TG, ↓TC, ↓LDL, ↑HDL	Olayaki et al. (2015)
Methanolic leaf extract 250 and 500 mg/kg bw	STZ, male Wistar rats (54 days)	Improved weight loss, ↓polydipsia, improved glucose tolerance, ↓FBG, ↑plasma insulin, ↓plasma ALT, AST and GGT, ↑plasma albumin, improved hepatic histopathological aberrations caused by diabetes	Muzumbukilwa, Nlooto and Owira (2019)

High-fat diet (HFD); Streptozotocin (STZ); Fasting blood glucose (FBG); oral glucose tolerance test (OGTT); blood glucose (BG); carnitine palmitoyltransferase 1 (CPT1); cluster of differentiation 36 (CD36); inducible nitric oxide synthase (iNOS); glycated haemoglobin (HbA1c); interleukin-1b (IL-1b); hepatocyte nuclear factor 4 α (HNF-4α); peroxisomal acyl-CoA oxidase 1 (ACOX1); fatty-acid synthase (FAS); peroxisome proliferator-activated receptor γ (PPARγ); interleukin-6 (IL-6); triglycerides (TG); gamma-glutamyltransferase (GGT); glutathione peroxidase (GPx); low-density lipoprotein cholesterol (LDL-C); glutathione S-transferase (GST); glucose transporter type 4 (GLUT4); superoxide dismutase (SOD); hormone-sensitive lipase (HSL); vitamin E (Vit-E); reduced glutathione (GSH); vitamin c (Vit-C); acetyl-CoA carboxylase (ACC); forkhead box protein O1 (FOXO1); interleukin-12 (IL-12); malondialdehyde (MDA); 6-phosphofructo-2-kinase/fructose-2 ,6-biphosphatase 3 enzyme (PFKFB3); sterol regulatory element-binding protein 1c (SREBP1c); 5' adenosine monophosphate-activated protein kinase (AMPK); total cholesterol (TC); catalase (CAT); homeostasis model assessment for insulin resistance (HOMA-IR); sirtuin 1 (SirT1); protein kinase B (Akt); high-density lipoprotein cholesterol (HDL-C); peroxisome proliferator-activated receptor gamma coactivator 1-α (PGC-1α).

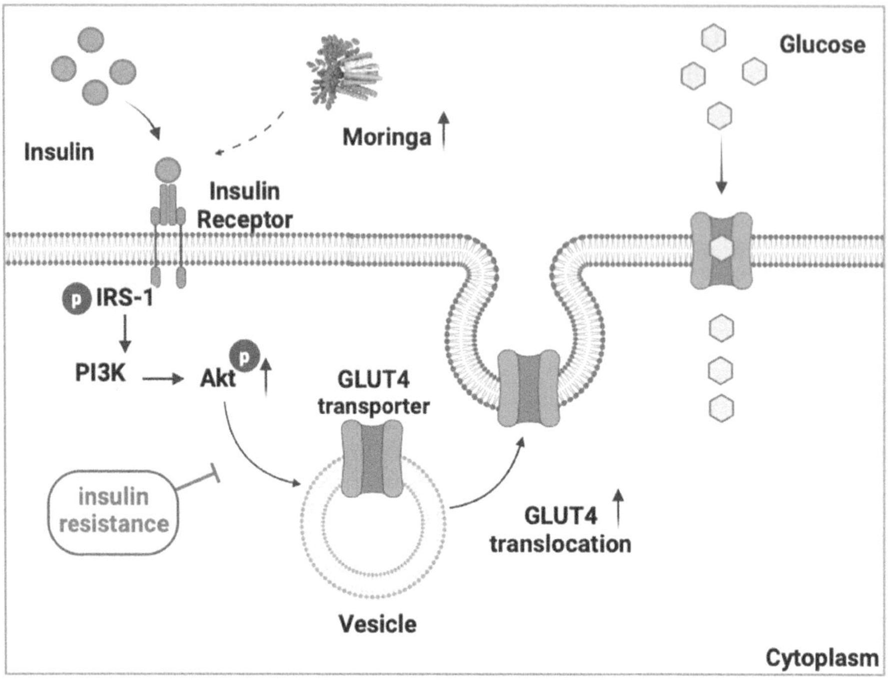

FIGURE 6.1 Proposed Mechanism of Action of *M. oleifera* against Diabetes. Image created with BioRender .com

2007). Yassa and Tohamy (2014) reported that the administration of moringa leaves to diabetic rats improved the cellular density of β-cells and restored the ultrastructure of the islets of Langerhans and pancreas. A different study found that the consumption of moringa seed powder by diabetic rats restored the normal histoarchitectural structure of the pancreatic islet and acini cells (Al-Malki and El Rabey 2015). Oral administration of methanolic extract of moringa pods to diabetic rats also improved islet β-cell density (Gupta et al. 2012) as well as moringa stem extracts, preventing apoptosis in islet β-cells (Jiang et al. 2025). Leaves or pod extracts have been shown to restore and regenerate pancreatic β-cells in diabetic rats (Abd El Latif et al. 2014; Gupta et al. 2012) and resolve inflammation and oxidative stress caused by a high-fat diet, thus deactivating the NF-κB pathway (El-Shehawi et al. 2021). Bioactive compounds such as glucomoringin, quercetin, kaempferol and phenolic acids found in moringa have been shown to scavenge free radicals and exert modulatory effects on oxidative stress markers, hence contributing to its hypoglycaemic, antidyslipidaemic and anti-inflammatory effects (Al-Malki and El Rabey 2015).

6.4 CLINICAL STUDIES ON THE GLUCOSE CONTROL EFFECT OF *MORINGA OLEIFERA*

Although preliminary research has explored the efficacy of moringa therapy in prediabetic subjects and patients with type 2 diabetes (T2D) (Table 6.3), further human studies are necessary to determine its potential in preventing or treating T2D. A clinical trial has highlighted the possible role of moringa as an antidiabetic agent for T2D patients. The investigators treated diabetic patients with a capsule of moringa, containing 98.34% leaf powder, twice daily for 90 days, in addition to their oral hypoglycaemic medications. The study found that postprandial glucose levels decreased significantly over time, from a baseline of 210±48.83 mg/dL to 191±48.64 at the first month ($p<0.0088$), 174±36.77 at the second month ($p<0.035$) and 150±21.10 mg/dL at the third month ($p<0.0098$), respectively. Additionally, glycated haemoglobin decreased from 7.81±0.51% to 7.40±0.63%

TABLE 6.3
Clinical Studies on the Effects of *M. oleifera* on T2D and Prediabetic Subjects

Intervention	Study design	Duration	Experimental outcome	Reference
Leaf capsules 0.5 g three times daily	Prospective quasi-experimental study (T2DM)	12 weeks	Improvement in blood HbA1c (0.6%) and a reduction in hsCRP	Mozo and Caole-Ang (2015)
Leaf capsules 4 g twice daily	Randomised, placebo-controlled trial (T2DM)	Four weeks	No significant effect on blood glucose	Taweerutchana et al. (2017)
Leaf capsules 3 g and 6 g twice daily	Randomised, placebo-controlled trial (T2DM)	12 weeks	A higher dose of the treatment improved glycaemic response as well as blood pressure in the studied group	Hameed et al. (2023)
Leaf capsules 0.8 g three times a day	Randomised, placebo-controlled, double-blind trial (prediabetes)	12 weeks	Improved glycaemic response by improving HbA1c levels in 58% of subjects after 12 weeks, compared with baseline in the moringa group and only 38% in the placebo group	Gómez-Martínez et al. (2022)
Leaf powder 10 g/day + OHA	Randomised, controlled, unblinded trial (T2DM)	12 weeks	Reduced body fat and improved glycated haemoglobin levels	Leone et al. (2025)
Leaf capsules 0.8 g three times a day	Randomised, placebo-controlled double-blind trial	12 weeks	Results indicate that moringa did not significantly affect inflammatory and cardiometabolic status markers	Díaz-Prieto et al. (2022)

($p<0.0076$) over the three-month period in patients given moringa, whereas it remained relatively stable in the placebo group, changing from 7.38±0.60% to 7.36±0.59% ($p<0.088$) (Veeranan et al. 2011). A randomised placebo-controlled study evaluated glucose control in 32 treatment-naive T2D patients following a four-week treatment with moringa (4 g twice/day) or a placebo. Both treatment groups showed a negligible decrease of 0.2 to 0.3% in HbA1c relative to the baseline, suggesting that monitoring blood glucose may have offered feedback that helped them make lifestyle changes, ultimately improving glycaemic control (Taweerutchana et al. 2017). In a study carried out by Leone and co-authors, 45 participants with T2D undergoing treatment with oral hypoglycaemic medications were administered 10 g of moringa leaves per day. In the experimental groups, the authors observed a significant decrease in glycated haemoglobin levels (up to −0.59%, $p<0.001$) and body fat (up to −1.5%, $p<0.001$) after three months of treatment, whereas the no intervention group showed no changes in these metrics (Leone et al. 2025). Gómez-Martínez and co-investigators examined the hypoglycaemic effect of moringa leaves on prediabetic subjects, revealing that 58% of participants in the treated group and 38% in the placebo group showed improved HbA1c levels relative to baseline after 12 weeks of intervention. The authors suggest that the marginally significant effects on HbA1c and glycaemia observed between the moringa and placebo groups may be attributed to the limited amounts of bioactive compounds present in the supplied dosage (2.4 g/day) (Gómez-Martínez et al. 2022). The researchers further evaluated the effect of moringa leaves on blood and faecal inflammatory markers, antioxidant capacity, blood pressure and lipid profile of prediabetic subjects but found no significant effects, which the authors attributed to the dosage and individual differences in the baseline biomarker values (Díaz-Prieto et al. 2022). Most clinical trials have shown that moringa has several benefits for diabetic patients. However, these studies are constrained by several factors, including small sample sizes, few trials, short durations and a lack of consideration for factors such as dietary control, disease span, moringa's phytochemical composition and comorbidities; thus, more comprehensive studies are required to fully elucidate moringa's effects on patients with type 2 diabetes.

6.5 SAFETY CONSIDERATIONS, CHALLENGES AND FUTURE PROSPECTS

Based on the results of these studies, moringa may have a positive impact on diabetes; nevertheless, additional preclinical and clinical studies are necessary to thoroughly examine its safety and effective doses before it can be adopted for glycaemic control management. Standardised methods are needed for extract preparation, including plant age at harvest, phytochemical composition, dosage and intervention duration. Results from observational studies and available clinical trials on diabetes have shown that moringa did not cause adverse effects on renal and hepatic functions after its consumption. However, Taweerutchana et al. (2017) reported that 25% of subjects who received moringa reported diarrhoea that was resolved within a few days. Leone et al. (2018) also reported a decrease in ratings of taste and colour after consumption of a moringa-supplemented meal among diabetic patients. A diabetic patient was also shown to exhibit Stevens–Johnson syndrome 14 hours after eating a meal supplemented with moringa leaves (Witharana, Wijetunga and Wijesinghe 2018). Consequently, further research is vital to identify the safety and effectiveness of utilising moringa as an antidiabetic treatment in humans.

REFERENCES

Abd El Latif, Amira, Badr El Said El Bialy, Hamada Dahi Mahboub, and Mabrouk Attia Abd Eldaim. 2014. "Moringa Oleifera Leaf Extract Ameliorates Alloxan-Induced Diabetes in Rats by Regeneration of β Cells and Reduction of Pyruvate Carboxylase Expression." *Biochemistry and Cell Biology* 92 (5): 413–19. doi:10.1139/bcb-2014-0081.

Adepoju, Feyisayo O., and Irina S. Selezneva. 2024. "Production and Evaluation of Set-Type Yoghurt Fortified with Chlorella Vulgaris and Moringa Oleifera Powder." *Plant Foods for Human Nutrition* 79 (2): 531–38. doi:10.1007/S11130-024-01194-7.

Adepoju, Feyisayo O., Irina S. Selezneva, and Charles O. R. Okpala. 2024. "Yoghurt Fortification with Moringa Oleifera: Nutritional and Production Aspects." *Mljekarstvo* 74 (1): 3–21. doi:10.15567/mljekarstvo.2024.0101.

Ahmad, Shadab, Alka Raj Pandey, Amit K. Rai, Suriya P. Singh, Pawan Kumar, Sushmita Singh, Farah Gulzar, Ishbal Ahmad, SashidharaKoneni V. , and Akhilesh K. Tamrakar. 2023. "Moringa Oleifera Impedes Protein Glycation and Exerts Reno-Protective Effects in Streptozotocin-Induced Diabetic Rats." *Journal of Ethnopharmacology* 305: 116117. doi:10.1016/j.jep.2022.116117.

Aljazzaf, Badriyah, Sassia Regeai, Sana Elghmasi, Nadia Alghazir, Amal Balgasim, Ismail M. Hdud Ismail, Areej A. Eskandrani, et al. 2023. "Evaluation of Antidiabetic Effect of Combined Leaf and Seed Extracts of Moringa Oleifera (Moringaceae) on Alloxan-Induced Diabetes in Mice: A Biochemical and Histological Study." *Oxidative Medicine and Cellular Longevity* 2023 (1): 9136217. doi:10.1155/2023/9136217.

Al-Malki, Abdulrahman L., and Haddad A. El Rabey. 2015. "The Antidiabetic Effect of Low Doses of Moringa Oleifera Lam. Seeds on Streptozotocin Induced Diabetes and Diabetic Nephropathy in Male Rats." *BioMed Research International* 2015 (1): 381040. doi:10.1155/2015/381040.

Anwer, Tarique, Mohammed M. Safhi, Hafiz A. Makeen, Saeed Alshahrani, Rahimullah Siddiqui, S. M. Sivakumar, Emad S. Shaheen, and Mohammad Firoz Alam. 2021. "Antidiabetic Potential of Moringa Oleifera Lam. Leaf Extract in Type 2 Diabetic Rats, and Its Mechanism of Action." *Tropical Journal of Pharmaceutical Research* 20 (2): 97–104. doi:10.4314/tjpr.v20i1.15.

Arise, Rotimi Olusanya, Oluwaseun Ruth Aburo, Samuel Tobi Farohunbi, and Adenike Adebola Adewale. 2016. "Antidiabetic and Antioxidant Activities of Ethanolic Extract of Dried Flowers of Moringa Oleifera in Streptozotocin-Induced Diabetic Rats." *Acta Facultatis Medicae Naissensis* 33 (4): 259–72. doi:10.1515/afmnai-2016-0028.

Attakpa, E. S., M. M. Sangaré, G. J. Béhanzin, J. M. Ategbo, B. Seri, and N. A. Khan. 2017. "Moringa Olifeira Lam. Stimulates Activation of the Insulin-Dependent Akt Pathway. Antidiabetic Effect in a Diet-Induced Obesity (DIO) Mouse Model." *Folia Biologica (Czech Republic)* 63 (2): 42. doi:10.14712/fb2017063020042.

Bamagous, Ghazi A., Saeed S. Al Ghamdi, Ibrahim Abdel Aziz Ibrahim, Amal M. Mahfoz, Mohamed A. Afify, Mahdi H.M. Alsugoor, Ahmed Ali Shammah, Palanisamy Arulselvan, and Thamaraiselvan Rengarajan. 2018. "Antidiabetic and Antioxidant Activity of Ethyl Acetate Extract Fraction of Moringa Oleifera Leaves in Streptozotocin-Induced Diabetes Rats via Inhibition of Inflammatory Mediators." *Asian Pacific Journal of Tropical Biomedicine* 8 (6): 320–27. doi:10.4103/2221-1691.235327.

Bao, Yifan, Jianbo Xiao, Zebin Weng, Xinyi Lu, Xinchun Shen, and Fang Wang. 2020. "A Phenolic Glycoside from Moringa Oleifera Lam. Improves the Carbohydrate and Lipid Metabolisms through AMPK in Db/Db Mice." *Food Chemistry* 311: 125948. doi:10.1016/j.foodchem.2019.125948.

Chigurupati, Sridevi, Atheer Al-murikhy, Suliman A. Almahmoud, Yosif Almoshari, Amira Saber Ahmed, Shantini Vijayabalan, Shatha Ghazi Felemban, and Vasanth Raj Palanimuthu. 2022. "Molecular Docking of Phenolic Compounds and Screening of Antioxidant and Antidiabetic Potential of Moringa Oleifera Ethanolic Leaves Extract from Qassim Region, Saudi Arabia." *Saudi Journal of Biological Sciences* 29 (2): 854–59. doi:10.1016/j.sjbs.2021.10.021.

Díaz-Prieto, Ligia E., Sonia Gómez-Martínez, Iván Vicente-Castro, Carlos Heredia, Elena A. González-Romero, María Del Carmen Martín-Ridaura, Mercedes Ceinos, María J. Picón, Ascensión Marcos, and Esther Nova. 2022. "Effects of Moringa Oleifera Lam. Supplementation on Inflammatory and Cardiometabolic Markers in Subjects with Prediabetes." *Nutrients* 14 (9): 1937. doi:10.3390/nu14091937.

El-Shehawi, Ahmed M., Mohamed Alkafafy, Samir El-Shazly, Samy Sayed, Samy Farouk, Saqer Alotaibi, Doaa A. Madkour, Hanem K. Khalifa, and Mohamed M. Ahmed. 2021. "Moringa Oleifera Leaves Ethanolic Extract Ameliorates High Fat Diet-Induced Obesity in Rats." *Journal of King Saud University - Science* 33 (6): 101552. doi:10.1016/j.jksus.2021.101552.

FAO. 2020. "Moringa | FAO | Food and Agriculture Organization of the United Nations." http://www.fao.org/traditional-crops/moringa/en/.

Gómez-Martínez, Sonia, Ligia E. Díaz-Prieto, Iván Vicente Castro, César Jurado, Nerea Iturmendi, Maria Carmen Martín-Ridaura, Nuria Calle, et al. 2022. "Moringa Oleifera Leaf Supplementation as a Glycemic Control Strategy in Subjects with Prediabetes." *Nutrients* 14 (1): 57. doi:10.3390/nu14010057.

Gopalakrishnan, Lakshmipriya, Kruthi Doriya, and Devarai Santhosh Kumar. 2016. "Moringa Oleifera: A Review on Nutritive Importance and Its Medicinal Application." *Food Science and Human Wellness* 5 (2): 49–56. doi:10.1016/j.fshw.2016.04.001.

Gupta, Rajnish, Manas Mathur, Vijay K. Bajaj, Pawan Katariya, Sunita Yadav, Raka Kamal, and Radhey S. Gupta. 2012. "Evaluation of Antidiabetic and Antioxidant Activity of Moringa Oleifera in Experimental Diabetes." *Journal of Diabetes* 4 (2): 164–71. doi:10.1111/j.1753-0407.2011.00173.x.

Hameed, Mishal, Alok Bharadwaj, Mamoona Mumtaz, Mumtaz Hussain, Mayank Goyal, Nida Fatima, Ashima Juyal, et al. 2023. "Evaluating the Effectiveness of Moringa Oleifera Leaf Capsules in Controlling Glycemic and Hypertension Levels in Type 2 Diabetes Patients." *Pakistan Journal of Pharmaceutical Sciences* 36 (4): 1343–47. doi:10.36721/PJPS.2023.36.4.SP.1343-1347.1

Hong, Zishan, Jing Xie, Huili Hu, Yuying Bai, Xia Hu, Tingting Li, Jinlian Chen, Jun Sheng, and Yang Tian. 2023. "Hypoglycemic Effect of Moringa Oleifera Leaf Extract and Its Mechanism Prediction Based on Network Pharmacology." *Journal of Future Foods* 3 (4): 383–91. doi:10.1016/j.jfutfo.2023.03.009.

Horn, Lydia, Natalia Shakela, Marius K. Mutorwa, Erold Naomab, and Habauka M. Kwaambwa. 2022. "Moringa Oleifera as a Sustainable Climate-Smart Solution to Nutrition, Disease Prevention, and Water Treatment Challenges: A Review." *Journal of Agriculture and Food Research* 10: 100397. doi:10.1016/J.JAFR.2022.100397.

International Diabetes Federation. 2021. *IDF Diabetes Atlas 2021 _ IDF Diabetes Atlas. IDF Official Website.*

Jaja-Chimedza, Asha, Li Zhang, Khea Wolff, Brittany L. Graf, Peter Kuhn, Kristin Moskal, Richard Carmouche, Susan Newman, J. Michael Salbaum, and Ilya Raskin. 2018. "A Dietary Isothiocyanate-Enriched Moringa (Moringa Oleifera) Seed Extract Improves Glucose Tolerance in a High-Fat-Diet Mouse Model and Modulates the Gut Microbiome." *Journal of Functional Foods* 47: 376–85. doi:10.1016/j.jff.2018.05.056.

Jiang, X, J Tang, Z Gao, C Chen, W Li, … L Peng. 2024. "Exploring the Protection on Islet β-Cells and Hypoglycemic Effect of Moringa Oleifera Stem Based on Network Pharmacology and Experimental Validation." *Journal of Functional Foods* 122: 106499. doi:10.1016/j.jff.2024.106499.

Jimoh, Tajudeen O. 2018. "Enzymes Inhibitory and Radical Scavenging Potentials of Two Selected Tropical Vegetable (Moringa Oleifera and Telfairia Occidentalis) Leaves Relevant to Type 2 Diabetes Mellitus." *Revista Brasileira de Farmacognosia* 28 (1): 73–79. doi:10.1016/j.bjp.2017.04.003.

Joung, Hyunchae, Bobae Kim, Hyunjoon Park, Kyuyeon Lee, Hee Hoon Kim, Ho Cheol Sim, Hyun Jin Do, Chang Kee Hyun, and Myoung Sool Do. 2017. "Fermented Moringa Oleifera Decreases Hepatic Adiposity and Ameliorates Glucose Intolerance in High-Fat Diet-Induced Obese Mice." *Journal of Medicinal Food* 20 (5): 439–47. doi:10.1089/jmf.2016.3860.

Kamrul-Hasan, A. B. M., Samir Kumar Talukder, Lakshmi Nagendra, Muhammad Shah Alam, Fatema Tuz Zahura Aalpona, Deep Dutta, and Shahjada Selim. 2023. "Effect of Moringa Oleifera Leaf Extract on Glycemic Parameters in Patients with Type 2 Diabetes Mellitus and Prediabetes: A Systematic Review and Meta-Analysis." *Bangladesh Journal of Endocrinology and Metabolism* 2 (3): 134–45. doi:10.4103/bjem.bjem_26_23.

Kaneto, Hideaki, Naoto Katakami, Dan Kawamori, Takeshi Miyatsuka, Ken'ya Sakamoto, Taka Aki Matsuoka, Munehide Matsuhisa, and Yoshimitsu Yamasaki. 2007. "Involvement of Oxidative Stress in the Pathogenesis of Diabetes." *Antioxidants and Redox Signaling* 9 (3): 355–66. doi:10.1089/ars.2006.1465.

Khan, Washim, Rabea Parveen, Karishma Chester, Shabana Parveen, and Sayeed Ahmad. 2017. "Hypoglycemic Potential of Aqueous Extract of Moringa Oleifera Leaf and in Vivo GC-MS Metabolomics." *Frontiers in Pharmacology* 8: 577. doi:10.3389/fphar.2017.00577.

Leone, A., Simona Bertoli, Sara Di Lello, Angela Bassoli, Stefano Ravasenghi, Gigliola Borgonovo, Fabio Forlani, and Alberto Battezzati. 2018. "Effect of Moringa Oleifera Leaf Powder on Postprandial Blood Glucose Response: In Vivo Study on Saharawi People Living in Refugee Camps." *Nutrients* 10 (10): 1494. doi:10.3390/nu10101494.

Leone, A., S. Di Lello, S. Bertoli, S. Ravasenghi, R. De Amicis, F. Menichetti, G. Fico, et al. 2025. "Moringa Oleifera Leaf Powder Enhances Glycemic Control in Sahrawi Women with Type 2 Diabetes: Findings from a 3-Month Unblinded Randomized Controlled Trial." *PharmaNutrition* 31: 100434. doi:10.1016/j.phanu.2025.100434.

Magaji, U. F., O. Sacan, and R. Yanardag. 2020. "Alpha Amylase, Alpha Glucosidase and Glycation Inhibitory Activity of Moringa Oleifera Extracts." *South African Journal of Botany* 128: 225–30. doi:10.1016/j.sajb.2019.11.024.

Mozo, Ranier Nery, and Imelda Caole-Ang. 2015. "The Effects of Malunggay (Moringa Oleifera) Leaves Capsule Supplements on High Specificity C-Reactive Protein and Hemoglobin A1c Levels of Diabetic Patients in Ospital Ng Maynila Medical Center: A Prospective Cohort Study." *Philippine Journal of Internal Medicine* 53 (4): 1–10.

Muzumbukilwa, Willy Tambwe, Manimbulu Nlooto, and Peter Mark Oroma Owira. 2019. "Hepatoprotective Effects of Moringa Oleifera Lam (Moringaceae) Leaf Extracts in Streptozotocin-Induced Diabetes in Rats." *Journal of Functional Foods* 57: 75–82. doi:10.1016/j.jff.2019.03.050.

Natsir, Hasnah, Rugaiyah Arfah, Abdur Rahman Arif, Muhammad Nadir, Nur Rahmi, and Aulia Karimah. 2024. "In Vitro and In Silico Assessment of Methanol Extract from Moringa Oleifera Seeds as α-Amylase Inhibitor." *Indonesian Journal of Chemical Research* 12 (2): 79–88. doi:10.30598/ijcr.2024.12-has.

Ogundipe, Adebanke, Babatunde Adetuyi, Franklyn Iheagwam, Keleko Adefoyeke, Joseph Olugbuyiro, Oluseyi Ogunlana, and Olubanke Ogunlana. 2022. "In Vitro Experimental Assessment of Ethanolic Extract of Moringa Oleifera Leaves as an α -Amylase and α -Lipase Inhibitor." *Biochemistry Research International* 2022 (1): 4613109. doi:10.1155/2022/4613109.

Olasehinde, Oluwaseun Ruth, Olakunle Bamikole Afolabi, Benjamin Olusola Omiyale, and Oyindamola Adeniyi Olaoye. 2021. "In Vitro Inhibitory Potentials of Ethanolic Extract of Moringa Oleifera Flower against Enzymes Activities Linked to Diabetes." *Journal of HerbMed Pharmacology* 10 (4): 408–14. doi:10.34172/jhp.2021.48.

Olayaki, Luqman A., Justice E. Irekpita, Musa T. Yakubu, and Opeolu O. Ojo. 2015. "Methanolic Extract of Moringa Oleifera Leaves Improves Glucose Tolerance, Glycogen Synthesis and Lipid Metabolism in Alloxan-Induced Diabetic Rats." *Journal of Basic and Clinical Physiology and Pharmacology* 26 (6): 585–93. doi:10.1515/jbcpp-2014-0129.

Paikra, Birendra Kumar, Hemant Kumar, J. Dhongade, and Bina Gidwani. 2017. "Phytochemistry and Pharmacology of Moringa Oleifera Lam." *Journal of Pharmacopuncture* 20 (3): 194. doi:10.3831/KPI.2017.20.022.

Shori, Amal Bakr. 2015. "Screening of Antidiabetic and Antioxidant Activities of Medicinal Plants." *Journal of Integrative Medicine* 13 (5): 297–305. doi:10.1016/S2095-4964(15)60193-5.

Su, Xinyue, Guanzheng Lu, Liang Ye, Ruyu Shi, Maomao Zhu, Xinming Yu, Zhiyong Li, Xiaobin Jia, and Liang Feng. 2023. "Moringa Oleifera Lam.: A Comprehensive Review on Active Components, Health Benefits and Application." *RSC Advances* 13 (35): 24353–84. doi:10.1039/d3ra03584k.

Świderska, Ewa, Justyna Strycharz, Adam Wróblewski, Janusz Szemraj, Józef Drzewoski, and Agnieszka Śliwińska. 2020. "Role of PI3K/AKT Pathway in Insulin-Mediated Glucose Uptake." In *Blood Glucose Levels* 37. doi:10.5772/intechopen.80402.

Tang, Yujiao, Eun Ju Choi, Weon Cheol Han, Mirae Oh, Jin Kim, Ji Young Hwang, Pyo Jam Park, Sang Ho Moon, Yon Suk Kim, and Eun Kyung Kim. 2017. "Moringa Oleifera from Cambodia Ameliorates Oxidative Stress, Hyperglycemia, and Kidney Dysfunction in Type 2 Diabetic Mice." *Journal of Medicinal Food* 20 (5): 502–10. doi:10.1089/jmf.2016.3792.

Taweerutchana, Rutchaporn, Natchagorn Lumlerdkij, Sathit Vannasaeng, Pravit Akarasereenont, and Apiradee Sriwijitkamol. 2017. "Effect of Moringa Oleifera Leaf Capsules on Glycemic Control in Therapy-Naïve Type 2 Diabetes Patients: A Randomized Placebo Controlled Study." *Evidence-Based Complementary and Alternative Medicine* 2017 (1): 6581390. doi:10.1155/2017/6581390.

Tuorkey, Muobarak J. 2016. "Effects of Moringa Oleifera Aqueous Leaf Extract in Alloxan Induced Diabetic Mice." *Interventional Medicine and Applied Science* 8 (3): 109–17. doi:10.1556/1646.8.2016.3.7.

Veeranan, V., Arun Giridhari, D. Malathi, and K. Geetha. 2011. "Anti Diabetic Property of Drumstick (Moringa Oleifera) Leaf Tablets." *International Journal of Health and Nutrition* 2 (1): 1–5.

Waterman, Carrie, Patricio Rojas, Tugba Tumer, Peter Kuhn, Allison Richard, Shawna Wicks, Jacqueline Stephens, et al. 2015. "Isothiocyanate-Rich Moringa Oleifera Extract Reduces Weight Gain, Insulin Resistance and Hepatic Gluconeogenesis in Mice." *Molecular Nutrition and Food Research* 59 (6): 1013–24. doi:10.1002/mnfr.201400679

Wen, Ying, Yanyang Liu, Qi Huang, Rong Liu, Jing Liu, Fengyu Zhang, Shao Liu, and Yueping Jiang. 2022. "Moringa Oleifera Lam. Seed Extract Protects Kidney Function in Rats with Diabetic Nephropathy by Increasing GSK-3β Activity and Activating the Nrf2/HO-1 Pathway." *Phytomedicine* 95: 153856. doi:10.1016/j.phymed.2021.153856.

Witharana, E., W. Wijetunga, and S. Wijesinghe. 2018. "Stevens-Johnson Syndrome (SJS) Following Murunga Leaf (Moringa Oleifera) Consumption." *Ceylon Medical Journal*, no. 63 (4): 188–89. doi:10.4038/cmj.v63i4.8771.

World Health Organization (WHO). 2013. *WHO Traditional Medicine Strategy 2014–2023*. World Health Organization (WHO).

Yassa, Hanan Dawood, and Adel Fathy Tohamy. 2014. "Extract of Moringa Oleifera Leaves Ameliorates Streptozotocin-Induced Diabetes Mellitus in Adult Rats." *Acta Histochemica* 116 (5): 844–54. doi:10.1016/j.acthis.2014.02.002.

Yong-Bing, Xu, Chen Gui-Lin, and Guo Ming-Quan. 2019. "Antioxidant and Anti-Inflammatory Activities of the Crude Extracts of Moringa Oleifera from Kenya and Their Correlations with Flavonoids." *Antioxidants* 8 (8): 296. doi:10.3390/antiox8080296.

Zaib, Sumera, Imtiaz Khan, Laiba Babar, Nawal Naseer, Danial Nadeem, Ravish Fatima, DeraAyed A. , Fadi Jaber, and K. K. Mujeeb Rahman. 2025. "Isolation and Characterization of an Insulin-Like Protein From Moringa Oleifera and Estimation of Its Antidiabetic Potential." *ChemistrySelect* 10 (6): 202401991. doi:10.1002/SLCT.202401991.

7 *Moringa oleifera* in Traditional Systems of Medicine

Sunday Ene-Ojo Atawodi, Jonathan Ilemona Achika, Emmanuel Chukwudalu Ohaekenyem and Joy Cecilia Atawodi

7.1 INTRODUCTION

Moringa oleifera Lam., commonly known as the 'miracle tree', 'drumstick tree', 'horseradish', 'moringa' and 'benzolive' (Atawodi et al., 2010; USDA, 2023; El Bilali et al., 2024), has been revered for centuries across diverse cultures for its remarkable therapeutic and nutritional properties (Ritu Paliwal et al., 2011). Belonging to the class Magnoliopsida, division Magnoliophyta, order Capparales and family Moringaceae, it is a fast-growing, drought-resistant, deciduous tree that is native to the Indian sub-continent, but now cultivated globally in tropical and subtropical regions. First described in 1785 by Jean-Baptiste Lamarck, a French naturalist (El Bilali et al., 2024), it is now naturalised in other parts of the world, where it has found widespread use. At present, it is widely distributed in Africa, the Mediterranean, America, the West Indies, the Philippines, China, Southeast Asia, and Arabia, where the family comprises 13 additional tree and shrub species (Olson, 2023).

In 2000 BCE, *M. oleifera* was recognised as a medicinal herb in the sub-Himalayan areas of India, Pakistan, Bangladesh and Afghanistan. Its extensive use in traditional systems of medicine, including Ayurveda, Siddha, Unani and traditional African and Chinese medicine, underscores its significance as a versatile medicinal plant.

Following the ancient texts *Charaka Samhita* and *Sushruta Samhita*, *M. oleifera* is classified as a *Rasayana* in the Ayurvedic system of medicine, where it is credited with therapeutic efficacies against over 300 ailments. Similarly, the significance of *M. oleifera* in other traditional systems of medicine, such as Unani, traditional African medicine (TAM), South Asian traditional medicine, Latin American systems of medicine and the Caribbean system of traditional medicine, is monumental (Sonewane et al., 2022; Pareek et al., 2023). Modern ethnopharmacological analyses have attributed these therapeutic efficacies to a variety of phytochemicals, including alkaloids, glucosinolates, isothiocyanates, phenolic acids and flavonoids like quercetin, which are responsible for numerous pharmacological activities such as antioxidant, antimicrobial and anti-inflammatory activities.

7.2 *MORINGA OLEIFERA* IN AYURVEDIC MEDICINE

7.2.1 The Historical Background of *Moringa oleifera* in Ayurveda

Moringa oleifera, often referred to as *Shigru* or *Sahijan* in Sanskrit, has been one of the oldest constituents of Ayurveda, the most ancient holistic healing system known to mankind. Moringa enjoys a reputation as a rejuvenator (*Rasayana*) and a premium plant in ethnopharmacology, with many therapeutic effects (Govindarajan et al., 2005).

DOI: 10.1201/9781003617860-8

In the *Charaka Samhita* and *Sushruta Samhita* texts, from the 2nd century BCE, moringa is considered with respect to taste and potency, while the post-digestive effect is defined in Ayurvedic pharmacology with regard to the pungent herb (*Katu*) and bitter taste (*Tikta*), with heat-inducing potency (*Ushna Virya*) and pungent post-digestive effect (*Katu Vipaka*). In addition, it addresses various health conditions arising from toxic substances (*Ama*) or imbalances, as in the case of obesity, diabetes and diseases related to the respiratory system, such as asthma and bronchitis, where the anti-inflammatory property mediates the clearing of airways and general reduction in symptoms (Total Ayurveda, 2024). Moringa enhances health in various ways, including the management of joint pain and neurological disorders of aggravated *Vata dosha*, indicating how it maintains therapeutic importance with traditional medicine (Total Ayurveda, 2024). According to Meireles et al. (2020), various preparations such as powders, decoctions and medicated oils from the leaves, bark, flowers, seeds and roots of moringa have been effectively used in Ayurvedic medicine.

Moringa has innate anti-inflammatory and antibacterial properties that stand out as the 'cure-all' in the healing of wounds and management of anti-infectious agents, as attested to in *Sushruta Samhita*, a basic text in Ayurvedic surgery (Li et al., 2023). The other classical text, *Ashtanga Hridaya*, which draws its knowledge mainly from *Charaka* and *Sushruta Samhitas*, mentions the detoxifying effects of moringa in the blood and liver (Jhofi et al., 2021). It has a predominantly bitter and pungent taste, contributing to its detoxifying and metabolism-stimulating properties. Its heating potency enhances circulation and digestion, making it particularly beneficial for cold-related ailments (Sayed et al., 2022).

7.2.2 Traditional Therapeutic Uses of *Moringa oleifera* Parts in Ayurveda

Moringa oleifera remains a cornerstone of Ayurvedic medicine, valued for its detoxifying, anti-inflammatory and digestion-enhancing properties. Whether used in herbal decoctions, powders or topical applications, *Shigru* continues to bridge the gap between ancient Ayurvedic wisdom and modern scientific research, reinforcing its relevance in holistic health and wellness (Jayachandran et al., 2021). Thus, *M. oleifera* exhibits a wide array of therapeutic actions that have been recognised in both traditional and modern medicine (Table 7.1). Its rich composition of bioactive compounds contributes to its effectiveness in managing various health conditions (Chiş et al., 2023). No wonder, therefore, that *M. oleifera* has been deeply integrated into Ayurvedic medicine for its diverse healing properties. Various parts of the plant – including the leaves, seeds, bark and roots – are utilised in traditional remedies to address a wide range of health concerns. Each component of the plant uniquely contributes to restoring balance within the body, making it a staple in holistic healing practices.

Moringa seeds play a crucial role in stimulating digestion (*Deepana*) and enhancing metabolic function and anti-inflammatory effects (*Shothahara*). Ayurveda texts suggest that consuming moringa leaf juice or incorporating dried leaf powder into the diet will enhance red blood cell production and combat fatigue (Jhofi et al., 2021). There are several reports of Ayurvedic practitioners using moringa either as teas, dietary supplements, decoctions, powders or pastes/ointments as nutritional supplements, and to treat diabetes, headaches, joint pain, nervous complications and inflammatory conditions like arthritis, utilising its *dosha* balancing properties (Li et al., 2023; Gautam, 2023).

7.2.3 Ayurvedic Formulations Containing Moringa

Moringa oleifera has become one of the most versatile and widely sought-after plants in holistic medicine, being part of most Ayurvedic preparations (Table 7.2), where it often functions synergistically with other herbs, either as simple powders, decoctions or quite intense preparations like pastes and herbal oils. Some traditional herbal preparations containing moringa are described by Gautam (2023):

TABLE 7.1
Ethnomedicinal Uses of *M. oleifera* in Ayurveda, Their Scientific Validation and Their Mechanisms of Action

S/N	Disease/ condition	Part used	Method of preparation	Traditional dose	Ayurvedic classification	Validation by modern laboratory evaluation	Phytochemical responsible	Mechanism of action	References
1.	Detoxification, liver support	Leaves	Herbal decoction, juice, dried powder	1–2 tsp dried powder or 1 cup decoction daily	*Tikta* (bitter) and *Katu* (pungent) *Rasa*	Validated for hepatoprotective and detoxifying properties	Flavonoids, polyphenols	Enhances liver enzyme activity and detox pathways	Leone et al., 2015; Gajender et al., 2023
2.	Skin health (acne, eczema, boils)	Leaves	Consumed as a decoction or applied as a fresh paste	Topical paste applied 1–2 times daily; oral decoction	*Raktashodhak* (blood purifier)	Antimicrobial and anti-inflammatory effects confirmed	Quercetin, tannins	Reduces oxidative stress, inhibits bacterial growth	Leone et al., 2015; Gajender et al., 2023
3.	Anaemia	Leaves	Consumed as juice or dried leaf powder in the diet	2–3 tsp dried powder daily	*Rakta-Vardhak* (blood enhancer)	Validated for its high iron content and haematopoietic effects	Iron, vitamin C	Increases haemoglobin and RBC production	Jhofi et al., 2021
4.	Antimicrobial, water purification	Seeds	Crushed and infused in water	1–2 crushed seeds per litre of water	*Jantughna* (antimicrobial)	Confirmed antibacterial properties	Isothiocyanates	Disrupts bacterial cell walls, inhibits growth	Sayed et al., 2022
5.	Digestive issues, intestinal worms	Seeds	Consumed in small quantities, infused in warm water	1–2 seeds daily	*Krimighna* (anthelmintic)	Validated for antimicrobial and anthelmintic properties	Alkaloids, saponins	Paralyses worms, disrupts gut flora	Sayed et al., 2022
6.	Pain relief, fever reduction	Bark	Decoction taken orally	1 cup decoction 2 times daily	*Vedanasthapana* (analgesic)	Validated for antipyretic and analgesic effects	Alkaloids, tannins	Inhibits inflammatory mediators	Jamil, 2024

7.	Digestive aid, indigestion, colic pain	Roots	Small amount of extract mixed with honey or warm water	3–5 drops root extract	*Deepana* (digestive stimulant)	Validated for digestive stimulation and carminative effects	Alkaloids, saponins	Stimulates digestive enzymes, reduces bloating	Khattab et al., 2021
8.	Metabolic disorders (diabetes)	Leaves	Extracts used as supplements	500 mg extract daily	Medohara (fat and sugar regulator)	Validated for antidiabetic effects	Flavonoids, polyphenols	Enhances insulin sensitivity, reduces glucose absorption	

TABLE 7.2
Traditional Ayurvedic Formulations of *M. oleifera*, Their Ethnomedicinal Uses and Their Preparations

S/N	Ayurvedic formulations containing moringa	Traditional herbal mixtures	Disease/condition	Part used	Method of preparation	Validation by modern laboratory evaluation	Phytochemical responsible for medicinal effects	References
1.	Shigru Churna (moringa powder mixed with honey or water for daily consumption)	Moringa + turmeric + black pepper	Detoxification, liver support	Leaves	Herbal decoction, moringa powder	Validated: studies confirm antioxidant and liver-protective effects (Li et al., 2023)	Flavonoids, polyphenols, quercetin, kaempferol	Leone et al., 2015; Gajender et al., 2023
2.	Lepa (moringa paste) (crushed leaves mixed with turmeric and sandalwood for skin healing)	Moringa + sandalwood + neem	Skin health (acne, eczema, wound healing)	Leaves	Paste applied topically or consumed as a decoction	Validated: antimicrobial and anti-inflammatory effects confirmed (Li et al., 2023)	Alkaloids, saponins, flavonoids	Leone et al., 2015; Gajender et al., 2023
3.	Shigru Churna (moringa powder taken to improve iron levels)	Moringa + amla + ashwagandha	Anaemia	Leaves	Dried leaf powder mixed with food or taken with warm water	Validated: high iron content supports red blood cell production (Jhofi et al., 2021)	Iron, vitamin C, chlorophyll	Jhofi et al., 2021
4.	Shigru Kwath (decoction) (moringa seeds boiled for purification and antimicrobial benefits)	Moringa + neem + clove	Antimicrobial, intestinal worms	Seeds	Crushed and infused in warm water	Validated: antimicrobial properties against bacteria and parasites (Sayed et al., 2022)	Isothiocyanates, benzylamine	Sayed et al., 2022

5.	Shigru Kwath (decoction) (moringa decoction used for digestion and detoxification)	Moringa + ginger + liquorice	Digestive health, bloating	Seeds	Infusion in warm water, consumed orally	Validated: supports gut microbiota and digestion (Sayed et al., 2022)	Saponins, flavonoids	Sayed et al., 2022
6.	Shigru Kwath (decoction) (moringa bark boiled for pain relief)	Moringa + boswellia + turmeric	Pain relief, arthritis	Bark	Decoction from boiled bark	Validated: anti-inflammatory and analgesic effects (Khattab et al., 2021)	Tannins, alkaloids, terpenoids	Khattab et al., 2021
7.	Shigru Kwath (decoction used for fever relief)	Moringa + tulsi + guduchi	Fever reduction	Bark, Leaves	Herbal decoction	Validated: antipyretic effects confirmed (Jamil et al., 2024)	Flavonoids, polyphenols, terpenes	Jamil, 2024
8.	Shigru Churna (root powder taken for digestion)	Moringa + ajwain + fennel	Digestive stimulation, gas relief	Roots	Root extract mixed with honey or warm water	Validated: improves digestion and relieves bloating (Khattab et al., 2021)	Alkaloids, terpenoids	Khattab et al., 2021
9.	Shigru Taila (moringa oil) (used for massage and inflammation relief)	Moringa + sesame oil + castor oil	Muscle pain, inflammation	Seeds	Cold-pressed oil applied topically	Validated: anti-inflammatory and skin-healing effects (Gautam, 2023)	Oleic acid, sterols, polyphenols	Gautam, 2023
10.	Shigru Churna (used to regulate blood sugar and oxidative stress)		Antioxidant, antidiabetic	Leaves	Consumed as powder or herbal infusion	Validated: antioxidant and antidiabetic effects confirmed (Anjali, 2023; Kanyal, 2023)	Quercetin, chlorogenic acid, polyphenols	Anjali, 2023; Kanyal, 2023

- Moringa leaf powder (*Shigru Churna*): This formulation is used to improve digestion, boost immunity and support overall vitality. It is prepared by mixing moringa leaf powder with honey.
- Moringa decoction (*Shigru Kwath*): This is a herbal tea prepared by boiling moringa leaves or bark in water until the liquid is reduced to half. It is commonly used for detoxification, reducing joint pain and alleviating respiratory conditions.
- Moringa oil (*Shigru Taila*): Moringa seed oil is obtained through cold-pressing or infusion methods, and it is used in Ayurvedic massage therapy for relieving muscle pain, reducing inflammation, moisturising and improving skin elasticity.
- Moringa paste (*Lepa*): This is a thick paste made from crushed moringa leaves, mixed with turmeric and sandalwood. It is applied topically in the treatment of skin conditions, wound healing, inflammatory conditions and joint pain.

7.3 *MORINGA OLEIFERA* IN TRADITIONAL AFRICAN MEDICINE

Traditionally, moringa has been used in TAM to treat a wide range of illnesses, including helminthiasis, paralysis, skin infections, diabetes, cardiovascular disease and anaemia (Alia et al., 2022; Liu et al., 2022). The leaves, in the form of herbal teas or pastes, are used to treat a variety of ailments, such as malaria, typhoid fever, parasitic diseases, arthritis, swellings, cuts, skin diseases, genito-urinary diseases, hypertension and diabetes in TAM (Leone et al., 2015). It is also used to induce lactation and boost the immune system in HIV and AIDS-related symptoms (Anwar et al., 2007; Popoola and Obembe, 2013; Sivasankari et al., 2014), as well as cardiac stimulants and contraceptive remedies.

Numerous screening tests and ethnopharmacological investigations (Abe and Ohtani, 2013; Popoola and Obembe, 2013; Sivasankari et al., 2014) have validated the therapeutic efficacy of moringa against the aforementioned disorders. Therefore, this section provides details on the use of moringa parts in nutrition and the traditional management of communicable and non-communicable disorders in the African health system (Table 7.3).

7.3.1 *Moringa oleifera* in Traditional African Diets

More than half (57%) of African respondents from South Africa (76%), Nigeria (53%) and Ghana (45%) had never heard of or eaten moringa (Hedhili et al., 2022), probably due to disparities in food habits and cultural norms, and the fact that it was introduced to Africa only in the early 20th century. However, the growing body of scientific research on moringa (especially in Nigeria and South Africa) indicates that knowledge about this plant remains in the diffusion and exploitation stage (Fahey, 2005).

The leaves of moringa are the most important part of the plant that features in dietary recipes in Africa (Ntila et al., 2020). In a survey by Hedhili et al. (2022), African respondents mostly drank moringa as tea. A few Africans who only recently learned about moringa from medical experts expressed negative attitudes about the use of moringa, despite their belief in its nutritional and health benefits. This is not surprising, since it is very challenging to alter people's eating habits (Hedhili et al., 2022). Generally, like other vegetables, Africans, depending on geographical location, nationality and regional customs, have varied but low consumption rates for moringa, because of poor familiarity (Faber et al., 2010; Glover-Amengor, 2015; Olusanya, 2018) and the unfavourable perception that vegetables are food for the underprivileged or socially backwards (Mwangi and Kimathi, 2006).

Despite the low knowledge and interest of many Africans in the consumption or inclusion of moringa in their diets, research has shown that the leaves contain appreciable amounts of the provitamin A carotenoid called beta-carotene and have an exceptional ability to take up and accumulate minerals like selenium and sulphur, even when grown on soils deficient in these minerals. Thus,

TABLE 7.3
Phytochemistry and Therapeutic, Nutritional and Ethnomedicinal Uses of *M. oleifera* Reported in TAM

S/N	Disease	Plant part	Method of preparation	Country of report	Validation of reported ethnomedicinal use	Phytochemical responsible for observed medicinal effects	Reference
1.	Metabolic and infectious diseases	L	Leaves and root decoction	Benin	Antidiabetic	No report	Agoyi et al., 2015
2.	Urinary tract infections	L	Leaf dietary inclusion	Cameroon	Antidiabetic	Tannins, phenols, saponins, sulfamethoxazole, trimethoprim, penicillin and chloramphenicol	Ngemenya et al., 2024
3.	Cancer	L	Leaf dietary inclusion	Cameroon	Anti-inflammatory	Flavonoids, glycoside, saponins, tannins, phenol, steroids and alkaloids	Djiogue et al., 2022
4.	Malnutrition	L and S	Lead and seed powder	Egypt	Food fortification	Saturated and unsaturated fatty acids	Abdalla., et al. 2022
5.	Heart disease and erectile dysfunction	L	Aqueous leaf extract mixed with ginger, milk honey or porridge	Ghana	Antidiabetic	Proteins and vitamins	Asigbaase et al., 2024
6.	Malnutrition	L, S and F	Leaf powder	Guinea-Bissau	Antidiabetic	Tannins, lycopene, polyphenols and peptides	Bancessi et al., 2020
7.	Stomach ailments	B	Barks are boiled in water and soaked in alcohol	Namibia	Antiulcer	a1-(4′-hydroxybenzyl)-4,5-dimethyl-1,3-dihydro-2*H*-imidazole-2-thione	Horn et al , 2022
8.	Malnutrition	L	Leaf powder	Nigeria	Food fortification	No report	Famakinwa et al., 2023
9.	Infectious diseases	L	Leaf powder	Nigeria	Abtibacteria activity	Flavonoids, alkaloids and tannins	Enerijiof et al., 2021
10.	Metabolic diseases	L, S, P, F and FL	Leaf powder	Nigeria	Antitumor	Alkaloids, flavonoids, steroids, glycosides, gums, phenols, tannings, terpenes, terpenoids and saponins	Omokpariola et al., 2020

(Continued)

TABLE 7.3 (CONTINUED)
Phytochemistry and Therapeutic, Nutritional and Ethnomedicinal Uses of *M. oleifera* Reported in TAM

S/N	Disease	Plant part	Method of preparation	Country of report	Validation of reported ethnomedicinal use	Phytochemical responsible for observed medicinal effects	Reference
11.	Malnutrition	L and F	Cooked leaf	Nigeria	Nutritional evaluation	Thiamine, riboflavin and niacin	Zaku et al., 2015
12.	Metabolic diseases	S	Ground seed	Nigeria	Anti-haemorrhagic activity	Vitamins (e.g. B, C, K, beta-carotene/provitamin A)	El Bilali et al., 2024
13.	Bacterial infections	L	Leaf powder	Nigeria	Antibacterial activity	Alkaloids, flavonoids, steroids, terpenoids, phenols, glycosides, tannins and saponins	Abdulmalik et al., 2024
14.	Diabetes	L, ST, R and S	Leaf decoctions	Nigeria	Antidiabetic	Contains alkaloids, tannins, saponins, flavonoids, phenol, terpenes and steroids	Akoko et al., 2022
15.	Haemorrhoids. and rheumatism.	S	Cream of moringa seed oil	Nigeria	Wound-healing activities	Flavonoids, quercetin, kaempferol and glycosides	Isimi et al., 2021
16.	Flu and fever.	R	Root concoction .	Nigeria	Antimalaria	No report	Abubakar et al., 2022
17.	Cancer	L, R and SB	Alcoholic root, leans and stembark extract	Nigeria	Antioxidant	Chlorogenic acid, rutin, quercetin glucoside and kaempferol rhamnoglucoside	Atawodi et al., 2010
18.	Malnutrition iron deficiency,	L	Leaf dietary inclusion	Nigeria	Antioxidants	Ascorbic acid, tocopherols and carotenoids	Oyeyinka et al., 2018

19.	Food poisoning.	L	Aqueous leaf extract	Nigeria	Anti-inflammatory	Steroids, tannins, terpenoids, probationers and saponins are all present	Oladeji et al., 2020
20.	Malnutrition	L	Leaf powder	Nigeria.	Antidiabetic	Benzyl amine	Obunda et al., 2023
21.	Cancer	L	Leaf dietary inclusion	Nigeria	Anti-inflammatory	Contains alkaloids, phenol, glycoside, saponins, tannins, flavonoids and steroids	Usman et al., 2022
22.	Cardiovascular ailments	L and S	Leaf and seed extract	Nigeria.	Anti-hypertensive	Flavonoids, polyphenol, lycopene, beta-carotene, hydrogen peroxide and hydroxyl	Oyeleye et al., 2023
23.	Infection and metabolic diseases	L	Leaf powder	Nigeria	Antimicrobial	Saponin, cardiac glycosides, alkaloids, terpenoids and phlorotannins	Akpor et al., 2021
24.	Male fertility and female infertility	L, P and S	Aqueous leaf extract	South Africa	Anti-inflammatory and anticancer qualities.	Quercetin, kaempferol, niazimicin, sterols and flavonoids	Moremane et al., 2023
25.	Malnutrition	L	Leaf powder	South Africa	Food fortification	Alkaloids, tannins, saponins, flavonoids, phenol, terpenes and steroids	Sokhela et al., 2023
26.	Malnutrition	L	No report	South Africa	Nutritional evaluation	The dried leaves had crude protein levels of 30.3% and 19 amino acids	Moyo et al., 2011
27.	Cancer	L	Leaf powder	South Africa	Anticancer activity	Gallic acid, orientin, quercetin, kaempferol and catechin	Bopape et al., 2023
28.	Diabetes and anaemia	B and R	Leaf powder and boiled roots	South Africa	antiulcer	No report	Mudau et al., 2022
29.	Cancer	L	Laef capsule	Tanzania	Anticancer	Polyphenols, flavonoids, tannins and terpenoids are present in the leaf	Wanjiru et al., 2022

L, leaves; S, stem; R, root; RB, root bark; P, pod; F, fruit; FL, flower.

Moringa leaves have enormous potential to help address the deficiencies of these nutrients, which are common in sub-Saharan Africa, with the attendant health and medical implications (Hedhili et al., 2022; Lyons et al., 2017). Malnutrition has been effectively treated with moringa, particularly in children and nursing mothers (Tamilselvi et al., 2019), suggesting that moringa can be crucial to achieving the UN's Sustainable Development Goal 2, 'zero hunger'. Keatinge et al. (2017), especially since it can serve as an all-year-round, climatically adaptable and nutrient-dense food source (FAO, 2023; El Bilali et al., 2024).

7.3.2 Moringa and TAM

According to Pareek et al. (2023), people worldwide have included *M. oleifera* in their diets since ancient times due to its vital therapeutic value. It is reported to have ethnomedicinal properties, with almost every part (leaf, pod, bark, gum, flower, seed, seed oil and root) of the plant being used to treat diseases ranging from hypertension, anxiety and diarrhoea; it is also used as a diuretic and for preparing dermal ointments for the management of skin conditions in ancient Egypt.

In West Africa, particularly in countries like Nigeria, Ghana and Senegal, *M. oleifera* is commonly used to manage conditions such as hypertension, diabetes and malnutrition. The leaves are often consumed as a vegetable or brewed into teas to boost immunity and improve overall health (Mbikay, 2012). Traditional healers in these regions also prescribe moringa leaf extracts to alleviate symptoms of anaemia and fatigue due to their high iron and vitamin content (Fahey, 2005). Additionally, the seeds are used to purify water, addressing the dual challenges of waterborne diseases and limited access to clean water in rural areas (Ndabigengesere et al., 1995). It is also reported that taking 1.5 teaspoons of powdered Moringa leaf for 90 days increases blood antioxidant levels. This is not surprising since one cup of freshly sliced pods (100 g) contains 157% of the daily vitamin C requirement (Obunda et al., 2023). The leaves are cooked or dried over a screen for several days and ground into a powder that is added to almost any food as a nutrient supplement, such as pap and cereal drinks, to improve their nutritive value. The roots are boiled with red potash to treat hepatitis or yellow fever, meningitis, smallpox, monkeypox, COVID-19 and poliomyelitis, while three cups of sieved boiled leaves, taken orally, three times daily, are also considered a treatment for fever, hepatitis, COVID-19 and yellow fever. The boiled leaves (dried or fresh), barks or roots are also used in the treatment of diabetes, anaemia, body pain, rheumatism, diarrhoea, catarrh fever, hypertension and stomach pain (Abubakar et al., 2022).

Nigerians have been reported to use the leaves to treat HIV infections and cancer (Popoola and Obembe, 2013), but published works on these specific uses are rather scant. Barks are boiled in water or soaked in alcohol to obtain decoction and extract, respectively, that can be used to treat stomach ailments (ease stomach pain and ulcers and aid in digestion), poor vision, joint pain, diabetes, anaemia, hypertension (Abe and Ohtani, 2013; Popoola and Obembe, 2013), toothaches, haemorrhoids and uterine disorders (Popoola and Obembe, 2013).

Moringa oleifera is widely used to treat a number of ailments in East Africa, from the treatment of communicable diseases in Kenya and Tanzania to the use of the decoction of the roots to treat skin diseases, wounds and respiratory diseases. Furthermore, various parts of the plant have reportedly been used for the treatment of bowel disorders and prostate and other forms of cancer, and to manage infections caused by *Staphylococcus aureus* and *Escherichia coli* (Kasolo et al., 2010; Chollom et al., 2012; Faehy, 2005). Similarly, the use of *M. oleifera* in the management of inflammation, arthritis, joint pains, swellings, gastrointestinal disorders, oxidative stress and liver disorders in southern African countries like Zimbabwe has been reported (Anwar et al., 2007; Stohs & Hartman, 2015), further confirming the importance of the plant in TAM (Mudau et al., 2022).

7.3.3 *Moringa oleifera* in African Ethnopharmacology

The importance of *M. oleifera* in African folkloric medicine is captured in Table 7.3. The extensive use of *M. oleifera* in TAM for the treatment and management of diseases, such as kidney disorders, cancer, obesity, inflammation, liver conditions, hyperglycaemia and dyslipidaemia, and to prevent the development of atherosclerotic plaques has been validated in well-orchestrated scientific experiments in animal models (Atawodi et al., 2010; Padayachee and Baijnath, 2020; Tiloke et al., 2018; Olounlade et al., 2021).

7.4 *MORINGA OLEIFERA* IN TRADITIONAL CHINESE MEDICINE

The harmony and balance of the internal system of the body are the guiding principles of traditional Chinese medicine (TCM). Accordingly, various parts of *M. oleifera*, commonly referred to as 'Lamu' or 'Mandarin' in South Chinese provinces, have been utilised, alone or in combination with other herbs, in order to create harmony and balance in the internal system of the body, either for its medicinal importance or nutritional significance (Zhang and Hengshan, 2001; Yuan et al., 2016).

7.4.1 Medicinal Uses of *M. oleifera* in TCM

Temperature and taste are important underlying principles in TCM. Accordingly, bitter-tasting medications can release, dry and clear the heat and purge the body, stimulate venting, reverse energy (*qi*) and move stools; sweet-tasting drugs can refill, reconcile, enhance smelting, nourish, tonify and soothe pain (Meireles et al., 2020). With their laxative properties that reduce the possibility of infectious diseases with their cooling effect, the majority of medicinal plants that store yin (an organic material) have a bitter taste. According to its bitter flavour, *M. oleifera* leaves can also help people lose weight and reduce blood fat (Meireles et al., 2020).

Research on *M. oleifera* categorised the plant under the cold or hot nature TCM concept (Olson et al., 2016). Based on scientific research recommendations and the taste theory of TCM, it has been inferred that *M. oleifera* leaves have a cold property and a sweet and bitter taste (Dong et al., 2019). In this regard, the role of *M. oleifera* in blood production, memory and muscle control, in addition to its functional properties arising from its high contents of amino acids, antioxidants and minerals, has been well documented in TCM (Kumar et al., 2024).

Various parts of *M. oleifera* are reported to exhibit a range of pharmacological characteristics, such as anti-inflammatory, antibacterial, anticancer and antioxidant effects (Atawodi et al., 2010; Chikara et al., 2021), owing to their constituents, including alkaloids, phenolic acids, flavonoids and glucosinolates. The characteristics of the phytoconstituents and the nutritional components, including minerals, in *M. oleifera* parts have been demonstrated to align closely with the potential of these constituents to balance body constituents towards enhancement of energy and the body's inherent healing potential (Kumar et al., 2024; Olson et al., 2016). It is generally accepted that the majority of cold-natured medicinal plants have a bitter taste and pharmacological effects that include antiviral, antibacterial and antipyretic properties (Ali et al., 2019).

Although moringa was introduced to China only recently, TCM professionals acknowledge its value and classify it based on their principles, allowing it to be used clinically as a natural TCM medication. In TCM, *M. oleifera*'s roots and bark are used to treat constitutional deficiencies because they are thought to be acrid and warm. They can be used to treat overactive energetic syndromes like hypertension, fever, diabetes, constipation and hyperthyroidism (Meireles et al., 2020). Its leaves, on the other hand, are rich in nutrients because, according to TCM principles, they have a unifying effect on the stomach and spleen, increasing the absorption and transport of nutrients. According to Meireles et al. (2020), this is indicated by the lack of energy and internal dampness, which are beneficial for tiredness, malnutrition, digestive issues, low appetite and anaemic states (Table 7.4).

TABLE 7.4
Phytochemistry and Therapeutic, Nutritional and Ethnomedicinal Uses of *M. oleifera* Reported in TCM

S/N	Disease	Plant part	Method of preparation	Country of report	Validation of reported ethnomedicinal use	Phytochemical responsible for observed medicinal effects	Reference
1. 1.	Cancer	L	Boiled leaf	China	Anticancer activity	Citric acid, pyruvic acid, succinic acid, lactic acid, amino acid, glycosides and phenols	Gaobotse et al., 2023
2.	Constipation	L	Leaf dietary inclusion	China	Anticonstipation	Flavonoids, amino acids, nucleotides, lipids and alkaloids	Gao et al., 2023
3.	Diabetes	L	Aqueous leaf extract	China	Antihyperglycaemic agents	Tricyclohumuladiol, 2-phenethyl-β-D-glucoside, benzyl-β-D-glucopyranoside and isorhamnolin-3-O-β-D-glucopyranoside	Jiang et al., 2024
4.	Diabetes	L	Alcoholic leaf extract	China	Antidiabetic	Amino acids, flavonoids and alkaloids	Tang et al., 2021
5.	Diabetes	S	Compound	China	Antioxidant	Spiroleiferthione A, with a 2-thiohydantoin	Jiang et al., 2023
6.	Diabetes	L	Ethanolic leaf extract	China	Antidiabetic	Alkaloids and polyphenols	Hong et al., 2023
7.	Diabetes	FL	Methanol extract	China	Antioxidant	Apigenin, acacetin	
8.	Hypertension, blood sugar ailments	L	Leaf decoctions	China	Anti-inflammatory	Flavonoids and polyphenols, oxalic acid, citric acid, pyruvic acid, succinic acid, lactic acid, amino acid, glycosides and phenols	Zheng et al., 2023

9.	Insomnia	S	Ethanolic seed extract	China	Antiinsomnia	Long-chain fatty acids, medium-chain fatty acids	Liu et al., 2022
10.	Malnutrition	L	Leaf dietary inclusion	China	Antioxidant	Proteins and vitamins	Shen et al., 2021
11.	Malnutrition	S	Seed milk	China	Food fortification	Phosphoserine, glutamic acid	Li et al., 2023
12.	Metabolic diseases	L	Leaf dietary inclusion	China	Anti-osteoporotic	No report	Hu et al., 2023
13.	Metabolic diseases	L	Leaf powder	China	Anti-inflammatory	Quercetin and kaempferol derivatives	Zhang et al., 2024
14.	Obesity	L	Leaf dietary inclusion	China	Anti-obesity	Galacturonic acid, glucuronic acid and protein	Li et al , 2022
15.	Polycystic ovary syndrome	L	Leaf dietary inclusion	China	Anti-inflammatory	Methyl linolenate, neoxanthin	Wu et al., 2023
16.	Prostate cancer	L	Compound	China	Anti-prostate cancers	Alkaloid	Xie et al., 2020
17.	Ulcerative colitis	L	Leaf dietary inclusion	China	Anti-inflammatory	Fucose, arabinose, rhamnose, galactose, glucose, xylose, mannose, galactose uronic acid, glucuronic acid, glucose uronic acid and mannose uronic acid	Mohamed-Husien et al., 2022

L, leaves; S, stem; R, root; RB, root bark; P, pod; F, fruit; FL, flower.

According to TCM principles, the primary action of *M. oleifera* leaf is to treat excess syndrome diseases. This is primarily supported by its bitter taste and cooling temperature effect, which removes excess heat and phlegm from the liver meridians. At the same time, the sweet taste can strengthen the spleen and stomach meridians and have a nourishing effect on the body (Meireles et al., 2020).

7.4.2 Therapeutic Applications of *M. oleifera* in TCM

In Chinese traditional medicine, the application of *M. oleifera* is based on the principles of tonifying vital energy called *qi*, as in the case of the leaves, and nourishment of the body's cooling and moistening properties, called *yin*. These have been attributed to their vitamin, mineral and antioxidant compound content. The use of moringa for health enhancement aligns with the use of other plants like *Astragalus membranaceus* and *Panax ginseng* for similar purposes and reasons, just as it is combined with herbs such as *Lilium brownii* and *Dendrobium nobile* in Chinese traditional medical practice in order to enhance its potency (Gopalakrishnan, 2016). Similarly, the usefulness of *M. oleifera* in Chinese traditional medicine has been attributed to properties such as the heat-clearing and detoxifying properties of the seeds and roots, the detoxifying effects and strengthening of the body's essence, called *jing*, and the cooling properties of the leaves, which may not be applicable to all individuals (Gopalakrishnan, 2016).

According to the TCM concept, the leaves of *M. oleifera* are a powerful rebuilder of the body's marrow, essentially restoring life force energy at an extraordinarily deep level (Gopalakrishnan, 2016), while the fruit and seeds of the tree tonify the spleen and eliminate dampness, making them suitable for people with low appetite, bloating and a heavy sensation in the limbs (Brown, 2002). The roots are utilised as food because of their warm nature and bitter, pungent flavour.

The root and bark help strengthen the spleen and stomach, eliminate air and encourage urine. Headaches, light-headedness, gastrointestinal issues, flatulence, hypertension and skin issues are among the conditions for which it is advised (Brown, 2002).

In TCM, the leaves of Moringa are the most popular part of the plant and are believed to possess the capacity to reverse severe malnutrition, remove heat and moisture and rid the body of toxins; it is also suggested for gynaecological issues like vaginitis and excessive discharge. It is also rich in iron, so it can be used to treat anaemia by tonifying the blood (Brown, 2002).

7.4.3 Scientific Validation of *M. oleifera* in TCM

A number of studies have validated the use of moringa in TCM (Table 7.4): the antidiabetic effect (Al-Malki et al., 2015); bacteriostatic and antimicrobial effects (Uddin et al., 2021); the preventive effect on pro-oxidative degeneration (Fejér et al., 2019); anti-inflammatory effects (Ndiaye et al., 2002) and the suppression of COX-2 and pro-inflammatory cytokines (IL-1β, IL-6, PGE2 and TNF-α) (Chen et al., 2021); blood vessel relaxing (Odii et al., 2022); and antioxidant effects (Atawodi et al., 2010; Gao et al., 2023).

7.5 *MORINGA OLEIFERA* IN SOUTHEAST ASIAN TRADITIONAL PRACTICES

Moringa oleifera has been a cornerstone of traditional medicine across Southeast Asia for centuries. Known locally as *Malunggay* in the Philippines, *Marum* in Thailand, *Daun Kelor* in Indonesia and *Chùm Ngây* in Vietnam, each name reflects the deep cultural and medicinal value of this plant. It is popular for its galactagogue properties in the Philippines, where it is used to boost milk production in nursing mothers when leaves are added to soup, and as an immune booster and modulator of metabolic disorders and hypertension and as a liver detoxifier in Indonesia (*Jamu medicine*), Thailand and Vietnam, respectively, where the leaf is a key component of the traditional medicine system (Pham, 2023; Poluan et al., 2023).

There is a wide diversity in the way moringa products are produced for consumption in South Asian traditional medicine. For instance, in the Philippines, the leaves of moringa, called malungay, are peculiar components of chicken soup, locally called 'tinola', and mung bean stew, commonly called 'ginisang mungo', while in Thailand and Indonesia, treatment of hypertension, diabetes and digestive disorders is effected by chewing the fresh leaves (alone or in combination with lemongrass or ginger), blended or steeped in hot water to produce moringa tea. On the other hand, for the treatment of wounds, pains (arthritis) and inflammation, a paste of the crushed leaves is used for dressing in Indonesia, while for skin diseases and environmental stressors like acne, eczema and psoriasis, moringa seed oil or ointment prepared from coconut oil and moringa leaves (Mudzviti et al., 2012) can be applied (Table 7.5). In Vietnamese and Indonesian traditional medicine, called *Jamu*, the use of moringa leaves, fermented with turmeric, tamarind or ginger or integrated into herbal brews and wines, for the treatment of digestive disorders, boosting immunity and the management of liver health and disorders has been reported (Pradana, 2022; Younis et al., 2020).

7.6 *MORINGA OLEIFERA* IN OTHER TRADITIONAL SYSTEMS OF MEDICINE

This chapter mainly discussed the utilisation of moringa in traditional systems of medicine in Africa, China and Southeast Asia, as well as in Ayurveda. However, the use of moringa is not limited to these traditional systems of medicine, but rather extends to other cultures, including Australia, the Caribbean, Latin America and the Pacific Islands (Vergara-Jimenez et al., 2017).

The use of moringa in Latin American systems of medicine largely aligns with its medicinal use in other cultures. For instance, in Mexico, Brazil and Peru, infusions of leaves, seeds and roots are used to treat inflammation, diabetes, hypertension, skin challenges, respiratory infections and gastrointestinal disorders. Similarly, in the folk medicines of Caribbean countries like Jamaica and Haiti, *M. oleifera* leaves, which are rich in flavonoids and phenolic acids (Atawodi et al., 2010), are prepared as tea or poultices and used in the management of arthritis, fatigue, inflammation, joint pain and anaemia in pregnant women (Nouman et al., 2014).

However, in Pacific Islands like Fiji and Samoa, as well as in Australia, moringa leaves are used as dietary supplements in boosting immunity because of their high nutrient density and the management of diabetes, dyslipidaemia and constipation, while extracts of the roots and seeds are respectively deployed in the treatment of urinary infections and kidney stones (Mbikay, 2012; Saini et al., 2014).

Additionally, the anti-inflammatory and antimicrobial properties of moringa are a key ingredient in functional foods designed to support gut health and immune function (Wang et al., 2016).

7.7 CONCLUSION

Moringa oleifera Lam. is widely used in nutritional supplements, tonics, detoxifiers and therapeutic agents in various traditional systems of medicine, including Ayurveda, TCM, South Asian traditional medicine, TAM, Australian traditional medicine, Latin American medicine and the Caribbean system of traditional medicine. The abundance of important phytochemicals in *M. oleifera* provides a scientific basis for its extensive traditional uses, as these bioactive compounds exert pharmacological activities, including antioxidant, anti-inflammatory and immunomodulatory effects, among others. Modern scientific research continues to validate many of these traditional uses, reinforcing the potential of moringa as a valuable resource in both traditional and contemporary medicine. However, further research is needed to fully elucidate the mechanisms of action of its abundant phytochemical constituents in order to optimise its therapeutic potential with respect to different diseases.

Acknowledgement: We thank the Tertiary Education Trust Fund (TETfund) for the award of a research grant.

TABLE 7.5
Ethnomedicinal Uses of *M. oleifera*: Preparation and Scientific Validation in Various Southeast Asian Countries

S/N	Disease/condition	Part used	Method of preparation	Country of report	Validation by modern laboratory evaluation	Phytochemical responsible for medicinal effects	References
1.	Lactation enhancement	Leaves	Added to soups and teas	Philippines	Supported by local practices and some studies	Flavonoids, phytosterols	Padma et al., 2022
2.	Diabetes, hypertension	Leaves	Brewed into herbal tea	Thailand	Supported by studies on hypoglycaemic and cardiovascular effects	Quercetin, chlorogenic acid	Leone et al., 2015
3.	Immune system boost	Leaves	Consumed as tea or in herbal tonics	Indonesia	Immunomodulatory properties confirmed in studies	Polyphenols, flavonoids	Nfambi et al., 2015; Poluan et al., 2023
4.	Liver protection	Leaves	Consumed in herbal brews	Vietnam	Hepatoprotective effects confirmed in studies	Kaempferol, quercetin	Pham, 2023
5.	Wound healing, anti-inflammatory	Leaves	Crushed into a poultice	Indonesia	Antibacterial and anti-inflammatory effects confirmed	Isothiocyanates, flavonoids	Elgamily et al., 2016
6.	Skin conditions (acne, eczema)	Seed oil	Applied as an ointment or balm	Thailand	Confirmed for antimicrobial and skin-protective properties	Oleic acid, tocopherols	Mudzviti et al., 2012
7.	Arthritis, joint pain	Leaves	Leaf paste applied externally	Vietnam	Anti-inflammatory effects supported by research	Quercetin, polyphenols	Ijarotimi et al., 2013
8.	Digestive health, detoxification	Leaves	Fermented with turmeric, tamarind or ginger	Indonesia	Fermented moringa extracts shown to enhance the bioavailability of compounds	Polyphenols, flavonoids	Pradana et al., 2022
9.	Kidney function, urinary tract infections	Leaves	Fermented into herbal tonics	Thailand	Some evidence supporting kidney-protective effects	Alkaloids, phenolic compounds	Younis et al., 2020
10.	Liver function, immune system boost	Leaves	Fermented into herbal wines and medicinal brews	Vietnam	Some studies confirm liver-supportive effects	Saponins, flavonoids	Zeng et al., 2018

REFERENCES

Abdalla, H. A., Ali, M., Amar, M. H., Chen, L., & Wang, Q. F. 2022. Characterization of phytochemical and nutrient compounds from the leaves and seeds of *Moringa oleifera* and *Moringa peregrina*. *Horticulture*, *8*(11), 1081.

Abdulmalik, U., Halliru, Z., Umar, A., Musa, M., & Adam, A. S. 2024. Phytochemical screening, GCMS analysis and antibacterial activity of *Moringa oleifera* ethanolic and aqueous leaf extracts against some clinical isolates. *UMYU Journal of Microbiology Research*, *9*(1), 34–45.

Abe, R., & Ohtani, K. 2013. An ethnobotanical study of medicinal plants and traditional therapies on Batan Island, the Philippines. *Journal of Ethnopharmacology*, *145*(2), 554–565.

Abubakar, I. B., Kankara, S. S., Malami, I., Danjuma, J. B., Muhammad, Y. Z., Yahaya, H., ... Nurudeen, Q. O. 2022. Traditional medicinal plants are used for treating emerging and re-emerging viral diseases in northern Nigeria. *European Journal of Integrative Medicine*, *49*, 102094.

Agoyi, E. E., Okou, F. A. Y., Assogbadjo, E. A., & Sinsin, B. 2015, November). Medicinal uses of *Moringa oleifera* in southern Benin (West Africa). In *I* international symposium *on Moringa* (Vol. 1158, pp. 303–308). Acta Horticulturae.

Akoko, S., Aleme, B. M., & Uahomo, P. O. 2022. The effect of the addition of extracts of Vernonia amygdalina and Moringa oleifera in the nutrition of alloxan-induced diabetic wistar rats. *International Journal of Pharma Research and Health Sciences*, *10*(4), 3455–3462.

Akpor, O. B., Ndakotsu, J., Evbuomwan, I. O., Olaolu, T. D., & Osemwegie, O. O. 2021. Bacterial growth inhibition and antioxidant potentials of leaf infusions of (Moringa oleifera), locust beans (*Parkia biglobosa*) and bitter leaf (*Vernonia amygdalin*). *Scientific African*, *14*, e01001.

Ali, A., Yusof, A., Chin, L., Ibrahim, M. N., & Muneer, S. 2019. Development and standardization of *Moringa oleifera* leaves as a natural dietary supplement. *Journal of Dietary Supplements*, *16*(1), 66–85. https://doi. org/10.1080/19390211.2018.1429517

Alia, F., Putri, M., Anggraeni, N., & Syamsunarno, M. R. A. A. 2022. The potency of Moringa oleifera Lam. as protective agent in cardiac damage and vascular dysfunction. *Frontiers in Pharmacology*, *12*, 724439.

Al-Malki, A. L., & El Rabey, H. A. 2015. The antidiabetic effect of low doses of Moringa oleifera Lam. seeds on streptozotocin-induced diabetes and diabetic nephropathy in male rats. *BioMed Research International*, *2015*(1), 381040.

Anjali, P. P., Krishnan, S. K., Madhusoodhanan, A., & Hima, V. H. 2023. A review on Kottam Thagaradi Agada Lepa in Mandalivisha Sopham. *Journal of Ayurveda and Integrated Medical Sciences*, *8*(4), 96–98. https://doi.org/10.21760/jaims.8.4.16

Anwar, F., Latif, S., Ashraf, M., & Gilani, A. H. 2007. *Moringa oleifera*: A food plant with multiple medicinal uses. *Phytotherapy Research*, *21*(1), 17–25.

Asigbaase, M., Adusu, D., Musah, A. A., Anaba, L., Nsor, C. A., Abugre, S., & Derkyi, M. 2024. Ethnobotanical and ethnopharmacological survey of medicinal tree species used in the treatment of diseases by forest-fringe communities of Southwestern Ghana. *Heliyon*, *10*(1). e23645. https://doi.org/10.1016/j.heliyon .2023.e2364

Atawodi, S. E., Atawodi, J. C., Idakwo, G. A., Pfundstein, B., Haubner, R., Wurtele, G., ... Owen, R. W. 2010. Evaluation of the polyphenol content and antioxidant properties of methanol extracts of the leaves, stem, and root barks of *Moringa oleifera* Lam. *Journal of Medicinal Food*, *13*(3), 710–716.

Bancessi, A., Bancessi, Q., Baldé, A., & Catarino, L. 2020. Present and potential uses of *Moringa oleifera* as a multipurpose plant in Guinea-Bissau. *South African Journal of Botany*, *129*, 206–208.

Bopape, M., Tiloke, C., & Ntsapi, C. 2023. *Moringa oleifera* and Autophagy: Evidence from *In Vitro* studies on chaperone-mediated autophagy in HepG2 cancer cells. *Nutrition and Cancer*, *75*(10), 1822–1847.

Brown, D. 2002. *The royal horticultural society- encyclopedia of herbs and their uses*. Dorling Kindersley Limited.

Chen, G., Xu, Y., Wu, J., Li, N., & Guo, M. 2021. Anti-inflammatory properties and potential bioactive components from *Moringa oleifera* leaves revealed by affinity ultrafiltration LC–MS and molecular docking. *ACS Food Science & Technology*, *1*(10), 1953–1962.

Chikara, N., Kaur, A., Mann, S., Garg, M. K., Sofi, S. A., & Panghal, A. 2021. Bioactive compounds, associated health benefits and safety considerations of *Moringa oleifera* L.: An updated review. *Nutrition & Food Science*, *51*(2), 255–277.

Chiş, A., Noubissi, P. A., Pop, O. L., Mureşan, C. I., Fokam Tagne, M. A., Kamgang, R., ... Suharoschi, R. 2023. Bioactive compounds in *Moringa oleifera*: Mechanisms of action, focus on their anti-inflammatory properties. *Plants*, *13*(1), 20.

Chollom, S. C., Agada, G. O. A., Gotep, J. G., Mwankon, S. E., Dus, P. C., Bot, Y. S., ... Okwori, A. E. J. 2012. Investigation of aqueous extract of *Moringa oleifera* Lam seed for antiviral activity against newcastle disease virus in ovo. *Journal of Medicinal Plants Research*, *6*(22), 3870–3875.

Djiogue, S., Youmbi, K. I., Seke, E. P. F., Kandeda, K. A., Zemo, G. F., Motoum, T. F., ... Dieudonne, N. 2022. Neuroprotective effects of the aqueous extract of leaves of Moringa oleifera (Moringaceae) in scopolamine-treated rats. *Bio Med*, *14*, 486.

Dong, X., Guo, X., Hu, Y., et al. 2019. Summary of studies on traditional Chinese medicine properties of overseas plant medicine *Moringa* leaves. *Global Traditional Chinese Medicine, 12*(1), 149–153. https://doi.org/10.3969/j.issn.1674-1749.2019.01.050

El Bilali, H., Guimbo, I. D., Nanema, R. K., Falalou, H., Kiebre, Z., Rokka, V.-M., Tietiambou, S. R. F., et al. 2024. Research on *Moringa oleifera* Lam. in Africa. *Plants, 13*(12), 1613.

Elgamily, H., Moussa, A., Elboraey, A., Hoda, E. S., Al-Moghazy, M., & Abdalla, A. 2016. Microbiological assessment of Moringa oleifera extracts and its incorporation in novel dental remedies against some oral pathogens. *Open access Macedonian Journal of Medical Sciences*, *4*(4), 585.

Faber, M., Van Jaarsveld, P. J., Wenhold, F. A. M., & Van Rensburg, J. 2010. African leafy vegetables are consumed by households in the Limpopo and KwaZulu-Natal provinces in South Africa. *South African Journal of Clinical Nutrition*, *23*(1), 30–38.

Fahey, J. W. 2005. *Moringa oleifera*: A review of the medical evidence for its nutritional, therapeutic, and prophylactic properties. *Trees for Life Journal*, *1*(5), 1–15.

Famakinwa, A., Shuttleworth, A., Lubisi, S., Olubi, O., Oguntibeju, O. O., & Obilana, A. 2023. Chemical and functional properties of snacks produced from wheat flour fortified with Moringa oleifera leaf powder. *African Journal of Food, Agriculture, Nutrition and Development*, *23*(9), 24467–24486.

FAO. 2023. *Traditional crops—Moringa*. https://www.fao.org/traditional-crops/moringa/en.

Fejér, J., Kron, I., Pellizzeri, V., Pľuchtová, M., Eliašová, A., Campone, L., ... Gruľová, D. 2019. First report on evaluation of basic nutritional and antioxidant properties of *Moringa oleifera* Lam. from the Caribbean Island of Saint Lucia. *Plants*, *8*(12), 537.

Gajender, Mazumder, A, Sharma, A, Azad, Md. A. K., 2023.A Comprehensive Review of the Pharmacological Importance of Dietary Flavonoids as Hepatoprotective Agents, Evidence-Based Complementary and Alternative Medicine, 4139117, 17 pages, 2023. https://doi.org/10.1155/2023/4139117

Gao, X., Yang, W., Li, S., Liu, S., Yang, W., Song, S., ... Tian, Y. 2023. *Moringa oleifera* leaf alleviates functional constipation via regulating the gut microbiota and the enteric nervous system in mice. *Frontiers in Microbiology*, *14*, 1315402.

Gaobotse, G., Venkataraman, S., Brown, P. D., Masisi, K., Kwape, T. E., Nkwe, D. O., ... Makhzoum, A. 2023. The use of African medicinal plants in cancer management. *Frontiers in Pharmacology*, *14*, 1122388.

Gautam. 2023. Pharmacognostical and phytochemical study on stem bark of Shigru (*Moringa oleifera* Lam). *Journal of Pharmacognosy and Phytochemistry*. doi: 10.22271/phyto.2023.v12.i4b.14704

Glover-Amengor, M. 2015. *Moringa Oleifera leaf supplementation on vitamin A status of children in Ada-East district of Ghana* (Doctoral dissertation), University of Ghana, Legon.

Gopalakrishnan, L., Doriya, K., & Kumar, D. S. 2016. *Moringa oleifera*: A review on nutritive importance and its medicinal application. *Food Science in Human Wellness, 5*, 49–56.

Govindarajan, R., Vijayakumar, M., & Pushpangadan, P. 2005. Antioxidant approach to disease management and the role of 'Rasayana'herbs of Ayurveda. *Journal of Ethnopharmacology*, *99*(2), 165–178.

Hedhili, A., Akinyemi, B. E., Otunola, G. A., Ashie-Nikoi, P. A., Kulkarni, M., Husson, F., & Valentin, D. 2022. *Moringa oleifera* Lam.: A comparative survey on consumer knowledge, usage, attitude and belief in Africa and India. *South African Journal of Botany*, *147*, 153–162.

Hong, Z., Xie, J., Hu, H., Bai, Y., Hu, X., Li, T., ... Tian, Y. 2023. Hypoglycemic effect of Moringa oleifera leaf extract and its mechanism prediction based on network pharmacology. *Journal of Future Foods*, *3*(4), 383–391.

Horn, L., Shakela, N., Mutorwa, M. K., Naomab, E., & Kwaambwa, H. M. 2022. *Moringa oleifera* as a sustainable climate-smart solution to nutrition, disease prevention, and water treatment challenges: A review. *Journal of Agriculture and Food Research*, *10*, 100397.

Hu, X. H., Yang, X. Y., Lian, J., Chen, Y., Zheng, C. Y., Tao, S. Y., ... Jiang, G. J. 2023. *Moringa oleifera* leaf attenuates osteoporosis in ovariectomized rats by modulating gut microbiota composition and MAPK signalling pathway. *Biomedicine & Pharmacotherapy*, *161*, 114434.

Ijarotimi, O. S., Fagbemi, T. N., & Osundahunsi, O. F. 2013. Comparative study of nutritional profiles and phytochemical components of raw, blanched and fermented flour from the leaves of *Moringa oleifera* Lam. *Malaysian Journal of Nutrition*, *19*(3), 371–382.

Isimi, C. Y., John-Africa, L. B., Ekere, K. E., Olayemi, O. J., Aremu, O. I., & Emeje, M. O. 2021. Formulation, evaluation and anti-hemorroidal activity of suppositories containing *Moringa oleifera* lam. seed oil. *ACTA Pharmaceutica Sciencia*, *59*(1), 647–665.

Jamil, S., Turabi, T. H., Ahmad, S., Riaz, M., Wariss, H. M., & Akter, Q. S. 2024. Comparative investigation of the nutritional profiling and antipyretic activity of *Moringa oleifera* leaves, bark, and root from different sites of Punjab, Pakistan. *Food Science & Nutrition*, *13*(1). https://doi.org/10.1002/fsn3.4706

Jayachandran, S., Kavitha, S., Gayathri, R., Vishnupriya, V., & Jayaraman, S. 2021. Evaluation of hypolipidemic potential of aqueous seed extract of *Moringa oleifera* –an in vitro study. *Journal of Pharmaceutical Research International*, 2722–2729. https://doi.org/10.9734/jpri/2021/ v33i60b34934

Jhofi, M., Husni, E., & Hamidi, D. 2021. Anticancer and antioxidant activity of asam kandis (*Garcinia cowa* roxb) leaf extract and fraction. *Advances in Health Sciences Research*. https://doi.org/10.2991/ahsr.k.211105.032

Jiang, X., Tang, J., Gao, Z., Chen, C., Li, W., Peng, L., ... Zhao, Y. 2024. Exploring the protection of islet β-cells and hypoglycemic effect of Moringa oleifera stem based on network pharmacology and experimental validation. *Journal of Functional Foods*, *122*, 106499.

Jiang, Y., Liu, R., Huang, L., Huang, Q., Liu, M., Liu, S., & Li, J. 2023. Spiroleiferthione A and oleiferthione a: Two unusual isothiocyanate-derived thioketone alkaloids from moringa oleifera Lam. *Seeds Pharmaceuticals*, *16*(3), 452.

Kanyal, K., & Shankar Pandey, V. 2023. A study of role of sharerika and mansika dosha in vyadhikshamatava. *International Ayurvedic Medical Journal*, *11*(10), 2497–2503. https://doi.org/10.46607/iamj1711102023

Kasolo, J. N., Bimenya, G. S., & Ojok, L. 2010. Phytochemicals and uses of *Moringa oleifera* leaves in Ugandan rural communities. *Journal of Medicinal Plants Research, 4*, 753–757.

Keatinge, J. D. H., Ebert, A. W., Hughes, J., Yang, R.-Y., & Curaba, J. 2017. Seeking to attain the UN's Sustainable Development Goal 2 worldwide: The important role of *Moringa oleifera*. *Acta Horticulturae*, 1–10.

Khattab, R. R., Alshamari, A. K., Hassan, A. A., Elganzory, H. H., El-Sayed, W. A., Awad, H. M., … Hassan, N. A. 2021. Click chemistry based synthesis, cytotoxic activity and molecular docking of novel triazole-thienopyrimidine hybrid glycosides targeting egfr. *Journal of Enzyme Inhibition and Medicinal Chemistry, 36*(1), 504–516. https://doi.org/10.1080/14756366. 2020.1871335

Kumar, S., Murti, Y., Arora, S., Akram, W., Bharadwaj, H., Gupta, K., Sachdev, A., ... Saha, S. 2024. Exploring the Therapeutic Potential of *Moringa oleifera* Lam. in traditional chinese medicine: A comprehensive review. *Pharmacological Research-Modern Chinese Medicine*, *12*, 100473.

Leone, A., Spada, A., Battezzati, A., Schiraldi, A., Aristil, J., & Bertoli, S. 2015. Cultivation, genetic, ethnopharmacology, phytochemistry and pharmacology of *Moringa oleifera* leaves: An overview. *International Journal of Molecular Sciences*, *17*(12), 2064.

Li, X., Fu, Y., Wang, Y., Xian-bin, S., Chen, H., Chen, T., … Chen, H. 2023. Synthesis, structure–activity relationship, and biological evaluation of quinolines for development of anticancer agents. Archiv Der Pharmazie, *356*(7). https://doi.org/10.1002/ardp.202200673

Liu, W. L., Wu, B. F., Shang, J. H., Wang, X. F., Zhao, Y. L., & Huang, A. X. 2022. *Moringa oleifera* seed ethanol extract and its active component kaempferol potentiate pentobarbital-induced sleeping behaviours in mice via a GABAergic mechanism. *Pharmaceutical Biology*, *60*(1), 810–824.

Lyons, G., Gondwe, C., Banuelos, G., Mendoza, C., Haug, A., Christophersen, O., & Ebert, A. W. 2017. Drumstick tree (*Moringa oleifera*) leaves as a source of dietary selenium, sulphur and pro-vitamin A. *Acta Hortic*, *1158*, 287–292. https://doi.org/10.17660/ActaHortic.2017.1158.32

Mbikay, M. 2012. Therapeutic potential of *Moringa oleifera* leaves in chronic hyperglycemia and dyslipidemia: A review. *Frontiers in Pharmacology*, *3*, 24.

Meireles, D., Gomes, J., Lopes, L., Hinzmann, M., & Machado, J. 2020. A review of properties, nutritional and pharmaceutical applications of *Moringa oleifera*: Integrative approach on conventional Asian medicine. *Advances in Traditional Medicine*, *20*(4), 495–515.

Mohamed-Husien, H., Peng, W., Su, H., Zhou, R., Tao, Y., Huang, J., ... Li, J. 2022. *Moringa oleifera* leaf polysaccharide alleviates experimental colitis by inhibiting inflammation and maintaining intestinal barrier. *Frontiers in Nutrition*, *9*, 1055791.

Moremane, M. M., Abrahams, B., & Tiloke, C. 2023. *Moringa oleifera*: A review on the antiproliferative potential in breast cancer cells. *Current Issues in Molecular Biology*, *45*(8), 6880–6892.

Moyo, B., Masika, P. J., Hugo, A., & Muchenje, V. 2011. Nutritional characterization of Moringa (*Moringa oleifera* Lam.) leaves. *African journal of biotechnology*, *10*(60), 12925–12933.

Mudau, T. E., Olowoyo, J. O., & Amoo, S. O. 2022. Ethnobotanical assessment of medicinal plants used traditionally for treating diabetes in Vhembe district, Limpopo Province, South Africa. *South African Journal of Botany*, *146*, 304–324.

Mudzviti, T., Maponga, C. C., Khoza, S., Ma, Q., & Morse, G. D. 2012. The impact of herbal drug use on adverse drug reaction profiles of patients on antiretroviral therapy in zimbabwe. *AIDS Research and Treatment*, *2012*, 1–4. https://doi.org/10.1155/2012/434171

Mwangi, S., & Kimathi, M. 2006, October. African leafy vegetables evolve from underutilized species to commercial cash crops. In *Research workshop on collective action and market access for smallholders* (Vol. 2, pp. 2–5).

Ndabigengesere, A., Narasiah, K. S., & Talbot, B. G. 1995. Active agents and mechanism of coagulation of turbid waters using *Moringa oleifera*. *Water Research*, *29*(2), 703–710.

Ndiaye, M., Dieye, A. M., Mariko, F., Tall, A., & Faye, B. 2002. Contribution to the study of the anti-inflammatory activity of *Moringa oleifera* (Moringaceae). *Dakar Medical*, *47*(2), 210–212.

Nfambi, J., Bbosa, G. S., Sembajwe, L. F., Gakunga, J. N., & Kasolo, J. 2015. Immunomodulatory activity of methanolic leaf extract of *Moringa oleifera* in wistar albino rats. *Journal of Basic and Clinical Physiology and Pharmacology*, *26*(6), 603–611. https://doi.org/10.1515/jbcpp-2014-0104

Ngemenya, M. N., Itoe, L. O., Awah, L. A., Asongana, R., & Ndip, R. A. 2024. Antibacterial activity against multidrug-resistant Salmonella, toxicity and biochemical effects of *Moringa oleifera* leaf extracts in mice. *Asian Journal of Natural Product Biochemistry*, *22*(1), 43–50.

Nouman, W., Siddiqui, M. T., Basra, S. M. A., Afzal, I., & Rehman, H. U. 2014. Enhancement of emergence potential and stand establishment of *Moringa oleifera* Lam. by seed priming. *Turkish Journal of Agriculture and Forestry*, 38(2), 227–235.

Ntila, S. L., Ndhlala, A. R., Mashela, P. W., Kolanisi, U., & Siwela, M. 2020. Supplementation of a complementary white maize soft porridge with Moringa oleifera powder as a promising strategy to increase nutritional and phytochemical values: A research note. *South African Journal of Botany*, *129*, 238–242.

Obunda, N. J., Ghavwan, O. R., Nwachukwu, C. V., Fasulu, P. E., Ogunwola, O. O., Achonwa, C. V., … Uchendu, I. K. 2023. Medicinal plant adverse effects with a focus on *Moringa oleifera* and *Telfairia occidentalis*. *Journal of Pharmacy and Allied Medicine, 1*(1), 57–64.

Odii, E. C., Okoro, J. O., Okorie, A. N., Asogwa, C. N., Chukwuka, C. O., Okoye, C. O., Okeke, E. S., Ejere, V. C., Eyo, J. E. 2022. Antihypertensive effect of *Moringa oleifera* (Moringaceae) Methanolic Leaf Extract (MoMLE) on Cricetomys gambianus (Muridae). *Tropical Journal of Natural Product Research*, *6*(10), 1695–1700. https://doi.org/10.26538/tjnpr/v6i10.23

Oladeji, O. S., Odelade, K. A., & Oloke, J. K. 2020. Phytochemical screening and antimicrobial investigation of Moringa oleifera leaf extracts. *African Journal of Science, Technology, Innovation and Development*, *12*(1), 79–84.

Olounlade, P. A., Konmy, B. S. B., Azando, E., Allou, S., & Baba-Moussa, L. 2021. *Moringa oleifera, Ocimum gratissimum* and *Vernonia amygdalina* as a natural antiparasitic alternative in growing rabbits. *Livestock Research for Rural Development*, *33*, 1–7.

Olson, M. E., Sankaran, R. P., Fahey, J. W., Grusak, M. A., Odee, D., & Nouman, W. 2016. Leaf protein and mineral concentrations across the "Miracle Tree" genus Moringa. *PLoS One*, *11*(7), 1–17. https:// doi .org/10.1371/journal.pone.0159782

Olson, M. E. 2023. *Moringaceae—Moringa*. http://www.efloras.org/florataxon.aspx?flora_id=1&taxon_id =200009759.

Olusanya, R. N. 2018. *The nutritional composition and acceptability of Moringa Oleifera Leaf Powder (MOLP)-supplemented Mahewu: A maize meal-based beverage for improved food and nutrition security*. Thesis in University of KwaZulu-Natal, Chapter 5, pp. 86–112.

Omokpariola, D. O., Precious-Egere, S. C., Omokpariola, P. L., & Okechukwu, V. U. 2021. Phytochemical and anti-microbial analysis of metabolites in seeds of Moringa oleifera grown in Nigeria. *Progress in Chemical and Biochemical Research*, *4*(3), 268–277.

Oyeleye, I. S., Ogunsuyi, O. B., Oluokun, O. O., & Oboh, G. 2023. Seeds of moringa (Moringa oleifera) and mucuna (*Mucuna pruriens* L.) modulate biochemical indices of L-NAME-induced hypertension in rats: a comparative study. *Journal of Agriculture and Food Research*, *12*, 100624.

Oyeyinka, A. T., & Oyeyinka, S. A. 2018. *Moringa oleifera* as a food fortificant: Recent trends and prospects. *Journal of the Saudi Society of Agricultural Sciences*, *17*(2), 127–136.

Padayachee, B., & Baijnath, H. J. S. A. J. O. B. 2020. An updated comprehensive review of the medicinal, phytochemical and pharmacological properties of Moringa oleifera. *South African Journal of Botany*, *129*, 304–316.

Padma, A., Indhuleka, A., Janet, J., & Ragavi, V. 2022. *Moringa oleifera*: Nutritional and medicinal properties for preventive health care. *Journal of University of Shanghai for Science and Technology*, *24*(1), 72–92. https://doi.org/10.51201/jusst/21/121020

Pareek, A., Pant, M., Gupta, M. M., Kashania, P., Ratan, Y., Jain, V., ... Chuturgoon, A. A. 2023. *Moringa oleifera*: an updated comprehensive review of its pharmacological activities, ethnomedicinal, phyto-pharmaceutical formulation, clinical, phytochemical, and toxicological aspects. *International Journal of Molecular Sciences*, *24*(3), 2098.

Pham, N. K., Bauerova, K., & Manh Cuong, N. 2023. Studies on hepatoprotective effects of Vietnamese medicinal plants. *Vietnam Journal of Science and Technology*, *61*(5), 725–750. https://doi.org/10.15625/2525-2518/17426

Poluan, J. C., Zubair, M. S., Ramadani, A. P., & Hayati, F. 2023. Narrative review: Potential of flavonoids from moringa (*Moringa oleifera* lam.) leaves as immunomodulators. *Jurnal Farmasi Galenika (Galenika Journal of Pharmacy)* (E-Journal), *9*(2), 270–283. https://doi.org/10.22487/j24428744.2023.v9.i2.16265

Popoola, J. O., & Obembe, O. O. 2013. Local knowledge, use pattern and geographical distribution of *Moringa oleifera* Lam.(Moringaceae) in Nigeria. *Journal of Ethnopharmacology*, *150*(2), 682–691.

Pradana, D. L. C., Rahmi, E. P., & Muti, A. F. 2022. The lethality dose and antioxidant activity of *Moringa oleifera* leaves extract. *IOP Conference Series: Earth and Environmental Science*, *1104*(1), 012017. https://doi.org/10.1088/1755-1315/1104/1/012017

Ritu Paliwal, R. P., Veena Sharma, V. S., & Pracheta, P. 2011. A review on horse radish tree (*Moringa oleifera*): A multipurpose tree with high economic and commercial importance. *Asian Journal of Biotechnology*, *3*(4), 317–328.

Saini, R. K., Sivanesan, I., & Keum, Y. S. 2014. Phytochemicals of *Moringa oleifera*: A review of their nutritional, therapeutic, and industrial significance. *3 Biotech*, *6*(2), 203.

Sayed, A. R., Shaaban, S., Gomha, S. M., & Al-Faiyz, Y. S. 2022. Design, synthesis, and biological evaluations of novel azothiazoles based on thioamide. *Current Issues in Molecular Biology*, *44*(7), 2956–2966. https://doi.org/10.3390/cimb44070204

Shen, M., Li, T., Qu, L., Wang, K., Hou, Q., Zhao, W., & Wu, P. 2021. Effect of dietary inclusion of *Moringa oleifera* leaf on productive performance, egg quality, antioxidant capacity and lipid levels in laying chickens. *Italian Journal of Animal Science*, *20*(1), 2012–2021.

Sivasankari, B., Anandharaj, M., & Gunasekaran, P. 2014. An ethnobotanical study of indigenous knowledge on medicinal plants used by the village peoples of Thoppampatti, Dindigul district, Tamilnadu, India. *Journal of Ethnopharmacology*, *153*(2), 408–423.

Sokhela, H., Govender, L., & Siwela, M. 2023. Complementary feeding practices and childhood malnutrition in South Africa: The potential of *Moringa oleifera* leaf powder as a fortificant: A narrative review. *Nutrients*, *15*(8), 2011.

Sonewane, K., Chouhan, S. S., Rajan, M., Chauhan, N. S., Rout, O. P., Kumar, A., ... Gupta, P. K. 2022. Pharmacological, ethnomedicinal, and evidence-based comparative review of *Moringa oleifera* Lam. (Shigru) and its potential role in the management of malnutrition in Tribal Regions of India, especially Chhattisgarh. *World Journal of Traditional Chinese Medicine*, *8*(3), 314–338.

Stohs, S. J., & Hartman, M. J. 2015. Review of the safety and efficacy of *Moringa oleifera*. *Phytotherapy Research*, *29*(6), 796–804.

Tamilselvi, N. A., & Arumugam, T. 2019. Health benefits, therapeutic and pharmacological properties of Moringa—A review. *Journal of Pharmaceutical Research International*, *25*, 1–11.

Tang, S. Q., Du, Q. H., & Fu, Z. 2021. Ultrasonic treatment on physicochemical properties of water-soluble protein from *Moringa oleifera* seed. *Ultrasonics Sonochemistry*, *71*, 105357.

Tiloke, C., Anand, K., Gengan, R. M., & Chuturgoon, A. A. (2018). Moringa oleifera and their phytonanoparticles: Potential antiproliferative agents against cancer. *Biomedicine & Pharmacotherapy*, *108*, 457–466. https://doi.org/10.1016/j.biopha.2018.09.060

Total Ayurveda. 2024. *The health benefits of Moringa- The modern superfood.* https://www.totalayurveda.in/benefits-of-moringa/

Uddin, T. M., Chakraborty, A. J., Khusro, A., Zidan, B. R. M., Mitra, S., Emran, T. B., ... Koirala, N. 2021. Antibiotic resistance in microbes: History, mechanisms, therapeutic strategies and future prospects. *Journal of Infection and Public Health*, *14*(12), 1750–1766.

USDA Taxon. *Moringa Oleifera lam.* Retrieved November 17, 2023, from https://npgsweb.ars-grin.gov/gringlobal/taxon/taxonomydetail?id=24597

Usman, A., Kawu, M. U., Shittu, M., Saleh, A., Jolayemi, K. O., Ibrahim, N. B., ... Okoronkwo, M. O. 2022. Comparative effects of methanol leaf extract of *Moringa oleifera* and ascorbic acid on haematological and histopathological changes induced by sub-chronic lead toxicity in male Wistar rats. *Pharmacological Research-Modern Chinese Medicine*, *2*, 100031.

Vergara-Jimenez, M., Almatrafi, M. M., & Fernandez, M. L. 2017. Bioactive components in *Moringa oleifera* leaves protect against chronic disease. *Antioxidants*, *6*(4), 91.

Wang, L., Chen, X., & Wu, A. 2016. Mini review on antimicrobial activity and bioactive compounds of *Moringa oleifera*. *Medicinal Chemistry*, *6*(9). https://doi.org/10.4172/2161-0444.1000402

Wanjiru, J., Gathirwa, J., Sauli, E., & Swai, H. S. 2022. Formulation, optimization, and evaluation of *Moringa oleifera* leaf polyphenol-loaded phytosome delivery system against breast cancer cell lines. *Molecules*, *27*(14), 4430.

Wu, Y., Yang, X., Hu, Y., Hu, X., Zhang, Y., An, T., ... Jiang, G. 2023. *Moringa oleifera* leaf supplementation relieves oxidative stress and regulates intestinal flora to ameliorate polycystic ovary syndrome in letrozole-induced rats. *Food Science & Nutrition*, *11*(9), 5137–5156.

Xie, J., Luo, F. X., Shi, C. Y., Jiang, W. W., Qian, Y. Y., Yang, M. R., ... Sheng, J. 2020. *Moringa oleifera* alkaloids inhibited PC3 cells growth and migration through the COX-2 mediated wnt/β-catenin signaling pathway. *Frontiers in Pharmacology*, *11*, 523962.

Younis, S. H., Obeid, R. F., & Ammar, M. M. 2020. Subsurface enamel remineralization by Lyophilized Moringa leaf extract loaded varnish. *Heliyon*, *6*(9), e05054. https://doi.org/10.1016/j.heliyon.2020.e05054

Yuan, H., Ma, Q., Ye, L., & Piao, G. 2016. Traditional medicine and modern medicine from natural products. *Molecules*, *21*(5), 559.

Zaku, S. G., Emmanuel, S., Tukur, A. A., & Kabir, A. 2015. *Moringa oleifera*: An underutilized tree in Nigeria with amazing versatility: A review. *African Journal of Food Science*, *9*(9), 456–461.

Zeng, B., Sun, J. J., Chen, T., Sun, B. L., He, Q., Chen, X. Y., ... Xi, Q. Y. 2018. Effects of *Moringa oleifera* silage on milk yield, nutrient digestibility and serum biochemical indexes of lactating dairy cows. *Journal of Animal Physiology and Animal Nutrition*, *102*(1), 75–81.

Zhang, Q, & Hengshan, S. 2001. *A practical Chinese-English dictionary of traditional Chinese medicine*. Shandong Science and Technology Press.

Zhang, M., Guo, M., Chen, N., Tang, Z., Xiang, J., Yang, L., ... Li, H. 2024. An integration of UPLC-Q-TOF-MS, GC-MS, electronic nose, electronic tongue, and molecular docking for the study of the chemical properties and flavor profiles of *Moringa oleifera* Leaves. *Chemosensors, 12*(9), 199.

Zheng, L., Lu, X., Yang, S., Zou, Y., Zeng, F., Xiong, S., ... Zhou, W. 2023. The anti-inflammatory activity of GABA-enriched *Moringa oleifera* leaves produced by fermentation with Lactobacillus plantarum LK-1. *Frontiers in Nutrition*, *10*, 1093036.

8 *Moringa oleifera* in Oral Care

Chandini Rajkumar, Saranya Ramsridhar, Indu Bharkavi S K and Murali Balasubramaniam A

8.1 INTRODUCTION

Oral diseases are among the most prevalent medical conditions worldwide, posing significant health and economic challenges for many nations (Peres et al., 2019). Approximately 3.5 billion people worldwide are estimated to suffer from various oral conditions, including periodontal disease, dental caries, tooth loss and oral cancers (Kassebaum et al., 2017). Most contemporary dental treatments have known side effects, and, due to their superior natural biocharacteristics, greater availability, reduced cost and broader safety profile, plant-derived substances are often used in preference to conventional drugs (Tartaglia et al., 2019). Across the globe, medicinal plants and plant extracts are commonly used for multiple therapeutic applications. They can be used to treat, regulate and manage numerous diseases due to their richness in bioactive compounds. Any plant that contains bioactive compounds used for treatment or as a raw material for drug manufacturing is medicinal (Abdelazim et al., 2024).

Besides being called the 'tree of life' or the 'miracle tree', *Moringa oleifera* is a major herbal plant because of its remarkable medical and non-medical advantages. Also known as the horseradish tree, ben oil tree or drumstick tree, it belongs to the Moringaceae family. It is mostly cultivated in the Himalayan foothills and has even extended to some regions of the United States, South America, East and West Africa, the Arabian Peninsula and Southeast and West Asia (Koul & Chase, 2015). *M. oleifera*'s seeds, flowers, pods and leaves are all valuable plant parts (Muzumbukilwa et al., 2019). The value of the plant as far as contribution to health and the environment goes back to approximately 2000 years; in 150 BCE, it was discovered in Indian Ayurvedic texts (Razis et al., 2014), suggesting that it was used to treat numerous illnesses, such as cholera, bronchitis, anaemia, respiratory infections, blood contamination and skin infections.

Terpenes, alkaloids, steroids, glucosinolates, flavonoids, phenolic acids and tannins are some of the bioactive compounds of *M. oleifera* (Leone, Spada, et al., 2015). It has some medicinal properties, including anti-inflammatory, antioxidant, antibacterial and hypoglycaemic activity, as well as flavonoids such as quercetin, rutin, kaempferol, myricetin, rhamnetin and apigenin. In addition, they help in wound healing and tissue regeneration (Mbikay, 2012; Mohanty et al., 2021; Muhammad et al., 2013). The leaves also contain various phenolic acids, including chlorogenic acid, caffeic acid and gallic acid (Leone, Fiorillo, et al., 2015; Leone, Spada, et al., 2015). We provide evidence that animals fed *M. oleifera* exhibit hypotensive and antimicrobial effects, and that the glucosinolates in *M. oleifera* are chemopreventive against cancer (Brunelli et al., 2010; Karim Nurul Ashikin Abdul, 2016). *M. oleifera* is also used to cure neurological disorders. The antibacterial, anticancer and antihepatotoxic activities of the plant are believed to be supported by tannins. Moringa contains alkaloids that also block calcium channels, and hence, it is a suitable drug against hypertension. Furthermore, leaf and seed extracts of the plant have been reported as powerful natural coagulants for water purification (Akhtar & Ahmad, 1995; Alam et al., 2020; Anwar & Bhanger, 2003; Dangi et al., 2002; Jaafaru et al., 2018; Ndhlala et al., 2014). Documenting the traditional Ayurvedic knowledge of *M. oleifera*, along with its sustainable practices for oral health care and integration with modern dentistry, can drive innovative developments in oral health care. There

DOI: 10.1201/9781003617860-9

is a significant lack of information on the use of *M. oleifera* in oral health care. This chapter will describe the current role of *M. oleifera* in dental health.

8.2 THE ANTI-INFLAMMATORY POTENTIAL OF *M. OLEIFERA*

M. Oleifera contains high levels of anti-inflammatory compounds that could help in oral inflammation. Different studies have examined its effectiveness in reducing/relieving oral inflammation and in treating gingivitis and periodontitis. Its high concentration of bioactive compounds, such as flavonoids, with anti-inflammatory and antioxidant properties, helps in the neutralisation of free radicals that cause cell damage and inflammation. The polyphenols interfere with the enzymes responsible for the inflammatory pathways. Therefore, these plant compounds reduce inflammation by inhibiting the activity of pro-inflammatory enzymes and modifying the synthesis of cytokines (Chiş et al., 2024)

The bioactive constituents that are found within the leaf extract are the prime target of modulation of the inflammatory reactions. Further, it is made evident through studies that the constituents effectively block the synthesis of primary pro-inflammatory cytokines such as tumour necrosis factor-alpha (TNF-α), interleukin-6 (IL-6), interleukin-8 (IL-8) and cyclooxygenase-2 (COX-2). These cytokines are typically seen elevated during inflammatory processes, particularly in immune cells such as monocyte-derived macrophages (MDMs) when stimulated by lipopolysaccharides (LPS).

M. oleifera primarily suppresses the nuclear factor-kappa B (NF-κB) signalling system, a crucial regulator of immunological and inflammatory processes, to provide its anti-inflammatory properties. To keep NF-κB in an inactive state within the cytoplasm, such extracts prevent the degradation of the inhibitor protein IκB-α. Furthermore, it inhibits the NF-κB p65 subunit's translocation into the nucleus, which stops the expression of the genes that produce inflammatory mediators. Through the disruption of this pathway, *M. oleifera* efficiently reduces inflammation and may help control inflammatory disorders (Kooltheat et al., 2014).

Periodontitis is a gum condition that changes the integral structure of the periodontium, leading to bone resorption and tooth support loss. Pro-inflammatory markers such as TNF-α and IL-6 play a role in osteoclast differentiation, which ultimately drives bone resorption and prevents new bone formation. In a study, the anti-inflammatory effects of *M. oleifera* leaf extract in managing periodontal inflammation caused by *Porphyromonas gingivalis*, which is a major pathogen in periodontitis, were explored. Results revealed that even at low concentrations, the moringa extract effectively reduced inflammatory responses by inhibiting the activity of the microorganism. Therefore, this study suggests that *M. oleifera* may serve as an indigenous therapeutic product in controlling periodontal inflammation, thereby supporting overall gum health (Ida et al., 2023; Luetragoon et al., 2020).

A clinical trial sought to ascertain the potential benefits for oral hygiene that come with using herbal lozenges containing leaf extracts from *M. Oleifera* and *Cyanthillium cinereum*. The research team discovered that the herbal lozenges exhibited robust antioxidant and anti-inflammatory effects in subjects with gingivitis, periodontitis and oral inflammation. Other studies focused on the anti-inflammatory properties of *M. Oleifera* leaf extract in a model of periodontitis induced by *P. gingivalis*. The research proved that the use of the extract significantly lowered IL-6 levels, the inflammation-promoting cytokine responsible for the worsening of periodontitis. The capacity of *M. Oleifera* to downregulate IL-6 was found to protect the gingiva through reductions in inflammatory reaction in the tissue and thus protecting the tissues from periodontal damage (Sugiharto et al., 2022).

Dos Santos et al. (2018) studied the effects of a natural isothiocyanate from *M. oleifera* for inflammatory hypernociception in the temporomandibular joint in a rat model. MC-D7 and MC-D9 reduced formalin-induced inflammation and nociception, and MC-H was more effective than indomethacin in reducing serotonin-induced hypernociception. In an in vivo study, male

Wistar rats were used to assess the analgesic effects of MC-H on TMJ pain relief. The findings indicated that MC-H exhibited anti-inflammatory and anti-nociceptive activity when administered orally. Such activities were attributed to peripheral mechanisms, including activation of the heme oxygenase-1 pathway and inhibition of intercellular adhesion molecule expression. Also, analgesic effects in the brain were due to the activation of opioid receptors (μ and δ) (Dos Santos et al., 2018).

Activity against periodontitis was tested using phenolic compounds extracted from the leaves of *M. oleifera* through a pharmacology and molecular docking approach. The results of an investigation involving both in vitro studies with RAW 264.7 macrophage cell cultures and an in vivo study of ligature periodontitis in rats suggested that moringa leaf extract provided its therapeutic action by modulating the classical p38α/MAPK14-OPG/RANKL pathway. The extract reduced the level of alveolar bone loss, elevated anti-inflammatory cytokines and reduced pro-inflammatory cytokines in the blood of the animal model (Wang et al., 2021).

The combined antioxidant and anti-inflammatory mechanisms of *M. oleifera* simultaneously protect oral tissue from damage and promote its healing, thereby supporting dental and oral hygiene. These *M. oleifera* features prevent further tissue damage and, at the same time, speed up the healing process, as they are useful for maintaining oral health (Shafiq & Mahdee, 2023).

8.3 THE ANTIMICROBIAL ACTIVITY OF *M. OLEIFERA*

Aqueous leaf juice and moringa seed extract suppress the growth of *Pseudomonas aeruginosa* and *Staphylococcus aureus*. Moringa seed ethanol extracts also suppress bacteria such as *Vibrio cholerae*, *S. aureus* and *Escherichia coli*. In addition, there are other antibacterial activities in moringa leaves and seeds that are useful for oral health (Padayachee & Baijnath, 2020). *M. oleifera*'s antiviral activity has been studied by a number of studies. The studies determined *M. oleifera*'s antibacterial activity using a variety of methods, such as disc agar diffusion and agar well diffusion. The leaves and seeds of this plant were extracted with the solvents of ethanol and chloroform and screened against all the bacterial species.

Moringa was compared with a battery of bacteria (disc agar diffusion test) using moringa against six Gram-negative bacteria: *Salmonella typhimurium*, *Enterobacter aerogenes*, *Shigella* spp., *E. coli*, *P. aeruginosa* and *Salmonella typhi* and a Gram-positive, *Staphylococcus aureus*. Ethanol extracts of moringa leaves proved to be inhibitory of four of the bacterial isolates: *E. coli*, *S. aureus*, *Enterobacter* spp. and *P. aeruginosa*. *Shigella* species, *S. typhimurium* and *S. typhi* were highly resistant to the ethanol extract of moringa leaves at all the concentrations tested. However, the moringa seed extract was active against *S. typhi*, *S. aureus* and *E. coli*, of which *S. aureus* and *E. coli* were resistant to other tested antibiotics, but was ineffective against *Shigella* spp., *Enterobacter* spp., *P. aeruginosa* and *S. typhimurium* bacterial isolates. Although not effective against other bacterial strains under study, the moringa leaf chloroform (MLC) extract was effective against *E. coli*, *S. typhi* and *S. typhimurium*. Likewise, the moringa seed chloroform (MSC) extract was not effective against other bacterial strains tested, but it was effective against *S. typhimurium* and *E. coli* (Bukar et al., 2010).

The bactericidal property of moringa seed extracts was assessed in Lar et al. (2011) through the agar well diffusion method. The relative potency of ethanol and aqueous moringa seed extracts was tested against three Gram-negative bacterial species: *S. typhi*, *Shigella flexneri* and *E. coli*. The aqueous extract, at subtest concentrations, was not active. The ethanol extract was also effective against *S. flexneri* and *E. coli*. No activity against *S. typhi* in ethanol and aqueous seed extracts was observed. Experiments demonstrated that aqueous, methanol, ethanol and petroleum ether extracts of moringa leaves have exhibited varying patterns of antibacterial inhibition. It was also found that extracts from various parts of the plant, such as leaves, flowers, pulp and seeds, exhibited activity against *E. coli* and *S. aureus* (Dodiya et al., n.d.).

Based on these findings, leaf extracts of *M. oleifera* can be used, either singly or in association with other antibiotics, to treat a variety of infectious diseases. To determine which organic solvent had the greatest antibacterial activity, Elgamily et al. (2016) carried out a study involving the preparation of *M. oleifera* plant extracts using ethanol, acetone and acetyl acetate. In accordance with earlier studies indicating that ethanolic leaf extract has no antifungal activity against *Candida albicans*, the current study demonstrated no antifungal activity among the tested extracts against this specific fungus. However, the results of the current study contradict another study that demonstrated antifungal activity of leaf extracts in ethanol and acetone against *C. albicans*. These discrepancies may be due to differences in seasonal conditions, the physiological status of the plant at the time of leaf collection or environmental conditions under which the plant was grown.

Cariogenic biofilms consisting of an overgrowth and accumulation of *Streptococcus mutans* are key players in the formation of dental caries. Bacterial biofilms possess a unique structure consisting of a glucan exopolymer matrix that serves as a barrier, rendering them resistant to antimicrobial drugs and preventing the entry of antibiotics. Recent research has shown that, based on the experimental results against this cariogenic biofilm, moringa and its extracts have been favourable.

The extracts were done with distilled water and ethyl alcohol. The results showed that the extracts of ethyl alcohol have significant inhibition of biofilm formation, removal of biofilms and mild bactericidal activity. Despite these positive aspects, however, distilled water did not completely remove the biofilm because it showed only moderate biofilm-preventing bactericidal activity, not as potent as calf blood serum. The results indicate that moringa extracts can be an effective way to prevent dental caries (Jwa, 2019).

8.4 THE ROLE OF *M. OLEIFERA* AS A DENTIFRICE AND MOUTHWASH

Besides being an antimicrobial, *M. oleifera* is rich in calcium, which results in denser bones and teeth. Recently, however, other studies have proved its effectiveness as an ingredient of a dentifrice formulation (Anas et al., 2021). This formula used much more than the control group to raise the calcium level in teeth. *M. oleifera* has done an excellent job in establishing itself in oral care systems, as well as the structural side of teeth, remineralising the tooth, if not being used against disease, and providing long-term oral health benefits. *M. oleifera* is a good therapeutic synergist and is useful as an agent in modern oral health practice. In addition to treating diseases, *M. oleifera* is an international and efficient method for increasing calcium uptake by teeth to prevent dental caries and maintain dental hygiene (Meena et al., 2021). A strong antimicrobial agent, chlorhexidine, is used for oral hygiene purposes to inhibit plaque formation, manage gingivitis and prevent oral infection. It is useful for bacteria and is usually recommended after dental surgery.

In vitro, the extract of *M. oleifera* (MOE) was evaluated as a possible source of antibiotics against *P. gingivalis*, *Tannerella forsythia* and *Treponema denticola* bacteria associated with periodontitis development. The purpose of this experiment was to determine the potential plant-based drug of *M. oleifera* as an investigational drug in the treatment of oral disease using the comparative efficiency of MOE to chlorhexidine mouthwash and control groups (i.e. phosphate-buffered saline). The disc diffusion assay, a traditional antimicrobial susceptibility testing method, was used to establish the antimicrobial property of MOE. *M. oleifera* was tested in aqueous solution and ethanol solution, and greater activity was proved in the latter. Chlorhexidine mouthwash produced the greatest zones of inhibition, but the aqueous and ethanolic extracts of *M. oleifera* were more effective against Gram-positive and Gram-negative bacteria, respectively. The microbicidal activity of the *M. oleifera* extract was attributed to its antibiotic compounds with broad antimicrobial activity and a natural antioxidant effect. *M. oleifera* extracts were stable for three weeks.

In vivo findings indicated that both chlorhexidine and *M. oleifera* mouthwashes significantly reduced plaque accumulation. There was a decline in plaque observed with the use of the *M. oleifera* mouthwash, which was on par with chlorhexidine, leading to its promising potential as a substitute treatment regimen. Though chlorhexidine showed a little greater potential, it was shown that *M. oleifera* mouthwash was a reasonable natural antimicrobial compound in dentistry (Sultana et al., 2024).

Another interesting study evaluated the antibacterial action of 5% *M. oleifera* mouthwash containing silver nanoparticles (AgNPs) against oral aerobic pathogens. AgNPs, when utilised as an antibacterial agent, emerged as a likely solution for reducing dental biofilm, dental caries and periodontal disease. *M. oleifera*-silver nanoparticle mouthwash showed dose-dependent antibacterial activity against bacteria like *S. aureus*, *S. mutans* and *Enterococcus faecalis* and maximum antimicrobial activity against plaque-forming bacteria. Moreover, it exhibited superior healing of wounds compared with chlorhexidine mouthwash in rabbit palatal wounds. These results point out the possibility of synergistic effects of phytomedicine and nanomedicine for safer plaque-associated gingival disease management (Kishore Kumar et al., 2022).

Another study evaluated the efficacy of mouthwashes with ethanolic leaf extracts of *M. oleifera* and *Azadirachta indica* in managing gingival inflammation. Over a period of two weeks, the participants using these herbal mouthwashes experienced a significant reduction in gingival and plaque indices. Additionally, there was a marked reduction in the numbers of *Staphylococcus* and *Candida* species in those using a chlorhexidine gluconate mouthwash. These findings indicate that *A. indica* and *M. oleifera* extracts could be useful alternatives in oral healthcare to reduce microbial load and gingivitis (Buakaew et al., 2021).

Duarte et al. (2022), in another randomised clinical crossover study, evaluated the effectiveness of a commercial dentifrice containing *M. oleifera* in subjects with mild to moderate gingivitis. The results indicated a significant decrease in the gingival and plaque indices in the dentifrice made from the moringa tree, compared with a dentifrice from the miswak (American tamarind) tree. Nevertheless, the research had several limitations, particularly a short experimental duration and a small study sample of 20 (Duarte et al., 2022). An investigation by AE and CP (2022) developed a novel antimicrobial toothpaste using *M. oleifera* root essential oil. The oil, incorporated at 2 ml per 100 g, gave a peppermint taste and contained highly active flavonoids, which are known for their antimicrobial activities. A new toothpaste was developed using 0.5 ml of the extract per 120 g. The efficacy of this toothpaste was subsequently tested against a range of bacteria and fungi associated with dental plaque (*S. mutans*, *Lactobacillus*, *S. aureus* and *C.albicans*) and compared with a well-known commercial toothpaste, using the agar well diffusion assay. The moringa toothpaste exhibited higher antimicrobial activity than the control toothpaste, with larger inhibition zones against all microorganisms tested. Particularly, it inhibited *S. mutans* (31 mm versus 22 mm) and *Lactobacillus* (33 mm versus 28 mm) more. This increased efficacy is possibly caused by the high content of flavonoid and phenolic groups in moringa extract, indicating that *M. oleifera* root essential oil possesses antibacterial activity against strains of bacteria responsible for dental plaque (AE & CP, 2022).

8.5 THE ANTICANCER POTENTIAL OF *M. OLEIFERA*

Oral cancer is a multifactorial disease likely caused by a combination of variables with no single causative agent. The majority of oral cancer occurrences are attributed to smoking. Cigarette smoking has several harmful by-products that are known to cause cancer, including polycyclic aromatic hydrocarbons (PAH). Among these, a well-known carcinogen is benzo[a]pyrene, a compound that is formed when organic matter is burnt incompletely. It binds to DNA, disrupting the transcription process and causing mutations that begin the process of transforming healthy cells into cancerous ones. The accumulation of altered DNA results in the development of cancer. Besides killing cancer cells, chemotherapy is harmful to healthy cells, and this has numerous effects.

Traditional cancer therapy systems like chemotherapy tend to destroy healthy cells along with cancerous cells, which results in unfavourable complications. Affected patients experience side effects like nausea, vomiting, dry mouth (xerostomia), difficulty in swallowing (dysphagia) and fatigue. These side effects adversely impact the patient's quality of life, in particular, their emotional well-being. As a result, many patients seek treatment options that can complement traditional therapy without producing significant side effects.

One promising alternative to the use of herbal plants relates to *M. oleifera*. The leaves of this miracle plant are said to contain anticancer bioactive molecules like flavonoids and isothiocyanates. Flavonoids proffer antioxidant and anti-inflammatory benefits. Likewise, isothiocyanates have been shown to induce apoptosis (programmed cell death) in cancerous cells. These properties make the moringa leaves a good natural medicinal source for cancerous diseases and minimise side effects (Budhy et al., 2023). *M. oleifera* leaf extract has displayed impressive anticancer potential. It triggers apoptosis in tumour cells by increasing the production of reactive oxygen species (ROS), which results in DNA fragmentation. Similar findings favouring apoptosis were observed with elevated production of caspase 3 and cell structural changes. Also, a drop in nuclear area factor (NAF) reading, which corresponds to a smaller or less circular nucleus that signifies cellular abnormality, is noted (Rajkumar et al., 2024).

Research shows that *M. oleifera* leaf extract has the potential to decrease cell viability in several cancer cell lines, including SCC15. Another study revealed that aqueous leaf extracts of *M. oleifera* demonstrated potent antioxidant properties. Similarly, stems, seeds and roots fractions of 70% ethanol extracts had good antioxidant activity. Additionally, the extracts were also found to have anticancer effects in head and neck cancer cells (Wang et al., 2020). Another advantage of *M. oleifera*'s leaf extract is that it has stronger efficacy when fused with chemotherapeutic drugs such as cisplatin. Proof of this finding comes from in vitro and animal research, yet more human trials are needed to corroborate these results, as well as standardised solutions to develop moringa extracts and solvents that give rise to the same anticancer outcome.

For instance, studies have also shown that *M. oleifera* seed oil possesses anticancer properties, namely, against oral cancer. It has been found that the seed oil has favourable anti-proliferative and cytotoxic properties against oral cancer cell lines CAL27 and SCC15, and heptacosane, the major component of this seed oil, exhibits anticancer properties. As a substrate, it blocks P-glycoprotein function, and as an inhibitor of P-glycoprotein, it increases the bioavailability of cytotoxic drugs and prevents their expulsion from cells following drug uptake. Thus, heptacosane can be used as an adjunct to chemotherapeutic drugs in order to enhance their efficacy. Such anticancer effects have previously been demonstrated in *M. oleifera* seed oil for oral cancer cell lines, but more in vivo studies and quantitative testing are needed to fully understand its therapeutic potential (Das et al., 2024).

8.6 THE EFFECT OF *M. OLEIFERA* IN REGENERATION AND HEALING

Research has shown that *M. oleifera* extract increases bone growth and improves the integration of intraosseous implants. They found that plant leaf extracts have osteo-conductive or osteo-inductive properties when applied to demineralised freeze-dried bovine bone xenografts. With the rise in transforming growth factor-beta one (TGF-β1) and osteocalcin, this preparation was found to exhibit enhanced bone regeneration and socket preservation following tooth extraction in an animal model of *Cavia cobaya* (Kresnoadi et al., 2019). Prior to their insertion into the femora and tibia of the rabbits, titanium dental implants were treated with a hydrophilic gel loaded with acemannan and *M. oleifera* extract. Hydrophilic coating created hydrophilic implant surfaces that increased de novo bone growth, promoted osseointegration and reduced fibrosis, inflammation, degeneration and necrotic activities in the altered bone (Pachimalla et al., 2020).

The bone-forming bioactivities of *M. oleifera* extracts need further research, especially in diseases such as diabetes mellitus, post-radiation treatment and osteoporosis, where systemic obstacles to bone formation complicate issues further for patients' health. To test the healing effects of *M. oleifera* leaf extract in gel form on oral excisional lesions of the palate in Sprague Dawley rats at two doses, 2% and 4%, another study used 10% iodine gel as a control (Amaliya et al., 2019). Histopathological analysis was performed on moringa-treated samples versus control samples, and in the early stage of palatal wound healing, the deposition of fibroblasts and collagen fibres was observed to have a higher activity compared with that of the control group. Results showed that the extract of *M. oleifera* leaf stimulated oral wound healing due to its antibacterial and anti-inflammatory properties.

The outcomes and consequences of this information and its application are, for now, not clearly seen and therefore need to be investigated. Additionally, the optimum therapeutic dose and assurance of the formulation's physical stability are to be determined prior to its application in clinical settings. Besides, in vitro studies have indicated that moringin, an isothiocyanate derived from *M. Oleifera* seeds, was capable of inducing hPDLSCs to neural progenitor, adipogenic and osteogenic differentiation without any evident tumourigenic effects (Romeo et al., 2018). This other set of investigations concentrated on the genomic aspects of mitophagy in moringin-pre-treated hPDLSCs. Their study has shown significant downregulation of almost all genes compared with untreated cells. However, this phenomenon can positively impact medicine by supplying validation that moringin might enhance the therapeutic efficacy of hPDLSCs under oxidative stress, most notably in neurodegenerative conditions (Chiricosta et al., 2019).

As per reports, *M. oleifera* water extracts increased human dermal fibroblast cell motility, viability and proliferation in vitro (Muhammad et al., 2013). Further, in a model organism, *M. oleifera* aqueous extract healed wounds in the presence of ongoing hyperglycaemia by diminishing the wound area, enhancing tissue repair and contraction, suppressing inflammatory mediators and increasing vascular endothelial growth factor (VEGF) (Muhammad et al., 2016). Without stimulating the development of malignant cancer cells, *M. oleifera* flower water extracts have exhibited promising activities in stimulating rat fibroblast cell proliferation, enhancing the differentiation of bone marrow-derived mesenchymal stem cells and stimulating angiogenesis. In contrast, aqueous leaf extract showed strong cytotoxic activity against MCF-7 breast cancer cell lines but a weakened capacity to induce cell development (Fernandes et al., 2016).

In another study, aqueous *M. oleifera* leaf extract was also shown to cause a decrease in skin wound area in diabetic rats, along with the enhancement of macrophage and fibroblast numbers. The result was a significantly increased degree of wound contraction over the control group treated with normal saline (Azevedo et al., 2018). An n-Hexane extract and *M. oleifera* seeds showed wound-healing activity by inducing the regeneration of tissues, inhibiting inflammatory cell infiltration and facilitating vascularity in the generated skin. These activities were observed in mouse models of incision and excision wound healing (Ali et al., 2021).

The use of a gel layer that contained *M. oleifera* leaf extract-loaded chitosan microparticles formed a barrier over the wound while permitting the delivery of active components over an extended period. Through an ex vivo model of pig skin, it was determined that the formulation caused the growth of keratinocytes and hence might be used in treating exuding wounds (Pagano et al., 2020). An exuding wound dressing with 0.5% *M. oleifera* leaf extract-charged film illustrated good drug-releasing characteristics, augmenting wound repair in streptozotocin-induced type II diabetic rat models (Chin et al., 2019). Another *M. oleifera* gum polysaccharide carbopol-based hydrogel dressing was created to provide moist conditions over the wound bed, facilitating re-epithelialisation and accelerating healing. Alginate-PEG methyl ether methacrylate scaffolds loaded with *M. oleifera* and *Aloe vera* greatly enhanced the viability of human skin fibroblast cells over control scaffolds in cell-free and cell-based assays (Rubio-Elizalde et al., 2019; Singh & Kumar, 2020).

In addition, 1-O-(4-hydroxymethylphenyl)-α-L-rhamnopyranoside (MPG), a phenolic glycoside extracted from *M. oleifera* seeds, was tested for its hepatoprotective activity against CCl_4-induced hepatotoxicity in mice and L02 cell lines (immortalised human hepatocyte cell line L02). Through its antioxidant activity, modulation of inflammatory mediators and apoptosis inhibition, MPG effectively prevents acute CCl_4-induced hepatotoxicity, as demonstrated in both in vitro and in vivo studies (Sun et al., 2019). In the biosynthesis of metal and metal oxide nanoparticles, extracts from various regions of *M. oleifera* have served as stabilisers and reducing/oxidising agents. It is known that many phytochemicals from the plant, in the form of tannins, saponins, flavonoids, alkaloids, vitamins and so on, present in the media support the development and stability of nanoparticles (Gopalakrishnan et al., 2016; Jadhav et al., 2022).

Gold nanoparticles were biosynthesised using *M. Oleifera* leaf extract as a reducing and stabilising agent. Low cytotoxicity of the blood cells and biological activity of the synthesised nanoparticles were observed. They further demonstrated that the nanoparticles enhanced neuronal cell regeneration in Swiss albino mice, evidenced by improved brain tissue architecture. It was reported that carriers with nanoparticles were biosynthesised with gold chloride ($HAuCl_4 \cdot 3H_2O$) and *M. Oleifera* leaf extract as a stabiliser. Some stirring at 75°C was conducted on the mixture; then it was filtered and centrifuged for the removal of excess heavy contaminants (Boruah et al., 2021). Some researchers considered using a polysaccharide extract from *M. Oleifera* seeds as a substitute in nanocomposite synthesis. For the synthetic process, the extract and $AgNO_3$ solution in distilled water were mixed, and the products were isolated by centrifugation. Afterwards, the nanoparticles were heated in an oven at 80°C to accomplish nanoparticle synthesis. These nanoparticles were non-toxic to mouse fibroblast L929 cells, promoting their in vitro migration while exhibiting antimicrobial activity towards pathogens responsible for wound infections. Furthermore, in a rat wound-healing model in male rats, they showed superior biocompatibility, facilitating infected wound healing by mitigating inflammation, eliminating bacteria and inducing complete epithelialisation within nine days (Mehwish et al., 2021). Increasing evidence points to extracts from various parts of *M. oleifera* that promote normal stem cell proliferation and other cell types, facilitating wound healing and tissue regeneration and inducing angiogenesis (Ramamurthy et al., 2021). Given their antimicrobial activity, these extracts of *M. oleifera* are therapeutic candidates for regenerating the dentin-pulp complex following dental injury and warrant further study.

8.7 THE SIDE EFFECTS OF *M. OLEIFERA*

Although *M. oleifera* is generally regarded as safe, it is essential to be careful of its adverse consequences. Few studies report the possible adverse side effects of moringa consumption. To date, there have been no notable negative effects reported concerning moringa consumption.

In a study, it was found that the phytochemical constituents responsible for the plant's bitter taste include the sulphur-containing compound glucosinolates and other alkaloids (Chiş et al., 2024; Chodur et al., 2018).

In addition to skin problems, breathing difficulties and swelling of the tongue, moringa powder has been reported to cause toxicity in the body (Sagrera et al., 2021).

8.8 CONCLUSION

Overall, *M. oleifera* may be a good candidate for a natural oral remedy for healthy gums, plaque prevention and battling disease-causing microorganisms taking hold in the mouth, thanks to its high nutrient content, antioxidants and antibacterial activities. It is easy to incorporate the plant into oral care owing to its use in powder, oil and extract forms. It also has minimal side effects, but more research will be needed to demonstrate that it does not cause adverse health effects after prolonged use. Due to additional knowledge about its regenerative properties, which give it consideration as a general as well as an oral health candidate, further research is needed on its application. As scientific research advances, *M. oleifera* may become a major component of a natural, practical and effective dental care product for enhancing overall health.

REFERENCES

Abdelazim, A. M., Afifi, M., Abu-Alghayth, M. H., & Alkadri, D. H. (2024). Moringa oleifera: Recent insights for its biochemical and medicinal applications. *Journal of Food Biochemistry, 2024*. https://doi.org/10.1155/2024/1270903

AE, A., & CP, E. (2022). Production of herbal toothpaste using Moringa root essential oil extract. *International Journal of Advanced Biochemistry Research*, *6*(2). https://doi.org/10.33545/26174693.2022.v6.i2a.134

Akhtar, A. H., & Ahmad, K. U. (1995). Anti-ulcerogenic evaluation of the methanolic extracts of some indigenous medicinal plants of Pakistan in aspirin-ulcerated rats. *Journal of Ethnopharmacology*, *46*(1). https://doi.org/10.1016/0378-8741(94)01220-T

Alam, M. W., Pandey, P., Khan, F., Souayeh, B., & Farhan, M. (2020). Study to investigate the potential of combined extract of leaves and seeds of moringa oleifera in groundwater purification. *International Journal of Environmental Research and Public Health*, *17*(20). https://doi.org/10.3390/ijerph17207468

Ali, A., Garg, P., Goyal, R., Kaur, G., Li, X., Negi, P., Valis, M., Kuca, K., & Kulshrestha, S. (2021). A novel herbal hydrogel formulation of moringa oleifera for wound healing. *Plants*, *10*(1). https://doi.org/10.3390/plants10010025

Amaliya, A., Muhaimina, R. K., Susanto, A., & Sutjiatmo, A. B. (2019). Histological assessment of palatal donor site wound healing after application of Moringa oleifera lamarck leaf extract in rats. *European Journal of Dentistry*, *13*(2). https://doi.org/10.1055/s-0039-1695065

Anas, R., Kirana Mattulada, I., Akbar, F. H., Indriany, L., Asmah, N., Irawati, E., & Dharma Utama, M. (2021) Effectiveness of paste-based *Moringa oleifera* to increase calcium levels in human tooth (in vitro). *Volatiles & Essential Oils, 8*(4), 15193-15201.

Anwar, F., & Bhanger, M. I. (2003). Analytical characterization of Moringa oleifera seed oil grown in temperate regions of Pakistan. *Journal of Agricultural and Food Chemistry*, *51*(22). https://doi.org/10.1021/jf0209894

Azevedo, Í. M., Araújo-Filho, I., Teixeira, M. M. A., Moreira, M. D. F. de C., & Medeiros, A. C. (2018). Wound healing of diabetic rats treated with moringaoleifera extract. *Acta Cirurgica Brasileira*, *33*(9). https://doi.org/10.1590/s0102-865020180090000008

Boruah, J. S., Devi, C., Hazarika, U., Bhaskar Reddy, P. V., Chowdhury, D., Barthakur, M., & Kalita, P. (2021). Green synthesis of gold nanoparticles using an antiepileptic plant extract: In vitrobiological and photocatalytic activities. *RSC Advances*, *11*(45), 28029–28041. https://doi.org/10.1039/d1ra02669k

Brunelli, D., Tavecchio, M., Falcioni, C., Frapolli, R., Erba, E., Iori, R., Rollin, P., Barillari, J., Manzotti, C., Morazzoni, P., & D'Incalci, M. (2010). The isothiocyanate produced from glucomoringin inhibits NF-kB and reduces myeloma growth in nude mice in vivo. *Biochemical Pharmacology*, *79*(8). https://doi.org/10.1016/j.bcp.2009.12.008

Buakaew, W., Sranujit, R. P., Noysang, C., Sangouam, S., Suphrom, N., Thongsri, Y., Potup, P., & Usuwanthim, K. (2021). Evaluation of mouthwash containing citrus hystrix dc., moringa oleifera lam. and azadirachta indica a. juss. leaf extracts on dental plaque and gingivitis. *Plants*, *10*(6). https://doi.org/10.3390/plants10061153

Budhy, T. I., Naori, P., Ridhatillah, D., Dwiamanda, L., & Evriany, M. (2023). The potency of chitosan-based Moringa oleifera leaves extract nanoparticles as anti-cancer agent. *Research Journal of Pharmacy and Technology*, *16*(1). https://doi.org/10.52711/0974-360X.2023.00007

Bukar, A., Uba, A., & Oyeyi, T. (2010). Antimicrobial profile of moringa oleifera lam. Extracts against some food – borne microorganisms. *Bayero Journal of Pure and Applied Sciences*, *3*(1). https://doi.org/10.4314/bajopas.v3i1.58706

Chin, C. Y., Ng, P. Y., & Ng, S. F. (2019). Moringa oleifera standardised aqueous leaf extract-loaded hydrocolloid film dressing: in vivo dermal safety and wound healing evaluation in STZ/HFD diabetic rat model. *Drug Delivery and Translational Research*, *9*(2). https://doi.org/10.1007/s13346-018-0510-z

Chiricosta, L., Gugliandolo, A., Diomede, F., Pizzicannella, J., Trubiani, O., Iori, R., Tardiolo, G., Guarnieri, S., Bramanti, P., & Mazzon, E. (2019). Moringin pretreatment inhibits the expression of genes involved in mitophagy in the stem cell of the human periodontal ligament. *Molecules*, *24*(18). https://doi.org/10.3390/molecules24183217

Chiş, A., Noubissi, P. A., Pop, O. L., Mureşan, C. I., Fokam Tagne, M. A., Kamgang, R., Fodor, A., Sitar-Tăut, A. V., Cozma, A., Orăşan, O. H., Hegheş, S. C., Vulturar, R., & Suharoschi, R. (2024). Bioactive compounds in Moringa oleifera: Mechanisms of action, focus on their anti-inflammatory properties. *Plants*, *13*(1). https://doi.org/10.3390/plants13010020

Chodur, G. M., Olson, M. E., Wade, K. L., Stephenson, K. K., Nouman, W., Garima, & Fahey, J. W. (2018). Wild and domesticated Moringa oleifera differ in taste, glucosinolate composition, and antioxidant potential, but not myrosinase activity or protein content. *Scientific Reports*, *8*(1). https://doi.org/10.1038/s41598-018-26059-3

Dangi, S. Y., Jolly, C. I., & Narayanan, S. (2002). Antihypertensive activity of the total alkaloids from the leaves of Moringa oleifera. *Pharmaceutical Biology*, *40*(2). https://doi.org/10.1076/phbi.40.2.144.5847

Das, M., Mohanty, J. N., Bhuyan, S. K., & Bhuyan, R. (2024). Anti-cancer activity of Moringa oleifera Lam. seed oil on oral cancer. *Beni-Suef University Journal of Basic and Applied Sciences*, *13*(1). https://doi.org/10.1186/s43088-024-00475-z

Dodiya, B., Amin, B., Kamlaben, S., & Patel, P. (n.d.). Antibacterial activity and phytochemical screening of different parts of Moringa oleifera against selected gram positive and gram negative Bacteria. *Journal of Pharmaceutical, Chemical and Biological Sciences*, (3), 1-27.

dos Santos, A. O., do Val, D. R., da Silveira, F. D., Gomes, F. I. F., Freitas, H. C., de Assis, E. L., de Almeida, D. K. C., da Silva, I. I. C., Barbosa, F. G., Mafezoli, J., da Silva, M. R., de Castro Brito, G. A., Clemente-Napimoga, J. T., de Paulo Teixera Pinto, V. de P. T., Filho, G. C., Bezerra, M. M., & Chaves, H. V. (2018). Antinociceptive, anti-inflammatory and toxicological evaluation of semi-synthetic molecules obtained from a benzyl-isothiocyanate isolated from Moringa oleifera Lam. in a temporomandibular joint inflammatory hypernociception model in rats. *Biomedicine and Pharmacotherapy*, *98*. https://doi.org/10.1016/j.biopha.2017.12.102

Duarte, K., Thomas, B., Varma, S. R., Kamath, V., Shetty, B., Kuduruthullah, S., & Nambiar, M. (2022). Antiplaque efficacy of a novel Moringa oleifera dentifrice: A randomized clinical crossover study. *European Journal of Dentistry*, *16*(4). https://doi.org/10.1055/s-0041-1736418

Elgamily, H., Moussa, A., Elboraey, A., El-Sayed, H., Al-Moghazy, M., & Abdalla, A. (2016). Microbiological assessment of Moringa oleifera extracts and its incorporation in novel dental remedies against some oral pathogens. *Open Access Macedonian Journal of Medical Sciences*, *4*(4). https://doi.org/10.3889/oamjms.2016.132

Fernandes, E., Pulwale, A., Patil, G., & Moghe, A. (2016). Probing regenerative potential of Moringa oleifera aqueous extracts using in vitro cellular assays. *Pharmacognosy Research*, *8*(4). https://doi.org/10.4103/0974-8490.188877

Gopalakrishnan, L., Doriya, K., & Kumar, D. S. (2016). Moringa oleifera: A review on nutritive importance and its medicinal application. *Food Science and Human Wellness*, *5*(2). https://doi.org/10.1016/j.fshw.2016.04.001

Ida, M. J., Shetty, B., Khan, S. F., Yadalam, U., & Nambiar, M. (2023). Development and in vitro characterization of a mucoadhesive gel with Moringa oleifera extract for periodontal drug delivery. *Journal of Indian Society of Periodontology*, *27*(2). https://doi.org/10.4103/jisp.jisp_710_21

Jaafaru, M. S., Nordin, N., Shaari, K., Rosli, R., & Abdull Razis, A. F. (2018). Isothiocyanate from Moringa oleifera seeds mitigates hydrogen peroxide-induced cytotoxicity and preserved morphological features of human neuronal cells. *PLoS One*, *13*(5). https://doi.org/10.1371/journal.pone.0196403

Jadhav, V., Bhagare, A., Ali, I. H., Dhayagude, A., Lokhande, D., Aher, J., Jameel, M., & Dutta, M. (2022). Role of Moringa oleifera on green synthesis of metal/metal oxide nanomaterials. *Journal of Nanomaterials*, *2022*. https://doi.org/10.1155/2022/2147393

Jwa, S. K. (2019). Efficacy of moringa oleifera leaf extracts against cariogenic biofilm. *Preventive Nutrition and Food Science*, *24*(3). https://doi.org/10.3746/pnf.2019.24.3.308

Karim NA, Ibrahim MD, Kntayya SB, Rukayadi Y, Hamid HA, Razis AF. Moringa oleifera Lam: Targeting Chemoprevention. Asian Pac J Cancer Prev. 2016;17(8):3675-86. PMID: 27644601.

Kassebaum, N. J., Smith, A. G. C., Bernabé, E., Fleming, T. D., Reynolds, A. E., Vos, T., Murray, C. J. L., Marcenes, W., Abyu, G. Y., Alsharif, U., Asayesh, H., Benzian, H., Dandona, L., Dandona, R., Kasaeian, A., Khader, Y. S., Khang, Y. H., Kokubo, Y., Kotsakis, G. A., … Yonemoto, N. (2017). Global, regional, and national prevalence, incidence, and disability-adjusted life years for oral conditions for 195 countries, 1990–2015: A systematic analysis for the global burden of diseases, injuries, and risk factors. *Journal of Dental Research*, *96*(4). https://doi.org/10.1177/0022034517693566

Kishore Kumar, G., Ramamurthy, S., Ulaganathan, A., Varghese, S., Praveen, A. A., & Saranya, V. (2022). Moringa oleifera mouthwash reinforced with silver nanoparticles - Preparation, characterization and its efficacy against oral aerobic microorganisms - in vitro study. *Biomedical and Pharmacology Journal*, *15*(4). https://doi.org/10.13005/bpj/2542

Kooltheat, N., Pankla Sranujit, R., Chumark, P., Potup, P., Laytragoon-Lewin, N., & Usuwanthim, K. (2014). An ethyl acetate fraction of Moringa oleifera Lam. inhibits human macrophage cytokine production induced by cigarette smoke. *Nutrients*, *6*(2). https://doi.org/10.3390/nu6020697

Koul, B., & Chase, N. (2015). Moringa oleifera Lam.: panacea to several maladies. *Journal of Chemical and Pharmaceutical Research*, *7*(6), 687-707.

Kresnoadi, U., Rahmania, P. N., Caesar, H. U., Djulaeha, E., Agustono, B., & Ari, M. D. A. (2019). The role of the combination of Moringa oleifera leaf extract and demineralized freeze-dried bovine bone xenograft (xenograft) as tooth extraction socket preservation materials on osteocalcin and transforming growth factor-beta 1 expressions in alveolar bone of Cavia cobaya. *The Journal of Indian Prosthodontic Society*, *19*(2). https://doi.org/10.4103/jips.jips_251_18

Lar, P. M., Ojile, E. E., Dashe, E., & Oluoma, J. N. (2011). Antibacterial activity of moringa oleifera seed extracts on some gram negative bacterial isolates. *African Journal of Natural Sciences*, *14*, 57-62.

Leone, A., Fiorillo, G., Criscuoli, F., Ravasenghi, S., Santagostini, L., Fico, G., Spadafranca, A., Battezzati, A., Schiraldi, A., Pozzi, F., Di Lello, S., Filippini, S., & Bertoli, S. (2015). Nutritional characterization and phenolic profiling of moringa oleifera leaves grown in chad, sahrawi refugee camps, and Haiti. *International Journal of Molecular Sciences*, *16*(8). https://doi.org/10.3390/ijms160818923

Leone, A., Spada, A., Battezzati, A., Schiraldi, A., Aristil, J., & Bertoli, S. (2015). Cultivation, genetic, ethnopharmacology, phytochemistry and pharmacology of Moringa oleifera leaves: An overview. *International Journal of Molecular Sciences*, *16*(6). https://doi.org/10.3390/ijms160612791

Luetragoon, T., Sranujit, R. P., Noysang, C., Thongsri, Y., Potup, P., Suphrom, N., Nuengchamnong, N., & Usuwanthim, K. (2020). Anti-cancer effect of 3-hydroxy-β-ionone identified from moringa oleifera Lam. leaf on human squamous cell carcinoma 15 cell line. *Molecules*, *25*(16). https://doi.org/10.3390/molecules25163563

Mbikay, M. (2012). Therapeutic potential of Moringa oleifera leaves in chronic hyperglycemia and dyslipidemia: A review. *Frontiers in Pharmacology*, *3*. https://doi.org/10.3389/fphar.2012.00024

Meena, R., Prajapati, S. K., Nagar, R., Porwal, O., Nagar, T., Tilak, V. K., Jayakumararaj, R., Arya, R. K. K., & Dhakar, R. C. (2021). Application of Moringa oleifera in dentistry. *Asian Journal of Dental and Health Sciences*, *1*(1). https://doi.org/10.22270/ajdhs.v1i1.5

Mehwish, H. M., Liu, G., Rajoka, M. S. R., Cai, H., Zhong, J., Song, X., Xia, L., Wang, M., Aadil, R. M., Inam-Ur-Raheem, M., Xiong, Y., Wu, H., Amirzada, M. I., Zhu, Q., & He, Z. (2021). Therapeutic potential of Moringa oleifera seed polysaccharide embedded silver nanoparticles in wound healing. *International Journal of Biological Macromolecules*, *184*. https://doi.org/10.1016/j.ijbiomac.2021.05.202

Mohanty, M., Mohanty, S., Bhuyan, S. K., & Bhuyan, R. (2021). Phytoperspective of Moringa oleifera for oral health care: An innovative ethnomedicinal approach. *Phytotherapy Research*, *35*(3). https://doi.org/10.1002/ptr.6896

Muhammad, A. A., Arulselvan, P., Cheah, P. S., Abas, F., & Fakurazi, S. (2016). Evaluation of wound healing properties of bioactive aqueous fraction from Moringa oleifera Lam on experimentally induced diabetic animal model. *Drug Design, Development and Therapy*, *10*. https://doi.org/10.2147/DDDT.S96968

Muhammad, A. A., Pauzi, N. A. S., Arulselvan, P., Abas, F., & Fakurazi, S. (2013). In vitro wound healing potential and identification of bioactive compounds from Moringa oleifera Lam. *BioMed Research International*, *2013*. https://doi.org/10.1155/2013/974580

Muzumbukilwa, W. T., Nlooto, M., & Owira, P. M. O. (2019). Hepatoprotective effects of Moringa oleifera Lam (Moringaceae) leaf extracts in streptozotocin-induced diabetes in rats. *Journal of Functional Foods*, *57*. https://doi.org/10.1016/j.jff.2019.03.050

Ndhlala, A. R., Mulaudzi, R., Ncube, B., Abdelgadir, H. A., Du Plooy, C. P., & Staden, J. Van. (2014). Antioxidant, antimicrobial and phytochemical variations in thirteen moringa oleifera lam. Cultivars. *Molecules*, *19*(7). https://doi.org/10.3390/molecules190710480

Pachimalla, P. R., Mishra, S. K., & Chowdhary, R. (2020). Evaluation of hydrophilic gel made from Acemannan and Moringa oleifera in enhancing osseointegration of dental implants. A preliminary study in rabbits. *Journal of Oral Biology and Craniofacial Research*, *10*(2). https://doi.org/10.1016/j.jobcr.2020.01.005

Padayachee, B., & Baijnath, H. (2020). An updated comprehensive review of the medicinal, phytochemical and pharmacological properties of Moringa oleifera. *South African Journal of Botany*, *129*. https://doi.org/10.1016/j.sajb.2019.08.021

Pagano, C., Perioli, L., Baiocchi, C., Bartoccini, A., Beccari, T., Blasi, F., Calarco, P., Ceccarini, M. R., Cossignani, L., di Michele, A., Ortenzi, R., Scuota, S., & Ricci, M. (2020). Preparation and characterization of polymeric microparticles loaded with Moringa oleifera leaf extract for exuding wound treatment. *International Journal of Pharmaceutics*, *587*. https://doi.org/10.1016/j.ijpharm.2020.119700

Peres, M. A., Macpherson, L. M. D., Weyant, R. J., Daly, B., Venturelli, R., Mathur, M. R., Listl, S., Celeste, R. K., Guarnizo-Herreño, C. C., Kearns, C., Benzian, H., Allison, P., & Watt, R. G. (2019). Oral diseases: A global public health challenge. *The Lancet*, *394*(10194). https://doi.org/10.1016/S0140-6736(19)31146-8

Rajkumar, C., Ramsridhar, S., Veeraraghavan, V. P., Francis, A. P., Purushotham, M., & Mageshwari, U. (2024). Anticancer effect of Moringa oleifera in oral squamous cell carcinoma: A systematic review. *Discover Oncology*, *15*(1). https://doi.org/10.1007/s12672-024-01557-1

Ramamurthy, S., Varghese, S., Sudarsan, S., Muruganandhan, J., Mushtaq, S., Patil, P. B., Raj, A. T., Zanza, A., Testarelli, L., & Patil, S. (2021). Moringa oleifera: Antioxidant, anticancer, anti-inflammatory, and related properties of extracts in cell lines: A review of medicinal effects, phytochemistry, and applications. *Journal of Contemporary Dental Practice*, *22*(12). https://doi.org/10.5005/jp-journals-10024-3187

Razis, A. F. A., Ibrahim, M. D., & Kntayya, S. B. (2014). Health benefits of Moringa oleifera. *Asian Pacific Journal of Cancer Prevention*, *15*(20), 8571–8576. https://doi.org/10.7314/APJCP.2014.15.20.8571

Romeo L, Diomede F, Gugliandolo A, Scionti D, Lo Giudice F, Lanza Cariccio V, Iori R, Bramanti P, Trubiani O, Mazzon E. Moringin Induces Neural Differentiation in the Stem Cell of the Human Periodontal Ligament. Sci Rep. 2018 Jun 14;8(1):9153. doi: 10.1038/s41598-018-27492-0. PMID: 29904155; PMCID: PMC6002387.

Rubio-Elizalde, I., Bernáldez-Sarabia, J., Moreno-Ulloa, A., Vilanova, C., Juárez, P., Licea-Navarro, A., & Castro-Ceseña, A. B. (2019). Scaffolds based on alginate-PEG methyl ether methacrylate-Moringa oleifera-Aloe vera for wound healing applications. *Carbohydrate Polymers, 206.* https://doi.org/10.1016/j.carbpol.2018.11.027

Sagrera, A., Montenegro, T., & Borrego, L. (2021). Cutaneous toxicity due to Moringa oleifera. *Actas Dermo-Sifiliograficas, 112*(10). https://doi.org/10.1016/j.ad.2020.03.011

Shafiq, N. E., & Mahdee, A. F. (2023). Moringa oleifera use in maintaining oral health and its potential use in regenerative dentistry. *Scientific World Journal, 2023.* https://doi.org/10.1155/2023/8876189

Singh, B., & Kumar, A. (2020). Graft and crosslinked polymerization of polysaccharide gum to form hydrogel wound dressings for drug delivery applications. *Carbohydrate Research, 489.* https://doi.org/10.1016/j.carres.2020.107949

Sugiharto, S., Ramadany, S., Handayani, H., Achmad, H., Mutmainnah, N., Inayah, N. H., Sesioria, A., & Thamrin, A. Z. (2022). Assessment of the anti-inflammatory activities of the moringa leaf extract in periodontitis cases through IL-6 Cytokine analysis in wistar (rattus novergicus). *Open Access Macedonian Journal of Medical Sciences, 10.* https://doi.org/10.3889/oamjms.2022.8381

Sultana, A., Tasneem, A., Krishnappa, P., & Shwetha, K. M. (2024). Effectiveness of Moringa oleifera mouthwash in young adults as an anti plaque agent – an interventional study. *Journal of Indian Association of Public Health Dentistry, 22*(2). https://doi.org/10.4103/jiaphd.jiaphd_242_23

Sun, C., Li, W., Liu, Y., Deng, W., Adu-Frimpong, M., Zhang, H., Wang, Q., Yu, J., & Xu, X. (2019). In vitro/in vivo hepatoprotective properties of 1-O-(4-hydroxymethylphenyl)-α-L-rhamnopyranoside from Moringa oleifera seeds against carbon tetrachloride-induced hepatic injury. *Food and Chemical Toxicology, 131.* https://doi.org/10.1016/j.fct.2019.05.039

Tartaglia, G. M., Tadakamadla, S. K., Connelly, S. T., Sforza, C., & Martín, C. (2019). Adverse events associated with home use of mouthrinses: A systematic review. *herapeutic Advances in Drug Safety, 10.* https://doi.org/10.1177/2042098619854881

Wang, F., Long, S., & Zhang, J. (2021). Moringa oleifera Lam. leaf extract safely inhibits periodontitis by regulating the expression of p38α/MAPK14-OPG/RANKL. *Archives of Oral Biology, 132.* https://doi.org/10.1016/j.archoralbio.2021.105280

Wang, F., Long, S., Zhang, J., Yu, J., Xiong, Y., Zhou, W., Qiu, J., & Jiang, H. (2020). Antioxidant activities and anti-proliferative effects of Moringa oleifera L. extracts with head and neck cancer. *Food Bioscience, 37,* 100691. https://doi.org/10.1016/J.FBIO.2020.100691

9 *Moringa oleifera* as an Antithrombotic and Thrombolytic

Sawetaji, Jyoti Sehrawat and Kamal Krishan Aggarwal

9.1 INTRODUCTION

Natural products have demonstrated their antithrombotic and thrombolytic effects in both experimental and clinical settings, offering a valuable alternative and supplement for pharmaceutical treatments (Chen et al., 2015). New insights to advance human health will come from our understanding of the biological processes involved in natural products as well as the mechanisms of thrombus formation. Currently, much effort has been made towards the discovery and development of natural products as anticoagulant agents in the replacement of available antithrombotic drugs (Newman & Cragg, 2016) and thrombolytic agents. Experimental and clinical phases have revealed that natural compounds, traditional medicines, functional foods and certain unique animal components exhibit antithrombotic properties (Chen et al., 2015). The primary rationale behind using natural products in the treatment of thrombotic illnesses is that they contain a variety of molecules, each of which may have many targets. These constituents may exhibit pleiotropic and synergistic actions that serve to increase therapeutic efficacy (Yang et al., 2023).

Every year, ten million people worldwide die from arterial thrombosis. Pharmaceutical industries have taken note of herbal medicines as alternative therapies due to the rising prevalence of thromboembolic diseases, and investigations are being conducted on new antithrombotic drugs with fewer harmful side effects (Subramani & Sathiyarajeswaran, 2022). Recent studies on the antiplatelet, fibrinolytic and anticoagulant qualities of plant extracts or natural compounds have shown a growing interest in antithrombotic research (Lamponi, 2021). These products have potential applications as complementary or alternative therapies. They may also serve as a starting point for the identification of novel antithrombotic compounds (Lamponi, 2021).

Moringa oleifera Lam. is a widely grown species in India and is locally known as the 'drumstick' plant, as seen in Figure 9.1. This plant is known for its nutritional and medicinal properties (Table 9.1). *Moringa* sp. is used in the treatment of different diseases, including heart-related problems (Bijina et al., 2011). *Moringa* sp. leaf powder is known for its heart health benefits, particularly in blood lipid control as well as in the prevention of plaque formation in the arteries (Thapa et al., 2019).

Moringa oleifera is highly regarded for its exceptional nutritional content and the presence of bioactive phytochemicals with medicinal importance. The edible parts of the plant are rich in nutrients, essential and non-essential amino acids, minerals, vitamins, antioxidants and various phenolic compounds (Boopathi et al., 2021). *Moringa oleifera* has gained considerable recognition for its thrombin-inhibitory and antithrombotic properties, with different plant parts showcasing unique bioactive effects.

DOI: 10.1201/9781003617860-10

FIGURE 9.1 A Specimen of *M. oleifera* Lam. (Source: http://hdl.handle.net/10603/585924).

TABLE 9.1
Medicinal/Pharmaceutical Applications of *M. oleifera* Lam.

	Medicinal/pharmaceutical uses	
Plant parts	**Uses**	**References**
Leaves	Antioxidant, wound healing, antibacterial, antifungal, antitumor, antiulcer, antidiabetic, hepatoprotective and cardiovascular activity	Nadkarni, 2009; Caceres et al., 1991; Doughari et al., 2007; Chuang et al., 2007; Murakami et al., 1998; Mishra et al., 2011; Ndong et al., 2007; Pari & Kumar, 2002; Faizi et al., 1995
Bark	Abortifacient, antifungal, antibacterial and anti-inflammatory activity	Mishra et al., 2011; Nadkarni, 2009; Caceres et al., 1991
Pods and seeds	Antipyretic, anthelmintic, antimicrobial, anti-inflammatory, antiarthritic, anticancer, antiallergic and hepatoprotective activity	Khare, 2007; Caceres et al., 1991; Mishra et al., 2011; Jabeen et al., 2008; Guevara et al., 1999; Hamza, 2007
Roots	Stimulant, diuretic, antilithic, anti-inflammatory, antimicrobial, antifertility and hepatoprotective activity	Caceres et al., 1991; Mishra et al., 2011; Nikkon et al., 2003; Prakash et al., 1987
Flowers	Cholagogue, stimulant, diuretic, anti-inflammatory and antiurolithiatic activity	Nadkarni, 2009; Caceres et al., 1991; Mishra et al., 2011

9.2 THROMBOSIS

Thrombosis is a process of blood clot (thrombus) formation in veins or arteries that interferes with blood flow. Clots formed in arteries may lead to ischaemic strokes (Kim et al., 2010). Blood clot formation in the veins causes venous thrombosis, which is a common cause of pulmonary embolism (Wolberg et al., 2015). According to Centers for Disease Control and Prevention (CDC) data (2024), up to 900,000 people are affected by venous thromboembolism (VTE), and approximately 60,000 to 100,000 people die each year due to VTE in the United States, especially after hospitalisation during cancer treatment or pregnancy. According to National Academy of Medical Sciences (NAMS) data (2024), about 30% of individuals with symptomatic VTE have pulmonary embolism (PE), whereas others have deep vein thrombosis (DVT). Approximately 31.8 million people in India are affected by thrombosis (Wendelboe et al., 2016). The occurrence of DVT in India is estimated at 0.1% of the population (Mann & Lorand, 1993; Tabassum et al., 2023). The disease is one of the main reasons for death and disability globally, causing one out of four deaths worldwide in 2010 (Davie, 1995). Older people are more affected by thrombosis; however, the age-dependent risk of thrombosis has not been properly studied (Greenberg & Davie, 2001).

Thrombin is a key enzyme in blood clotting and haemostasis. It converts fibrinogen to fibrin, which is involved in clot formation and its stabilisation by further processing. Thrombin, a serine protease, functions both as a procoagulant and an anticoagulant and regulates its own production through the activation of coagulation factors V, VIII and even XI. It activates platelets as a procoagulant by binding to its receptor on the platelets. Thrombus formation in the blood has been attributed to the imbalance between procoagulant and anticoagulant factors. Risk factors for thrombus formation may be acquired and genetic. The acquired factors have been identified as obesity, heart-related disorders, cancer, old age, trauma and so on (Walsh & Ahmad, 2002). Some of the genetic factors responsible for thrombus formation are increased levels of blood clotting factors and antithrombin protein C deficiency (Govindarajan et al., 2016). The pathophysiology of thrombosis and the biology of thrombus formation have greatly advanced, but the pharmacological agents that were once used for prevention or treatment are now outdated and offer only slight improvements over older medications (Lichota et al., 2020).

9.3 MORINGA AS A SOURCE OF BIOACTIVES AGAINST THROMBOSIS

The therapeutic potential of *M. oleifera* has been demonstrated for its haemostatic, antifibrotic and cardioprotective effects (Kunwar et al., 2022). It has been suggested as a natural alternative to commercial haemostatic agents for promoting wound healing. With its strong inhibitory effect on blood coagulation time, leaf extract has been considered as a potential local haemostatic agent (Kunwar et al., 2022). It has been reported that extracts of *M. oleifera* improve survival rates in thrombosis-induced animal models and prolonged bleeding times, suggesting their potential as effective natural anticoagulants. These beneficial effects are attributed to bioactive compounds present in the plant that may play crucial roles in modulating platelet aggregation. These compounds also influence multiple coagulation pathways, thereby contributing to the anticoagulant properties of *M. oleifera* extracts (Saputri et al., 2022).

Moringa oleifera leaves have been reported to contain bioactive compounds that have the potential to inhibit platelet aggregation. This anti-aggregation effect of the leaf extract is influenced by the concentration of the extract and the incubation duration with platelets. These compounds may interfere with thrombin-induced platelet aggregation through the modulation of key biochemical pathways (Arabshahi-Delouee et al., 2009). Studies have identified a thrombin inhibitor and a protease from *M. oleifera* leaves that exhibit non-competitive inhibition of thrombin and fibrinolytic potential, respectively (Sawetaji & Aggarwal, 2023, 2024). The thrombin inhibitor isolated from *M. oleifera* showed significant thrombin inhibition activity over a pH range of 6.2 to pH 8.2 and a temperature range of 25°C to 37°C. The isolated inhibitor exhibits an IC_{50} value of 4.23 μg. The inhibition constant (Ki) was calculated as 4.35×10^{-7} M. PT and APTT assays of the isolated inhibitor showed its inhibitory effect on clotting factors involved in both extrinsic and intrinsic blood clotting pathways by prolonging the incubation time for clot formation as compared with heparin (Sawetaji & Aggarwal, 2023). Thus, bioactive compounds in both the leaves and roots of *M. oleifera* have effects on thrombin activity and platelet function. The inhibitory effects of aqueous leaf extracts on platelet aggregation reinforce the antithrombin potential of *M. oleifera* (Satish et al., 2012). Hexane and ethyl acetate extracts of the root have shown significant inhibition of thrombin activity and platelet aggregation, highlighting their potential as a natural anticoagulant (Kwon et al., 2019). A compound ferulic acid, from *M. oleifera* leaves, was shown to have antithrombotic activity (Pareek et al., 2023). Kwon et al. (2019) also showed anti-thrombosis activity with the root extract of *M. oleifera*. Thrombolytic activity of methanolic extracts of *M. oleifera* leaves has been reported (Kunwar et al., 2022). These findings emphasise the therapeutic potential of *M. oleifera* in the development of alternative or adjunctive therapies for thrombotic diseases.

Seeds of the *M. oleifera* plant are known to possess anticoagulant properties, as their extract has been shown to prevent platelet aggregation and thrombosis. Phytochemicals that are present in the seeds of moringa most probably act synergistically and thereby modulate coagulation (Singnap et al., 2019). The bioactive compounds present in the seeds interfere with thrombin-induced platelet aggregation and modulate key enzymes involved in the clotting cascade (Singnap et al., 2019). A lectin, cMoL, was isolated from the seeds of *M. oleifera*, which was shown to reduce blood coagulation under *in vitro* conditions. The isolated protein significantly delayed blood clotting time (de Andrade Luz et al., 2013). *Moringa oleifera* seed powder was shown to reduce nocturnal heart rate and improve cardiac diastolic function. The cardioprotective properties of *M. oleifera* seeds as dietary inclusions have been documented in animal models. Extracts of seeds were shown to prolong bleeding and clotting time, indicating that the seeds may have an antagonistic effect on haemostatic activities (Singnap et al., 2019). Phytochemicals with the potential to impact blood coagulation and thrombin activity have been documented from the flowers of moringa. The flower extracts were shown to solubilise blood clots in a dose-dependent manner (Sudha et al., 2022; Kunwar et al., 2022).

9.4 IN VITRO AND IN VIVO THROMBOLYTIC ACTIVITY OF *MORINGA OLEIFERA*

Moringa oleifera has been shown to possess antithrombin and fibrinolytic properties. A fibrin-hydrolysing protease has been reported from the leaves of *M. oleifera*. The isolated protease solubilised 96.41% of a blood clot over 96 hours, suggesting that *M. oleifera* possesses significant fibrinolytic potential and could be explored as a natural, alternative therapy for managing thrombosis. (Sawetaji & Aggarwal, 2024). A proteinaceous inhibitor was also reported from moringa that inhibits thrombin activity and possibly interferes with the clotting process (Sawetaji & Aggarwal, 2023). The antithrombotic and blood clot solubilisation activity of moringa was also demonstrated using animal models (Nurhayati et al., 2023). A methanolic extract of moringa was administered orally for seven days prior to thrombosis induction. The results showed that *M. oleifera* prevented paralysis and death caused by collagen-epinephrine-induced pulmonary thrombosis. The extract significantly prolonged bleeding time, indicating its potential antithrombotic activity (Saputri et al., 2022). The beneficial role of moringa has been shown to improve cardiometabolic abnormalities and cardiotoxicity conditions (Louisa et al., 2022). Thus, *M. oleifera* may act as a source of thrombolytic activity, thrombin inhibition and fibrin solubilisation activity. Moringa holds potential as an alternative or supplementary therapy to current thrombolytic therapies (Sawetaji & Aggarwal, 2023).

9.5 CONCLUSION

Thrombosis is a global life-threatening disease, and available synthetic and chemical drugs for the treatment of thrombosis are known to cause serious side effects. The documented blood clot solubilisation and the clot-delaying potential of molecules and compounds isolated from different parts of the plant make them suitable molecules for the development of natural therapeutics against thrombosis. As *M. oleifera* is an edible plant and consumption of it traditionally for its anticoagulant and thrombolytic properties, it could be considered as a nutraceutical supplement to reduce the risk of thrombosis. However, more clinical experiments at the molecular level are needed to consider *M. oleifera* as a drug against thrombosis.

REFERENCES

Arabshahi-Delouee, S., Aalami, M., Urooj, A. and Krishnakantha, T. P. 2009. *Moringa oleifera* leaves as an inhibitor of human platelet aggregation. *Pharmaceutical Biology* 47(8): 734–739.

Bijina, B., Chellappan, S., Basheer, S. M., Elyas, K. K., Bahkali, A. H. and Chandrasekaran, M. 2011. Protease inhibitor from *Moringa oleifera* leaves: Isolation, purification, and characterization. *Process Biochemistry* 46(12): 2291–2300.

Boopathi, N.M., Raveendran, M. (2021). Moringa and Its Importance. In: Boopathi, N.M., Raveendran, M., Kole, C. (eds) The Moringa Genome. Compendium of Plant Genomes:1-9, Springer, Cham.

Caceres, A., Cabrera, O., Morales, O., Mollinedo, P. and Mendia, P. 1991. Pharmacological properties of *Moringa oleifera*. 1: Preliminary screening for antimicrobial activity. *Journal of Ethnopharmacology* 33(3): 213–216.

Chen, C., Yang, F. Q., Zhang, Q., Wang, F. Q., Hu, Y. J. and Xia, Z. N. 2015. Natural products for antithrombosis. *Evidence-Based Complementary and Alternative Medicine* 2015(1): 876426.

Chuang, P. H., Lee, C. W., Chou, J. Y., Murugan, M., Shieh, B. J. and Chen, H. M. 2007. Anti-fungal activity of crude extracts and essential oil of *Moringa oleifera* Lam. *Bioresource Technology* 98(1): 232–236.

Davie, E. W. 1995. Biochemical and molecular aspects of the coagulation cascade. *Thrombosis and Haemostasis* 74(01): 001–006.

de Andrade Luz, L., Silva, M. C. C., da Silva Ferreira, R., Santana, L. A., Silva-Lucca, R. A., Mentele, R., Oliva, M. L. V., Paiva, P. M. G. and Coelho, L. C. B. B. 2013. Structural characterization of coagulant *Moringa oleifera* Lectin and its effect on hemostatic parameters. *International Journal of Biological Macromolecules* 58: 31–36.

Doughari, J. H., Pukuma, M. S. and De, N. 2007. Antibacterial effects of *Balanites aegyptiaca* L. Drel. and *Moringa oleifera* Lam. on *Salmonella typhi*. *African Journal of Biotechnology* 6(19): 2212–2215.

Faizi, S., Siddiqui, B. S., Saleem, R., Siddiqui, S., Aftab, K. and Gilani, A. H. 1995. Fully acetylated carbamate and hypotensive thiocarbamate glycosides from *Moringa oleifera*. *Phytochemistry* 38(4): 957–963.

Govindarajan, V., Rakesh, V., Reifman, J. and Mitrophanov, A. Y. 2016. Computational study of thrombus formation and clotting factor effects under venous flow conditions. *Biophysical Journal* 110(8): 1869–1885.

Greenberg, D. L. and Davie, E. W. 2001. Blood coagulation factors: Their complementary DNAs, genes and expression. In *Hemostasis and Thrombosis: Basic Principles and Clinical Practice*. 4th ed. Philadelphia, PA: Lippincott Williams & Wilkins, 21–57.

Guevara, A. P., Vargas, C., Sakurai, H., Fujiwara, Y., Hashimoto, K., Maoka, T., Kozuka, M., Ito, Y., Tokuda, H. and Nishino, H. 1999. An antitumor promoter from *Moringa oleifera* Lam. *Mutation Research/Genetic Toxicology and Environmental Mutagenesis* 440(2): 181–188.

Hamza, A. A. 2007. *Curcuma longa, Glycyrrhiza glabra* and *Moringa oleifera* ameliorate diclofenac-induced hepatoxicity in rats. *American Journal of Pharmacology and Toxicology* 2: 80–88.

Jabeen, R., Shahid, M., Jamil, A. and Ashraf, M. 2008. Microscopic evaluation of the antimicrobial activity of seed extracts of *Moringa oleifera*. *Pakistan Journal of Botany* 40(4): 1349–1358.

Khare, CP. 2007. *Indian Medicinal Plants*. Springer, 422–423.

Kim, S. Y., Koo, Y. K., Koo, J. Y., Ngoc, T. M., Kang, S. S., Bae, K., Kim, Y. S. and Yun-Choi, H. S. 2010. Platelet anti-aggregation activities of compounds from *Cinnamomum cassia*. *Journal of Medicinal Food* 13(5): 1069–1074.

Kunwar, B., Jain, V. and Verma, S. K. 2022. In vitro thrombolytic activity of *Moringa oleifera*. *Nusantara Bioscience* 14(1): 63–69.

Kwon, C. S., Sung, H. J. and Sohn, H. Y. 2019. Anti-thrombosis activities of the root extract of *Moringa oleifera* Lam. *Microbiology and Biotechnology Letters* 47(1): 20–24.

Lamponi, S. 2021. Bioactive natural compounds with antiplatelet and anticoagulant activity and their potential role in the treatment of thrombotic disorders. *Life* 11(10): 1095.

Lichota, A., Szewczyk, E. M. and Gwozdzinski, K. 2020. Factors affecting the formation and treatment of thrombosis by natural and synthetic compounds. *International Journal of Molecular Sciences* 21(21): 7975.

Louisa, M., Patintingan, C. G. H. and Wardhani, B. W. K. 2022. *Moringa oleifera* Lam. in cardiometabolic disorders: a systematic review of recent studies and possible mechanism of actions. *Frontiers in Pharmacology* 13: 792794.

Mann, K. G. and Lorand, L. 1993. Introduction: blood coagulation. *Methods in Enzymology* 222: 1–10.

Mishra, G., Singh, P., Verma, R., Kumar, S., Srivastav, S., Jha, K. K. and Khosa, R. L. 2011. Traditional uses, phytochemistry and pharmacological properties of *Moringa oleifera* plant: An overview. *Der Pharmacia Lettre* 3(2): 141–164.

Murakami, A., Kitazono, Y., Jiwajinda, S., Koshimizu, K. and Ohigashi, H. 1998. Niaziminin, a thiocarbamate from the leaves of *Moringa oleifera*, holds a strict structural requirement for inhibition of tumor-promoter-induced Epstein-Barr virus activation. *Planta Medica* 64(04): 319–323.

Nadkarni, K. M. 2009. Indian materia medica. *Bombay Popular Prakashan* 1: 811–816.

Ndong, M., Uehara, M., Katsumata, S. I. and Suzuki, K. 2007. Effects of oral administration of *Moringa oleifera* Lam on glucose tolerance in Goto-Kakizaki and Wistar rats. *Journal of Clinical Biochemistry and Nutrition* 40(3): 229–233.

Newman, D. J. and Cragg, G. M. 2016. Natural products as sources of new drugs from 1981 to 2014. *Journal of Natural Products* 79(3): 629–661.

Nikkon, F., Saud, Z. A., Rahman, M. H. and Haque, M. E. 2003. In vitro antimicrobial activity of the compound isolated from chloroform extract of *Moringa oleifera* Lam. *Pakistan Journal of Biological Sciences* 6(22): 1888–1890.

Nurhayati, T., Fathoni, M. I., Fatimah, S. N., Tarawan, V. M., Goenawan, H. and Dwiwina, R. G. 2023. Effect of *Moringa oleifera* leaf powder on hematological profile of male Wistar rats. *Journal of Blood Medicine 14*: 477–485.

Pareek, A., Pant, M., Gupta, M. M., Kashania, P., Ratan, Y., Jain, V. and Chuturgoon, A. A. 2023. *Moringa oleifera*: An updated comprehensive review of its pharmacological activities, ethnomedicinal, phytopharmaceutical formulation, clinical, phytochemical, and toxicological aspects. *International Journal of Molecular Sciences* 24(3): 2098.

Pari, L. and Kumar, N. A. 2002. Hepatoprotective activity of *Moringa oleifera* on antitubercular drug-induced liver damage in rats. *Journal of Medicinal Food* 5(3): 171–177.

Prakash, A. O., Pathak, S., Shukla, S. and Mathur, R. 1987. Uterine histoarchitecture during pre and post-implantation periods of rats treated with aqueous extract of *Moringa oleifera* Lam. *Acta Europaea Fertilitatis* 18(2): 129–135.

Saputri, F. C., Andriani, A. and Azmi, N. U. 2022. *Imperata cylindrica* and *Moringa oleifera*: Antithrombotic effect on pulmonary thromboembolism in mice. *Pharmacognosy Journal* 14(1): 148–153.

Satish, A., Sairam, S., Ahmed, F. and Urooj, A. 2012. *Moringa oleifera* Lam.: Protease activity against blood coagulation cascade. *Pharmacognosy Research* 4(1): 44–49.

Sawetaji, and Aggarwal, K. K. 2023. A non-competitive serpin-like thrombin inhibitor isolated from *Moringa oleifera* exhibit a high affinity for thrombin. *The Protein Journal* 42(4): 305–315.

Sawetaji, and Aggarwal, K. K. 2024. A protease from *Moringa oleifera* Lam. exhibits in-vitro blood clot solubilization and fibrin hydrolysis. *The Protein Journal* 43(4): 923–934.

Singnap, C. L., Sabo, A. M., Idah, O. O. V., Lekshak, B. N., Yakubu, T. P., Samuel, O. O., Hamza, W. R., Musa, T. F., Olisemeke, E. U., Dandam, V. R. and Lugos, M. D. 2019. Ethanolic extract of *Moringa oleifera* seed prolongs blood coagulation in Wistar albino rats. *International Journal of Hematology and Blood Disorders* 4(2): 1–6.

Subramani, B. and Sathiyarajeswaran, P. 2022. Current update on herbal sources of antithrombotic activity—a comprehensive review. *The Egyptian Journal of Internal Medicine* 34(1): 26.

Sudha, K.M., Praveen, S., Murali, N., Raghu, N. and Suresh, P. 2022. Evaluation of in-vitro thrombolytic activity of *Moringa oleifera* flower extract. *International Journal of Scientific Research in Science and Technology* 9(10): 230–238.

Tabassum, H., Chakraborty, R. and Chatterjee, N. S. 2023. A review of venous thromboembolism in India. *Indian Journal of Medical Research* 157(4): 281–292.

Thapa, K., Poudel, M. and Adhikari, P. 2019. *Moringa oleifera*: A review article on nutritional properties and its prospect in the context of Nepal. *Acta Scientific Agriculture* 3(11): 47–54.
Walsh, P. N. and Ahmad, S. S. 2002. Proteases in blood clotting. *Essays in Biochemistry* 38: 95–111.
Wendelboe, A. M. and Raskob, G. E. 2016. Global burden of thrombosis: epidemiologic aspects. *Circulation Research* 118(9): 1340–1347.
Wolberg, A. S., Rosendaal, F. R., Weitz, J. I., Jaffer, I. H., Agnelli, G., Baglin, T. and Mackman, N. 2015. Venous thrombosis. *Nature Reviews Disease Primers* 1(1): 1–17.
Yang, H., Zahan, M. N., Yoon, Y., Kim, K., Hwang, D. H., Kim, W. H., Rho, I. R., Kim, E. and Kang, C. 2023. Unveiling the potent fibrino(geno)lytic, anticoagulant, and antithrombotic effects of papain, a cysteine protease from *Carica papaya* latex using κ-carrageenan rat tail thrombosis model. *International Journal of Molecular Sciences* 24(23):16770.

Part 2

Phytochemistry, Technology and Applications

10 *Moringa oleifera* Leaves
Chemistry, Technology, Functionality and Applications

Muhammad Junaid Anwar, Aneela Hameed, Aimen Mazhar, Sahar Qundeel and Muhammad Hammad Anwar

10.1 INTRODUCTION

Moringa oleifera (MO) is well known for its fast growth. MO belongs to the family Moringaceae and is a perennial angiosperm tree with a height of 7–15 m and a diameter of 20–40 cm at chest height (Alegbeleye, 2018; Mbikay, 2012). It is widely cultivated in most regions, i.e. Asia, Africa and the Middle East (Farid & Hegazy, 2020; Hisam et al., 2018). MO is native to the sub-Himalayan tracts of Pakistan, India, Afghanistan and Bangladesh with different regional names, i.e. suhanjana, malunggay, marango, horseradish tree, drumstick tree, kelor and benzolive. It has attracted socioeconomic and medical attention in various tropical and subtropical regions, i.e. Tropical Asia, Eastern, Western and Southern Africa, the Pacific Islands, the Caribbean and Latin America, where MO is cultivated widely and widely accepted. It is famous for its easy cultivation, low nutrient demand and its ability to grow in drought and deprived soils. It is capable of tolerating diverse ranges of rainfall, with an estimated minimum annual rainfall of 250 mm to a maximum above 3000 mm (Anwar et al., 2007; El Bilali et al., 2024; Mashamaite et al., 2024). The morphological features of MO are given in Figure 10.1.

MO was promoted by the World Health Organization (WHO) in 1998 as an alternative source of supplements to cure malnutrition (Mahmood et al., 2010). MO is utilised as a vegetable, medicine and a source of cooking oil in underdeveloped countries. MO has been used for centuries in more than 80 countries due to its well-known functional, therapeutic and medicinal properties, as it contains a unique nutritional and phytochemical composition and various bioactive compounds. MO was explored for the treatment of 300 diseases, as shown in the ancient Ayurvedic scriptures (Dixit et al., 2016). Healthcare professionals have utilised its various phytochemicals in various medicines or in their natural form to treat and protect human beings against various illnesses. Generally, phytochemicals present in vegetables are divided into five major categories: sulphur-containing compounds, terpenoids, alkaloids, carotenoids and polyphenols (Ma et al., 2020). The majority of these are available in different parts of the MO tree. Its biological activities and potential to prevent disease are attributed to phytochemicals. According to estimates, 80% of the population in underdeveloped countries use MO phytochemicals as a conventional medicine for health (Camilleri et al., 2024; Ma et al., 2020).

The technology for processing MO leaves involves various methods to extract beneficial compounds, enhance nutritional value and ensure product stability. These methods range from the extraction and purification of specific compounds to innovative drying techniques. Each method is designed to improve the health benefits and usability of MO leaves in different applications, such as health supplements and food products. The process for MO flavones extraction involves drying, crushing and swelling the leaves, followed by ethanol extraction and purification using macroporous adsorption resin. This process maximises the purity and stability of the flavone product, making it

DOI: 10.1201/9781003617860-12

FIGURE 10.1 Morphological features of *Moringa oleifera*. A) Full grown MO tree (B) Tripinnately compound with pinnate with multiple leaflets (C) Upper surface of a MO leaf (D) Lower surface of the MO leaf

suitable for industrial production (Ma, 2018). A detoxifying method uses ultrasonic rinsing with citric or acetic acid to remove pungent tastes and toxins, followed by drying and grinding. This process retains the nutritional value while making the leaves more palatable (Tang, 2014). A dual-energy solar-powered dryer, controlled by a microcontroller, efficiently reduces the moisture content of MO leaves from 18.5% to 4% without compromising nutritional value. This method is more efficient than traditional drying, achieving better quality leaves in a shorter time (Jalil et al., 2022).

MO is a highly nutritious food plant and a well-known functional food as all its parts are edible (Fahey, 2005; Cajuday et al., 2010). The seeds, fruits, sap, flowers, leaves, bark and roots of MO have therapeutic, prophylactic and nutritional properties. Different studies have reported its antidiabetic, diuretic, antiepileptic, antipyretic, anti-obesity, thyroid hormone regulatory, radioprotective, anticancer, hypoglycaemic and hypotensive properties (Chen et al., 2017; Metwally et al., 2017; Ritu Paliwal et al., 2011; Sharma et al., 2012). It is also used as a complementary medicinal option to prevent or provide relief against various diseases (Alegbeleye, 2018; Mbikay, 2012; Ritu Paliwa et al., 2011). MO leaves contain high concentrations of bioactive compounds, such as carotenoids, vitamins, saponins, tannins, isothiocyanates, glucosinolates, flavonoids and phenolics, which contribute to their remarkable potential as a promising candidate for medicinal applications (Hisam et al., 2018). The numerous benefits, i.e. antihypertensive, anti-inflammatory and antioxidant activity and hepatoprotective attributes, are linked to MO leaves (Farid & Hegazy, 2020). Due to its countless health benefits, it is known as the 'tree of life' and the 'miracle tree'. It is also known for its beneficial impact on water cleanliness (Padayachee & Baijnath, 2020).

This chapter covers the nutritional and phytochemical composition of MO, various extraction techniques (i.e. conventional methods, ultrasound-assisted extraction, microwave-assisted extraction and supercritical fluid extraction), drying technologies (i.e. shade drying, oven drying, freeze-drying) and functional properties (i.e. antioxidant, antimicrobial, anticancer, antidiabetic, anti-inflammatory and neuroprotective properties). Moreover, the applications of MO for different purposes, such as functional foods and nutraceuticals, fortification in food products, encapsulation and delivery systems and therapeutic and pharmaceutical applications, are also highlighted. The conclusion discusses the safety, toxicology and regulatory aspects of MO and its safe consumption.

10.1.1 Nutritional and Phytochemical Composition of MO Leaves

MO is studied worldwide due to its unique composition, which includes higher concentrations of macronutrients, i.e. fats, dietary fibres and proteins. MO can serve as a valuable tool for treating diet deficiencies and improving health, particularly during malnutrition. MO plays a prominent role in major dietary components as an alternative source of animal protein for populations where the availability of animal protein is limited. The leaves of MO are rich in protein with balanced essential amino acids (Yang et al., 2023). The protein composition of MO leaves complies with WHO standards and enhances its potential to combat protein-energy malnutrition (Kashyap et al., 2022). There are 17 essential and non-essential amino acids quantified. Higher concentrations of lysine (69.13 mg/100 g) and leucine (94.36 mg/100 g) were observed in MO leaves. The amino acids, i.e., valine, isoleucine and leucine, present in MO leaves are effective in muscle repair and development, suggesting it is a better option for rehabilitation patients and athletes (Ntshambiwa et al., 2023). MO leaves were investigated as a protein supplement to increase the range of functional foods, i.e. fortified snacks, multi-beneficial powders and protein bars. These innovations explored the bioavailability of plant-based high-protein to meet the nutritional requirements among various age groups, i.e. the elderly, youngsters and children (Arora & Arora, 2021; Liang et al., 2023).

10.1.1.1 Macronutrients and Micronutrients

MO leaves reveal an extraordinary nutritional composition, with high concentrations of lipids, proteins, carbohydrates, ash and fibre content with percentile values of 7.76, 9.38, 56.33, 4.56 and 11.23%, respectively (Hodas et al., 2021). MO leaves are a rich source of micronutrients, i.e. vitamins and minerals. MO is considered a major source of vital minerals, i.e. manganese (Mn), calcium (Ca), copper (Cu), zinc (Zn), iron (Fe), phosphorus (P), magnesium (Mg), potassium (K) and sodium (Na). MO leaves have a strong vitamin profile as they contain the vitamin A precursor (β-carotene), vitamin B (i.e., B1, B2 and B3), vitamin C, different forms of vitamin E (α-tocopherol, β-tocopherol, γ-tocopherol and δ-tocopherol) and vitamin K (El Sohaimy et al., 2015; Hodas et al., 2021).

10.2 PHYTOCHEMICALS AND BIOACTIVE COMPOUNDS

MO contains polyphenols such as phenolic acids, tannins and flavonoids in high concentrations (Owon et al., 2021). The leaves of the MO plant contain the highest quantity of polyphenols in comparison with the other parts. The phenolic content of MO is significantly affected by the environmental circumstances of growth and geography. The most common flavonoids present in various parts of the MO plant, except seeds and roots, are kaempferol glycosides (malonyl glucoside, rutinoside and glucoside) and quercetin. The other flavonols of the MO plant include epicatechin, rutin and myricetin. Flavonoid levels varied among different varieties depending on geographical differences. MO contains numerous phenolic acids such as ellagic acid, coumaric acid, chlorogenic acid, ellagic acid, caffeic acid and gallic acid (Ariani et al., 2023; Chiş et al., 2023; George, 2021). MO leaves also contain higher concentrations of complex polyphenols (tannins), which have the ability to bind to proteins. The concentrations of tannin vary in all parts of the MO tree, but the highest levels are present in leaves (Ma et al., 2020).

Various phenolic acid derivatives such as kaempferol acetyl hexose, isorhamnetin-hexose, cyanidin hexose, kaempferol-3-hexose, luteolin-7-O-glucoside, kaempferol-3-O-rutinoside, quercetin-3-acetyl-hexose, quercetin-3-hexose, quercetin-7-hexose, apigenin-C-hexose, quercetin-3-O-rutinoside, 4-feruloylquinic acid, 4-p-coumaroylquinic acid, C-diglycoside, apigenin acetyl 4-(a-L-rhamnopyranosyloxy) benzyl glucosinolate, 4-caffeoylquinic acid, 3-p-coumaroylquinic acid, 3-caffeoyl quinic acid and glucomoringin were detected by using ultra-high-performance liquid chromatography with quadrupole time-of-flight mass spectrometry (UPLC-Q-TOF-MS) (Teclegeorgish et al., 2021). Liquid chromatography electrospray ionisation tandem mass spectrometry (LC-ESI-MS) was used to identify kaempferol 3-glucoside, quercetin-3-acetylglucoside, quercetin-3-glucoside, kaempferol acetyl glycoside and rutin (Hamed et al., 2019). The phytochemical components such as phytol, palmitic acid, ethyl ester, hexadecenoic acid and 9, 12, 15-octadecatrienoic acid with concentrations of 9.81, 14.70, 16.71 and 39.66% were detected in the ethanolic extract of MO leaves by using gas chromatography-mass spectrometry (GC-MS) (Adeyemi et al., 2021).

10.2.1 Alkaloids

Alkaloids are included in the nitrogen-containing compounds category and are well known for their prominent potential in the medical field. They exhibit various health benefits such as anti-inflammatory, antibacterial and analgesic properties. A study reported that MO leaves can reduce sleep and pain due to the presence of alkaloids (Roni et al., 2021). MO is well known for its ability to treat various neurological injuries, cancers and cardiovascular diseases because of its antioxidant scavenging activity (Hu et al., 2023; Guzmán-Maldonado et al., 2020).

10.2.2 Flavonoids

Flavonoids are a group of phytochemicals present in MO leaves and are well known for their antioxidant and anti-inflammatory attributes. Flavonoids provide protection to the cells against oxidative and inflammatory damage, which are linked with various degenerative diseases. They lower blood glucose levels due to their antibacterial properties, providing relief for diabetic patients (Olusanya et al., 2020).

10.2.3 Glucosinolates

Glucosinolates are classified as sulphur-containing compounds. The ability to be involved in cell cycle-linked enzymes and detoxification processes has made them capable of preventing the development of cancerous cells. Glucosinolates in MO leaves enhance immunity and also prevent inflammatory reactions in the body (Shang et al., 2024; Chhikara et al., 2021).

10.2.4 Isothiocyanates

Isothiocyanates are also included in the category of sulphur-containing components that are linked to the treatment of various types of cancers. These components in MO leaves expedite the detoxification phenomenon in the body, which prevents cancerous cell formation. Moreover, these compounds also assist the liver in the detoxification process, which improves the body's ability to combat toxins (Kashyap et al., 2022).

10.2.5 Polyphenols

There is a wide range of naturally occurring phytochemicals, particularly polyphenols, showing antioxidant properties (Li et al., 2024). The leaves of MO reduce oxidative stress in the human body, which aids in the prevention of various chronic diseases such as cardiac issues, cancer and diabetes (Owon et al., 2021). It also exhibits anti-inflammatory properties, helping prevent cellular damage.

10.3 FUNCTIONAL PROPERTIES OF MO

10.3.1 Antioxidant Properties

Cellular macromolecules are oxidatively damaged by reactive oxygen species (ROS), contributing to diseases like hypertension and diabetes, whereas natural antioxidants from MO leaves, such as quercetin, kaempferol and polyphenols, offer potent protective effects (Gönenç et al., 2013; Kashyap et al., 2017). Various solvent extracts, particularly ethyl acetate and methanolic extracts, exhibit strong antioxidant activity correlated with high phenolic content, surpassing standard antioxidants such as ascorbic acid (Vongsak et al., 2012; Atawodi et al., 2010; Charoensin, 2014; Rakesh & Singh, 2011). Moringa leaf preparations have demonstrated efficacy in reducing oxidative stress in animal models, improving antioxidant enzyme levels and mitigating tissue damage caused by toxic agents (Khalofah et al., 2020; Gouda et al., 2018). Additionally, topical and dietary applications enhance skin hydration and systemic antioxidant status (Table 10.1).

10.3.2 Antimicrobial Properties

The rise of antibiotic-resistant pathogens has intensified interest in plant-based antimicrobials, with MO exhibiting broad-spectrum activity against bacteria such as *Staphylococcus aureus*, *E. coli*, *Salmonella typhi*, *Pseudomonas aeruginosa* and *Bacillus subtilis*, though it showed resistance to aqueous extracts (Abalaka et al., 2012; Ezugwu & Chukwubike, 2014). Bioactive compounds like benzyl isothiocyanates and glucosinolates disrupt microbial membranes and inhibit enzyme synthesis, contributing to the strong antimicrobial and antifungal effects of methanolic, ethanolic and acetone extracts (Tshabalala et al., 2020). Ethanol extracts demonstrated larger inhibition zones than aqueous extracts, with efficacy increasing dose-dependently (Mursyid et al., 2019; Abadallah & Ali, 2019). Comparative studies further confirm MO's superior activity against resistant strains and various fungal pathogens, including *Candida* spp. (Moyo et al., 2012).

TABLE 10.1
Antioxidant Properties of *Moringa oleferia*

Sample Type	Outcome Summary	Mechanism	Citations
Methanol, acetone, ethyl acetate, etc. extracts of MO leaves	Ethyl acetate extract showed strongest superoxide anion ($O_2^{\bullet-}$) scavenging	Activity linked to high phenolic content	Verma et al., 2009; Atawodi et al 2010; Charoensin., 2014
Methanolic leaf extract	IC_{50} = 49.86 µg/mL (better than ascorbic acid, IC_{50} = 56.44 µg/mL)	High polyphenol content	Rakesh & Singh., 2011
MO tea	81% DPPH inhibition vs. 76.5% for vitamin C	Strong radical scavenging capacity	Fombang et al., 2016; Posmontier, 2011
Hydroalcoholic extract (gel and nanoparticle forms)	High antioxidant activity	Presence of flavonoids and phenolic acids; suitable skin formulation	Álvarez-Román et al., 2020; Lopez-Teros et al., 2017
Freeze-dried Moringa Oleferia leaves (aqueous, aqueous ethanol extract)	89.7-92.0% lipid peroxidation inhibition	Linoleic acid β-carotene model; dose-dependent superoxide radical scavenging	Siddhuraju & Becker, 2003
MO seed oil	Antioxidative and hypoglycemic effects	Improved hepatic, renal, and pancreatic function	Sajid et al., 2020
MO leaf and fruit extracts	Potent antioxidant activity	Validated across different regions	Luqman et al., 2012
Leaf and seed extract	Contains >40 natural antioxidants	Flavonoids, polyphenols act synergistically	Pakade et al., 2013; Unuigbe et al., 2014; Niziot-Tukaszewska et al., 2020

10.3.3 Anticancerous Properties

MO shows significant anticancer potential due to bioactive compounds such as niazimicin, quercetin, isothiocyanates and kaempferol, which induce antiproliferative and pro-apoptotic effects through mitochondrial dysfunction and ROS generation (McGuire, 2016; Gopalakrishnan et al., 2016). In vitro studies demonstrate efficacy against multiple cancer cell lines (HeLa, HepG2, DU145, A549, MDA-MB-231 and HCT-8) by modulating apoptosis-related genes, including enhancement of caspase-3 and BAX and downregulation of BCL-2 and Notch-1. In vivo, moringa extracts reduce tumour growth and suppress carcinogenesis in hepatocellular and colorectal cancer models in a dose-dependent manner (Cuellar-Nuñez et al., 2018; Sadek et al., 2017), with up to 78.69% tumour volume reduction reported. Additionally, methyl isothiocyanate modulates tumour suppressive and antioxidant gene expression via epigenetic pathways (Wang et al., 2019), supporting MO's potential in cancer therapy and functional food applications.

10.3.4 Antidiabetic Properties

MO has shown significant antidiabetic properties through both in vivo and in vitro studies. In clinical trials, diabetic patients supplemented with moringa leaf powder showed notable reductions in postprandial glucose and HbA1c levels (Momoh et al., 2013; Kumar & Mandapaka, 2013). Animal studies using streptozotocin-induced diabetic rats confirmed moringa's ability to reduce blood glucose, improve insulin levels, restore antioxidant enzyme activity (CAT, SOD, GPx) and protect pancreatic β-cells from oxidative damage (Khan et al., 2017; Li et al., 2020). Moringa leaf extracts also inhibit key enzymes like α-glucosidase and α-amylase, suggesting a mechanism for delayed carbohydrate digestion and absorption (Jimoh, 2018). Methyl isothiocyanate from moringa reversed high-glucose-induced transcriptomic and epigenetic changes in diabetic nephropathy models (Li et al., 2020). Histological studies in diabetic rats demonstrated tissue recovery in the parotid gland, along with reduced DNA damage following moringa administration (Abo Baker & Moawad, 2020). In a human study, moringa reduced postprandial glucose spikes and α-amylase activity, attributed to its high fibre and bioactive compound content. These findings support moringa's potential as a natural adjunct therapy for diabetes management.

10.3.5 Anti-inflammatory Properties

The bioactive phytochemical and phenolic compound content of MO is very high, which accounts for its potent anti-inflammatory activity. In animal models, ethanolic and aqueous extracts significantly decreased proteinase activity, granuloma formation, formaldehyde- and carrageenan-induced paw oedema, demonstrating efficacy that is comparable to standard drugs like dexamethasone and diclofenac (Saleem et al., 2020; Mittal et al., 2017). The regulation of the NF-κB signalling pathway and the inhibition of pro-inflammatory cytokines due to leaf extract reduce inflammation at the molecular level (Luetragoon et al., 2020). Moringa demonstrated significant results in drug-induced liver inflammation and asthma due to a reduction in histamine level, leukocyte infiltration and improvement in antioxidant enzyme activity (Suresh et al., 2020). These findings support the role of MO as an anti-inflammatory remedy.

10.3.6 Neuroprotective Effects

MO exhibits neuroprotective effects by reducing oxidative stress, preserving mitochondrial function and modulating vitagene expression (Amara et al., 2021; Igado & Olopade, 2016). It has been shown to counteract lead-induced neurotoxicity by enhancing detoxifying enzymes (Alqahtani & Albasher, 2021) and may mitigate Alzheimer's-related damage by influencing brain monoamines and electrical activity (Ganguly & Guha, 2008). These properties are largely attributed to its high polyphenolic and antioxidant content (Luqman et al., 2012; Calabrese et al., 2007) (Table 10.2).

10.4 EXTRACTION TECHNIQUES

Extraction of the flavonoids from MO leaves requires optimised techniques to prevent degradation of these sensitive compounds during processing (Chávez-González et al., 2020). While basic methods, such as maceration and Soxhlet, are employed, more advanced methods, such as ultrasonic-assisted extraction (UAE), microwave-assisted extraction (MAE) and supercritical fluid extraction (SFE), greatly improve yields due to superior mass transfer mechanisms (Vinatoru et al., 2017; Vergara-Jimenez et al., 2017). Meticulous sample preparation is of utmost importance, and careful quantification is critical for evaluating the plant's nutraceutical potential (Lin et al., 2018).

10.4.1 Conventional Methods

Some traditional extraction methods for MO include grinding, pouring, boiling, soaking, hydro-distillation and Soxhlet extraction (Hamed et al., 2019). The choice of solvents can also affect yield results; some include ethanol, methanol, chloroform and acetate (Chávez-González et al., 2020). Both Soxhlet and reflux methods are high-temperature continuous techniques. However, Soxhlet can lower the concentration of flavonoids and phenolic compounds in moringa extract (Vergara-Jimenez et al., 2017). On the other hand, maceration requires only that the sample be stirred at room temperature, which helps break down cell walls and cause the phytochemicals to dissolve. Ecoction and infusion are variations that use boiling or cold water for specific durations (Owon et al., 2021). However, traditional methods are often time-consuming, energy-intensive and involve toxic solvents (Worku Dadi et al., 2020).

10.4.2 Ultrasound-Assisted Extraction

Ultrasound-assisted extraction (UAE) enhances flavonoid recovery by inducing acoustic cavitation, which reduces particle size, disrupts cell walls and improves solvent penetration (Tariq et al., 2023). UAE yields higher extraction efficiency than conventional methods, with reduced solvent use and time and better bioactive content, including quercetin and kaempferol derivatives from moringa leaves (Vergara-Jimenez et al., 2017; Worku Dadi et al., 2020). Its efficiency depends on variables like solvent type, ultrasonic power, temperature and ethanol concentration. Optimisation using response surface methodology (RSM) revealed that deep eutectic solvents (DES) and 50% ethanol, and 300 rpm agitation, maximise flavonoid and phenolic content (Hamed et al., 2019). Despite UAE's advantages, under certain low-temperature conditions, maceration may yield higher TFC/TPC (Worku Dadi et al., 2020; Zahari et al., 2020). However, excessive ultrasound frequency (>20 kHz) can degrade sensitive phytochemicals.

10.4.3 Microwave-Assisted Extraction

Microwave-assisted extraction (MAE) is an efficient and inexpensive method for the extraction of flavonoids from plant material with advantages, including low extraction time, solvent utilisation and energy consumption (Rodríguez-Pérez et al., 2016; Bagade et al., 2021). Microwaves penetrate the plant cell, causing it to burst, which ejects bioactive compounds like quercetin, kaempferol and their glycosides, all of which are antioxidants that involve anti-inflammatory and anticancer properties (Rodríguez-Pérez et al., 2016). MAE yielded higher amounts of extracts and antioxidant activity than pressurised liquid extraction (PLE) under the best conditions of 42% ethanol, 158°C and 20

TABLE 10.2
Functional Properties of *Moringa oleferia*

Activity Type	Sample/Plant Part	Mechanism/Summary	Reference
Anti-inflammatory	Moringa leaf extracts (aqueous, ethanolic, methanolic, ethyl acetate)	Exhibited anti-inflammatory, anti-asthmatic and hepatoprotective effects by reducing pro-inflammatory cytokines (e-g- TNF-α, ILs) stabilizing membranes and enhancing antioxidant activity	Rakesh & Singh 2011; Saleem et al., 2020; Edoga et al., 2013; Mittal et al., 2017; Luetragoon et al., 2020; Suresh et al., 2020; El-Hadary & Ramadana., 2019
Anti-diabetic	Dehydrated leaf powder	Significantly reduced postprandial glucose, HbA1c, serum glucose, glycemic index, and α-amylase activity; effects attributed to fibre, phenolics and antioxidants	Momoh et al., 2013, Kumar & Mandapaka., 2013, Leone et al., 2018
Anti-hypertensive & cardioprotective	Aqueous/ethyl/methyl extract	Lowered blood glucose and inflammation; protected β-cells; enhanced antioxidant enzymes; inhibited α-amylase/α-glucosidase; improved organ function and reduced DNA damage	Berkovich et al., 2013; Khan et al., 2017; Kamalakkannan & 2006; Muzumbukilwa et al., 2019; Villarruel-López et al., 2018; Jimoh, 2018; Abo Baker & Moawad., 2020
Anti-hypertensive & cardio-protective	Aqueous and methanolic extracts	Lowered blood pressure, heart rate, glucose, triglycerides and oxidative stress; improved antioxidant enzyme activity, histopathology, and heart function markers	Aju et al., 2019; Aekthammarat et al., 2019; Mabrouki et al., 2020; Azeez et al., 2017
Cardiovascular risk reduction	Capsule (moringa + bryophyllum + vitamin C)	Reduced diastolic BP and blood glucose; improved HDL in males; hypotensive and antioxidant effects attributed to plant glycosides and vitamin C	Sparman, 2017
Neuro-protective, Anticonvulsant, CNS depressant	Ethanolic/methanolic/ aqueous leaf extract	Enhances serotonin level, modulates redox genes, reduces tau and Aβ pathology, restores cerebellar structure	Abo-Bakr et al., 2013; Bhattacharya et al., 2014; Al-Abri et al., 2018; Mahaman et al., 2018
Anti-nociceptive, memory enhancer	Leaf extracts	Improves memory, reduces pain and oxidative stress, protects against neurotoxicity in hypoxia, HHcy, quinine and behavioural stress	Ray et al., 2004; Mohamed et al., 2019; Umoh et al., 2018
Antimicrobial	Leaf extracts (Aqueous, ethanol, methanol)	Broad spectrum antibacterial and antifungal activity via membrane disruption and enzyme inhibition	Abalaka et al., 2012; Kekuda et al., 2010; Ezugwu & Chukwubike., 2014; Bukar et al., 2010; Pandey et al., 2012; Atef et al., 2019; Bancessi et al., 2020; Fouad et al., 2019
Antifungal	Steam distillate, methanol, aqueous extract	Effective against *Aspergillus, Penicillium, Fusarium, Candida* spp. Due to phytochemical richness	Abalaka et al., 2012; Ishnava et al., 2012; Al-husnan & Alkahtani, 2016

Antibacterial	Ethanol, acetone, water leaf extracts	Active against *E.coli, S.aureus, B.subtilis*, etc., with mic as low as 0.39 mg/ml; activity is solvent dependent	Tshabalala et al., 2020; Prabakaran et al., 2018; Abadallah & Ali.m, 2019; Bancessi et al., 2020; Fouad et al., 2019
Anticancer	Leaf phytochemicals and extracts	Induce apoptosis, inhibit proliferation, modulate Bcl-2/Bax expression, restore redox and signaling pathways	Anwar et al 2007; Saini et al., 2016; Balusamy et al., 2019; Al-Asmari et al., 2015; Sadek et al., 2017; Wang et al., 2019
Chemo-protective	Leaf powder or aqueous extract	Suppress tumour formation, reduce oxidative damage, and improve live health in in-vivo models	Cuellar-Nuñez et al., 2018; Jung et al., 2015; Barhoi et al., 2021

minutes (Rodríguez-Pérez et al., 2016). It also matched the efficiency of UAE performed over 60 minutes and conventional methods over 24 hours in just five minutes (Bagade et al., 2021).

10.4.4 Supercritical Fluid Extraction

Supercritical fluid extraction (SFE), particularly using CO_2 (critical point: 7.4 MPa, 32°C), efficiently isolates bioactives such as flavonoids by combining gas-like diffusivity with liquid-like solvating power (Abhari & Khaneghah, 2020). The addition of the GRAS-approved cosolvents, such as ethanol, enhances the recovery of the mid-polar compounds like polyphenols (Tariq et al., 2023). SFE applied to MO seeds yielded significantly higher antioxidant-rich oil (396±23 mg/g) with reduced thermal degradation compared WITH Soxhlet extraction (Chávez-González et al., 2020). Subcritical ethanol extraction achieved a 2.60% flavonoid yield, improving efficiency by 26.7% over conventional methods (Dinesha et al., 2018; Tariq et al., 2023). Overall, SFE provides selective, energy-efficient extraction with superior bioactive compound recovery.

10.5 DRYING METHODS

In the post-harvest processing of plant materials, drying is an essential unit activity, especially for those used in functional foods and nutraceuticals. The drying process can greatly influence how well the ingredients' colour, smell and quality are kept in your food. Even with all these drying methods, the most common ways to process herbs, leafy vegetables and plants, such as MO, mint and basil, are shade drying, oven drying and freeze-drying. Here, we examine these methods by looking at their science, what they can do well, the issues they have and how preparatory approaches can improve their results.

10.5.1 Shade Drying

Shade drying was adopted as a cheap and effective process for preserving the sensitive, fragrant parts of plants. Ensure that the place where you store your materials is out of direct light, well ventilated and shielded from the sun. Flavonoids, phenolic acids, chlorophyll and essential oils are some of the sensitive phytochemicals that are better protected by cold and low UV light (Alakali et al., 2015; Foline et al., 2011).

It has been found through comparative investigations that shadow drying retains the original colour and fragrance of plant products, which cannot be sustained during sun or high-heat drying methods. Overall, this method results in green vegetables and herbs, such as moringa, basil and mint, retaining their important volatile compounds and chlorophyll (Ali et al., 2017). It is important to remember that drying is also influenced by air climate, such as air temperature, humidity and air circulation. Drying food outside in shaded areas requires a well-ventilated area; otherwise, the food can dry slowly or cause microbial contamination (Setiaboma et al., 2021). Blanching, osmotic dehydration or enzymatic inactivation are methods that address these problems by enhancing safety and allowing faster drying processes. In addition, using shade drying, together with air movement brought in by the sun, increases the drying rate, provides uniform results and helps retain the quality of the phytochemicals (as noted by Mbah et al., 2012).

10.5.2 Oven Drying

One typical way to carry out thermal dehydration is oven drying, in which air heated from below by a continuous fan is used to dry the food evenly. The most important feature of oven drying is that it regulates temperature and airflow perfectly, allowing for reliable and equal drying. This technique is commonly used by industrialists and researchers to decrease moisture, microbes and enzymes (Kshirsagar et al., 2017). Oven drying takes little effort and time, but it often degrades vitamins, antioxidants and essential oils. Similar to the open-air drying of herbs, the application of normal temperatures results in notable losses of chlorophyll and volatile ingredients in leafy greens (Tetteh et al., 2019; Alakali et al., 2015).

It has been found that using blanching and antioxidant dips results in less oxidative and enzymatic damage after oven drying. If you blanch the vegetables, the enzymes polyphenol oxidase and peroxidase are effectively stopped, which lowers browning and phenol loss (Aremu & Akintola, 2014). Still, oven drying is not suitable for many high-value or delicate plants because it takes a lot of energy and may cause nutrients to be lost. European researchers have studied the use of hybrid techniques that combine oven drying with thermal energy from the sun or from infrared light (Mbah et al., 2012).

10.5.3 Freeze Drying

The term lyophilisation, which is the same as freeze-drying, is known for its outstanding effectiveness in preserving biological components and sensory qualities. The process is divided into three steps: freezing, primary drying and secondary drying. Carrying out freeze-drying at low temperatures with less pressure helps reduce thermal damage and save both chemical components and sensitive nutrients (Manju et al., 2021; Kashaninejad et al., 2021). It is particularly suitable for processing high-value goods, namely pharmaceuticals, probiotics, powders made from plants and fruits full of antioxidants. Whereas standard drying may cause materials to shrink and fade in colour, items that are freeze-dried shrink less, keep their shades, absorb water well and remain stable for a longer time (Dadi et al., 2020; Reubun et al., 2021).

Before freeze-drying, some methods, for example, pulping and osmotic dehydration or using maltodextrin as a carrier, are used to increase effectiveness. Applying these techniques increases the mixing of powders, stops them from sticking, helps to transfer more mass and protects the bioactives involved during storage and processing (Shobhit et al., 2022). Even though freeze-drying requires a lot of energy and money, it is useful in situations where maintaining high product quality is important.

10.6 STANDARDISATION AND QUALITY CONTROL

Before a product is released onto the market, the final stages of standardisation and quality control ensure its consistency, safety and effectiveness. During this stage, plants' botanical status is changed into trusted functional foods, nutraceuticals or herbal products for the market. This is mostly carried out in laboratories.

10.6.1 Analytical Methods: The Guardians of Accuracy

High-performance liquid chromatography (HPLC) is known as a reliable tool for working with proteins and their fragments. HPLC functions as a detective, painstakingly isolating each compound in the plant extract and identifying them as alkaloids, flavonoids and phenolic acids. Chromatography matters in research and industry for its unique resolution, precise results and overall repeatability, even though it is also known for its ability to separate molecules (Uddin et al., 2022). Phenolic compound analysis depends on RP-HPLC, which is performed with UV or DAD detection. It helps to accurately follow the product's biological qualities ('fingerprint') (de Magalhães et al., 2020).

Experts use GC-MS to analyse fatty acids and essential oils, which are highly volatile. It uses mass spectrometry's structure to identify compounds and combines this with the high resolution of gas chromatography to recognise and detect the smallest parts of crucial molecules (Cheung et al., 2021). Plant-derived chemicals are often poorly suited for GC-MS analysis when they are too unstable or magnetic, so they have to be first converted with a derivatisation process (Ali et al., 2017).

UV-visual (UV-vis) spectrophotometry is the next technique, and it is as useful as the chromatographic methods, even though it is simpler. For determining rapid insights into the levels of flavonoids and antioxidants and the total phenolic content, UV-vis takes the lead in quality tests. It gives reliable results when used for batch comparisons and is still preferred for many tests despite its lower potential accuracy at the molecular level (Pudziuvelyte et al., 2022). UV-vis is usually combined with the aluminium chloride test or Folin–Ciocalteu method, enabling scientists to soon test for phytochemicals before starting detailed research.

10.6.2 Marker Compounds and Chemical Profiling: The Signatures of Trust

To standardise, it is important to focus on the marker compounds that set a plant extract's personality and strength. The evidence these substances provide helps confirm their legitimacy and therapeutic value, just like a signature at the bottom of a picture. The chemicals are carefully selected according to their importance in medicine, ease of detection, stability and how consistently they appear in different batches (Pan et al., 2018; Srivastava et al., 2022). It is important for a product's reliability in regulated fields that these indicators exist, appear in the same amount and are repeatable.

In addition, chemical profiling involves more than just particular indicators. There are modern methods for examining phytochemicals in a sample, such as nuclear magnetic resonance (NMR) spectroscopy and liquid chromatography-mass spectrometry (LC-MS/MS) (Alavijeh et al., 2014). They give us helpful data, both numerical and non-numerical, about everything present in the extract.

Additionally, chromatographic fingerprinting helps people perform work in this field. They are usually applied to confirm a product's quality, rather than to simply analyse it. They support making all batches alike, check for errors and guarantee the botanicals come from true sources (Das et al., 2020). Strong chemical markers confirm product quality, effectiveness and location within an industry with both rising customer trust and government regulations (Chen et al., 2017).

10.7 TECHNOLOGICAL APPLICATIONS IN FOOD SYSTEMS

10.7.1 Functional Foods and Nutraceuticals

Numerous investigations have emphasised the nutritional and therapeutic benefits of MO leaves (Kashyap et al., 2022; Agamou et al., 2015). The leaves exhibited an adequate nutritional profile, including substantial amounts of carbohydrates, fibre, protein, important amino acids (particularly lysine), minerals (calcium, iron, potassium, magnesium) and vitamins (A, B complex, C and E). HPLC analysis showed 70% methanol was the most efficient solvent for phenolic extraction. Strong antioxidant properties in the leaves supported their possible application as a functional food additive (El Sohaimy et al., 2015). Although the key nutrients in dried MO leaves are retained, the resulting powder remains a great source of proteins, vitamins, minerals and essential fatty acids that promote general health and well-being (Arshad et al., 2025). Dehydration techniques have been used to enhance the shelf life and availability of MO leaf powder. The blanched and unblanched MO leaf powders can be stored for a longer period by reducing the moisture content and water activity at isothermal drying temperature (50–80°C) (Sultana, 2020).

10.7.2 Fortification in Food Products

In the food industry, MO leaf powder and extracts are being utilised in functional foods due to their high nutritional value. MO leaf powder has been widely used in a variety of edible functional foods, i.e. cookies, snacks and wheat bread (Joshi & Varma, 2017). Compared with the control sample, bread prepared with wholewheat flour and 5% MO leaf powder had a considerably higher protein content and a higher crude fibre content, with increases of 54% and 56%, respectively (Sengev et al., 2013). Another study showed that fortifying MO leaf powder (2.5% to 10%) into wholewheat bread enhanced its nutritional value and antioxidant and mineral profile, i.e. Ca, Mg and β-carotene (vitamin A) (Khan et al., 2023). Moreover, fortifying MO leaf powder (5% to 20%) into pan bread also showed significant improvements in nutritional content, i.e. protein, fibre and essential minerals including Ca, Mg and Fe (El-Gammal et al., 2016).

The fortification of 10% MO leaf powder retained the flavour, texture and colour of cookies and significantly increased the protein, iron and calcium content (Dachana et al., 2010). Manaois et al. (2013) observed that rice cookies supplemented with freshly collected MO leaves (5%) substantially increased protein content (26%), while 5% supplemented with dried MO leaf powder increased protein content by 14%. Wheat flour biscuits fortified with blanched MO leaves showed higher protein and β-carotene content. However, dehydrated MO leaves fortification significantly increased the protein, fibre, ash and mineral contents. These findings indicate that MO leaves have the potential to be used as a functional component in baking products (Navarathna and Rathnayake, 2017).

MO is a multifunctional plant used in a wide range of beverages and has positive aspects, including better digestion, increased immunity and antioxidant support (Pareek et al., 2023). A blended health drink containing beetroot, MO leaf extract and pineapple in a 25:15:60 ratio was found suitable for beverage preparation. The beverage remained safe for ten weeks at 4±2°C in the refrigerator and six weeks at room temperature when sodium benzoate was added (Kharode et al., 2019). MO leaves are often blended with other organic components like green tea extract, ginseng or caffeine to produce energy drinks that offer extra vitality without the use of synthetic stimulants (Rodrigues et al., 2023). MO leaves fortification in orange or pomegranate fruit juices improves the vitamins, minerals and antioxidant potential. These beverages have anti-ageing, anti-inflammatory and detoxifying properties and also boost immunity, digestion and overall health. Some formulations also contain fibre or probiotics for additional benefits (Rodrigues et al., 2023; Perumalsamy et al., 2024).

MO has become widely used in dairy and plant-based dairy sectors due to its high protein and calcium content, and is a useful additive to milk and yoghurt products. Studies have shown that MO leaf extract fortification enhances the nutritional profiling of yoghurt without compromising its flavour and palatability (Salem et al., 2013). Additionally, adding MO leaf extract improves the health benefits of plant-based milk substitutes, such as almond and soy, by increasing their mineral and vitamin content (Matabura & Rweyemamu, 2022).

10.7.3 Encapsulation and Delivery Systems

MO plants are highly susceptible to degradation despite being a rich source of bioactive compounds and antioxidants. Microencapsulation is a useful technique for improving bioavailability and stability. MO leaf extract was microencapsulated using soy and pea protein isolates by spray-drying, which enhanced the stability of bioactive components and improved antioxidant retention. Compared with 25°C, phenolic content and antioxidant activity were better maintained at 4°C (Osamede Airouyuwa & Kaewmanee, 2019). Encapsulating MO leaf extract using maltodextrin, gum arabic and their combination formed microcapsules that revealed excellent techno-functional qualities and enhanced thermal stability, solubility and storage stability. The findings support their possible application as fortificants in processed foods (George et al., 2021). MO leaf polyphenols encapsulated in phytosomes demonstrated higher cytotoxicity, encapsulation efficiency and selectivity against 4T1 breast cancer cells, confirming their promise for breast cancer treatment (Wanjiru et al., 2022). In comparison with bulk extract, green-synthesised nanoparticles from MO leaf extract (size 141.6 nm, PDI 0.32) showed strong antidiabetic and anticancer potential, increased antioxidant activity and a greater phenolic/flavonoid content. These results demonstrate their potential in functional remedies and nanomedicine (Virk et al., 2023).

10.8 THERAPEUTIC AND PHARMACEUTICAL APPLICATIONS

MO, known as the 'miracle tree', has long been used in traditional medicine systems, including Ayurveda, Traditional Chinese Medicine (TCM) and African medicine, for its therapeutic properties (Klimek-Szczykutowicz et al., 2024). In TCM, the plant's leaves are used to treat colds, gynaecological disorders and anaemia and to restore vitality (Pareek et al., 2023; Hairi et al., 2025; Balamurugan et al., 2018). Traditionally, MO leaves have been used to treat several ailments, including typhoid fever, malaria, parasitic diseases, inflammations, arthritis, skin diseases, hypertension and diabetes and to support lactation, immunity (including HIV-related symptoms) and reproductive health (Figure 10.2) (Anwar et al., 2007; Abe & Ohtani, 2013; Yabesh et al., 2014; Leone et al., 2015).

MO has potent anti-inflammatory and antioxidant properties because of its abundance of bioactive compounds, including polyphenols, flavonoids, glucosinolates and isothiocyanates, which assist in preventing oxidative stress and inflammation, two major contributors to the onset of chronic diseases like diabetes, cancer and cardiovascular disorders (Pop et al., 2022; Ramamurthy et al., 2021). The different molecular pathways of the MO leaf are shown in Figure 10.3. The anti-inflammatory potential of MO has been well documented. MO leaves contain flavonoids such as kaempferol and quercetin, which have shown substantial anti-inflammatory effects. PKM1 and PKM2 significantly reduced nitric oxide (NO) generation in RAW 264.7 macrophages. MO leaf extract and its key components, especially quercetin, effectively mitigated inflammation and oxidative stress in high-fat diet-fed rats by scavenging H_2O_2, superoxide and other ROS, indicating their potential as alternative anti-inflammatory agents (Coppin et al., 2013; Das et al., 2013). The comparative research used a carrageenan-induced paw oedema model in albino Wistar rats to assess the anti-inflammatory effects of MO ethanolic leaf extract at dosages of 50, 100, 200 and 400 mg/kg versus aspirin (200 mg/kg). Paw oedema was considerably ($p<0.01$) decreased by the extract at 100, 200 and 400 mg/kg, presumably as a result of the inhibition of prostaglandin precursors. The anti-inflammatory

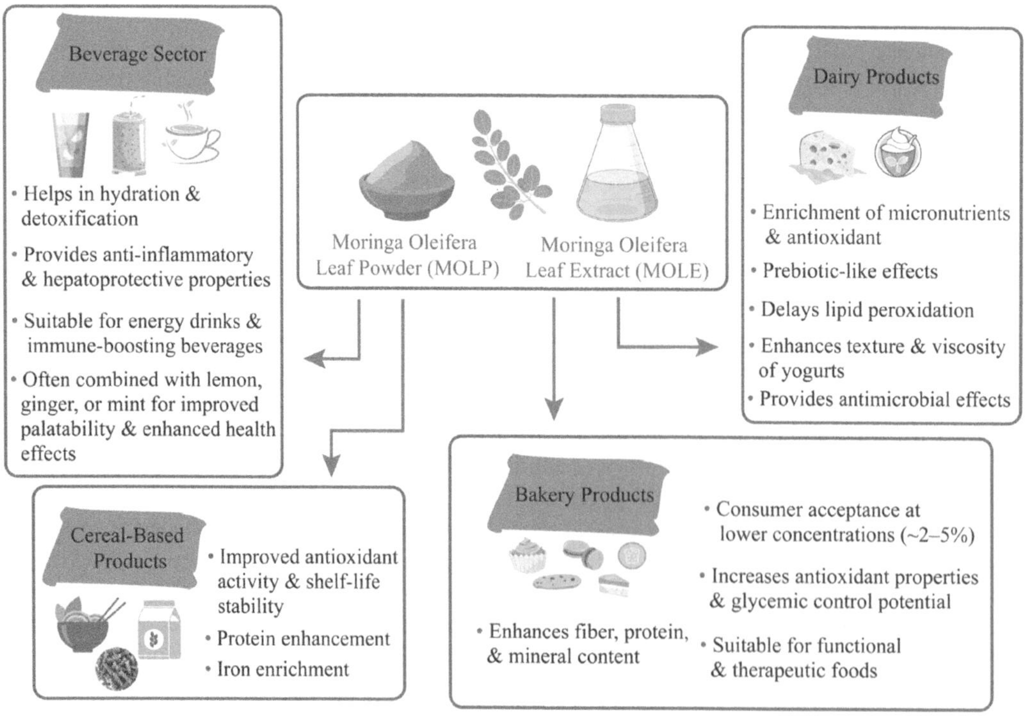

FIGURE 10.2 Molecular pathways of *Moringa oleifera* Leaf in managing metabolic disorders

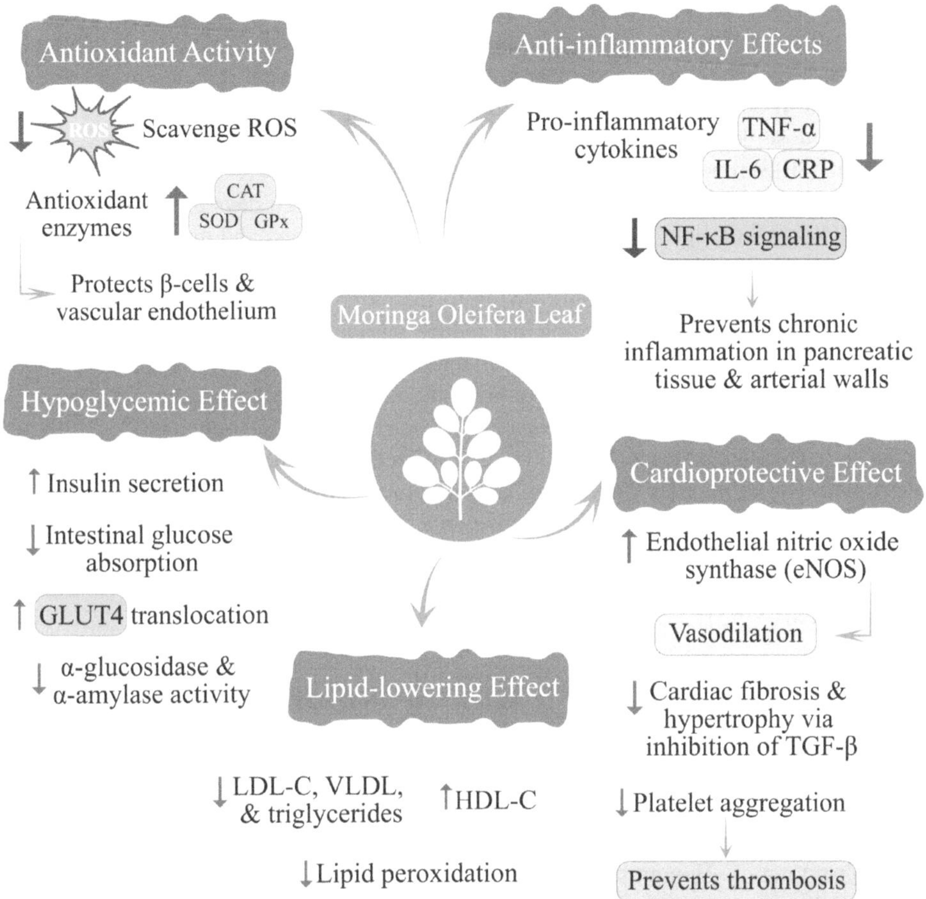

FIGURE 10.3 ***Moringa oleifera*** **leaves powder and extract in food fortification, highlighting benefits in beverage, dairy, cereals and bakery sectors.**

activity was attributed to phytoconstituents, including vanillin, 4-hydroxymellein, β-sitosterol and flavonoids (Bhattacharya et al., 2014).

MO leaf extracts have demonstrated notable antidiabetic benefits in diabetic animal models. The aqueous extract of MO leaves alleviated both type 1 (STZ-induced) and type 2 (insulin-resistant) diabetes by reducing insulin resistance and increasing insulin levels, lowering hyperglycaemia (Khan et al., 2017; Tuorkey, 2016). Furthermore, a methanolic extract of MO leaves improved glucose uptake, glycogen content, glucose tolerance and glycogen synthetase activity in alloxan-induced diabetic rats (Olayaki et al., 2015). Human clinical investigations have shown that supplementing with powdered or extracted MO leaves could reduce fasting blood glucose. It could potentially be a useful strategy for managing diabetes (Hamza et al., 2023).

In human tumour (KB) cells, phenolic compounds of MO leaf extract, including quercetin and kaempferol, showed potent antiproliferative and pro-apoptotic actions through DNA fragmentation and ROS production (Sreelatha et al., 2011). Furthermore, MO leaf extract showed potent anticancer activity against intestinal and breast adenocarcinoma cell lines (Al-Asmari et al., 2015), inhibited the proliferation of B16F10 melanoma cells (Gismondi et al., 2013) and suppressed NF-κB signalling in pancreatic cancer cells (Berkovich et al., 2013). It has been observed that isothiocyanates in MO leaves, including moringin, cause cell cycle arrest and significantly decrease the viability of cancer cells (Wu et al., 2021).

Das et al. (2012) showed that MO leaf extract suppressed lipid peroxidation (LPO), ALT, AST, ALP and liver histopathological damage while improving glutathione (GSH) levels in rats fed on a high diet, indicating hepatoprotective effects. In *Plasmodium berghei*-infected mice, aqueous MO leaf extract (100–1000 mg/kg) effectively reduced AST and ALT levels and restored serum albumin, indicating dose-dependent hepatoprotective effects (Dondee et al., 2016).

Panda et al. (2013) reported the cardioprotective effect of N,α-1-VR, an alkaloid produced from MO leaves, in isoproterenol-induced cardiotoxic mice. After seven days of oral administration of this alkaloid (40 mg/kg), lipid peroxidation, serum cardiac markers and myocardial necrosis were reduced while antioxidant levels increased. MO leaves possess antihypertensive effects by decreasing systolic and diastolic blood pressure owing to their rich potassium, magnesium and calcium content, which promote vascular function and reduce resistance (Alia et al., 2022). Its antioxidant properties also suppress oxidative stress and endothelial dysfunction while enhancing endothelial nitric oxide synthase activity, which helps to lower blood pressure (Hassan et al., 2021).

Several studies have proven the pharmacological activities of MO. Gupta et al. (2010) reported that an MO ethanolic leaf extract (125–500 mg/kg) effectively reduced cyclophosphamide-induced immunosuppression in mice by improving haematological parameters and organ weight, presumably by alleviating myelosuppression and leukopenia. In animal studies, the methanolic extract of MO leaves promoted both humoral and cellular immunity, while the aqueous and ethanolic extracts increased antibody response, delayed-type hypersensitivity and phagocytic activity (Sudha et al., 2010; Nfambi et al., 2015).

10.9 SAFETY, TOXICOLOGY AND REGULATORY ASPECTS

MO leaf extract and powder are widely used in commercial products such as teas, soups, cereals and herbal remedies, and global market trends suggest their significant growth. While in vitro and in vivo studies support the safety of MO leaves, prolonged overconsumption may raise concerns due to nutrient excess and contamination with heavy metals such as lead and arsenic, depending on the geographical region. Human studies have revealed no harmful effects when consumed within acceptable limits; however, a maximum daily consumption of 70 g is advised to avoid cumulative toxicity (Asiedu-Gyekye et al., 2014; Chhikara et al., 2021).

According to studies on acute toxicity, MO leaf powder is usually safe. In Sprague Dawley rats, oral administration of dried leaf powder up to 2000 mg/kg showed no negative effects (Adedapo et al., 2009). Similarly, studies revealed that aqueous MO leaf extract is generally safe, with no fatalities reported in male Wistar mice up to 6400 mg/kg. However, prolonged use beyond 1600 mg/kg can lead to symptoms of sluggishness and dullness without altering metabolic or haematological markers (Awodele et al., 2012). In another study, Stohs and Hartman (2015) reported no genotoxic effects of MO leaf extract at 1000 mg/kg in rats, while a lethal dose of 1585 mg/kg was observed.

The toxicological profile of MO leaf powder and extracts has been assessed in a number of in vitro and in vivo studies. Although dose-dependent cytotoxicity was observed at higher doses, results from normal human cell lines indicate overall safety. In rats, in vivo studies found no substantial toxicity, except at extremely high dosages (≥3000 mg/kg). Insufficient information is available on the toxicity of leaf extracts in humans; however, human studies using MO leaf powder have not revealed any negative effects. Thus, more toxicological analysis must be conducted to determine the long-term safety of MO leaf-based supplements (Gomes et al., 2023).

10.10 CONCLUSION

Moringa oleifera has emerged as a highly valued plant due to its exceptional nutritional, medicinal and functional properties. Its leaves, in particular, are rich in essential nutrients, including proteins, vitamins, minerals and a wide array of bioactive compounds such as flavonoids, polyphenols, glucosinolates and isothiocyanates. These compounds contribute to its potent antioxidant, anti-inflammatory, antidiabetic, antimicrobial and anticancer activities. The integration of MO leaves into food systems – through fortification, encapsulation and as functional ingredients – has demonstrated significant potential in addressing malnutrition and enhancing public health, especially in resource-limited settings. Furthermore, advanced processing and extraction technologies have improved the stability and bioavailability of its key components, expanding its application in nutraceuticals and pharmaceuticals. Despite its wide-ranging benefits, attention must be paid to the standardisation, safety and regulatory aspects to ensure consumer protection and product consistency. Overall, MO represents a promising, sustainable and versatile bioresource that supports both nutrition security and therapeutic innovation.

REFERENCES

Abadallah, M. S., and M. Ali. 2019. Antibacterial activity of *Moringa oleifera* leaf extracts against bacteria isolated from patients attending general Sani Abacha specialist hospital damaturu. *Journal of Allied Pharmaceutical Sciences i, 1*, pp.61–66.

Abalaka, M. E., S. Y. Daniyan., S. B. Oyeleke, and S. O. Adeyemo. 2012. The antibacterial evaluation of *Moringa oleifera* leaf extracts on selected bacterial pathogens. *Journal of Microbiology Research, 2*(2), pp.1–4.

Abe, R., and K. Ohtani. 2013. An ethnobotanical study of medicinal plants and traditional therapies on Batan Island, the Philippines. *Journal of Ethnopharmacology, 145*(2), pp.554–565.

Abhari, K., and A. M. Khaneghah. 2020. Alternative extraction techniques to obtain, isolate and purify proteins and bioactive from aquaculture and by-products. In *Advances in food and nutrition research* (Vol. 92, pp. 35–52). Academic Press.

Abo Baker, S. H., and A. A. Moawad. 2020. Anti-diabetic effect of *Moringa oleifera* extract on parotid gland of albino rats. *Egyptian Dental Journal, 66*, pp.187–196.

Abo-Bakr, A. M., H. M. Hassan, H. H. Temirek, and A. M. Mosallam. 2013. Synthesis of new dipyrimidine and dialkyl terephthalate derivatives from 2,6-diphenyl-sulphonyloxy-1H,3H,5H,7H-pyrrolo[3,4-f] isoindole-1,3,5,7-tetrone. *Journal of Chemistry, 56*(2), pp.112–127.

Adedapo, A. A., O. M. Mogbojuri, and B. O. Emikpe. 2009. Safety evaluations of the aqueous extract of the leaves of *Moringa oleifera* in rats. *Journal of Medicinal Plants Research, 3*(8), pp.586–591.

Adeyemi, S., R. Larayetan, A. D. Onoja, A. Ajayi, A. Yahaya, O. O. Ogunmola, A. O. Adeyi, and O. Chijioke. 2021. Anti-hemorrhagic activity of ethanol extract of *Moringa oleifera* leaf on envenomed albino rats. *Scientific African, 12*, p.e00742.

Agamou, J. A. A., E. N. Fombang, and C. M. Mbofung. 2015. Particular benefits can be attributed to *Moringa oleifera* Lam leaves based on origin and stage of maturity. *Journal of Experimental Biology and Agricultural Sciences, 3*(6), pp.541–555.

Aju, B. Y., R. Rajalakshmi, and S. Mini, 2019. Protective role of *Moringa oleifera* leaf extract on cardiac antioxidant status and lipid peroxidation in streptozotocin induced diabetic rats. *Heliyon, 5*(12), e02935.

Alakali, J. S., C. T. Kucha, and I. A. Rabiu. 2015. Effect of drying temperature on the nutritional quality of *Moringa oleifera* leaves. *African Journal of Food Science, 9*(7), pp.395–399.

Al-Asmari, A. K., S. M. Albalawi, M. T. Athar, A. Q. Khan, H. Al-Shahrani, and M. Islam. 2015. *Moringa oleifera* as an anti-cancer agent against breast and colorectal cancer cell lines. *PLoS One, 10*(8), p.e0135814.

Alegbeleye, O. O. 2018. How functional is *Moringa oleifera*? A review of its nutritive, medicinal, and socioeconomic potential. *Food and Nutrition Bulletin, 39*(1), pp.149–170.

Ali, M. A., Y. A. Yusof, N. L. Chin, and M. N. Ibrahim. 2017. Processing of Moringa leaves as natural source of nutrients by optimization of drying and grinding mechanism. *Journal of Food Process Engineering, 40*(6), p.e12583.

Alia, F., M. Putri, N. Anggraeni, and M. R. A. A. Syamsunarno. 2022. The potency of *Moringa oleifera* Lam. as protective agent in cardiac damage and vascular dysfunction. *Frontiers in Pharmacology, 12*, p.724439.

Alqahtani, W. S., and G. Albasher. 2021. *Moringa oleifera* Lam. extract rescues lead-induced oxidative stress, inflammation, and apoptosis in the rat cerebral cortex. *Journal of Food Biochemistry, 45*(1), p.e13579.

Álvarez-Román, R., P. G. Silva-Flores, S. A. Galindo-Rodríguez, A. A. Huerta-Heredia, W. Vilegas, and D. Paniagua-Vega. 2020. Moisturizing and antioxidant evaluation of *Moringa oleifera* leaf extract in topical formulations by biophysical techniques. *South African Journal of Botany, 129*, pp.404–411.

Alavijeh, N. S., S. Ramezanpour, M. S. Alavijeh, S. Balalaie, F. Rominger, A. Misra, and H. R. Bijanzadeh. 2014. Synthesis and lipophilicity evaluation of some novel indole-containing pseudopeptides. *Monatshefte für Chemie – Chemical Monthly*, *145*(2), pp.349–356.

Amara, I., M. L. Ontario, M. Scuto, G. M. Lo Dico, S. Sciuto, V. Greco, S. Abid-Essefi, A. Signorile, A. T. Salinaro, and V. Calabrese. 2021. *Moringa oleifera* protects SH-SY5YCells from DEHP-induced endoplasmic reticulum stress and apoptosis. *Antioxidants*, *10*(4), p.532.

Anwar, F., S. Latif, M. Ashraf, and A. H. Gilani. 2007. *Moringa oleifera*: A food plant with multiple medicinal uses. *Phytotherapy Research: An International Journal Devoted to Pharmacological and Toxicological Evaluation of Natural Product Derivatives*, *21*(1), pp.17–25.

Aremu, A. K., and A. Akintola. 2014, October. Effects of some drying methods on nutritional characteristics of moringa (*Moringa oleifera*) seeds. In *4th International conference on biotechnology and environment management* (IPCBEE Vol. 75, pp. 66–71), IACSIT Press, Singapore.

Ariani, L. N., T. Estiasih, W. B. Sunarharum, and A. Khatib. 2023. Potential of moringa (*Moringa oleifera*) leaf powder for functional food ingredients: A review. *Czech Journal of Food Sciences*, *41*(1), pp.8–20.

Arora, S. and S. Arora. 2021. Nutritional significance and therapeutic potential of *Moringa oleifera*: The wonder plant. *Journal of Food Biochemistry*, *45*(10), p.e13933.

Arshad, M. T., S. Maqsood, A. Ikram, and K. T. Gnedeka. 2025. Recent perspectives on the pharmacological, nutraceutical, functional, and therapeutic properties of *Moringa oleifera* plant. *Food Science & Nutrition*, *13*(4), p.e70134.

Asiedu-Gyekye, I. J., S. Frimpong-Manso, C. Awortwe, D. A. Antwi, and A. K. Nyarko. 2014. Micro- and macroelemental composition and safety evaluation of the nutraceutical *Moringa oleifera* leaves. *Journal of Toxicology*, *2014*(1), p.786979.

Atawodi, S. E., J. C. Atawodi, G. A. Idakwo, B. Pfundstein, R. Haubner, G. H. Wurtele, Bartsch, and R. W. Owen. 2010. Evaluation of the polyphenol content and antioxidant properties of methanol extracts of the leaves, stem, and root barks of *Moringa oleifera* Lam. *Journal of Medicinal Food*, *13*(3), pp.710–716.

Awodele, O., I. A. Oreagba, S. Odoma, J. A. T. da Silva, and V. O. Osunkalu. 2012. Toxicological evaluation of the aqueous leaf extract of *Moringa oleifera* Lam.(Moringaceae). *Journal of Ethnopharmacology*, *139*(2), pp.330–336.

Bagade, S. B., and M. Patil. 2021. Recent advances in microwave assisted extraction of bioactive compounds from complex herbal samples: A review. *Critical Reviews in Analytical Chemistry*, *51*(2), pp.138–149.

Balamurugan, S., S. Vijayakumar, S. Prabhu, and J. E. Morvin Yabesh. 2018. Traditional plants used for the treatment of gynaecological disorders in Vedaranyam taluk, South India-An ethnomedicinal survey. *Journal of Traditional and Complementary Medicine*, *8*(2), pp.308–323.

Berkovich, L., G. Earon, I. Ron, A. Rimmon, VexlerA. , and S. Lev-Ari. 2013. *Moringa oleifera* aqueous leaf extract down-regulates nuclear factor-kappaB and increases cytotoxic effect of chemotherapy in pancreatic cancer cells. *BMC Complementary and Alternative Medicine*, *13*, pp.1–7.

Bhattacharya, A., M. R. Naik, D. Agrawal, K. Rath, S. Kumar, and S. S. Mishra. 2014. Antipyretic, anti-inflammatory and analgesic effects of leaf extract of drumstick tree. *Journal of Young Pharmacists*, *6*(4), p.20.

Bhattacharya, A., R. Behera, D. Agrawal, P. K. Sahu, S. Kumar, and S. S. Mishra 2014. Antipyretic effect of ethanolic extract of *Moringa oleifera* leaves on albino rats. *Tanta Medical Journal*, *42*(2), 74–78.

Cajuday, L. A. and G. L. Pocsidio. 2010. Effects of *Moringa oleifera* Lam.(Moringaceae) on the reproduction of male mice (Mus musculus). *Journal of Medicinal Plants Research*, *4*(12), pp.1115–1121.

Calabrese, V. 2007, October. Redox regulation of cellular stress response in ageing and neurodegenerative disorders: role of vitagenes and modulation by antioxidants. In *Free radical research* (Vol. 41, Suppl. 1, pp.S10). Taylor & Francis Ltd.

Camilleri, E., and R. Blundell. 2024. A comprehensive review of the phytochemicals, health benefits, pharmacological safety and medicinal prospects of *Moringa oleifera*. *Heliyon, 10*(6), e27807.

Charoensin, S. 2014. Antioxidant and anticancer activities of *Moringa oleifera* leaves. *Journal of Medicinal Plants Research*, *8*(7), pp.318–325.

Chávez-González, M. L., L. Sepúlveda, D. K. Verma, H. A. Luna-García, L. V. Rodríguez-Durán, A. Ilina, and C. N. Aguilar. 2020. Conventional and emerging extraction processes of flavonoids. *Processes*, *8*(4), p.434.

Chen, C., B. Zhang, Q. Huang, X. Fu, and R. H. Liu. 2017. Microwave-assisted extraction of polysaccharides from *Moringa oleifera* Lam. leaves: Characterization and hypoglycemic activity. *Industrial Crops and Products*, *100*, pp.1–11.

Cheung, S. M., E. Husain, V. Mallikourti, Y. Masannat, S. Heys, and J. He. 2021. Intra-tumoural lipid composition and lymphovascular invasion in breast cancer via non-invasive magnetic resonance spectroscopy. *European Radiology*, *31*(6), pp.3703–3711.

Chhikara, N., A. Kaur, S. Mann, M. K. Garg, S. A. Sofi, and A. Panghal. 2021. Bioactive compounds, associated health benefits and safety considerations of *Moringa oleifera* L.: An updated review. *Nutrition & Food Science*, *51*(2), pp.255–277.

Chiş, A., P. A. Noubissi, O. L. Pop, C. I. Mureşan, M. A. Fokam Tagne, R. Kamgang, A. Fodor, A. V. Sitar-Tăut, A. Cozma, O. H. Orăşan, and S. C. Hegheş. 2023. Bioactive compounds in *Moringa oleifera*: mechanisms of action, focus on their anti-inflammatory properties. *Plants*, *13*(1), p.20.

Coppin, J. P., Y. Xu, H. Chen, M. H. Pan, C. T. Ho, R. Juliani, R., J. E. Simon, and Q. Wu. 2013. Determination of flavonoids by LC/MS and anti-inflammatory activity in *Moringa oleifera*. *Journal of Functional Foods*, *5*(4), pp.1892–1899.

Cuellar-Nuñez, M. L., I. Luzardo-Ocampo, R. Campos-Vega, M. A. Gallegos-Corona, E. G. De Mejía, and G. Loarca-Piña 2018. Physicochemical and nutraceutical properties of moringa (*Moringa oleifera*) leaves and their effects in an in vivo AOM/DSS-induced colorectal carcinogenesis model. *Food Research International*, *105*, pp.159–168.

Dachana, K.B., J. Rajiv, D. Indrani, and J. Prakash. 2010. Effect of dried moringa (*Moringa oleifera* lam) leaves on rheological, microstructural, nutritional, textural and organoleptic characteristics of cookies. *Journal of Food Quality*, *33*(5), pp.660–677.

Dadi, D. W., S. A. Emire, A. D. Hagos, and J. B. Eun 2020. Physical and functional properties, digestibility, and storage stability of spray-and freeze-dried microencapsulated bioactive products from Moringa stenopetala leaves extract. *Industrial Crops and Products*, *156*, p.112891.

Das, N., K. Sikder, S. Bhattacharjee, S. B. Majumdar, S. Ghosh, S. Majumdar, and S. Dey. 2013. Quercetin alleviates inflammation after short-term treatment in high-fat-fed mice. *Food & Function*, *4*(6), pp.889–898.

Das, N., K. Sikder, S. Ghosh, B. Fromenty, and S. Dey. 2012. *Moringa oleifera* Lam. leaf extract prevents early liver injury and restores antioxidant status in mice fed with high-fat diet. *Indian Journal of Experimental Biology*, *50*(6), p.404.

Das, P. E., A. F. Majdalawieh, I. A. Abu-Yousef, S. Narasimhan, and P. Poltronieri. 2020. Use of a hydroalcoholic extract of *Moringa oleifera* leaves for the green synthesis of bismuth nanoparticles and evaluation of their antimicrobial and antioxidant activities. *Materials*, *13*(4), p.876.

de Magalhães, B. E. A., D. A. de Andrade Santana, I. M. de Jesus Silva, L. A. C. Minho, M. A. Gomes, J. R. G. da Silva Almeida, and W. N. L. dos Santos. 2020. Determination of phenolic composition of oilseed whole flours by HPLC-DAD with evaluation using chemometric analyses. *Microchemical Journal*, *155*, 104683.

Dinesha, B. L., U. Nidoni, C. T. Ramachandra, N. Naik, and K. B. Sankalpa. 2018. Effect of extraction methods on physicochemical, nutritional, antinutritional, antioxidant and antimicrobial activity of Moringa (*Moringa oleifera* Lam.) seed kernel oil. *Journal of Applied and Natural Science*, *10*(1), pp.287–295 .

Dixit, S., A. Tripathi, and P. Kumar. 2016. Medicinal properties of *Moringa oleifera*: A review. *International Journal of Education and Science Research Review*, *3*(2), pp.173–185.

Dondee, K., P. Borkaew, C. Klubsri, P. Bootprom, B. Saiphet, and V. Somsak. 2016. The protective effect of *Moringa oleifera* leaf extract on liver damage in mice infected with Plasmodium berghei ANKA. *Journal of Coastal Life Medicine, 4*(9), pp.742–746.

Edoga, C. O., O. O. Njoku, E. N. Amadi, and J. J. Okeke 2013. Blood sugar lowering effect of *Moringa oleifera* Lam in albino rats. *International Journal of Science and Technology*, *3*(1), pp.88–90.

El Bilali, H., I. Dan Guimbo, R. K. Nanema, H. Falalou, Z. Kiebre, V. M. Rokka, S. R. F. Tietiambou, J. Nanema, L. Dambo, F. Grazioli, and A. K. Naino Jika. 2024. Research on Moringa (*Moringa oleifera* Lam.) in Africa. *Plants*, *13*(12), p.1613.

El Sohaimy, S. A., G. M. Hamad, S. E. Mohamed, M. H. Amar, and R. R. Al-Hindi. 2015. Biochemical and functional properties of *Moringa oleifera* leaves and their potential as a functional food. *Global Advanced Research Journal of Agricultural Science*, *4*(4), pp.188–199.

El-Gammal, R. E., G. A. Ghoneim, and S. M. ElShehawy. 2016. Effect of moringa leaves powder (*Moringa oleifera*) on some chemical and physical properties of pan bread. *Journal of Food and Dairy Sciences*, *7*(7), pp.307–314.

Ezugwu, R. I., and C. Chukwubike. 2014. Evaluation of the antimicrobial activity of *Moringa oleifera* leaves extract on Helicobacter pylori. *IOSR Journal of Pharmacy and Biological Sciences*, *9*, pp.57–60.

Fahey, J. W. 2005. *Moringa oleifera*: A review of the medical evidence for its nutritional, therapeutic, and prophylactic properties. Part 1. *Trees for life Journal*, *1*(5), pp.1–15..

Farid, A. S. and A. M. Hegazy. 2020. Ameliorative effects of *Moringa oleifera* leaf extract on levofloxacin-induced hepatic toxicity in rats. *Drug and Chemical Toxicology*, *43*(6), pp.616–622.

Foline, O., A. F. Emenike, and B. I. Eunice. 2011. Comparative analysis of the nutritional composition of three different drying methods of *Moringa oleifera* leaves. *International Journal of Applied Agricultural Research*, *6*(2), pp.131–138.

Fombang, E. N. and W. R. Saa. 2016. Production of a functional tea from *Moringa oleifera* LAM leaf powder: Optimization of phenolic extraction using response surface methodology. *Journal of Nutrition & Food Sciences*, *6*, p.556.

Fouad, E. A., A. S. A. Elnaga, and M. M. Kandil. 2019. Antibacterial efficacy of *Moringa oleifera* leaf extract against pyogenic bacteria isolated from a dromedary camel (Camelus dromedarius) abscess. *Veterinary World*, *12*(6), p.802.

Ganguly, R., and D. Guha. 2008. Alteration of brain monoamines and EEG wave pattern in rat model of Alzheimer's disease and protection by *Moringa oleifera*. *Indian Journal of Medical Research*, *128*(6), pp.744–751.

George, T. T. 2021. *Characterisation and encapsulation of Moringa oleifera extracts* (Doctoral dissertation), Cape Peninsula University of Technology.

George, T. T., A. B. Oyenihi, F. Rautenbach, and A. O. Obilana. 2021. Characterization of *Moringa oleifera* leaf powder extract encapsulated in maltodextrin and/or gum arabic coatings. *Foods*, *10*(12), p.3044.

Gismondi, A., L. Canuti, S. Impei, G. Di Marco, M. Kenzo, V. Colizzi, and A. Canini. 2013. Antioxidant extracts of African medicinal plants induce cell cycle arrest and differentiation in B16F10 melanoma cells. *International Journal of Oncology*, *43*(3), pp.956–964.

Gomes, S. M., A. Leitão, A. Alves, and L. Santos. 2023. Incorporation of *Moringa oleifera* leaf extract in yoghurts to mitigate children's malnutrition in developing countries. *Molecules*, *28*(6), p.2526.

Gönenç, A., A. Hacışevki, Y. Tavil, ÇengelA. , and M. Torun. 2013. Oxidative stress in patients with essential hypertension: A comparison of dippers and non-dippers. *European journal of Internal Medicine*, *24*(2), pp.139–144.

Gopalakrishnan, L., K. Doriya, and D.S. Kumar. 2016. *Moringa oleifera*: A review on nutritive importance and its medicinal application. *Food Science and Human Wellness*, *5*(2), pp.49–56.

Gouda, A. S., N. A. El-Nabarawy, and S. F. Ibrahim. 2018. *Moringa oleifera* extract (Lam) attenuates Aluminium phosphide-induced acute cardiac toxicity in rats. *Toxicology Reports*, *5*, pp.209–212.

Gupta, A., M. K. Gautam, R. K. Singh, M. V. Kumar, C. V. Rao, R. K. Goel, and S. Anupurba. 2010. Immunomodulatory effect of *Moringa oleifera* Lam. extract on cyclophosphamide induced toxicity in mice.*Indian Journal of Experimental Biology*, *48*, pp.1157–1160.

Guzmán-Maldonado, S. H., M. J. López-Manzano, T. J. Madera-Santana, C. A. Núñez-Colín, C. P. Grijalva-Verdugo, A. G. Villa-Lerma, and J. R. Rodríguez-Núñez. 2020. Nutritional characterization of *Moringa oleifera* leaves, seeds, husks and flowers from two regions of Mexico. *Agronomía Colombiana*, *38*(2), pp.287–297.

Hairi, H. A., R. R. Jusoh, M. Z. Sadikan, W. N. W. Hasan, and A. N. Shuid. 2025. Exploring the potential of *Moringa oleifera* in managing bone loss: insights from preclinical studies. *International Journal of Medical Sciences*, *22*(4), p.819.

Hamed, Y. S., M. Abdin, H. M. S. Akhtar, D. Chen, P. Wan, G. Chen, and X. Zeng. 2019. Extraction, purification by macrospores resin and in vitro antioxidant activity of flavonoids from Moringa oliefera leaves. *South African Journal of Botany*, *124*, pp.270–279.

Hamza, M. A., M. Naimuzzaman, and S. K. Roy. 2023. Health benefits of *Moringa oleifera*: Used as an anti-diabetic agent. *International Journal of Agricultural Research, Innovation and Technology, 13*(1), pp.96–102.

Hassan, M. A., T. Xu, Y. Tian, Y. Zhong, F. A. Z. Ali, X. Yang, and B. Lu. 2021. Health benefits and phenolic compounds of *Moringa oleifera* leaves: A comprehensive review. *Phytomedicine*, *93*, p.153771.

Hisam, E. E. A., M. S. Rofiee, A. KHALID, A. F. Jalaluddin, M. I. M. Yusof, M. H. Idris, S. Ramli, R. J. James, W. J. Yoeng, T. L. Kek, and M. Z. Salleh. 2018. Combined extract of *Moringa oleifera* and Centella asiatica modulates oxidative stress and senescence in hydrogen peroxide-induced human dermal fibroblasts. *Turkish Journal of Biology*, *42*(1), pp.33–44.

Hodas, F., M. R. T. Zorzenon, and P. G. Milani. 2021. *Moringa oleifera* potential as a functional food and a natural food additive: A biochemical approach. *Anais da Academia Brasileira de Ciencias*, *93*, p.e20210571.

Hu, E., Z. Li, T. Li, X. Yang, R. Ding, H. Jiang, H. Su, M. Cheng, Z. Yu, H. Li, and T. Tang. 2023. A novel microbial and hepatic biotransformation-integrated network pharmacology strategy explores the therapeutic mechanisms of bioactive herbal products in neurological diseases: The effects of Astragaloside IV on intracerebral hemorrhage as an example. *Chinese Medicine*, *18*(1), p.40.

Igado, O. O., and J. O. Olopade. 2016. A review on the possible neuroprotective effects of *Moringa oleifera* leaf extract. *Nigerian Journal of Physiological Sciences*, *31*(2), pp.183–187.

Ishnava, K. B., K. H. Chauhan, and C. A. Bhatt. 2012. Screening of antifungal activity of various plant leaves extracts from Indian plants. *Archives of Phytopathology and Plant Protection*, *45*(2), pp.152–160.

Jalil, S. M., H. Husaini, R. Munadi, and I. D. Sara. 2022. Microcontroller based dual energy Moringa leaf dryer design and development. *International Journal of Power Electronics and Drive Systems*, *13*(2), p.1075.

Jimoh, T. O. 2018. Enzymes inhibitory and radical scavenging potentials of two selected tropical vegetable (*Moringa oleifera* and Telfairia occidentalis) leaves relevant to type 2 diabetes mellitus. *Revista brasileira de farmacognosia*, *28*(1), pp.73–79.

Joshi, P., and K. Varma. 2017. Preparation of Value Added Products from Dehydrated Drumstick Leaves. *Online International Interdisciplinary Research Journal, 7*, pp.41–48.

Jung, I. L., J. H. Lee, and S. C. Kang. 2015. A potential oral anticancer drug candidate, *Moringa oleifera* leaf extract, induces the apoptosis of human hepatocellular carcinoma cells. *Oncology Letters*, *10*(3), pp.1597–1604.

Kamalakkannan, N., and P. S. M. Prince. 2006. Antihyperglycaemic and antioxidant effect of rutin, a polyphenolic flavonoid, in streptozotocin-induced diabetic wistar rats. *Basic and Clinical Pharmacology and Toxicology*, *98*(1), pp.97–103.

Kashaninejad, M., B. Blanco, O. Benito-Román, S. Beltrán, S. M. Niknam, and M. T. Sanz. 2021. Maximizing the freeze-dried extract yield by considering the solvent retention index: Extraction kinetics and characterization of *Moringa oleifera* leaves extracts. *Food and Bioproducts Processing*, *130*, pp.132–142.

Kashyap, P., S. Anand, and A. Thakur. 2017. Evaluation of antioxidant and antimicrobial activity of Rhododendron arboreum flowers extract. *International Journal of Food and Fermentation Technology*, *7*(1), pp.123–128.

Kashyap, P., S. Kumar, C. S. Riar, N. Jindal, P. Baniwal, R. P. Guiné, P. M. Correia, MehraR. , and H. Kumar. 2022. Recent advances in Drumstick (*Moringa oleifera*) leaves bioactive compounds: Composition, health benefits, bioaccessibility, and dietary applications. *Antioxidants*, *11*(2), p.402.

Kekuda, T. P., N. Mallikarjun, D. Swathi, K. V. Nayana, M. B. Aiyar, and T. R. Rohini. 2010. Antibacterial and Antifungal efficacy of steam distillate of *Moringa oleifera* Lam. *Journal of Pharmaceutical Sciences and Research*, *2*(1), p.34.

Khalofah, A., N. A. Bokhari, H. M. Migdadi, and M. S. Alwahibi. 2020. Antioxidant responses and the role of *Moringa oleifera* leaf extract for mitigation of cadmium stressed Lepidium sativum L. *South African Journal of Botany*, *129*, pp.341–346.

Khan, D., A. N. Khan, and M. J. Zaki. 2022. Brief profile of doctor tree (*Moringa oleifera* Lam.). *International Journal of Biological Research*, *10*(1), pp.01–31.

Khan, M. A., S. Shakoor, K. Ameer, M. A. Farooqi, M. Rohi, M. Saeed, M. T. Asghar, M. B. Irshad, M. Waseem, S. Tanweer, and U. Ali. 2023. Effects of dehydrated moringa (*Moringa oleifera*) leaf powder supplementation on physicochemical, antioxidant, mineral, and sensory properties of whole wheat flour leavened bread. *Journal of Food Quality*, *2023*(1), p.4473000.

Khan, W., R. Parveen, K. Chester, S. Parveen, and S. Ahmad. 2017. Hypoglycemic potential of aqueous extract of *Moringa oleifera* leaf and in vivo GC-MS metabolomics. *Frontiers in Pharmacology*, *8*, p.577.

Kharode, R., Y. Chavan, M. Gaikwad, A. Hiremath, A. and A. Sahoo. 2019. Development of health drink from fruit and vegetables (Beetroot, Pineapple and Moringa leaves). *Journal of Pharmaceutical Innovation*, *8*, pp.776–780.

Klimek-Szczykutowicz, M., K. Gaweł-Bęben, A. Rutka, E. Blicharska, M. Tatarczak-Michalewska, K. Kulik-Siarek, W. Kukula-Koch, M.A. Malinowska, and A. Szopa. 2024. *Moringa oleifera* (drumstick tree)—nutraceutical, cosmetological and medicinal importance: A review. *Frontiers in Pharmacology, 15*, p.1288382.

Kshirsagar, R. B., A. R. Sawate, S. K. Sadawate, B. M. Patil, and M. A. Zaker. 2017. Effect of blanching and drying treatment on the proximate composition of *Moringa oleifera* leaves. *International Journal of Agricultural Engineering, 10 (1)*, 10-15.

Kumar, P. K., and R. T. Mandapaka. 2013. Effect of *Moringa oleifera* on blood glucose, ldl levels in types ii diabetic obese people. *Innovative Journal of Medical and Health Sciences*, *3*(1), pp.23–25.

Leone, A., A. Spada, A. Battezzati, A. Schiraldi, J. Aristil, and S. Bertoli. 2015. Cultivation, genetic, ethnopharmacology, phytochemistry and pharmacology of *Moringa oleifera* leaves: An overview. *International Journal of Molecular Sciences, 16*(6), pp.12791–12835.

Leone, A., S. Bertoli, S. Di Lello, A. Bassoli, S. Ravasenghi, G. Borgonovo, F. Forlani, and A. Battezzati. 2018. Effect of *Moringa oleifera* leaf powder on postprandial blood glucose response: In vivo study on Saharawi people living in refugee camps. *Nutrients, 10*(10), p.1494.

Li, S., W. Li, R. Wu, R. Yin, D. Sargsyan, I. Raskin, and A. N. Kong. 2020. Epigenome and transcriptome study of moringa isothiocyanate in mouse kidney mesangial cells induced by high glucose, a potential model for diabetic-induced nephropathy. *The AAPS Journal, 22*, pp.1–11.

Li, X., Han, H., Ma, Y., Wang, X., & Lü, X. 2024. Identification of phenolic compounds from fermented *Moringa oleifera* Lam. leaf supplemented with Fuzhuan brick tea and their volatile composition and anti-obesity activity. *Journal of Food Science, 89*(5), pp.3094–3109

Liang, M., T. Li, Y. Qu, J. Qin, Z. Li, X. Huang, Q. Zhang, Y. Zhang, Q. Guo, and Q. Wang. 2023. Mitigation mechanism of resveratrol on thermally induced trans-α-linolenic acid of trilinolenin. *LWT, 189*, p.115508.

Lin, M., J. Zhang, and X. Chen. 2018. Bioactive flavonoids in *Moringa oleifera* and their health-promoting properties. *Journal of Functional Foods, 47*, pp.469–479.

López-Martínez, L. X., J. A. Dominguez-Ávila, N. J. Salazar-Lopez, and G. A. Gonzalez-Aguilar. 2021. Quality Control and Safety of Moringa. In *Biological and pharmacological properties of the genus Moringa* (pp. 83–98). CRC Press.

Lopez-Teros, V., J. L. Ford, M. H. Green, G. Tang, M. A. Grusak, L. Quihui-Cota, M. Tawanda, Mariela Paz-Cassini, and Humberto Astiazaran-Garcia. 2017. Use of a "super-child" approach to assess the vitamin A equivalence of *Moringa oleifera* leaves, develop a compartmental model for vitamin A kinetics, and estimate vitamin A total body stores in young Mexican children. *The Journal of Nutrition, 147*(12), pp.2356–2363.

Luetragoon, T., R. Pankla Sranujit, C. Noysang, Y. Thongsri, P. Potup, N. Suphrom, N. Nuengchamnong, and K. Usuwanthim. 2020. Bioactive compounds in *Moringa oleifera* Lam. leaves inhibit the pro-inflammatory mediators in lipopolysaccharide-induced human monocyte-derived macrophages. *Molecules, 25*(1), p.191.

Luqman, S., S. Srivastava, R. Kumar, A. K. Maurya, and D. Chanda. 2012. Experimental assessment of *Moringa oleifera* leaf and fruit for its antistress, antioxidant, and scavenging potential using in vitro and in vivo assays. *Evidence-Based Complementary and Alternative Medicine, 2012*(1), p.519084.

Ma, Y. 2018. Technology for extracting and purifying *Moringa oleifera* flavone. https://patents.google.com/patent/CN104398542A/en

Ma, Z. F., J. Ahmad, H. Zhang, I. Khan, and S. Muhammad. 2020. Evaluation of phytochemical and medicinal properties of Moringa (*Moringa oleifera*) as a potential functional food. *South African Journal of Botany, 129*, pp.40–46.

Mabrouki, L., I. Rjeibi, J. Taleb, and L. Zourgui. 2020. Cardiac ameliorative effect of *Moringa oleifera* leaf extract in high-fat diet-induced obesity in rat model. *BioMed Research International, 2020*(1), p.6583603.

Mahaman, Y. A.R., F. Huang, M. Wu, Y. Wang, Z. Wei, J. Bao, M. T. M. Salissou, D. Ke, Q. Wang, R. Liu, and J. Z. Wang. 2018. *Moringa oleifera* alleviates homocysteine-induced Alzheimer's disease-like pathology and cognitive impairments. *Journal of Alzheimer's Disease, 63*(3), pp.1141–1159.

Mahmood, K. T., T. Mugal, and I. U. Haq. 2010. *Moringa oleifera*: A natural gift-A review. *Journal of Pharmaceutical Sciences and Research, 2*(11), p.775.

Manaois, R. V., A. V. Morales, and R. G. Abilgos-Ramos. 2013. Acceptability, shelf life and nutritional quality of moringa-supplemented rice crackers. *Philippine Journal of Crop Science, 38*(2), pp.1–8.

Manju, K. M., Rekha, Priyanka, and N. Kumar. 2021. Effect of fluidized-bed and freeze-drying techniques on physicochemical, nutritional, thermal, and structural properties of *Moringa oleifera* flowers, leaves, and seeds. *Journal of Food Processing and Preservation, 45*(9), p.e15719.

Mashamaite, C. V., M. N. Ramatsitsi, and A. Manyevere. 2024. *Moringa oleifera* Lam.: A versatile climate-smart plant for nutritional security and therapeutic usage in semi-arid regions. *Journal of Agriculture and Food Research, 16*, p.101217.

Matabura, V. V. and L. M. Rweyemamu. 2022. Formulation of plant-based food and characterisation of the nutritional composition: A case study on soy-moringa beverage. *Journal of Food Science and Technology, 59*(10), pp.3794–3805.

Mbah, B. O., P. E. Eme, and A. E. Paul. 2012. Effect of drying techniques on the proximate and other nutrient composition of *Moringa oleifera* leaves from two areas in Eastern Nigeria. *Pakistan Journal of Nutrition*, *11*(11), p.1044.

Mbikay, M. 2012. Therapeutic potential of *Moringa oleifera* leaves in chronic hyperglycemia and dyslipidemia: A review. *Frontiers in Pharmacology*, *3*, p.24.

McGuire, S. 2016. *World cancer report 2014*. Geneva, Switzerland: World Health Organization, international agency for research on cancer, WHO Press. 2015. *Advances in Nutrition*, *7*(2), p.418.

Metwally, F. M., H. M. Rashad, H. H. Ahmed, A. A. Mahmoud, E. R. A. Raouf, and A. M. Abdalla. 2017. Molecular mechanisms of the anti-obesity potential effect of *Moringa oleifera* in the experimental model. *Asian Pacific Journal of Tropical Biomedicine*, *7*(3), pp.214–221.

Mittal, A., M. Sharma, A. David, P. Vishwakarma, M. Saini, M. Goel, and K. K. Saxena. 2017. An experimental study to evaluate the anti-inflammatory effect of *Moringa oleifera* leaves in animal models. *International Journal of Basic & Clinical Pharmacology*, *6*(2), pp.452–457.

Mohamed, A. A. R., M. M. Metwally, S. R. Khalil, G. A. Salem, and H. A. Ali. 2019. *Moringa oleifera* extract attenuates the CoCl2 induced hypoxia of rat's brain: Expression pattern of HIF-1α, NF-kB, MAO and EPO. *Biomedicine and Pharmacotherapy*, *109*, pp.1688–1697.

Momoh, M. A., S. A. Chime, and F. C. Kenechukwu. 2013. Novel drug delivery system of plant extract for the management of diabetes: An antidiabetic study. *Journal of Dietary Supplements*, *10*(3), pp.252–263.

Moyo, B., P. J. Masika, and V. Muchenje. 2012. Effect of supplementing crossbred Xhosa lop-eared goat castrates with *Moringa oleifera* leaves on growth performance, carcass and non-carcass characteristics. *Tropical Animal Health and Production*, *44*, pp.801–809.

Mursyid, M., R. N. Annisa, I. Zahran, J. Langkong, and I. Kamaruddin. 2019, October. Antimicrobial activity of moringa leaf (*Moringa oleifera* L.) extract against the growth of Staphylococcus epidermidis. In *IOP conference series: Earth and environmental science* (Vol. 343, No. 1, p.012145). IOP Publishing.

Muzumbukilwa, W. T., M. Nlooto, and P. M. O. Owira. 2019. Hepatoprotective effects of *Moringa oleifera* Lam (Moringaceae) leaf extracts in streptozotocin-induced diabetes in rats. *Journal of Functional Foods*, *57*, pp.75–82.

Navarathna, S. B. and A. R. M. H. A. Rathnayake. 2017. Utilization of Moringa olifera leaves as a functional food ingredient in the bakery industry. *International Journal of Science and Research (IJSR)*, *6*(2), pp.339–344.

Nfambi, J., G. S. Bbosa, L. F. Sembajwe, J. Gakunga, and J. N. Kasolo. 2015. Immunomodulatory activity of methanolic leaf extract of *Moringa oleifera* in Wistar albino rats. *Journal of Basic and Clinical Physiology and Pharmacology*, *26*(6), pp.603–611.

Niziot-Tukaszewska, Z., D. Furman-Toczek, T. Bujak, WasilewskiT. , and Z. Hordyjewicz-Baran. 2020. *Moringa oleifera* L. extracts as bioactive ingredients that increase safety of body wash cosmetics. *Dermatology Research and Practice*, 8197902.

Ntshambiwa, K. T., E. Seifu, and G. Mokhawa. 2023. Nutritional composition, bioactive components and antioxidant activity of Moringa stenopetala and *Moringa oleifera* leaves grown in Gaborone, Botswana. *Food Production, Processing and Nutrition*, *5*(1), p.7.

Olayaki, L. A., J. E. Irekpita, M. T. Yakubu, and O. O. Ojo. 2015. Methanolic extract of *Moringa oleifera* leaves improves glucose tolerance, glycogen synthesis and lipid metabolism in alloxan-induced diabetic rats. *Journal of Basic and Clinical Physiology and Pharmacology, 26*(6), pp.585–593.

Olusanya, R. N., U. Kolanisi, A. Van Onselen, N. Z. Ngobese, and M. Siwela. 2020. Nutritional composition and consumer acceptability of *Moringa oleifera* leaf powder (MOLP)-supplemented mahewu. *South African Journal of Botany*, *129*, pp.175–180.

Osamede Airouyuwa, J., and T. Kaewmanee. 2019. Microencapsulation of *Moringa oleifera* leaf extracts with vegetable protein as wall materials. *Food Science and Technology International, 25*(6), pp.533–543.

Owon, M., M. Osman, A. Ibrahim, M. A. Salama, and B. Matthäus. 2021. Characterisation of different parts from *Moringa oleifera* regarding protein, lipid composition and extractable phenolic compounds. *OCL*, *28*, p.45.

Padayachee, B., and H. J. S. A. J. O. B. Baijnath. 2020. An updated comprehensive review of the medicinal, phytochemical and pharmacological properties of *Moringa oleifera*. *South African Journal of Botany*, *129*, pp.304–316.

Pakade, V., E. Cukrowska, and L. Chimuka. 2013. Comparison of antioxidant activity of *Moringa oleifera* and selected vegetables in South Africa. *South African Journal of Science*, *109*(3), pp.1–5.

Pan, D., A. Kobayashi, P. Jiang, L. Ferrari de Andrade, R. E. Tay, A. M. Luoma, D. Tsoucas, X. Qiu, K. Lim, P. Rao, and H. W. Long. 2018. A major chromatin regulator determines resistance of tumor cells to T cell–mediated killing. *Science*, *359*(6377), pp.770–775.

Panda, S., A. Kar, P. Sharma, and A. Sharma. 2013. Cardioprotective potential of N, α-l-rhamnopyranosyl vincosamide, an indole alkaloid, isolated from the leaves of *Moringa oleifera* in isoproterenol induced cardiotoxic rats: In vivo and in vitro studies. *Bioorganic & Medicinal Chemistry Letters, 23*(4), pp.959–962.

Pandey, A., R. D. Pandey, P. Tripathi, P. P. Gupta, J. Haider, S. Bhatt, and A. V. Singh. 2012. *Moringa oleifera* Lam. Sahijan)-a plant with a plethora of diverse therapeutic benefits: An updated retrospection. *Medicinal and Aromatic Plants, 1*(1), pp.1–8.

Pareek, A., M. Pant, M.M. Gupta, P. Kashania, Y. Ratan, V. Jain, A. Pareek, and A. A. Chuturgoon. 2023. *Moringa oleifera*: An updated comprehensive review of its pharmacological activities, ethnomedicinal, phytopharmaceutical formulation, clinical, phytochemical, and toxicological aspects. *International Journal of Molecular Sciences, 24*(3), p.2098.

Perumalsamy, H., S. R. Balusamy, J. Sukweenadhi, S. Nag, D. Mubarak Ali, M. El-Agamy Farh, H. Vijay, and S. Rahimi. 2024. A comprehensive review on *Moringa oleifera* nanoparticles: Importance of polyphenols in nanoparticle synthesis, nanoparticle efficacy and their applications. *Journal of Nanobiotechnology, 22*(1), p.71.

Pop, O. L., A. D. Kerezsi, and C. Ciont. 2022. A comprehensive review of *Moringa oleifera* bioactive compounds—cytotoxicity evaluation and their encapsulation. *Foods, 11*(23), p.3787.

Posmontier, B. 2011. The medicinal qualities of *Moringa oleifera. Holistic Nursing Practice, 25*(2), pp.80–87.

Prabakaran, M., S. H. Kim, A. Sasireka, ChandrasekaranM. , and I. M. Chung. 2018. Polyphenol composition and antimicrobial activity of various solvent extracts from different plant parts of *Moringa oleifera. Food Bioscience, 26*, pp.23–29.

Pudziuvelyte, L., A. Siauruseviciute, R. Morkuniene, R. Lazauskas, and J. Bernatoniene. 2022. Influence of technological factors on the quality of chitosan microcapsules with *Boswellia serrata* L. essential oil. *Pharmaceutics, 14*(6), p.1259.

Rakesh, S., and V. J. Singh. 2011. Anti-inflammatory activity of *Moringa oleifera* leaf and pod extracts against carrageenan induced paw edema in albino mice. *Journal of Pharmaceutical Science and Innovation, 1*, pp.22–24.

Ramamurthy, S., S. Varghese, S. Sudarsan, J. Muruganandhan, S. Mushtaq, P. B. Patil, A. T. Raj, A. Zanza, L. Testarelli, and S. Patil. 2021. *Moringa oleifera*: Antioxidant, anticancer, anti-inflammatory, and related properties of extracts in cell lines: A review of medicinal effects, phytochemistry, and applications. *Journal of Contemporary Dental Practice, 22*(12), pp.1483–1492.

Ray, K., R. Hazra, P. K. Debnath, and D. Guha. 2004. Role of 5-hydroxytryptamine in *Moringa oleifera* induced potentiation of pentobarbitone hypnosis in albino rats. *Indian Journal of Experimental Biology, 42*(6), pp.632–635.

Reubun, Y. T. A., S. Kumala, SetyahadiS. , and P. Simanjuntak. 2021. Freezed drying of Kelor leaves extract (*Moringa oleifera* lam.). *Jurnal Sains dan Kesehatan, 3*(4), pp.470–474.

Ritu Paliwal, R. P., V. S. Veena Sharma, and P. Pracheta. 2011. A review on horse radish tree (*Moringa oleifera*): A multipurpose tree with high economic and commercial importance.*Asian Journal of Biotechnology, 3*(4), pp.317–328.

Rodrigues, J. F., C. Soares, M. M. Moreira, M. J. Ramalhosa, N. F. Duarte, C. Delerue-Matos, and C. Grosso. 2023. *Moringa oleifera* Lam. Commercial beverages: A multifaceted investigation of consumer perceptions, sensory analysis, and bioactive properties. *Foods, 12*(11), p.2253.

Rodríguez-Pérez, C., J. A. Mendiola, R. Quirantes-Piné, E. Ibáñez, and A. Segura-Carretero. 2016. Green downstream processing using supercritical carbon dioxide, CO2-expanded ethanol and pressurized hot water extractions for recovering bioactive compounds from *Moringa oleifera* leaves. *The Journal of Supercritical Fluids, 116*, pp.90–100.

Roni, R. A., M. N. H. Sani, S. Munira, M. A. Wazed, and S. Siddiquee. 2021. Nutritional composition and sensory evaluation of cake fortified with *Moringa oleifera* leaf powder and ripe banana flour. *Applied Sciences, 11*(18), p.8474.

Sadek, K. M., T. K. Abouzed, R. Abouelkhair, and S. Nasr. 2017. The chemo-prophylactic efficacy of an ethanol *Moringa oleifera* leaf extract against hepatocellular carcinoma in rats. *Pharmaceutical Biology, 55*(1), pp.1458–1466.

Saini, R. K., I. Sivanesan, and Y. S. Keum. 2016. Phytochemicals of *Moringa oleifera*: a review of their nutritional, therapeutic and industrial significance. *3 Biotech, 6*, pp.1–14.

Sajid, A., L. Naz, A. Mughal, G. Yasmeen, N. I. Khan, and S. Sohail. 2020. Antioxidant effects of *Moringa oleifera* seed oil against oxidative stress induced by alloxan in rats. *International Journal of Endorsing Health Science Research*, pp.257–264.

Saleem, A., M. Saleem, and M. F. Akhtar. 2020. Antioxidant, anti-inflammatory and antiarthritic potential of *Moringa oleifera* Lam: An ethnomedicinal plant of Moringaceae family. *South African Journal of Botany, 128*, pp.246–256.

Salem, A. S., W. M. Salama, A. M. Hassanein, and H. M. El Ghandour. 2013. Enhancement of nutritional and biological values of Labneh by adding dry leaves of *Moringa oleifera* as innovative dairy products. *World Applied Sciences Journal, 22*(11), pp.1594–1602.

Sengev, A. I., J. O. Abu, and D. I. Gernah. 2013. Effect of *Moringa oleifera* leaf powder supplementation on some quality characteristics of wheat bread. *Food and Nutrition Sciences, 4*(3), p.270.

Setiaboma, W., D. Desnilasari, A. C. Iwansyah, D. P. Putri, W. Agustina, E. Sholichah, and A. Herminiati. 2021. Karakterisasi kimia dan uji organoleptik bakso ikan manyung (Arius thalassinus, Ruppell) dengan penambahan daun kelor (Moringa oleiferea Lam) segar dan kukus. *Biopropal Industri, 12*(1), pp.9–18.

Shang, K., C. Ge, Y. Zhang, J. Xiao, S. Liu, and Y. Jiang.2024. An evaluation of sex-specific pharmacokinetics and bioavailability of Kokusaginine: An in vitro and in vivo investigation. *Pharmaceuticals, 17*(8), p.1053.

Sharma, V., R. Paliwal, P. Janmeda, and S. Sharma. 2012. Chemopreventive efficacy of *Moringa oleifera* pods against 7, 12-dimethylbenz [a] anthracene induced hepatic carcinogenesis in mice. *Asian Pacific Journal of Cancer Prevention, 13*(6), pp.2563–2569.

Shobhit, A. Sharma, P. Kajla, S. P. Bangar, and J. M. Lorenzo. 2022. Drying kinetics of pretreated drumstick (*Moringa oleifera*) leaves during lyophillization. *Food Analytical Methods, 15*(12), pp.3334–3345.

Siddhuraju, P., and K. Becker. 2003. Antioxidant properties of various solvent extracts of total phenolic constituents from three different agroclimatic origins of drumstick tree (*Moringa oleifera* Lam.) leaves. *Journal of Agricultural and Food Chemistry, 51*(8), pp.2144–2155.

Sparman, A. 2017. Combination of *Moringa oleifera*, Bryophyllum pinnatum and Vitamin C in the management of key risk factors for cardiovascular disease. *Natural Products Chemistry & Research, 5*(10.4172), pp.2329–6836.

Sreelatha, S., A. Jeyachitra, and P. R. Padma. 2011. Antiproliferation and induction of apoptosis by *Moringa oleifera* leaf extract on human cancer cells. *Food and Chemical Toxicology, 49*(6), pp.1270–1275.

Srivastava, V., D. Davis, K. Kuns, P. Landry, S. Ballmer, M. Evans, et al. 2022. Science-driven tunable design of cosmic explorer detectors. *The Astrophysical Journal, 931*(1), p.22.

Stohs, S.J. and M. J. Hartman. 2015. Review of the safety and efficacy of *Moringa oleifera*. *Phytotherapy Research, 29*(6), pp.796–804.

Sudha, P., S. M., Asdaq, S. S. Dhamingi, and G. K. Chandrakala. 2010. Immunomodulatory activity of methanolic leaf extract of *Moringa oleifera* in animals. *Indian Journal of Physiololgy and Pharmacololgy, 54*(2), pp.133–140.

Sultana, S. 2020. Nutritional and functional properties of *Moringa oleifera*. *Metabolism Open, 8*, p.100061.

Suresh, S., A. S. Chhipa, M. Gupta, S. Lalotra, S. S. Sisodia, R. Baksi, and M. Nivsarkar. 2020. Phytochemical analysis and pharmacological evaluation of methanolic leaf extract of *Moringa oleifera* Lam. in ovalbumin induced allergic asthma. *South African Journal of Botany, 130*, pp.484–493.

Tang, G. 2014. *Moringa oleifera* leaf detoxifying method and application of *Moringa oleifera* leaf product. https://patents.google.com/patent/CN103583930A/en

Tariq, I., A. Yasmin, A. Imran, H. Akhtar, M. Afzaal, M. Azeem, R. Pant, J. Abid, F. Islam, T. Zahoor, and A. B. Kinki. 2023. A review on extraction technique and immune-boosting properties of *Moringa oleifera* Lam. *International Journal of Food Properties, 26*(1), pp.2493–2508.

Teclegeorgish, Z. W., Y. M. Aphane, N. S. Mokgalaka, P. Steenkamp, and V. J. Tembu. 2021. Nutrients, secondary metabolites and anti-oxidant activity of *Moringa oleifera* leaves and Moringa-based commercial products. *South African Journal of Botany, 142*, pp.409–420.

Tetteh, O. N. A., C. Ulrichs, S. Huyskens-Keil, I. Mewis, N. K. Amaglo, I. N. Oduro, C. Adarkwah, D. Obeng-Ofori, and N. Förster. 2019. Effects of harvest techniques and drying methods on the stability of glucosinolates in *Moringa oleifera* leaves during post-harvest. *Scientia Horticulturae, 246*, pp.998–1004.

Tshabalala, T., A. R. Ndhlala, B. Ncube, H. A. Abdelgadir, and J. Van Staden. 2020. Potential substitution of the root with the leaf in the use of *Moringa oleifera* for antimicrobial, antidiabetic and antioxidant properties. *South African Journal of Botany, 129*, pp.106–112.

Tuorkey, M. J. 2016. Effects of *Moringa oleifera* aqueous leaf extract in alloxan induced diabetic mice. *Interventional Medicine and Applied Science, 8*(3), pp.109–117.

Uddin, N., N. Muhammad, M. Nisar, N. Aisha, E. A. Ali, R. Ullah, A. A. Khan, I. U. Rahman, A. Khan, and A. Zeb. 2022. Distribution of polyphenolic compounds, antioxidant potential, and free amino acids in *Ziziphus* fruits extract; a study for determining the influence of wider geography. *Food Science & Nutrition, 10*(5), pp.1414–1430.

Umoh, I. U., I. A. Edagha, and A. N. Aquaisua. 2018. Neuroprotective effect of ethanol extract of *Moringa oleifera* leaf on the neurofibres of cerebellum of quinine-treated adult Wistar rats. *Asian Journal of Research in Medical and Pharmaceutical Sciences*, *4*, pp.1–10.

Unuigbe, C. A., H. A. Okeri, O. Erharuyi, E. E. Oghenero, and D. A. Obamedo. 2014. Phytochemical and antioxidant evaluation of *Moringa oleifera* (Moringaceae) leaf and seed. *Journal of Pharmacy & Bioresources*, *11*(2), pp.51–57.

Vergara-Jimenez, M., M. M. Almatrafi, and M. L. Fernandez. 2017. Bioactive components in *Moringa oleifera* leaves protect against chronic disease. *Antioxidants*, *6*(4), p.91.

Verma, A. R., M. Vijayakumar, C. S. Mathela, and C. V. Rao. 2009. In vitro and in vivo antioxidant properties of different fractions of *Moringa oleifera* leaves. *Food and Chemical Toxicology*, *47*(9), pp.2196–2201.

Villarruel-López, A., D. A. López-De La MoraO. D. Vázquez-Paulino, , A. G. Puebla-Mora, M. R. Torres-Vitela, L. A. Guerrero-Quiroz, and K. Nuño. 2018. Effect of *Moringa oleifera* consumption on diabetic rats. *BMC complementary and Alternative Medicine*, *18*, pp.1–10.

Vinatoru, M., T. J. Mason, and I. Calinescu. 2017. Ultrasonically assisted extraction (UAE) and microwave assisted extraction (MAE) of functional compounds from plant materials. *TrAC Trends in Analytical Chemistry*, *97*, pp.159–178.

Virk, P., M. A. Awad, S. S. A. L. Alsaif, A. A. Hendi, M. Elobeid, K. Ortashi, R. Qindeel, M. F. El-Khadragy, H. M. Yehia, M. F. S. El-Din, and H. A. Salama. 2023. Green synthesis of *Moringa oleifera* leaf nanoparticles and an assessment of their therapeutic potential. *Journal of King Saud University-Science, 35*(3), p.102576.

Vongsak, B., P. Sithisarn, and W. Gritsanapan. 2012. HPLC quantitative analysis of three major antioxidative components of *Moringa oleifera* leaf extracts. *Planta Medica*, *78*(11), p.PJ15.

Wang, C., R. Wu, D. Sargsyan, M. Zheng, S. Li, R. Yin, S. Su, I. Raskin, and A. N. Kong. 2019. CpG methyl-seq and RNA-seq epigenomic and transcriptomic studies on the preventive effects of Moringa isothiocyanate in mouse epidermal JB6 cells induced by the tumor promoter TPA. *The Journal of Nutritional Biochemistry*, *68*, pp.69–78.

Wanjiru, J., J. Gathirwa, E. Sauli, and H. S. Swai. 2022. Formulation, optimization, and evaluation of *Moringa oleifera* leaf polyphenol-loaded phytosome delivery system against breast cancer cell lines. *Molecules, 27*(14), p.4430.

Wu, Y. Y., Y. M. Xu, and A. T. Lau. 2021. Anti-cancer and medicinal potentials of moringa isothiocyanate. *Molecules*, *26*(24), p.7512.

Yabesh, J. M., S. Prabhu, and S. Vijayakumar. 2014. An ethnobotanical study of medicinal plants used by traditional healers in silent valley of Kerala, India. *Journal of Ethnopharmacology*, *154*(3), pp.774–789.

Yang, M., L. Tao, X. R. Kang,Z. L. Wang, L. Y. Su, L. F. Li, F. Gu, C.C. Zhao, J. Sheng, and Y. Tian. 2023. *Moringa oleifera* Lam. leaves as new raw food material: A review of its nutritional composition, functional properties, and comprehensive application. *Trends in Food Science & Technology*, *138*, pp.399–416.

Zahari, N. A. A. R., G. H. Chong, L. C. Abdullah, and B. L. Chua. 2020. Ultrasonic-assisted extraction (UAE) process on thymol concentration from Plectranthus amboinicus leaves: Kinetic modeling and optimization. *Processes*, *8*(3), p.322.

11 *Moringa oleifera* Fruits
Chemistry, Technology, Functionality and Applications

Periyakaruppan Pradeepkumar, Varatharaj Rajapandian and Periyakaruppan Karuppasamy

11.1 INTRODUCTION

11.1.1 An Overview of *M. oleifera* Fruits

Moringa oleifera, generally known as the 'drumstick tree' or the 'miracle tree', has been hailed for its multifaceted benefits, including its nutritional, medicinal and environmental applications. While the leaves, seeds and roots have been extensively studied, the fruits of *M. oleifera*, commonly known as drumsticks, represent an underutilised component of this versatile plant. It is an Indian subcontinent native plant and is extensively cultivated in tropical and subtropical regions. *M. oleifera* has received much attention at the global level owing to its contributions to food security, health and sustainability (Anwar et al., 2007; Leone et al., 2015). The fruits, with their unique chemical composition and functionality, offer a wealth of opportunities across diverse industries, yet they remain inadequately explored in comparison with other plant parts. This chapter aims to provide a comprehensive examination of the chemical composition, technological applications, functional properties and potential industrial uses of *M. oleifera* fruits. By doing so, it argues for their broader utilisation as a resource to address global challenges in nutrition, health and sustainability.

The chemical profile of *M. oleifera* fruits underscores their nutritional and functional value. The pods are rich in important vitamins, for example, vitamin C, β-carotene and a spectrum of B-complex vitamins, which contribute to their antioxidant potential. Minerals such as calcium, potassium and magnesium are also abundant, making the fruits nutritionally dense (Gopalakrishnan et al., 2016). The amino acid profile, including essential amino acids, further enhances their appeal as a dietary resource, particularly in regions suffering from malnutrition. Beyond macronutrients, *M. oleifera* fruits are a source of various bioactive mixtures, including flavonoids, isothiocyanates and phenolic acids, which are prevalent in the fruit pulp and pods, offering anti-inflammatory, antimicrobial and antioxidant performances (Bennett et al., 2003). Additionally, the seeds within the fruits contain high-quality oil, unsaturated fatty acids and proteins, which show promising food as well as pharmaceutical applications (Leone et al., 2015).

Technological advancements have opened new pathways for the utilisation of *M. oleifera* fruits. Processing techniques such as freeze-drying, spray-drying and enzymatic extraction have been employed to create value-added products. These methods not only enhance the shelf life and bioavailability of nutrients but also expand the functionality of the fruits in various formulations (Kou et al., 2021). The fibrous pods can be processed into powders, pastes or extracts, finding applications in the food and beverage industry as natural thickeners, flavour enhancers and nutritional supplements. For instance, drumstick powders have been incorporated into soups, snacks and bakery products to improve their nutritional profiles (Reddy & Reddy, 2020). Furthermore, the oil extracted

DOI: 10.1201/9781003617860-13

from seeds embedded in the fruits is used in cosmetics and skincare formulations for its moisturising and anti-ageing properties (Waterman et al., 2014).

M. oleifera fruits exhibit an array of functional properties that support their use in health-promoting products. Their high antioxidant capacity is attributed to bioactive compounds such as flavonoids and phenolic acids that help mitigate oxidative stress and reduce the risk of chronic diseases (Vergara-Jimenez et al., 2017). The fruits also possess hypoglycaemic properties, making them valuable in managing conditions like diabetes [4]. Emerging studies suggest potential anticancer activities are ascribed to the presence of isothiocyanates and glucosinolates, which promote apoptosis in cancer cells [7]. Additionally, the dietary fibre content in the fruits aids gut health, further highlighting their functionality as a versatile food ingredient. Beyond their role in food and health, *M. oleifera* fruits offer significant industrial potential. The fibrous nature of the pods makes them suitable for biocomposites and bioplastic production, aligning with sustainable manufacturing practices [6]. The oil extracted from seeds has been explored as a biofuel source, offering an eco-friendly alternative to fossil fuels. The valourisation of *M.* fruits also addresses sustainability concerns by reducing agricultural waste. By converting fruit biomass into high-value products such as bioplastics, bioenergy and animal feed, industries can contribute to a circular economy, reducing environmental footprints while maximising resource efficiency [5].

11.1.2 The Importance of Studying *M. oleifera* Fruits

Despite their vast potential, the fruits of *M. oleifera* remain underutilised, primarily due to limited awareness and research. Most studies have focused on the leaves and seeds, overshadowing the opportunities presented by the fruits. This chapter argues that unlocking the potential of *M. oleifera* fruits could address pressing global challenges related to nutrition, health and sustainability. By leveraging their unique chemical composition, functional properties and technological versatility, *M. oleifera* fruits can emerge as a valuable resource for various industries.

11.2 THE CHEMISTRY OF *M. OLEIFERA* FRUITS

The chemistry of *M. oleifera* fruits is categorised by a complex matrix of essential nutrients and bioactive compounds, contributing to their diverse health benefits and functional applications. These fruits contain carbohydrates, proteins and dietary fibre, alongside an array of vitamins and minerals, including vitamin C, vitamin A, potassium, calcium and iron. They also contain significant amounts of phytochemicals, including phenolic acids, flavonoids (e.g. quercetin and kaempferol), carotenoids (e.g. β-carotene), alkaloids and isothiocyanates [1,9]. Phenolic acids and flavonoids are primarily responsible for the fruits' antioxidant activity, while isothiocyanates contribute to anti-inflammatory and antimicrobial properties [11]. Furthermore, fatty acids present in fruit seeds include oleic acid and other unsaturated fatty acids, making them valuable for oil production and cosmetic applications [10]. This diverse chemical composition underpins the fruits' nutritional, therapeutic and industrial significance, reinforcing their role as a versatile functional food. Figure 11.1 displays the structures of different phytochemicals that exist in *M. oleifera* fruits.

11.2.1 Functionalities Existing in *M. oleifera* Fruits

M. oleifera fruits, often referred to as drumsticks, exhibit a wide range of functionalities owing to their diverse nutrient composition and bioactive compounds. These functionalities make the fruits a valuable resource for nutrition, health and therapeutic applications. The beneficial properties stem from their antioxidant, antimicrobial, anti-inflammatory and nutraceutical potential, among others.

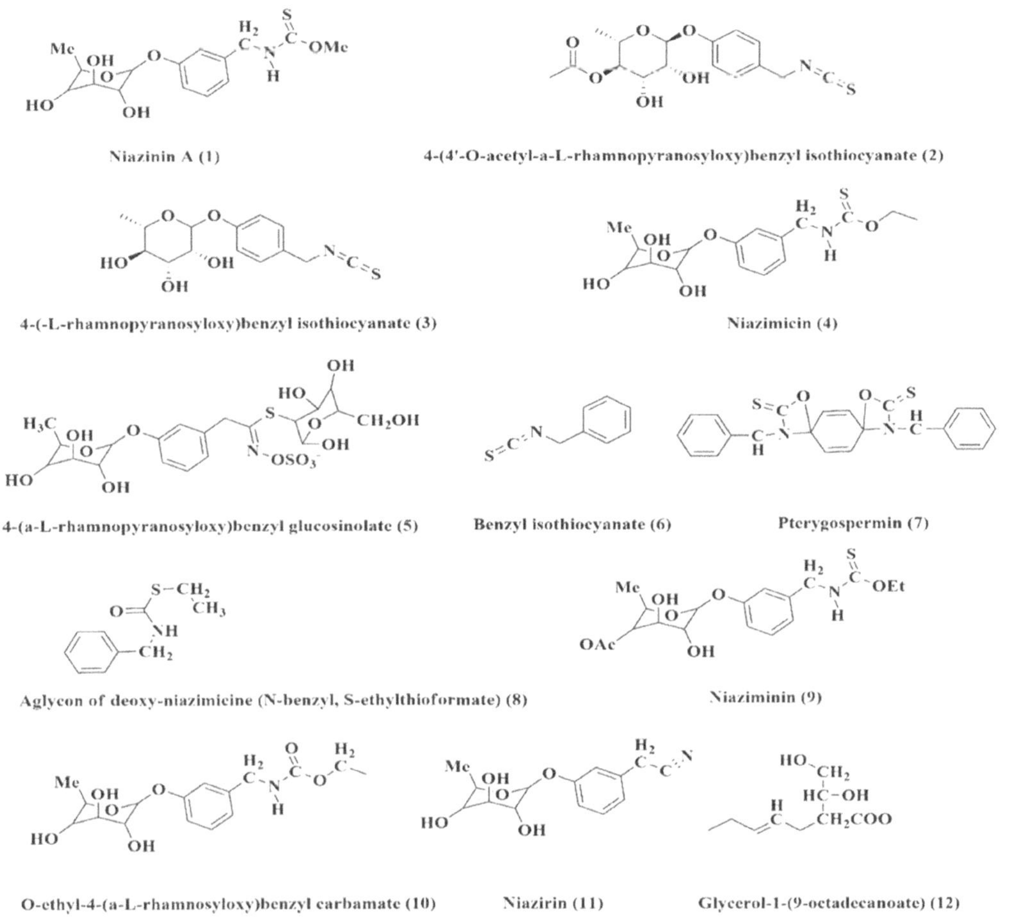

FIGURE 11.1 Structures of Certain Phytochemicals in *M. oleifera* Fruits: (1) Niazinin A, (2) 4-(4′-O-acetyl-α-L-rhamnopyranosyloxy)benzyl isothiocyanate, (3) 4-(-L-rhamnopyranosyloxy)benzyl isothiocyanate, (4) Niazimicin, (5) 4-(α-L-rhamnopyranosyloxy)benzyl glucosinolate, (6) Benzyl isothiocyanate, (7) Pterygospermin, (8) Aglycon of deoxy-niazimicine (N-benzyl, S-ethylthioformate), (9) Niaziminin, (10) O-ethyl-4-(α-L-rhamnosyloxy)benzyl carbamate, (11) Niazirin and (12) Glycerol-1-(9-octadecanoate) respectively.

11.2.1.1 Antioxidant Properties

The antioxidant properties of *M. oleifera* fruits are among their most significant functionalities, which are attributed to the existence of phenolic content, flavonoids and carotenoids. Antioxidants play a vital role in neutralising free radicals and reducing oxidative stress, which is an important factor in pathogenesis and age-related diseases such as cancer, diabetes and cardiovascular disorders (CVDs). The fruits are particularly rich in quercetin, chlorogenic acid and β-carotene, all of which contribute to their potent antioxidant capacity [1]. Studies have shown that the extracts from *M. oleifera* fruits exhibited significant free radical scavenging activity (DPPH (2,2-diphenyl-1-picrylhydrazyl)), indicating their potential to protect cellular structures from oxidative damage [11]. Antioxidant activity levels were analysed in an ethanolic plant extract, free of toxic effects, up to 100.0 mg kg^{-1}, as shown in Figure 11.2.

11.2.1.2 Anti-Inflammatory Functionality

M. oleifera fruits have shown substantial anti-inflammatory performance, making them a valued natural resource for handling inflammatory conditions. These properties are largely attributed to

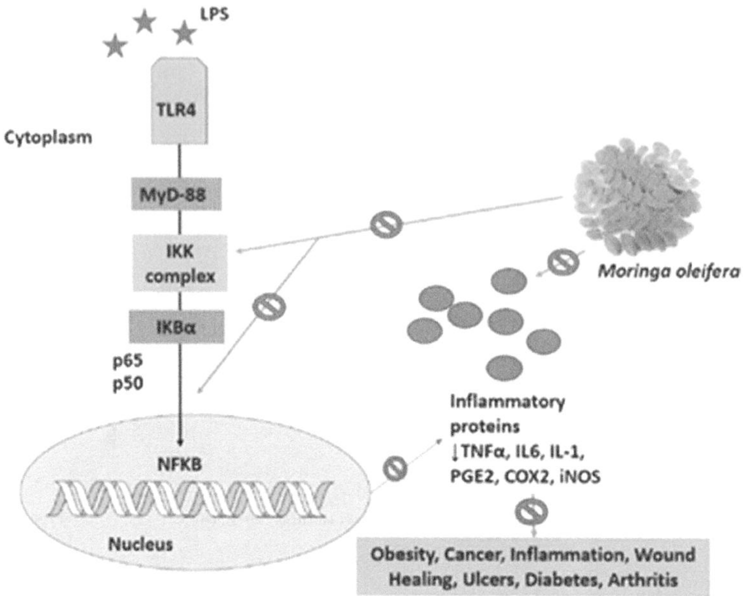

FIGURE 11.2 As an Oxidative and Inflammatory Marker, *M. oleifera* Prevents NFκB Suppression by Blocking IκB Phosphorylation. By blocking the nuclear translocation and dimerisation of IκB and NFκB, it lessens inflammation and cures conditions such as obesity, arthritis, cancer, diabetes and ulcers by preventing the production of inflammatory proteins like TNFα, COX-2, IL6 and iNos. Images were obtained from Pareek et al. (2023).

the existence of flavonoids, isothiocyanates and phenolic acids. These bioactive compounds inhibit pro-inflammatory mediators, such as cytokines and prostaglandins, in the pathogenesis of chronic inflammatory conditions [7]. Research has shown that extracts from *M. oleifera* fruits can suppress nuclear factor-kappa B (NF-κB) activity, regulate inflammation and reduce oxidative stress markers, further mitigating inflammatory responses [8]. The synergistic effects of antioxidants and anti-inflammatory agents in the fruits offer therapeutic potential for conditions such as arthritis, cardiovascular diseases and metabolic syndromes. Additionally, their high vitamin C content enhances the body's natural defence mechanisms against inflammation by scavenging free radicals and supporting immune function [3]. *M. oleifera* fruits have been used as traditional medicine to treat inflammatory ailments such as bronchitis and joint pain, highlighting their ethnopharmacological relevance. Modern studies support these traditional uses, demonstrating that the fruit extracts can serve as a natural alternative or complement to synthetic anti-inflammatory drugs, reducing side effects associated with long-term medication use [4].

11.2.1.3 Antimicrobial and Antifungal Properties

M. oleifera fruits exhibit potent antimicrobial and antifungal activities, attributed to their rich phytochemical composition. Bioactive compounds such as alkaloids, flavonoids and phenolic acids play a vital role in inhibiting the growth of pathogenic microorganisms. These compounds destroy microbial cell walls, control their enzymatic activity and restrict DNA replication, effectively combating a range of bacterial and fungal species [2]. Studies have demonstrated that extracts from *M. oleifera* fruits exhibit antimicrobial activity against both Gram-positive and Gram-negative bacteria, for instance, *Escherichia coli*, *Staphylococcus aureus* and *Pseudomonas aeruginosa* [5]. The antifungal properties are equally noteworthy, with efficacy against common fungi such as *Candida albicans* and *Aspergillus niger.* These findings accentuate that the antimicrobial potential of extracts of *M. oleifera* fruit acts as a preservative in food products and as an active ingredient in pharmaceutical formulations for treating microbial infections [3]. The antimicrobial efficacy of *M. oleifera* fruits is further enhanced by their synergistic interaction with conventional antibiotics. Research suggests that combining fruit extracts with antibiotics can improve the overall antimicrobial activity, reduce the dosage of synthetic drugs and mitigate the risk of antibiotic resistance [8]. This highlights their relevance in addressing the global challenge of multidrug-resistant pathogens. The practical applications of these properties extend to agriculture and the food industries. For instance, *M. oleifera* fruit extracts can be used as natural biocontrol agents to protect crops from microbial infestations. Similarly, their incorporation into edible coatings and packaging materials can prolong the shelf life of perishable food items by inhibiting microbial growth [6].

11.2.1.4 Antidiabetic Potential

The antidiabetic potential of *M. oleifera* fruits has received substantial consideration in recent years, largely owing to their capability for regulating glucose and insulin levels in the blood sensitively. These effects are attributed to the occurrence of various bioactive compounds like isothiocyanates, flavonoids and phenolic acids that modulate the key metabolic pathways involved in glucose homeostasis [3]. Studies have demonstrated that *M. oleifera* fruit extracts can reduce postprandial blood glucose levels by inhibiting the enzymatic activity of α-amylase and α-glucosidase enzymes, which are responsible for the digestion and absorption of carbohydrates. This inhibitive action of enzymes diminishes glucose release into the bloodstream, thereby avoiding swift spikes in blood sugar levels [5]. Additionally, the high fibre content in the fruits contributes to improved glycaemic control by delaying gastric emptying and enhancing satiety. The fruits also exhibit antioxidant properties towards oxidative stress, which is a crucial issue in the progression of insulin resistance and diabetes-associated problems. By neutralising reactive oxygen species (ROS) and reducing inflammation, *M. oleifera* fruits support pancreatic β-cell function and insulin secretion [2]. In animal models, supplementation with *M. oleifera* fruit extracts has been shown to improve glucose tolerance and lipid profiles, further highlighting their therapeutic potential. Clinical trials in humans have corroborated these findings, demonstrating significant reductions in fasting blood glucose levels and HbA1c in individuals with type 2 diabetes [8]. The integration of *M. oleifera* fruits with dietary products and functional foods offers a promising avenue for managing diabetes naturally. Their incorporation into beverages, snacks and meal replacements can provide a convenient and effective means of leveraging their antidiabetic properties. Moreover, the sustainability and affordability of *M. oleifera* fruits make them an accessible resource for addressing the growing global burden of diabetes, particularly in low-income regions [6].

11.2.1.5 Cardiovascular Health

M. oleifera fruits contribute to cardiovascular health through multiple mechanisms. The potassium (K^+) ion content in the fruits helps regulate blood pressure through the effects of sodium ions (Na^+). In addition, the antioxidant properties reduce oxidative stress on blood vessels, thereby preventing endothelial dysfunction and atherosclerosis. The bioactive compounds in the fruits also exhibit lipid-lowering effects, reducing low-density lipoprotein (LDL) cholesterol levels while increasing high-density lipoprotein (HDL) cholesterol levels [13]. These effects collectively diminish the threat of CVDs, hypertension and heart attacks.

11.2.1.6 Digestive Health

Dietary fibre in *M. oleifera* fruits plays a crucial role in promoting digestive health. Fibre enhances intestinal regularity, averts constipation and supports the development of useful gut bacteria. In addition, the fruits contain compounds with mild laxative properties, making them useful for individuals with digestive issues. Moreover, the anti-inflammatory and antioxidant characteristics of the fruits help to protect the gastrointestinal lining from harm produced by inflammation and oxidative stress, reducing the risk of ulcers and inflammatory bowel diseases [9].

11.2.1.7 Immunomodulatory Effects

The high vitamin C content in moringa fruits is a key factor in their immunomodulatory effects. Vitamin C enhances the manufacture of white blood cells, which provide immunity to the body for fighting infection. In addition, flavonoids and phenolic acids, bioactive compounds in the fruits, contribute to immune health by means of decreasing inflammation and counterbalancing destructive free radicals. These effects make *M. oleifera* fruits a valuable dietary component for boosting immunity, especially in populations vulnerable to infections and malnutrition [1].

11.2.1.8 Nutraceutical Applications

The diverse functionalities of *M. oleifera* fruits have led to their incorporation into nutraceuticals and functional foods. The fruits are processed into powders, capsules and extracts, which are marketed for their health-promoting properties. Nutraceutical products derived from moringa fruits are particularly popular for their antioxidant, anti-inflammatory and antidiabetic benefits. Additionally, the fruits are used in functional beverages, energy bars and dietary supplements to address specific health anxieties, for instance, malnutrition, chronic diseases and ageing [3].

11.2.1.9 Skin and Cosmetic Benefits

The bioactive compounds in *M. oleifera* fruits, particularly antioxidants and fatty acids, make them beneficial for skin health. These compounds protect the skin protein from oxidative impairment caused by ultraviolet (UV) radiation and harmful pollutants. They also help with the synthesis of collagen, which decreases the signs of ageing, for example, wrinkles and fine lines on the skin. As a result, moringa fruit extracts are increasingly used in the preparation of natural skincare creams, serums, masks and so on [10].

11.2.1.10 Neuropharmacological Applications

The neuropharmacological potential of *M. oleifera* fruits has been increasingly recognised, with studies highlighting their role in supporting cognitive function and protecting against neurodegenerative disorders due to the existence of anti-inflammatory and antioxidant characteristic alkaloids, flavonoids and phenolic acids [5]. Extracts from *M. oleifera* fruit have demonstrated neuroprotective effects by mitigating oxidative stress and reducing neuronal inflammation in age-related disease conditions. The high levels of vitamin C and β-carotene antioxidants in the fruits help to neutralise ROS that can damage neurons and impair cognitive function [3]. Animal studies have shown that supplementation with *M. oleifera* fruit extracts can improve memory retention, learning ability and overall cognitive performance. These effects are thought to result from enhanced neurotransmitter activity and improved cerebral blood flow [4]. In addition, the anti-inflammatory characteristics of the fruits help reduce neuroinflammation, which is a significant issue in the development of neurodegenerative disorders [2]. The potential applications of *M. oleifera* fruits in neuropharmacology extend to their use in functional foods, dietary supplements and pharmaceutical formulations. For example, incorporating *M. oleifera* fruit extracts into beverages or capsules could provide a natural approach to enhancing brain health and preventing age-related cognitive decline (Figure 11.3). Furthermore, their affordability and sustainability make them a viable option for addressing neurological health challenges, particularly in resource-limited settings [6].

11.3 ADVANCED TECHNIQUES FOR THE INVESTIGATION OF STRUCTURAL INSIGHTS OF *M. OLEIFERA* FRUITS

Understanding the structural composition of *M. oleifera* fruits is indispensable for exploiting their nutritional, therapeutic and industrial potential. Advanced techniques in structural analysis provide detailed insights into the structural aspects of (physico-chemical and functional) *M. oleifera* fruit components. These methods, ranging from spectroscopy to imaging and chromatography, enable

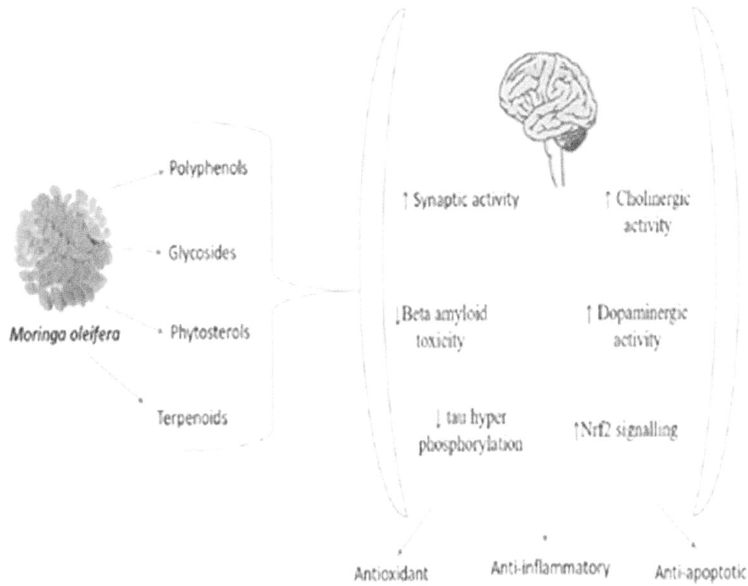

FIGURE 11.3 Numerous Neuroprotective Properties Are Attributed to Several Phytoconstituents Found in *M. oleifera*. *M. oleifera* is in charge of increasing dopaminergic, cholinergic and synaptic activity as well as signalling nuclear factor erythroid 2-related factor 2 and concurrently lowering tau protein phosphorylation and beta-amyloid toxicity.

precise characterisation of bioactive compounds, nutrient distribution and physical structure. This section explores key advanced techniques used in the investigation of *M. oleifera* fruits.

11.3.1 Spectroscopic Analysis

11.3.1.1 Fourier Transform Infrared Spectroscopy

Fourier transform infrared spectroscopy (FTIR) is a widely used technique for categorising functional groups and the chemical composition of *M. oleifera* fruits. By analysing the infrared absorption spectra, FTIR provides detailed information about the presence of carbohydrates, flavonoids, lipids, phenolic acids and proteins. For example, FTIR studies have confirmed the presence of polyphenolic compounds in *M. oleifera* fruit extracts, which are associated with their antioxidant properties [11].

11.3.1.2 Nuclear Magnetic Resonance Spectroscopy

Nuclear magnetic resonance (NMR) spectroscopy offers complete structural evidence of the compounds and their molecular arrangements in *M. oleifera* fruits. Using proton and carbon NMR, researchers can identify and quantify bioactive molecules such as isothiocyanates, alkaloids and other phytochemicals [1]. This technique is particularly useful for studying water-soluble components and their interactions in fruit tissues.

11.3.1.3 Raman Spectroscopy

Raman spectroscopy provides complementary insights into the vibrational modes of molecules. It is a non-destructive method used to analyse the presence of carotenoids and other pigments in *M. oleifera* fruits. Raman analysis enhances the understanding of the antioxidant potential of these compounds and their stability under various conditions [13].

11.3.2 Chromatographic Analysis

11.3.2.1 High-Performance Liquid Chromatography

High-performance liquid chromatography (HPLC) is one of the important effective techniques for the identification, separation and quantification of various bioactive compounds present in *M. oleifera* fruits. Together with various detectors like UV-visible and mass spectrometry (MS), HPLC allows precise analysis of phenolic acids, flavonoids and vitamins. For instance, anti-inflammatory and antioxidant characteristics of chlorogenic acid and quercetin compounds have been identified in *M. oleifera* fruit extracts using HPLC [9].

11.3.2.2 Gas Chromatography-Mass Spectrometry

Gas chromatography-mass spectrometry (GC-MS) is well matched for analysing volatile and semi-volatile organic compounds present in *M. oleifera* fruits. The GC-MS technique has been used to identify essential oils and fatty acid profiles in fruit seeds, revealing their potential applications in cosmetics and biofuel production [10]. The ability of GC-MS to provide both qualitative and quantitative data makes it invaluable for comprehensive chemical profiling.

11.3.2.3 Ultra-High-Performance Liquid Chromatography

Ultra-high-performance liquid chromatography (UHPLC) enhances the resolution and speed of compound separation compared with conventional HPLC. It is increasingly used to study complex mixtures in *M. oleifera* fruits, providing detailed insights into their polyphenolic and glycosidic content [5].

11.3.3 Imaging and Microscopic Techniques

11.3.3.1 Scanning Electron Microscopy

Scanning electron microscopy (SEM) provides high-resolution images of the surface morphology of *M. oleifera* fruits. The SEM technique is used to analyse the microstructure of the fruit's skin, seeds and pulp, shedding light on their physical properties and texture. SEM studies have revealed the porous nature of seed coats, which play a critical role in oil extraction and water purification applications [3].

11.3.3.2 Transmission Electron Microscopy

Transmission electron microscopy (TEM) offers insights into the internal ultrastructure of *M. oleifera* fruit cells. It is particularly useful for studying the arrangement of cellular organelles and the distribution of nutrient reserves. This technique has been employed to investigate the localisation of lipid bodies in seeds and their potential applications in nutraceutical formulations.

11.3.3.3 X-Ray Diffraction

X-ray diffraction (XRD) is used to study the crystalline structure of bioactive compounds in *M. oleifera* fruits. It is particularly valuable for analysing starch granules, cellulose fibres and other polysaccharides in the fruit pulp. XRD patterns have confirmed the semi-crystalline nature of these polysaccharides, which contributes to their functional properties such as water retention and gelling [11].

113.4 Mass Spectrometry-Based Techniques

11.3.4.1 Matrix-Assisted Laser Desorption/Ionisation Time-of-Flight Mass Spectrometry

Matrix-assisted laser desorption/ionisation time-of-flight mass spectrometry (MALDI-TOF MS) is a powerful tool for profiling complex biomolecules, including proteins, peptides and polysaccharides in *M. oleifera* fruits. The technique provides high-throughput and accurate mass measurements, enabling the identification of novel bioactive compounds with therapeutic potential [10].

11.3.4.2 Liquid Chromatography-Mass Spectrometry

Liquid chromatography-mass spectrometry (LC-MS) is the combination of the separation capabilities of LC with the precise quantitative detection of MS. LC-MS is used to identify and quantify trace compounds in *M. oleifera* fruits, such as vitamins, polyphenols and secondary metabolites. LC-MS has been instrumental in studying the pharmacokinetics of bioactive compounds derived from the fruits [9].

11.3.5 Thermal and Rheological Techniques

11.3.5.1 Differential Scanning Calorimetric Technique

The differential scanning calorimetric (DSC) technique is used to study the thermal properties of *M. oleifera* fruits, such as their melting and decomposition temperatures. This technique provides insights into the stability of bioactive compounds during processing and storage, which is crucial for developing functional foods and nutraceuticals.

11.3.5.2 Rheometric Technique

Rheological studies assess the flow and deformation properties of *M. oleifera* fruit pulp and extracts. These studies are essential for understanding the textural properties of products derived from the fruits, such as purees, beverages and gels. Rheological data contribute to optimising product formulations and processing conditions.

11.3.6 Advanced Spectrometry Techniques

11.3.6.1 X-Ray Fluorescence

X-ray fluorescence (XRF) is employed to analyse the elemental composition of *M. oleifera* fruits. This technique is particularly useful for detecting trace elements such as calcium, magnesium and potassium, which are vital for the nutritional value of the fruits.

11.3.6.2 Electron Paramagnetic Resonance

Electron paramagnetic resonance (EPR) spectroscopy provides insights into the paramagnetic properties of antioxidants in *M. oleifera* fruits. This method has been used to study the free radical scavenging activities of phenolic compounds, contributing to a better understanding of the fruits' antioxidant mechanisms [13].

11.4 NUTRITIONAL AND BIOCHEMICAL SIGNIFICANCE OF *M. OLEIFERA* FRUITS

M. oleifera has gained attention for its impressive nutritional and biochemical properties. While its leaves, seeds and flowers are widely studied, the fruits (commonly called drumsticks) are equally important due to their rich nutrient content and bioactive compounds. The fruits offer a sustainable source of nutrition and therapeutic benefits, making them valuable in addressing global health challenges like malnutrition and chronic diseases.

11.4.1 Nutritional Profile of *M. oleifera* Fruits

M. oleifera fruits are a dense source of essential nutrients, including vitamins, minerals, proteins and fibre. One of their notable features is their high vitamin C content, which supports immune function and acts as a potent antioxidant. A 100 g serving of moringa fruits provides approximately 50–60 mg of vitamin C, fulfilling a significant portion of the recommended daily intake [1]. The fruits are also rich in vitamin A, crucial for maintaining vision, immune health and skin integrity. In addition, they contain vital minerals such as calcium, potassium, magnesium and iron. Calcium supports bone health, while potassium helps regulate blood pressure. Iron content, though moderate, helps combat anaemia, especially when the fruits are consumed alongside vitamin C-rich foods that enhance iron absorption [3]. Protein content in the fruits, although not as high as in the seeds or leaves, still contributes significantly to dietary needs. They contain essential amino acids like leucine, isoleucine and valine, which are critical for muscle repair and metabolic functions. The presence of dietary fibre enhances gastrointestinal health by promoting bowel regularity and preventing constipation.

11.4.2 Bioactive Compounds in *M. oleifera* Fruits

M. oleifera fruits are a powerhouse of bioactive compounds, including flavonoids, phenolic acids, carotenoids and alkaloids. These compounds have been linked to numerous health benefits, primarily due to their antioxidant, anti-inflammatory and antimicrobial properties.

11.4.3 Flavonoids and Phenolic Acids

The fruits contain significant amounts of quercetin, kaempferol and chlorogenic acid, which are potent antioxidants. These compounds neutralise free radicals, reducing oxidative stress and preventing cellular damage. Chlorogenic acid, in particular, has been shown to regulate blood sugar levels by improving glucose metabolism, making the fruits beneficial for managing diabetes [9].

11.4.4 Carotenoids

The fruits are rich in β-carotene and other carotenoids, precursors of vitamin A. These compounds contribute to the fruits' antioxidant properties and support eye health by helping prevent conditions such as age-related macular degeneration and cataracts [5].

11.4.5 Isothiocyanates and Alkaloids

Isothiocyanates derived from *M. oleifera* fruits exhibit antimicrobial and anticancer activities. These compounds are known to inhibit the growth of pathogens such as *E. coli* and *S. aureus*, as well as reducing the risk of certain cancers by inducing apoptosis in malignant cells [10].

11.4.6 Combating Malnutrition

The nutrient density of *M. oleifera* fruits makes them an effective tool for addressing malnutrition, particularly in resource-limited settings. In regions where protein-energy malnutrition and micronutrient deficiencies are prevalent, *M. oleifera* fruits can serve as a cost-effective dietary supplement. Their high vitamin C and iron content combat anaemia, while the amino acids support growth and development in children [11]. The fruits are also easy to grow and harvest, making them accessible to vulnerable populations. Their incorporation into local diets, either fresh or in dried and powdered forms, can significantly enhance nutritional outcomes.

11.4.7 Antioxidant Properties and Their Health Implications

Oxidative stress, caused by an imbalance between free radicals and antioxidants, is a major contributor to chronic diseases such as cardiovascular disorders, diabetes and cancer. The antioxidant properties of *M. oleifera* fruits play a crucial role in mitigating these risks. Compounds like flavonoids and phenolics scavenge free radicals, protecting cells from damage and reducing inflammation. Studies have demonstrated that extracts from *M. oleifera* fruits exhibit strong antioxidant activity, comparable to standard antioxidants like ascorbic acid. Regular consumption of these fruits can help lower oxidative stress markers, improve lipid profiles and reduce the risk of cardiovascular diseases [13].

11.4.8 Anti-inflammatory and Antimicrobial Benefits

Chronic inflammation is a hallmark of many diseases, including arthritis, diabetes and neurodegenerative disorders. Bioactive compounds in *M. oleifera* fruits, such as phenolic acids and isothiocyanates, exhibit strong anti-inflammatory effects. These compounds inhibit the production of pro-inflammatory cytokines, thereby reducing inflammation and its associated complications. The antimicrobial properties of *M. oleifera* fruits further enhance their therapeutic potential. Studies have shown that fruit extracts can inhibit the growth of bacteria, fungi and viruses. This makes them valuable not only as dietary components but also in the development of natural antimicrobial agents for food preservation and medicinal use [1].

11.4.9 Functional Foods and Nutraceutical Applications

M. oleifera fruits are increasingly used in the development of functional foods and nutraceuticals due to their nutritional and biochemical significance. They are incorporated into a variety of products, including energy bars, health drinks and dietary supplements. The high antioxidant and anti-inflammatory properties of the fruits enhance the functional value of these products, appealing to health-conscious consumers. The powdered form of the fruits is often added to smoothies, soups and baked goods to boost their nutritional content. Functional beverages made from *M. oleifera* fruit extracts have also gained popularity, particularly for their detoxifying and immunity-boosting benefits [3].

11.5 ENVIRONMENTAL AND SUSTAINABILITY ASPECTS

In addition to their nutritional and biochemical benefits, *M. oleifera* fruits contribute to environmental sustainability. The tree is drought-resistant and thrives in arid and semi-arid regions, making it a valuable resource for food security in climate-stressed areas. The cultivation of *M. oleifera* fruits requires minimal inputs, reducing the environmental footprint compared with other crops. Moreover, the by-products of *M. oleifera* fruits, such as seed cake after oil extraction, are used as natural fertilisers and water purifiers. These applications highlight the holistic value of *M. oleifera* fruits in sustainable agricultural and environmental practices [10].

11.6 SUMMARY

M. oleifera fruit has numerous health and medicinal benefits. It exhibits a wide spectrum of properties, including antidepressant, antifungal, antibacterial, anti-inflammatory, antioxidant and antiviral characteristics, helping protect the body by improving the immune system. The fruit of *M. oleifera* encompasses numerous vitamins, fibres, minerals, fats, ascorbic acid, carbohydrates, β-sitosterol, β-nicotinic acid, oestrogenic compounds, tocopherol and proteins, which are utilised for controlling age-related diseases such as CVDs, Alzheimer's, Parkinson's and heart attacks. *M. oleifera* fruits are used successfully to decrease the HDL level of the liver, aorta and heart in rabbits. *M. oleifera* fruits, based on various commercial products like Rumalaya, Septilin, Orthoherb, Kupid Fort and Livospin, are available on the market to cure many disorders. *M. oleifera* fruit extract formulated markers are available to control oxidative stress and evaluate toxicity and antioxidant performances via in vitro and in vivo assays. The isothiocyanate derivatives present in *M. oleifera* fruits are responsible for the NO-inhibitory effect. The fibres or polyethylene terephthalate composites present in the fruits are used for packaging. In this chapter, the chemistry, functionalities and techniques available for analysing the structural insights and numerous applications of *M. oleifera* fruits were discussed.

REFERENCES

1. Anwar, F., Latif, S., Ashraf, M., & Gilani, A. H. (2007). Moringa oleifera: A food plant with multiple medicinal uses. *Phytotherapy Research, 21*(1), 17–25.
2. Leone, A., Spada, A., Battezzati, A., Schiraldi, A., Aristil, J., & Bertoli, S. (2015). Cultivation, genetic, ethnopharmacology, phytochemistry, and pharmacology of *Moringa oleifera* leaves: An overview. *International Journal of Molecular Sciences, 16*(6), 12791–12835.
3. Gopalakrishnan, L., Doriya, K., & Kumar, D. S. (2016). Moringa oleifera: A review on nutritive importance and its medicinal application. *Food Science and Human Wellness, 5*(2), 49–56.
4. Bennett, R. N., Mellon, F. A., Foidl, N., Pratt, J. H., DuPont, M. S., Perkins, L., & Kroon, P. A. (2003). Profiling glucosinolates and phenolics in vegetative and reproductive tissues of *Moringa oleifera*. *Journal of Agricultural and Food Chemistry, 51*(12), 3546–3553.
5. Kou, X., Li, B., Olayanju, J. B., Drake, J. M., & Chen, N. (2021). Nutraceutical or pharmacological potential of Moringa oleifera Lam. *Nutrients, 10*(3), 343–355.
6. Reddy, C. S., & Reddy, K. H. K. (2020). Utilization of *Moringa oleifera* pods as biomass in sustainable product development. *Sustainable Materials and Technologies, 25*, e00174.
7. Waterman, C., Cheng, D. M., Rojas-Silva, P., Poulev, A., Dreifus, J., Lila, M. A., & Raskin, I. (2014). Stable, water-extractable isothiocyanates from *Moringa oleifera* leaves attenuate inflammation in vitro. *Phytochemistry, 103*, 114–122.
8. Vergara-Jimenez, M., Almatrafi, M., & Fernandez, M. L. (2017). Bioactive components in *Moringa oleifera* leaves protect against chronic disease. *Antioxidants, 6*(4), 91.
9. Fahey, J. W. (2005). Moringa oleifera: A review of the medical evidence for its nutritional, therapeutic, and prophylactic properties. *Trees for Life Journal, 1*(5), 1–15.
10. Rockwood, J. L., Anderson, B. G., & Casamatta, D. A. (2013). Potential uses of Moringa oleifera and an examination of antibiotic efficacy conferred by Moringa oleifera seed and leaf extracts using crude extraction techniques available to underserved indigenous populations. *International Journal of Phytotherapy Research, 3*(2), 61–71.
11. Siddhuraju, P., & Becker, K. (2003). Antioxidant properties of various solvent extracts of total phenolic constituents from three different agroclimatic origins of drumstick tree (*Moringa oleifera Lam.*) leaves. *Journal of Agricultural and Food Chemistry, 51*(8), 2144–2155.
12. Pareek, A., Pant, M., Gupta, M. M., Kashania, P., Ratan, Y., Jain, V., Pareek, A., &-Chuturgoon, A. A. (2023). *Moringa oleifera*: An updated comprehensive review of its pharmacological activities, ethnomedicinal, phytopharmaceutical formulation, clinical, phytochemical, and toxicological aspects. *International Journal of Molecular Sciences, 24*(3), 2098.
13. Sharma, V., Paliwal, R., & Pracheta. (2018). A review on the ameliorative potential of Moringa oleifera on health complications: A functional food perspective. *Indian Journal of Traditional Knowledge, 17*(3), 546–558.

12 *Moringa oleifera* Flowers
Chemistry, Technology, Functionality and Applications

Reham S. Ibrahim, Reham S. Darwish, Hend M. Dawood and Mohamed A. Farag

12.1 INTRODUCTION

Moringa oleifera, commonly known as the drumstick tree, miracle tree or tree of life, is a fast-growing, drought-tolerant species native to the sub-Himalayan regions of northern India. It belongs to the family Moringaceae and is now widely cultivated throughout South and Southeast Asia due to its exceptional adaptability to arid and semi-arid environments (Lakshmidevamma et al., 2021).

Renowned for its resilience, *M. oleifera* can thrive in poor soils with minimal water, making it particularly valuable in regions affected by food insecurity and limited agricultural resources. Its rapid maturation, together with its high nutritive value and versatile utility, ranging from nutrition and medicine to water refinement, positions *M. oleifera* as a vital crop for sustainable development and climate-sturdy agriculture (Afrianto & Metanda, 2023). Almost every part of *M. oleifera*, including its leaves, flowers, seeds, pods and roots, serves a practical aim, whether in food, alternative medicine or industrial applications (Gopalakrishnan et al., 2016). Among these, the leaves are possibly the most renowned, illustrious for their high nutritional value. As well as macronutrients and minerals, *M. oleifera* is rich in bioactive phytoconstituents, such as flavonoids, phenolic acids and alkaloids. These compounds play a role in moringa's antioxidant, anti-inflammatory and antimicrobial properties, which underpin its prevalent use in both traditional remedies and emerging biomedical research (Anwar et al., 2007). While the leaves and seeds are the most commonly reported as beneficial, the flowers also warrant substantial attention. Moreover, *M. oleifera* flowers are crucial to the plant's reproduction, being fragrant and hermaphroditic to attract the pollinators (Vaknin et al., 2021).

Culinary uses for *M. oleifera* flowers include various traditional dishes: lightly cooked, incorporated into soups and curries or added fresh to salads. They are praised for their delicate taste and notable nutritional content, including essential amino acids and trace micronutrients (Oyeyinka & Oyeyinka, 2018). The flowers have long been valued in medicine for their reported diuretic and anti-inflammatory properties. Traditionally, they have been used to treat urinary tract infections and promote lactation (Anwar et al., 2007). The flowers make a more subtle but noticeable contribution than the highly nutritious leaves and the functionally varied seeds. The plant's reputation as an exceptionally multipurpose species – an endowment not only for individual health but also for wider environmental and community resilience – is highlighted by its distinctive culinary applications and traditional medicinal uses (Anwar et al., 2007).

The objective of this chapter is to present a comprehensive review of *M. oleifera* flowers, emphasising their phytochemistry, nutritional value and functional characteristics. It reveals extraction methods, bioactive potential and a wide range of uses in the food, medicine and cosmetics sectors. The chapter also addresses current issues and provides future directions for sustainable use and research.

DOI: 10.1201/9781003617860-14

12.2 PHYTOCHEMICAL COMPOSITION

M. oleifera flowers are a very important part of the plant, possessing numerous bioactive phytoconstituents that belong to different chemical classes. Flavonoids are the most common phytochemicals in flowers, making up about 7.12% of the dry weight. Flavonoids are well-known for their anticancer, anti-inflammatory and free-radical-scavenging abilities. *M. oleifera* flowers have higher flavonoid content than the leaves, making them more powerful antioxidants (Jaglan et al., 2024). Alkaloids are nitrogenous compounds that constitute up to 1.56% of dry weight. They have a broad spectrum of pharmacological effects, such as antibacterial and analgesic activities. Saponins, which make up 3.2% of the substance, help the immune system and lower cholesterol. Unlike flavonoids, alkaloids and saponins are present in a higher concentration in the leaves. Additionally, *M. oleifera* flowers contain a wealth of antioxidant compounds such as cyanogenic glycosides (2.6%), polyphenols, vitamin C and carotenoids (Sathanya et al., 2025). Vitamin C, or ascorbic acid, is particularly prevalent in the flowers (Monteiro et al., 2022).

Moringa flowers are noteworthy from a nutritive viewpoint because of their high protein content (21–22%) and a variety of essential amino acids, such as lysine, methionine and threonine, which are exceptional for edible flowers. Additionally, they offer dietary fibre, which facilitates digestion and is a good source of minerals such as calcium and potassium (Monteiro et al., 2022).

12.3 EXTRACTION AND PROCESSING TECHNOLOGIES

12.3.1 Conventional Extraction Techniques

Determining the presence of phytochemical components, including alkaloids, tannins, flavonoids, steroids, terpenoids and saponins, in plant extracts could be achieved using different solvents, such as ethanol, chloroform, hexane, acetone and ethyl acetate. From qualitative and quantitative screenings of moringa flowers, the majority of these components were predominant in methanol and ethanol extracts (Selvakumar, 2023). In the same context, different solvent extracts of the flowers were evaluated on RAW 264.7 macrophages using the MTT assay, which revealed that the hexane extract showed no significant cytotoxicity at 1 μg/mL and displayed a strong anti-inflammatory activity with an IC_{50} value of 274.55 μg/mL when compared with celecoxib. Moreover, it suppressed the expression levels of nitric oxide (NO), interleukin-6 (IL-6) and tumour necrosis factor-alpha (TNF-α) in these macrophages. Regarding the total phenolic and flavonoid content, the hexane extract was found to contain 12.51±0.28 mg GAE/g extract and 0.16±0.01 mg RuE/g extract, respectively (Fahmy et al., 2024). However, in another study, an 80% hydroethanolic extract of *M. oleifera* flower, with seven identified compounds, mostly phenolics, remarkably decreased the secretion and expression of some inflammatory mediators such as NO, prostaglandin E2 (PGE2), IL-6, IL-1β, TNF-α, nuclear factor-kappa B (NF-κB), inducible NO synthase (iNOS) and cyclooxygenase-2 (COX-2), while increasing the production of IL-10 and IκB-α (inhibitor of κB) in a dose-dependent manner (Tan et al., 2015).

Comparison of aroma composition of the fresh flowers of *M. oleifera* versus the hydrodistilled oil and hexane extract using GC-MS, aldehydes (mainly E-2-hexanal) constituted the major class in the flowers (64.75%); however, alkane hydrocarbons, monoterpenes and aldehydes were predominantly present in the hydrodistilled oil (>50%). The hexane extract contained exclusively alkane hydrocarbons, with heptacosane being the major component (46.2%). Owing to these constituents, it showed significant antioxidant activity as indicated by the in vitro 2,2-diphenyl-1-picrylhydrazyl (DPPH), 2,2′-azino-bis(3-ethylbenzothiazoline-6-sulfonic acid) (ABTS), Cupric Ion Reducing Antioxidant Capacity (CUPRAC), ferric reducing antioxidant power (FRAP) and phosphomolybdenum (PBA) assays (Fahmy et al., 2024).

One study employed by Dutta et al. investigated the effect of different extraction methods and sample preparation procedures on the yield of active constituents. Air-dried powdered flowers and macerated specimens with occasional Soxhlet extraction using distilled water, methanol, acetone, ethyl acetate, chloroform or hexane were tested. The maximum yield content was attained using water, followed by methanol, ethanol, acetone, ethyl acetate and finally hexane. The best extraction solvent was found to be methanol, as it constituted the most bioactive metabolites, concluding that the more polar the solvents, the more efficiently they can extract the major biologically active secondary metabolites (Kalappurayil & Joseph, 2017). For instance, the highest percentage inhibition of DPPH radical scavenging activity was attributed to aqueous extract (91.45%) (Dutta et al., 2024).

A study evaluating the influence of Soxhlet extraction treatment with three different solvents on inhibiting carrageenan-induced paw oedema revealed that petroleum ether and chloroform extracts were remarkably less effective than the methanolic extract when compared with indomethacin (reference standard) at a dose of 10 mg/kg body weight. Moreover, the methanol extract inhibited the oedema after the third hour of carrageenan injection by 29.02%. Results indicated that the plant's anti-inflammatory action was due to the predominance of some polar anti-inflammatory phytoconstituents (flavonoid, alkaloid, tannin) in interrupting arachidonic acid metabolism and production of reactive free radicals (Velaga et al., 2017).

Regarding the effect of different solvent extracts on antibacterial activity, 70% ethanol extract exhibited the highest inhibition zone against *Escherichia coli* (23 mm) and *Staphylococcus aureus* (17 mm) in flowers (Dodiya et al., 2015). In another study, bioactive compounds were extracted by the cold extraction method employing methanol, dichloromethane (DCM) and ethyl acetate as solvents; then the resulting extracts were assessed for their antibacterial activity using the agar well diffusion method. DCM extract exhibited significant antibacterial activity on most of the ocular pathogens tested (*S. aureus*, *Streptococcus agalactiae* and *Propionibacterium acnes*), revealing its promising potential as a pharmaceutical agent in treating eye infections (Sumitha et al., 2015)

The antioxidant power of extracts was found to be associated positively with the concentration of the extraction solvent. Experimental data indicated that 100% methanol exhibited a 7.40% yield, 32.20 amino acid equivalent (AAE/g) of antioxidant capacity, 6.46 mg/100 g of total phenolic content and 2.80 mg/100 g of total flavonoid content. FRAP assay on the optimised parameters included: 59.11% solvent concentration, 43 h drying time and a 10:00 am harvest time showed a 52.94 g/100g yield with 52.40% inhibition to oxidation, 16.42 AAE/g of IC_{50}, 4.94 mg/100 g phenolic acid content and 3.02 mg/100 g flavonoid content (Ukpe et al., 2024).

12.3.2 Advanced Extraction Techniques

Conventional methods, including maceration, Soxhlet extraction and percolation, are generally used. However, they suffer from many limitations, being long, time-consuming and expensive (owing to the use of flammable and non-sustainable solvent practices), in addition to their low yield. This provoked the use of modern technologies such as subcritical liquid extraction, enzyme-assisted extraction, ultrasonic-assisted extraction and high-pressure-assisted extraction. These extraction methods afforded more efficient and less time-consuming procedures in addition to shorter extraction times and higher yields (Jaglan et al., 2024). For instance, a study was undertaken to compare the antioxidant, anticancer and antimicrobial activities of different *M. oleifera* flower extracts using hexane (MOF-H), ethyl acetate (MOF-EA) and methanol (MOF-M) as solvents, assisted by ultrasound-assisted extraction (UAE). Optimum IC_{50} values for DPPH, ABTS and nitric oxide (33.54±1.13, 29.86±0.08 and 49.7±1.12 μg/mL, respectively) were retrieved from MOF-EA, making it the most effective solvent for the extraction of bioactive metabolites (Talath et al., 2022).

To assess the extraction efficacy comparing conventional Soxhlet with ultrasound-assisted extraction methods, solvents such as ethanol, methanol and water were used to extract bioactive compounds from *M. oleifera* flowers. This was followed by encapsulation using freeze-drying with different mixtures of wall materials such as gum arabic (GA) and maltodextrin (MD). Results revealed that ultrasound-assisted extraction using methanol was the most efficient method and was thus further subjected to encapsulation. The highest yield (98.08%) was achieved with a combination of GA and MD, which had the lowest water activity (0.26) and moisture content (5%), suggesting a better shelf life and higher antimicrobial activity. Furthermore, encapsulation with GA and MD displayed the highest antioxidant activity, total phenolic content (TPC) and total flavonoid content (TFC) (Sathanya et al., 2025).

Water is considered a universal solvent for all types of extraction processes, yet it is restricted to the extraction of polar and hydrophilic compounds. The presence of water-soluble impurities can also negatively affect the yield and purity of the extract. Therefore, ionic liquids acted as good alternatives, being non-flammable and thermally stable; however, their selectivity limited their use for certain extraction and purification procedures. This led to the use of deep eutectic solvents (DES) for the extraction of pure bioactive compounds. They shared the same physicochemical properties as ionic liquids, as both are biodegradable and considerably safe. DES are made by mixing two solvents, one of which is a hydrogen-bond donor (HBD), while the other is a hydrogen-bond acceptor (HBA), leading to the reduction of the melting points of the solvents. In a study conducted by Jaglan et al. (2024), optimisation of the extraction process was achieved using six different types of DES combinations of various ratios: ChCl-Gly, ChCl-MA, ChCl-LA, ChCl-Glu, Pro-Gly and Pro-LA. Total phenolic content, flavonoid content and antioxidant activity of *M. oleifera* flower extracts were calculated at different temperatures ranging from 50 to 80°C. The maximum yield of phytochemicals was attained at 70°C with Pro-Gly in a 1:2 M ratio with total phenolic content (32.4 mg GAE/g DW), total flavonoid content (42.1 mg QE/g DW) and antioxidant activity (65.82 %) (Jaglan et al., 2024).

A novel method for the extraction of flavonol aglycones (myricetin, quercetin and kaempferol) from the flowers was applied using a molecularly imprinted polymer employing quercetin as the imprinting compound. The results of this method were compared with those obtained after coupling with the heated reflux extraction method. Calculated adsorption capacities were 2.0, 3.4 and 3.7 mol/g for myricetin, quercetin and kaempferol, respectively. Recoveries of quercetin and kaempferol from *M. oleifera* methanol extracts of flowers ranged from 75 to 86%, respectively, indicating the effectiveness of this method in the quantitative clean-up of these flavonoids. Using heated reflux extraction coupled with molecularly imprinted polymers, higher concentrations of quercetin and kaempferol (845±32 mg/kg and 2802±157 mg/kg) were retrieved from *M. oleifera* flowers (Pakade et al., 2013).

Recent studies focused on protein isolated from *M. oleifera* flowers to be utilised in the preparation of protein bars. By the aid of physicochemical analysis like colour, caloric value, water activity, pH and sensory characteristics, evaluation of the nutritional and quality attributes of bars was carried out. Different concentrations of the isolated protein were used, with 15% and 10% protein isolates showing the highest caloric value (498.33 and 480.37 kcal), protein percentage (42.55 and 35.29%), satisfactory texture parameters (hardness 4.68 and 4.85 kg, firmness 3.88 and 3.67 kg, toughness 3.78 and 3.63 kg and work of shear 16.84 and 16.80 kg/mm). The obtained results concluded that the incorporation of *M. oleifera* protein isolate in bars was not restricted to an increase in their aesthetic value but also to their nutritional profile (Javed et al., 2021).

The effect of impregnating *M. oleifera* flower (MF) extract to develop nutritional chicken nuggets via formulating control (C), T1 (with 1% MF) and T2 (2% MF) groups alongside testing their physicochemical, nutritional and storage stability at 4°C up to 20 days was investigated. T1 and T2 groups exhibited higher cooking yield, emulsion stability, ash, protein, total phenolics and dietary fibre compared to the control. Moreover, the addition of MF extract significantly reduced lipid oxidation during storage time; thus, it could be considered an effective ingredient in cooked chicken nuggets (Madane et al., 2019). Fortification of fermented yellow maize and millet blend (Ogi) with MF powder at different ratios showed that blend 6 (20% flower extract) can be used as an alternative to weaning foods to enhance the nutritional value of food for children and help to limit protein malnutrition (A. K. Arise et al., 2014).

12.3.3 Post-Harvest Processing Techniques

M. oleifera has been described as one of the most amazing trees and is deemed to be a rich source of proteins, vitamins and minerals, including potassium, calcium, phosphorus, iron and folic acid, as well as β-carotene. However, the need for preservation of these metabolites is crucial owing to their medicinal and therapeutic properties. Flowers can be processed by cooking, oven drying or even incorporated into tea bags. In some parts of India, flowers are used to prepare pickles. Recent research reported that the flower ash is rich in potassium and calcium (Ravani et al., 2017).

Since the conventional drying process consumes a lot of time and energy that negatively affected the product quality and safety, optimisation and quality attributes in *M. oleifera* flowers using a freeze dryer (FD) were studied via response surface methodology (RSM) at different drying temperatures (−65 to −45°C), vacuum pressures (0.5 to 2.5 mmHg) and drying times (18 to 24 h). Results were interpreted in terms of vitamin C content, total antioxidant activity (TAA) and hygroscopicity (HS). RSM results defined the optimum parameters for freeze-drying that were found to be: −63.75°C for 18 hours under 0.55 vacuum pressure, which, in turn, yielded optimum values of vitamin C (285.84 mg/100g), total antioxidant activity (453.20 mg/100g) and hygroscopicity (1.57%) (Pandidurai et al., 2022). Another study focused on optimising the drying process of the flowers with a slight modification comprising steam blanching of fresh flowers, then sulfation at 0.1%, followed by dehydration at different temperatures (45, 55 and 65°C). The dried powder at 55°C exhibited maximum physicochemical properties, highest preservation of phytochemicals, especially β-carotene (55.41 mg), and optimum total antioxidant (460.18 mg), flavonoids (19.34 mg/RE), phenol (234.13 mg/GAE) and vitamin C (485.09 mg) contents (Pandidurai et al., n.d.; Selvakumar, 2023).

Upon studying the effect of processing operations such as fluidised bed drying (FBD) and freeze-drying (FD) on thermal, nutritional, physicochemical and structural properties of *M. oleifera* flowers, it was revealed that powders prepared by FBD at 50°C outlet temperature and FD at −50°C exhibited better nutritional values, phenolic content, ascorbic acid, crystallinity index and antioxidant activities, making the flowers good candidates as functional ingredients in the food and pharmaceutical industry (Manju et al., 2021).

Despite the fact that *M. oleifera* flowers are one of the richest sources of potassium, calcium, protein and dietary fibre, they may not be available throughout the year, particularly in South India. A combination of blanching and drying proved to be a fast, efficient and energy-saving way to enhance the flowers' yield and to increase their shelf life while retaining their nutritional value. The optimised conditions, including drying temperature (60.6°C), blanching temperature (79.4°C) and blanching time (2.85 min), were determined based on examination of the rehydration ratio, some chemical characteristics (polyphenols and flavonoid contents) and the antioxidant activity of the dried samples (Kumaraguruparaswami et al., 2021).

12.4 *M. OLEIFERA* FLOWER FUNCTIONAL PROPERTIES

12.4.1 Antioxidant Activity

M. oleifera flowers have effective antioxidant activity; this action happens through neutralising free radicals and relieving oxidative stress. The antioxidant action is mainly due to its content of bioactive compounds such as phenolic acids and flavonoids, which have an important role in the plant's health-enhancing properties (Madane et al., 2019). Previous studies have assessed the antioxidant properties of *M. oleifera* flowers using the DPPH assay. A study showed that the dried extract of *M. oleifera* flowers contained 19.31 mg/g of total phenolics, measured as gallic acid equivalents. Despite that phenolic content, it showed only moderate antioxidant activity (Alhakmani et al., 2013). Moreover, the flower extract possessed the most potent DPPH radical scavenging activity, with an IC_{50} value of 405 μg/mL, indicating its promising antioxidant potential (Vats & Gupta,

2017). Furthermore, the antioxidant effect of *M. oleifera* flower extracts when incorporated into chicken meat nuggets was assessed. At a concentration of 250 µg/mL, the aqueous ethanolic extract showed a 72.22% DPPH radical scavenging effect. This effect was significantly comparable to the effect of the synthetic antioxidant butylated hydroxytoluene (Madane et al., 2019). Another study illustrated that the flower's ethanolic extract possessed an effective radical scavenging potential. Based on the mentioned information, *M. oleifera* flowers can be considered as a natural potential source of antioxidants (Santos et al., 2012).

12.4.2 Anti-Inflammatory Activity

Due to its content of bioactive compounds such as phenolics, flavonoids and alkaloids, *M. oleifera* flower extracts have a strong anti-inflammatory potential. This action occurs through numerous mechanisms, mainly through inhibiting the production of important inflammatory mediators such as nitric oxide, prostaglandins and cytokines, as will be shown in the next subsection.

12.4.2.1 Suppression of NO, PGE2, TNF-α, IL-6, NF-κB and COX-2 Pathways in Macrophages

According to the literature, the *M. oleifera* flower hydroethanolic extract showed a strong anti-inflammatory activity. The mechanisms behind this action were inhibiting the secretion and expression of several inflammatory mediators such as NO, PGE2, IL-6, IL-1β, TNF-α, NF-κB, iNOS and COX-2 in lipopolysaccharide-stimulated RAW 264.7 macrophages. These results supported the anti-inflammatory action of *M. oleifera* flowers. Furthermore, the anti-inflammatory action may be due to preventing the degradation of IκB-α in the NF-κB signalling pathway. Previous studies linked the anti-inflammatory action to the flower's content of flavanols, quinic acid derivatives and fatty acids (Tan et al., 2015).

12.4.3 Antidiabetic Effect

A previous study showed that the ethanolic extract *of M. oleifera* flowers significantly inhibits two important enzymes: α-amylase and α-glucosidase, with IC_{50} values of 37.33 mg/mL and 38.30 mg/mL, respectively. These enzymes have important roles in the digestion of carbohydrates and the absorption of glucose, so their inhibition will help slow down glucose release and potentially decrease postprandial blood sugar levels. Moreover, the extract showed a strong antioxidant activity through the scavenging of DPPH, hydroxyl and hydrogen peroxide radicals, in addition to demonstrating FRAP, indicating its ability to inhibit oxidative stress linked to diabetes (Ruth et al., 2021).

Despite the numerous in vitro studies that showed the potentially promising antidiabetic action of *M. oleifera* flowers, in vivo research in this area is still limited. As a result, we need more studies, such as in vivo and clinical trials, to have comprehensive studies that can confirm the efficacy of the flower's extract, determine the safe dosage levels and evaluate their general safety. In vitro studies indicated that *M. oleifera* flowers have a promising antidiabetic action that occurs through numerous mechanisms, such as inhibition of amylase and glucosidase enzymes and reducing oxidative stress, in addition to regulating glucose metabolism pathways. However, more in vivo and clinical studies are needed for the validation of these mechanisms and to evaluate the effectiveness of the flower's extracts for the management of diabetes (Ruth et al., 2021).

12.4.4 Hepatoprotective Properties

The content of *M. oleifera* flowers of flavonoids, phenolics and alkaloids has a strong correlation to its hepatoprotective effects, which occur through numerous mechanisms (Omodanisi et al., 2017;

Saki et al., 2023; Younis et al., 2022), including its antioxidant action by neutralising free radicals and reducing oxidative stress. Meanwhile, the extract shows hepatoprotective activity by modulating the liver enzymes that are capable of detoxification. Moreover, the extract possesses strong anti-inflammatory action, so it helps in reducing the inflammation linked to liver damage. A previous study found that *M. oleifera* flower aqueous extract caused a significant inhibition in serum liver enzymes (ALT, ALP). Moreover, it resulted in enhanced antioxidant levels and the reconstruction of the normal structure of liver tissue in Wistar rats exposed to carbon tetrachloride (CCl_4), a well-known liver toxin (R. O. Arise et al., 2019). These results showed the effect of *M. oleifera* flower extract in stabilising the membrane of hepatocytes in addition to increasing the antioxidant defence of liver cells.

Another study evaluated the hepatoprotective activity of MF extract in a rat model of liver damage caused by obesity. The results showed a significant decrease in liver enzymes, in addition to strong antioxidant activity and the restoration of normal liver tissue structure (Wahyuniasri et al., 2024).

Although all the studies showed promising hepatoprotective action of *M. oleifera* flowers, more comprehensive studies, including clinical trials, are required to confirm these effects in humans and determine the dosages and the overall safety. *M. oleifera* flowers can be considered a potential source of natural hepatoprotective functional constituents.

12.4.5 Cardioprotective Properties

Antioxidant activity, anti-inflammatory properties and modulation of cardiac function are three mechanisms behind the potential cardioprotective action of *M. oleifera* flowers.

An in vitro study tested the ability of MF extract to dissolve blood clots in order to assess its thrombolytic potential. Results showed that the extract possessed a promising clot-dissolving activity, suggesting it could be used to prevent atherothrombotic cardiovascular conditions (Kunwar et al., 2022). Further research studies are needed to confirm such findings and determine overall safety for human use (Kunwar et al., 2022).

12.4.6 Traditional Uses for Treating Ulcers, Wounds and Hypertension

M. oleifera flowers are traditionally used in folk medicine in different countries for the treatment of ulcers, wounds and hypertension. In the Mijikenda community of Kilifi, Kenya, *M. oleifera* flowers are used for their medicinal effects, especially in the treatment of ulcers and wounds. They prepare the flowers by grinding them after shade-drying; then they add them to food or brew them into herbal teas (Mwami et al., 2024). Meanwhile, in Mijikenda, they use *M. oleifera* flowers to control high blood pressure by mixing flowers with leaves, tomatoes, amaranths or meat (Mwami et al., 2024).

12.5 NUTRITIONAL BENEFITS

M. oleifera flower is well-known for its nutritional value, besides its therapeutic efficacy. It is mainly used in tropical areas of developing countries as a solution for nutritional deficiencies (Mushtaq et al., 2021). Adding *M. oleifera* flowers to the diet is considered a promising approach to combat malnutrition and improve dietary diversity. They are known for their high nutritional content in addition to their use in various recipes. They are an important resource, especially in areas suffering from nutritional deficiencies. *M. oleifera* flowers are rich in various important nutrients such as protein, carbohydrates, dietary fibre and antioxidants like vitamins A and C. Moreover, they are an important source of minerals such as calcium, iron and potassium, and contain essential amino acids along with beneficial phytochemicals. Furthermore, Senegal and Benin use moringa to treat children (Madane et al., 2019; Madhumitha K, 2020).

12.6 APPLICATIONS

12.6.1 Food Industry

M. oleifera flowers are used in protein-enriched bars and beverages due to their nutritional value. Packing the flowers with protein, dietary fibre and antioxidants can provide a natural, plant-based substitute instead of conventional additives (Subramaniam & Basri, 2024). *M. oleifera* flowers, as mentioned before, are rich in protein (17.87% dry matter), dietary fibre (36.14%) and total phenolic compounds (18.34–19.49 mg GAE/g dry matter). These components are behind the flowers' use as functional constituents in numerous food products (Madane et al., 2019). A previous study showed that moringa flower protein isolate was added to energy bars. Bars that are fortified with 10–15% moringa protein exhibited elevated protein levels (35.29–42.55%) and high caloric content (480–498 kcal), in addition to the required texture characteristics, including hardness and toughness. These bars are effective in combating malnutrition and are considered a cost-effective protein source (Javed et al., 2021). The use of *M. oleifera* flower protein in beverages is limited; however, its protein and antioxidant content make it an important fortifying ingredient in protein shakes, smoothies and health drinks (A. K. Arise et al., 2014). When MF extract was incorporated into chicken meat nuggets, they increased the nuggets' odour quality and lipid stability, in addition to extending their refrigerated shelf life to up to 20 days (Madane et al., 2019). Adding *M. oleifera* flowers to different culinary recipes not only enhances the visual appeal of the dishes but also adds nutritional value.

12.6.2 Pharmaceuticals

Most pharmaceutical preparations contain *M. oleifera* leaves and seeds, whereas preparations that contain the flowers are rare. As mentioned in the previous subsections, flowers are rich in functional constituents associated with their anti-inflammatory, antimicrobial and antioxidant activities. Flowers are considered a promising candidate for various pharmaceutical products. A previous study showed that *M. oleifera* flower extract nanoparticles possessed anti-inflammatory properties. These may be due to modulating the NF-κB pathway by inhibiting the expression of inflammation-related proteins, resulting in low levels of inflammatory cytokines. At the same time, this study showed the cytotoxicity of *M. oleifera* flower extract nanoparticles to lung cancer cells; however, they were non-toxic to normal healthy cells. Another study illustrated the synthesis of *M. oleifera* flower extract silver nanoparticles (AgNPs) that showed effective antibacterial actions against a wide range of bacteria, such as *S. aureus*, *Bacillus subtilis*, *Micrococcus luteus*, *Salmonella paratyphi*, *Pseudomonas aeruginosa* and *Klebsiella pneumoniae* (Perumalsamy et al., 2024).

12.6.3 Cosmetics

Incorporating *M. oleifera* flower extracts into different cosmetic products has a wide range of benefits, such as antioxidant, anti-inflammatory and anti-ageing properties. As the demand for natural and effective skincare options increases, moringa flowers can be considered a promising ingredient for future development in cosmetic preparations. The flowers can help in reducing the signs of ageing, such as wrinkles and fine lines, due to their content of antioxidants and their ability to protect the skin cells from oxidative stress, giving it a youthful and healthy skin appearance (Fahmy et al., 2024). Adding MF extracts to cleansers and body washes provides antioxidant benefits beyond their effective cleansing, giving the skin vitality (Nizioł-Łukaszewska et al., 2020). Today, numerous skincare brands have introduced products containing flower extracts to obtain the benefits of the flowers' antioxidant and anti-ageing potentials.

12.6.4 Water Treatment

No research studies were found concerning the use of moringa flowers in water treatment; however, all studies discussed the seeds. Moringa seeds are rich in protein that can help in eliminating impurities and enhancing water clarity, as protein is considered an effective natural coagulant for water treatment. This action occurs as proteins that possess a positive charge attract the negatively charged particles, resulting in their aggregation and settling, reducing turbidity, bacteria and other contaminants (Abouzied et al., 2023; Ueda Yamaguchi et al., 2021). Compared with synthetic coagulants, moringa seeds provide a natural and cost-effective substitute for water treatment and purification, especially in developing nations (Darshan et al., 2020).

12.6.5 Agriculture

The flowers, as a part of the organic material of the plants, can be considered a potential feedstock for anaerobic digestion, and their decomposition can generate biogas that is primarily composed of methane (Darshan et al., 2020). Furthermore, no research studies showed that moringa flowers are applied directly as fertilisers, but other moringa plant parts, such as leaves, seeds and extracts, are extensively used in the production of organic fertilisers. Meanwhile, packing these parts with essential nutrients, minerals and plant growth hormones can effectively enrich soil quality in addition to support healthy plant development (Lawal et al., 2021). Furthermore, a study declared that the growth of moringa plants can result in the improvement of soil conditions in tropical and subtropical regions (Kustiyani et al., 2021).

12.7 CHALLENGES AND FUTURE PERSPECTIVES

Extensive studies were conducted on different parts of *M. oleifera*, particularly the flowers, owing to their superior nutritional value, thereby encouraging their biofortification in many diets. For instance, incorporating flowers into pregnant and nursing mothers' diets can significantly enhance their nutritional values. They can also be integrated into foods as natural antioxidants and antimicrobials. Yet, insufficient information is available concerning the nutritional composition of flowers; thus, it might serve as a platform for further therapeutic research. Meanwhile, the antinutritional components may hinder the absorption of certain essential minerals; therefore, the addition of suitable enzymes aiming to enhance nutrient availability and digestibility could be worth investigating (Arora & Arora, 2021).

Promising pharmacological and therapeutic properties have been reported for this plant. However, more double-blind research is needed to confirm these properties. Additionally, a vacancy exists in ascertaining the pharmacokinetics and bioavailability of its secondary metabolites, so detailed investigations and extensive clinical studies on their mechanism of action are needed. Some previous research also suggested the safety of *M. oleifera* flowers; however, comprehensive toxicological investigations are required if the entire plant or its isolated constituents are introduced for regular consumption to ensure their safety (Arora & Arora, 2021).

Standardisation of medicinal plants is an important issue that dramatically affects the efficacy of herbal drugs. It involves complete data on herbs such as appearance, organoleptic, macroscopical and microscopic characterisation together with fingerprint profiling of the active constituents present in the plant; however, several factors such as the source and quality of raw materials, temperature, water availability, light exposure, method type, period and time of collection, drying, packing, storage and plant age form obstacles against the exact assessment of the quality and stability of herbal remedies that consequently influence the therapeutic value of herbal medicines (Khan et al., 2015). As the commercialisation of plant crude drugs has occurred, it is crucial to ensure the efficacy, safety and quality of medicinal plants. Hence, there is an urge to explore more about the standardisation parameters and data of *M. oleifera* flower extracts, as information is scarce on this particular issue. Some techniques, such as thin-layer chromatography (TLC), could serve as good standardising tools and stability indicators for natural products by affording reliable values that could be used in extract analysis and stability tests. TLC fingerprinting of plant extracts can also be used for the identification of components based on their retention factor (Rf) value and colour spots (Khan et al., 2015).

Sustainable food production has been a major challenge in recent centuries due to global environmental problems, including climate change, soil degradation and biodiversity loss. Moringa plant species are incorporated in climate-smart agriculture (CSA) using an agroforestry approach that ensures high agricultural productivity, adaptive capacity and reduced greenhouse gas emissions. This is attributed to their characteristics of being fast-growing, having a deep root system with various uses and being high in climate change mitigation and adaptation liability (Horn et al., 2022). Additionally, moringa is highly drought-resistant, can be cultivated in a wide range of soil types and pH values and is also resistant to most pests and diseases. For all these reasons, this plant can be cultivated in many regions, including Saudi Arabia (Mridha, 2015). Moringa can participate in reforestation projects due to its fast growth and flexibility, making it effective in combating soil erosion and desertification. The major challenge now is the propagation of the plant among farmers through well-organised programmes and extension services (Adikuru et al., 2011).

Recently, several attempts have been made to use novel delivery systems, such as nanoparticles, nanoemulsions and nanoencapsulation, to formulate the bioactive constituents from *M. oleifera*. Such a rapidly developing area of research for scientists could be explored to combat human pathogens and integrate them into human and veterinary therapeutic fields (Khan et al., 2015).

12.8 CONCLUSION

The flowers of *M. oleifera* offer far more than ornamental value; they are a rich source of bioactive compounds, essential nutrients and functional properties that support their use across a wide range of industries. While the leaves and seeds have received considerable attention, the flowers stand out for their unique combination of nutritional, medicinal and technological potential. From antioxidant-rich extracts to protein-based food applications, their versatility is both scientifically compelling and practically valuable.

As interest in natural, sustainable resources continues to grow, *M. oleifera* flowers represent an untapped opportunity for innovation in food, health and environmental solutions. Moving forward, collaborative research that bridges disciplines, such as phytochemistry, nutrition science, pharmacology and food technology, will be key to unlocking their full potential and ensuring their responsible and effective use.

REFERENCES

Abouzied, A. H., Radwan, T. E. E., & Hassan, H. A. S. (2023). Applying Moringa oleifera extract in water treatment as a natural coagulant to remove turbidity and algae. *Egyptian Journal of Aquatic Biology and Fisheries*, *27*(5), 1121–1131. https://doi.org/10.21608/ejabf.2023.323750

Adikuru, N. C., Okafor, O. E., Ojiako, F. O., & Ibeawuchi, I. I. (2011). Prospects and challenges in the utilization of Moringa oleifera Lam. for agroecosystem sustainability in Nigeria. *International Journal of Agriculture and Rural Development*, *14*(2), 660–665.

Afrianto, W. F., & Metanda, A. A. (2023). Revealing the biocultural importance of Moringa oleifera (Moringaceae) in three villages, Kediri District, Indonesia. *Biodiversitas Journal of Biological Diversity*, *24*(12), 6942–6952.

Alhakmani, F., Kumar, S., & Khan, S. A. (2013). Estimation of total phenolic content, in-vitro antioxidant and anti-inflammatory activity of flowers of Moringa oleifera. *Asian Pacific Journal of Tropical Biomedicine*, *3*(8), 623–627. https://doi.org/10.1016/S2221-1691(13)60126-4

Anwar, F., Latif, S., Ashraf, M., & Gilani, A. H. (2007). Moringa oleifera: A food plant with multiple medicinal uses. *Phytotherapy Research*, *21*(1), 17–25. https://doi.org/10.1002/ptr.2023

Arise, A. K., Arise, R. O., Sanusi, M. O., Esan, O. T., & Oyeyinka, S. A. (2014). Effect of Moringa oleifera flower fortification on the nutritional quality and sensory properties of weaning food. *Croatian Journal of Food Science and Technology*, *6*(2), 65–71.

Arise, R. O., Idris, B. I., Aburo, O. R., & Adewale, A. A. (2019). Hepatoprotective and antioxidant activities of aqueous extract of *Moringa oleifera* flower on ccl4-induced toxicity in rats. *Ife Journal of Science*, *21*(1), 205. https://doi.org/10.4314/ijs.v21i1.18

Arora, S., & Arora, S. (2021). Nutritional significance and therapeutic potential of Moringa oleifera: The wonder plant. *Journal of Food Biochemistry*, *45*(10), e13933.

Darshan, C. M., Bharadwaj, K. R., Tejash, S. M., Negi, P. S., Haware, D. J., & Radha, C. (2020). Moringa oleifera seed protein isolate as an alternative for purifying turbid water. *Desalination and Water Treatment*, *203*, 129–136. https://doi.org/10.5004/dwt.2020.26188

Dodiya, B., Amin, B., Kamlaben, S., & Patel, P. (2015). Antibacterial activity and phytochemical screening of different parts of Moringa oleifera against selected gram positive and gram negative bacteria. *Journal of Pharmaceutical, Chemical and Biological Sciences*, *3*(3), 421–425.

Dutta, S., Kar, S., Yasmin, R., & Choubey, R. (2024). Evaluation of Bioactive Compounds and Antioxidant Activity of Various Parts (Pod, Flower and Leaves) of Drumstick (Moringa oleifera). *Bulletin of Environment, Pharmacology and Life Sciences*, *13*, 105–110.

Fahmy, N. M., Fayez, S., Mohamed, R. W., Elissawy, A. M., Eldahshan, O. A., Zengin, G., & Singab, A. N. B. (2024). Moringa oleifera flowers: Insights into their aroma chemistry, anti-inflammatory, antioxidant, and enzyme inhibitory properties. *BMC Complementary Medicine and Therapies*, *24*(1), 286. https://doi.org/10.1186/s12906-024-04579-y

Gopalakrishnan, L., Doriya, K., & Kumar, D. S. (2016). Moringa oleifera: A review on nutritive importance and its medicinal application. *Food Science and Human Wellness*, *5*(2), 49–56. https://doi.org/10.1016/j.fshw.2016.04.001

Horn, L., Shakela, N., Mutorwa, M. K., Naomab, E., & Kwaambwa, H. M. (2022). Moringa oleifera as a sustainable climate-smart solution to nutrition, disease prevention, and water treatment challenges: A review. *Journal of Agriculture and Food Research*, *10*, 100397.

Jaglan, P., Kumar, M., Kaushik, D., Kumar, A., Argyropoulos, D., Oz, F., & Proestos, C. (2024). Optimization of the extraction process of Moringa oleifera flower by using Deep Eutectic Solvents (DES). *Results in Chemistry*, *7*, 101445.

Javed, M. S., Amjad, A., Shah, M., Shah, F.-H., Sardar, H., Tariq, M. R., Khan, A. A., Sajid, M. W., Ali, U., & Amir, M. (2021). Isolation and characterization of moringa oleifera l. Flower protein and utilization in functional food bars. *Food Science and Technology*, *41*, 643–652.

Kalappurayil, T. M., & Joseph, B. P. (2017). A review of pharmacognostical studies on Moringa oleifera Lam. flowers. *Pharmacognosy Journal*, *9*(1), 1–7.

Khan, M., Rizwani, G. H., & Zahid, H. (2015). Standardization of dried flowers of Moringa oleifera (Lamk.) and Jasminum sambac (L.) Ait according to who guidelines. *International Journal of Pharmacy and Pharmaceutical Sciences*, *7*, 19–22.

Kumaraguruparaswami, M., Easwaramoorthy, S., Sengottuvelu, A., Mohanasundaram, D., & Nallamuthu, D. (2021). Impact on blanching and drying of Moringa oleifera flower and its process parameter optimization. *AIP Conference Proceedings*, *2387*(1), 080005.

Kunwar, B., Jain, V., & Verma, S. K. (2022). In vitro thrombolytic activity of Moringa oleifera. *Nusantara Bioscience*, *14*(1), 63–69. https://doi.org/10.13057/nusbiosci/n140108

Kustiyani, Isdiantoni, Ekawati, I., Faisal, M., Wati, H. D., & Prasetyo, E. N. (2021). Spatial arrangement in alley cropping of moringa and crops to optimize farming income on small Potteran Island, Sumenep Regency. *IOP Conference Series: Earth and Environmental Science*, *649*(1). https://doi.org/10.1088/1755-1315/649/1/012026

Lakshmidevamma, T. N., Ugalat, J., Apoorva, K. A., Suresh, S. P. G., Doddamani, M., Kadam, S., Nayana, R. S., Jagadeesha, R. C., & Fakrudin, B. (2021). Genetic diversity of moringa (Moringa oleifera Lam.). *The Moringa Genome*, 57–65.

Lawal, O. O., Yusuf, T. M., Aliyu, O. M., Olowoake, A., Subair, S. K., & Nofiu, N. B. (2021). Comparative economic analysis of organic and inorganic fertilizers on grain yield of tropical maize varieties. *Badeggi Journal of Agricultural Research and Environment*, *3*(1), 37–43. https://doi.org/10.35849/bjare2020030010

Madane, P., Das, A. K., Pateiro, M., Nanda, P. K., Bandyopadhyay, S., Jagtap, P., Barba, F. J., Shewalkar, A., Maity, B., & Lorenzo, J. M. (2019). Drumstick (Moringa oleifera) flower as an antioxidant dietary fibre in chicken meat nuggets. *Foods*, *8*(8). https://doi.org/10.3390/foods8080307

Madhumitha, K, K., & Prabha, V. (2020). Moringa oliefera flowers : A review on nutritive benefits, medicinal values and culinary uses. *International Journal of Food Science and Nutrition*, *5*(1), 113–115.
Manju, K. M., Rekha, P., & Kumar, N. (2021). Effect of fluidized-bed and freeze-drying techniques on physicochemical, nutritional, thermal, and structural properties of Moringa oleifera flowers, leaves, and seeds. *Journal of Food Processing and Preservation*, *45*(9), e15719.
Monteiro, J., Scotti-Campos, P., Pais, I., Figueiredo, A. C., Viegas, D., & Reboredo, F. (2022). Elemental composition, total fatty acids, soluble sugar content and essential oils of flowers and leaves of Moringa oleifera cultivated in Southern Portugal. *Heliyon*, *8*(12), e12647. https://doi.org/10.1016/j.heliyon.2022.e12647
Mridha, M. A. U. (2015). Prospects of moringa cultivation in Saudi Arabia. *Journal of Applied Environmental and Biological Sciences*, *5*(3), 39–46.
Mushtaq, B. S., Hussain, M. B., Omer, R., Toor, H. A., Waheed, M., Shariati, M. A., Sergey, P., & Heydari, M. (2021). Moringa Oleifera in malnutrition: A comprehensive review. *Current Drug Discovery Technologies*, *18*(2), 235–243. https://doi.org/10.2174/1570163816666191105162722
Mwami, B., Maňourová, A., Hendre, P. S., Muchugi, A., Verner, V., Kariuki, P., Sulaiman, N., & Polesny, Z. (2024). Traditional knowledge, use, and management of Moringa oleifera among the mijikenda community in Kilifi, Kenya. *Plants*, *13*(24). https://doi.org/10.3390/plants13243547
Nizioł-Łukaszewska, Z., Furman-Toczek, D., Bujak, T., Wasilewski, T., & Hordyjewicz-Baran, Z. (2020). Moringa oleifera L. extracts as bioactive ingredients that increase safety of body wash cosmetics. *Dermatology Research and Practice*, *2020*, 8197902. https://doi.org/10.1155/2020/8197902
Omodanisi, E. I., Aboua, Y. G., Chegou, N. N., & Oguntibeju, O. O. (2017). Hepatoprotective, antihyperlipidemic, and anti-inflammatory activity of Moringa oleifera in diabetic-induced damage in male wistar rats. *Pharmacognosy Research*, *9*(2), 182–187. https://doi.org/10.4103/0974-8490.204651
Oyeyinka, A. T., & Oyeyinka, S. A. (2018). Moringa oleifera as a food fortificant: Recent trends and prospects. *Journal of the Saudi Society of Agricultural Sciences*, *17*(2), 127–136.
Pakade, V., Cukrowska, E., Lindahl, S., Turner, C., & Chimuka, L. (2013). Molecular imprinted polymer for solid-phase extraction of flavonol aglycones from M oringa oleifera extracts. *Journal of Separation Science*, *36*(3), 548–555.
Pandidurai, G., Amutha, S., Kanchana, S., Vellaikumar, S., & Prabhakaran, K. (n.d.). *The effect of different drying temperatures, phytochemical and nutrient retention in freeze dried Moringa flowers (Moringa oleifera L.), YMER Digital,* 21(05):1336–1344.
Pandidurai, G., Amutha, S., Kanchana, S., Vellaikumar, S., & Prabhakaran, K. (2022). Optimization of freeze drying parameters for moringa (Moringa oleifera) flower powder by using response surface methodology and principal component analysis. *Journal of Horticultural Sciences*, *17*(2), 388–396.
Perumalsamy, H., Balusamy, S. R., Sukweenadhi, J., Nag, S., MubarakAli, D., El-Agamy Farh, M., Vijay, H., & Rahimi, S. (2024). A comprehensive review on Moringa oleifera nanoparticles: Importance of polyphenols in nanoparticle synthesis, nanoparticle efficacy and their applications. *Journal of Nanobiotechnology*, *22*(1), 71. https://doi.org/10.1186/s12951-024-02332-8
Ravani, A., Prasad, R. V., Gajera, R. R., & Joshi, D. C. (2017). Potentiality of Moringa oleifera for food and nutritional security-A review. *Agricultural Reviews*, *38*(3), 228–232.
Ruth, O. O., Bamikole, A. O., Olusola, O. B., & Adeniyi, O. O. (2021). In vitro inhibitory potentials of ethanolic extract of Moringa oleifera flower against enzymes activities linked to diabetes. *Journal of HerbMed Pharmacology*, *10*(4), 408–414. https://doi.org/10.34172/jhp.2021.48
Saki, M., De Villiers, H., Ntsapi, C., & Tiloke, C. (2023). The Hepatoprotective Effects of Moringa oleifera against Antiretroviral-Induced Cytotoxicity in HepG2 Cells: A Review. *Plants*, *12*(18). https://doi.org/10.3390/plants12183235
Santos, A. F. S., Argolo, A. C. C., Paiva, P. M. G., & Coelho, L. C. B. B. (2012). Antioxidant activity of Moringa oleifera tissue extracts. *Phytotherapy Research : PTR*, *26*(9), 1366–1370. https://doi.org/10.1002/ptr.4591
Sathanya, P. S., Raja Rajeshwari, E., Vignesh, S., Chidanand, D. V., & Baskaran, N. (2025). Comparative study of soxhlet and ultrasound-assisted extraction and encapsulation of bioactives from Moringa oleifera flower. *South African Journal of Botany*, *184*, 4–17. https://doi.org/10.1016/j.sajb.2025.05.035
Selvakumar, P. (2023). Drying diverse temperatures alters the physico-chemical and phytochemical properties of Moringa flowers. *Journal of Preventive, Diagnostic and Treatment Strategies in Medicine*, *2*(1), 60–66.
Subramaniam, S. M., & Basri, H. (2024). Enhanced knowledge in sciences and technology development of high-protein bar incorporated with Moringa oleifera flower. *Sciences and Technology*, *4*(1), 443–452.

Sumitha, J., Padmalatha, C., & Singh, A. R. (2015). Antibacterial efficacy of Moringa oleifera and Tabernaemontana divaricata flower extracts on occular pathogens. *International Journal of Current Microbiology and Applied Sciences*, *4*, 203–216.

Talath, S., Wali, A. F., & Menezes, G. A. (2022). Ultra-sound assisted extraction, anti-oxidant, cytotoxic and antimicrobial studies of Moringa oleifera flowers. *Pakistan Journal of Pharmaceutical Sciences*, *35*(4).

Tan, W. S., Arulselvan, P., Karthivashan, G., & Fakurazi, S. (2015). Moringa oleifera flower extract suppresses the activation of inflammatory mediators in lipopolysaccharide-stimulated RAW 264.7 macrophages via NF-κB pathway. *Mediators of Inflammation*, *2015*(1), 720171. https://doi.org/10.1155/2015/720171

Ueda Yamaguchi, N., Cusioli, L. F., Quesada, H. B., Camargo Ferreira, M. E., Fagundes-Klen, M. R., Salcedo Vieira, A. M., Gomes, R. G., Vieira, M. F., & Bergamasco, R. (2021). A review of Moringa oleifera seeds in water treatment: Trends and future challenges. *Process Safety and Environmental Protection*, *147*, 405–420. https://doi.org/10.1016/j.psep.2020.09.044

Ukpe, R. A., Udo, I. I., & Udofia, P. G. (2024). Evaluation and optimization of yield and antioxidant activity of Moringa oleifera flower extract. *Journal of Applied Sciences and Environmental Management*, *28*(12B Suppl.), 4321–4328.

Vaknin, Y., Eisikowitch, D., & Mishal, A. (2021). Floral and Pollen traits of Moringa oleifera Lam. and Moringa peregrina (Forssk.) fiori provide reproductive adaptations for arid conditions. *Agronomy*, *11*(6). https://doi.org/10.3390/agronomy11061090

Vats, S., & Gupta, T. (2017). Evaluation of bioactive compounds and antioxidant potential of hydroethanolic extract of Moringa oleifera Lam. from Rajasthan, India. *Physiology and Molecular Biology of Plants : An International Journal of Functional Plant Biology*, *23*(1), 239–248. https://doi.org/10.1007/s12298-016-0407-6

Velaga, V., Suryadevara, N., Chee, L. L., & Ismail, N. E. (2017). Phytochemical analysis and immuno-modulatory effect of Moringa oleifera flowers. *Journal of Pharmacy & Pharmaceutical Sciences*, *9*, 24–28.

Wahyuniasri, N. R., Nardi, L., & Sopacua, E. (2024). Effect of moringa flower extract on liver function and histopathology appearance of obesity model rat ' s liver. *Jurnal Pima Medika Sainsima Medika Sains*, *6*(1), 12–20.

Younis, N., Khan, M. I., Zahoor, T., & Faisal, M. N. (2022). Phytochemical and antioxidant screening of Moringa oleifera for its utilization in the management of hepatic injury. *Frontiers in Nutrition*, *9*, 1078896. https://doi.org/10.3389/fnut.2022.1078896

13 *Moringa oleifera* Extracts
Chemistry, Technology, Properties and Applications

Sitaraman Krishnan, Temitope Orimolade, Ngoc-Tram Le, Phong T. Huynh

13.1 INTRODUCTION

Moringa oleifera, a medium-sized tree of the Moringaceae family, is widely recognised for its nutritional and medicinal value (Ramachandran, Peter and Gopalakrishnan 1980). The family consists of a single genus, *Moringa*, characterised by parietal placentation, three-valved fruits, elongated non-dehiscent berries and winged seeds. Taxonomic classifications have varied, with historical placements in Rheoadales, Capparidales and Violales. Among the species, *M. oleifera* and *M. concanensis* are the best known, with *M. oleifera* distinguished by tripinnate leaves, smaller leaflets and a medium-sized stature.

The tree has corky, fissured bark, tuberous roots, and spirally arranged, long-petioled leaves. Its fragrant, bisexual flowers form axillary panicles with yellowish-white petals. The fruit is a long, ribbed capsule, typically 20–45 cm in length, containing numerous globular, three-winged seeds. Seed anatomy includes a thick testa (seed coat) with a reticulate parenchyma zone, a mesh-like network of parenchyma cells that aid in storage and structural support, fibre layers and an oil-rich endosperm. The cotyledons, which serve as the seedling's first leaves, store oils and proteins, contributing to the plant's nutritional, ethnomedical and pharmacological properties (Pareek et al. 2023; Saini, Sivanesan and Keum 2016).

The plant is mentioned in the *Āyurveda* treatise *Caraka Saṃhitā*, a redacted version of the earlier *Agniveśa Saṃhitā*, estimated to have been composed around 1000 BCE (Sharma and Dash 2009). In this text, it is referred to as *kṛṣṇagandhā* and *śigru*, identified as *Moringa ovalifolia* and *Moringa oleifera*, respectively, though botanical equivalences of such classical terms are not always uniformly interpreted.

Various parts of the *M. oleifera* plant, including leaves, seeds, root bark, flowers and pods, have the potential to exhibit diverse pharmacological properties due to their rich phytochemical composition, with flavonoids contributing to antidiabetic, antioxidant and anti-inflammatory effects; isothiocyanates and quercetin possessing anticancer and antiproliferative properties; vitamins and minerals supporting immune function; pterygospermin providing antimicrobial activity; alkaloids in the bark offering antiulcer and cardiac stimulant effects; and polyunsaturated fatty acids (PUFAs) in the pods potentially aiding in obesity management (Gopalakrishnan, Doriya and Kumar 2016). In particular, the seeds of *M. oleifera* exhibit significant pharmacological potential, offering antimicrobial, anticancer, antidiabetic, antioxidant, antihypertensive, anti-inflammatory and cardioprotective properties (Dzuvor et al. 2022). *M. oleifera* seeds are rich in proteins, lipids, vitamins and minerals, providing benefits for food technology (Saa et al. 2019). Moringa oil, stable during frying, outperforms groundnut oil and enhances deep-fried food products like chips (Boukandoul et al. 2023). Treatments such as germination, fermentation, roasting and boiling improve the seeds' nutritional properties, with defatted flour used in products like cakes and cookies. *M. oleifera* remains a highly

 DOI: 10.1201/9781003617860-15

valuable plant with diverse nutritional offerings, making it an important resource for food security and health supplementation worldwide. Seed cakes, by-products of oil extraction, are effective in water purification, particularly in effluent management, due to their ability to remove microbes and organic matter.

This chapter will focus on the chemistry, extraction technologies, functional properties and applications of *M. oleifera* in food, pharmaceuticals and environmental sectors.

13.2 THE COMPOSITION OF *M. OLEIFERA* EXTRACTS

Makkar and Becker (1997) conducted a comprehensive analysis of the chemical composition of various morphological parts of the *M. oleifera* tree. The crude protein (CP) content ($N \times 6.25$) was highest in the leaves (26 wt% dry matter), followed by the twigs (7 wt%) and stems (6 wt%). True protein accounted for 87% of the total CP in leaves, 60% in twigs and 53% in stems. The results of this analysis are presented in Table 13.1. The amino acid compositions of proteins from the leaves, whole seeds and deoiled seeds are summarised in Table 13.2. Notably, the leaves contain higher concentrations of all essential amino acids than the World Health Organization (WHO) recommended levels (g/100 g protein) for children aged three to ten years (WHO 2007), with values comparable to those found in soybeans (*Glycine max*) (Belitz, Grosch and Schieberle 2004).

The seed kernel (the dehulled inner seed) contained approximately 37% crude protein (CP) and 42% lipid, whereas the deoiled seed meal (the fat-free residue remaining after solvent extraction) contained 61% CP. Following extraction of both the kernel and the meal with distilled water, the resulting extracted-kernel and extracted-meal residues contained 35% and 70% CP, respectively; the CP in both residues was reported to consist entirely of true protein. The retention of substantial protein in these residues may reflect the use of distilled water under room-temperature, low-ionic-strength extraction conditions, which likely limited the solubilization of some seed proteins. During distilled-water extraction, 20.5% of kernel dry matter and 41.8% of meal dry matter were solubilized. However, only 23.7% and 33.4% of the original crude protein in the kernel and meal, respectively, were lost to the aqueous phase, indicating that much of the protein remained in the residual solids. In contrast, aqueous saline extraction systems can yield more concentrated protein extracts and greater protein solubilization (Orimolade et al., 2025). The extracted meal was virtually free of the antinutritional factors examined, except for phytate (6.7%). Seed proteins also exhibit high digestibility, attributed to the absence of trypsin inhibitor and urease activity (Leone et al. 2016).

TABLE 13.1
Chemical Composition of Different Fractions of *M. oleifera*

Parts/fractions	Crude protein (%)	Lipid (%)	Ash (%)	NDF (%)	ADF (%)	ADL (%)	ADIP (%)
Leaves	26.4	nd	8.87	15.1	9.2	1.1	0.83
Twigs	7.2	nd	9.30	56.4	51.5	8.3	1.12
Stems	6.2	nd	6.90	68.4	60.9	11.0	1.06
Shells	9.9	2.2	2.27	84.2	80.5	45.2	nd
Seed kernels	36.8	41.7	3.18	4.8	3.9	1.4	0.47
Deoiled seed kernels	61.4	nil	5.65	8.2	6.7	2.5	0.85

Composition values in wt% dry matter. nd, not determined; crude protein = total nitrogen × 6.25; NDF - neutral detergent fibre (total cell wall fraction: hemicellulose, cellulose, lignin); ADF - acid detergent fibre (cellulose + lignin fraction, inversely related to digestibility); ADL - acid detergent lignin (lignin fraction of ADF); ADIP - acid detergent insoluble protein (protein bound to fibre, largely unavailable).

TABLE 13.2
Amino Acid Composition of *M. oleifera* Leaves and Seeds Compared with Soybean Protein and the WHO Reference Protein

Amino acids	Leaves	Seed kernels	Deoiled kernels	Extracted kernels	Extracted deoiled kernels	Soybean protein	WHO reference
Lysine	5.60	1.41	1.47	1.62	1.48	6.4	4.8
Leucine	8.70	5.11	5.27	5.85	5.84	7.8	6.1
Isoleucine	4.50	2.99	3.05	3.45	3.49	4.5	3.1
Methionine	1.98	1.93	1.90	2.23	2.13	1.3	2.4*
Cysteine	1.35	4.13	4.22	4.82	4.72	1.3	–
Phenylalanine	6.18	3.83	3.97	4.30	4.29	4.9	4.1**
Tyrosine	3.87	1.44	1.50	1.52	1.41	3.1	–
Valine	5.68	3.40	3.47	3.72	3.63	4.8	4.0
Histidine	2.99	2.20	2.27	2.38	2.28	–	1.6
Threonine	4.66	2.15	2.25	2.41	2.28	–	2.5
Serine	4.12	2.64	2.75	3.05	2.85	–	–
Glutamic acid	10.2	19.5	19.4	20.7	19.6	–	–
Aspartic acid	8.83	3.80	3.97	3.96	3.76	–	–
Proline	5.43	5.27	5.52	6.37	6.04	–	–
Glycine	5.47	4.76	4.90	5.00	4.40	–	–
Alanine	7.32	3.67	3.77	4.15	4.05	–	–
Arginine	6.23	11.4	11.6	13.9	16.7	–	–
Tryptophan	2.10	nd	nd	nd	nd	1.3	0.66

Composition values in g/100 g crude protein. Amino acid content was determined following acid hydrolysis (6 M HCl, 100°C, 24 h, vacuum), with analysis performed using an amino acid analyser; nd, not determined; *sulphur amino acids (methionine+cysteine); **aromatic amino acids (phenylalanine+tyrosine).

Makkar and Becker (1997) also reported the neutral detergent fibre (NDF), acid detergent fibre (ADF) and acid detergent lignin (ADL) contents of different fractions of *M. oleifera* in the context of its use as animal feed. NDF measures the total structural fibre fraction – including cellulose, hemicellulose and lignin – which contributes to the bulkiness of plant material and influences voluntary feed intake in ruminants, as bulkier forages tend to limit intake by filling the digestive tract. ADF represents the less digestible fibre fraction, composed primarily of cellulose and lignin, and serves as an indicator of digestibility. ADL refers to the lignin fraction determined from the ADF residue after further acid treatment. ADL is considered a measure of the indigestible portion of plant fibre, as lignin resists microbial degradation in the rumen. Higher ADL content generally indicates lower digestibility and energy availability for herbivores. Additionally, the study reported acid detergent insoluble protein (ADIP), the protein associated with the ADF fraction, which is generally considered to have low nutritional availability.

The amino acid sequences of the MO2 and *Mo*-CBP_3 proteins (a 2S albumin) present in the aqueous extracts of *M. oleifera* seeds were reported by Gassenschmidt et al. (1995) and Freire et al. (2015). These proteins have relatively low molecular weights of 6,782 and 18,756 g/mol, respectively, and are cationic, with theoretical isoelectric points of 11.61 and 7.55 (Orimolade et al. 2025). Both contain a glutamine-rich peptide sequence, QQQGQVGPQQV. Their cationic nature promotes binding to negatively charged surfaces, flocculation activity and antimicrobial properties.

M. oleifera leaves are rich in bioactive compounds, including vitamins, carotenoids, polyphenols, flavonoids, alkaloids, glucosinolates, tannins and saponins. The chemical structures of some

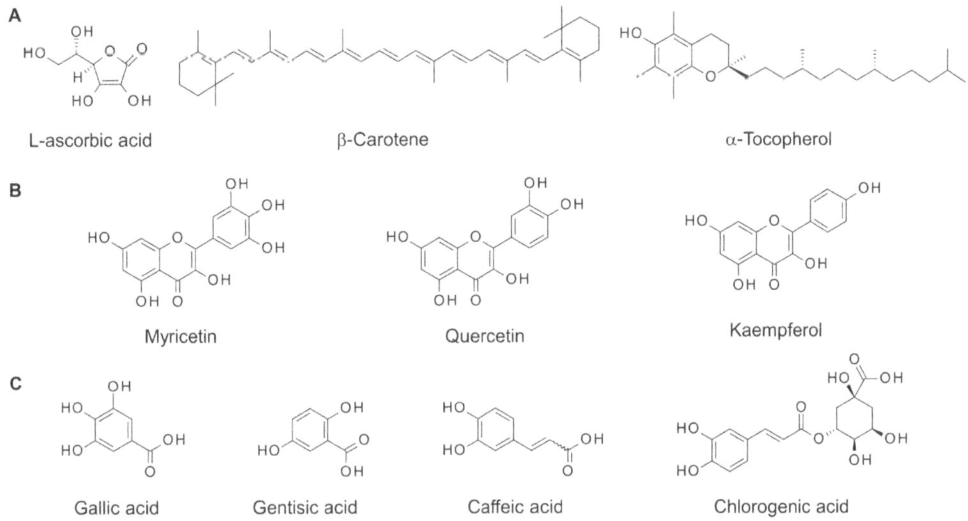

FIGURE 13.1 Representative Antioxidant and Polyphenolic Compounds in *M. oleifera*. (A) Antioxidant vitamins: L-ascorbic acid (vitamin C), β-carotene (provitamin A) and α-tocopherol (vitamin E). (B) Flavonoids that enhance radical-scavenging activity. (C) Phenolic acids, such as gallic and chlorogenic acids, recognised for their antimicrobial and antioxidant properties.

of these compounds are illustrated in Figures 13.1 and 13.2. Their concentrations vary depending on geographic and environmental factors, as well as the extraction method, as reviewed by Leone et al. (2015). The leaves provide significant amounts of β-carotene (6.6–6.8 mg/100 g fresh; up to 39.6 mg/100 g dried), vitamin C (200 mg/100 g fresh; 18.7–140 mg/100 g dried; up to 920 mg/100 g freeze-dried) and vitamin E (9 mg/100 g).

Flavonoids are abundant in *M. oleifera* leaves, with myricetin (5.8 mg/g), quercetin (up to 8 mg/g) and kaempferol (up to 4.6 mg/g) present in significant amounts. The leaves also contain phenolic acids, including gallic acid (≈1 mg/g), chlorogenic acid (up to 0.489 mg/g) and caffeic acid (up to 0.409 mg/g). Additionally, they are rich in alkaloids (e.g. *N*-(α-L-rhamnopyranosyl)vincosamide, niazirin, marumoside A), glucosinolates (primarily 4-*O*-(α-L-rhamnopyranosyloxy)benzyl glucosinolate, 21.84–59.4 mg/g), tannins (13.2–20.6 g TAE/kg; 5.0–12.0 g TAE/kg in freeze-dried leaves) and saponins (≈50 g DE/kg; 64–81 g DE/kg in freeze-dried leaves). All concentrations are as reported by Leone et al. (2015) on a dry weight basis, where TAE refers to tannic acid equivalents and DE to diosgenin equivalents.

Anti-nutrients, such as oxalates (430–1050 mg/100 g) and phytates (25–31 mg/100 g), are present in *M. oleifera* leaves at higher levels than in legumes and cereals (Leone et al. 2015), potentially reducing the bioavailability of essential minerals like calcium, magnesium, iron and zinc.

It is important to note that, in addition to geographical variations, seasonal differences in the phytochemical profile and bioactivities of *M. oleifera* leaf extracts are also evident. Baldisserotto et al. (2023) conducted a comprehensive multifunctional profiling of leaf extracts collected in February, March, April and May, revealing that hydroalcoholic extracts from the February harvests exhibited the highest total phenolic content (TPC). The various numerical findings presented in this chapter serve as representative benchmarks.

M. oleifera seeds contain 34–42% oil, which is more efficiently extracted using solvent extraction (*n*-hexane) than cold pressing, which recovers only about 70% of the oil (Leone et al. 2016). In rural areas, oil is traditionally extracted by boiling de-husked seeds in water and skimming the oil from the surface.

Moringa oil is golden yellow, liquid at room temperature, with a density of approximately 0.9 g/mL and a viscosity in the range of 43–103 cP (Leone et al. 2016). It has a saponification value (SV)

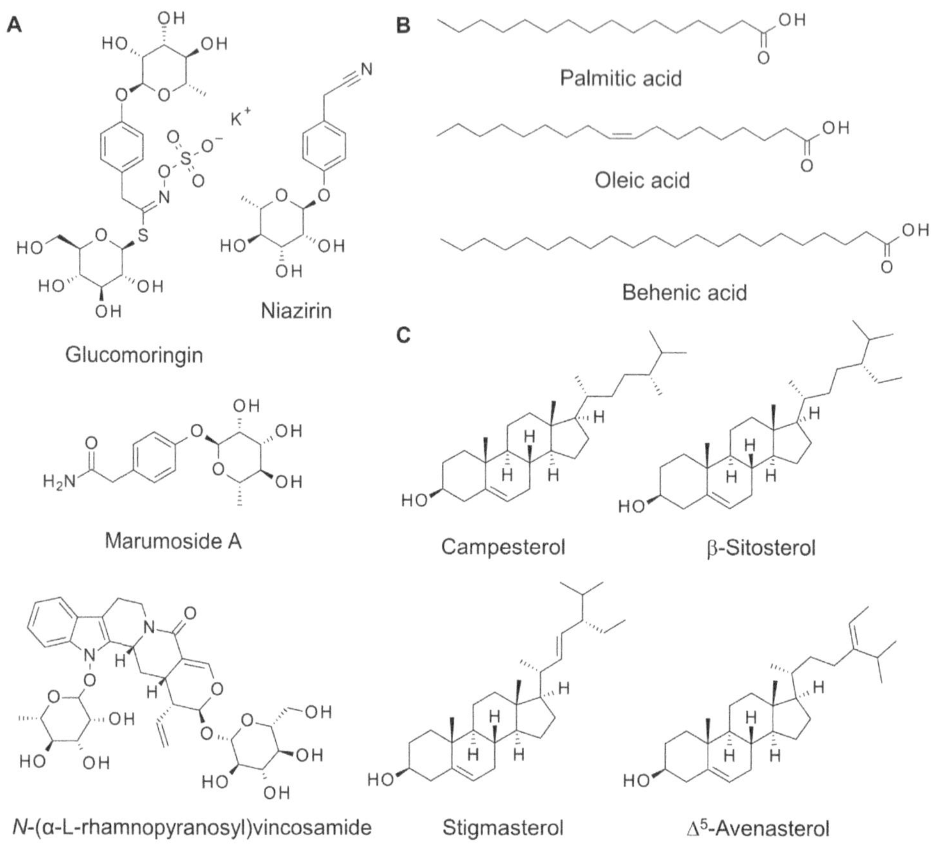

FIGURE 13.2 Glycosidic, Fatty Acid and Phytosterol Constituents of *M. oleifera*. (A) Representative glycosylated natural products and glucosinolates, including 4-*O*-(α-L-rhamnopyranosyloxy)benzylglucosinolate (glucomoringin). (B) Major fatty acids in *M. oleifera* seed oil, with oleic acid as the dominant monounsaturated component. (C) Phytosterols present in seeds and leaves, which are structurally similar to cholesterol and have been investigated for lipid-lowering potential.

of approximately 185 (mg of KOH needed to hydrolyse 1 g of oil) and an iodine value (IV) of ≈67 (g of halogen, as iodine, bound per 100 g of oil). In comparison, coconut oil has an SV of approximately 256 and IV of 9, olive oil has an SV of 190 and IV of 84 and peanut oil has an SV of 191 and IV of 95 (Belitz, Grosch and Schieberle 2004).

With a smoke point of approximately 200°C, at which its triacylglycerols begin to decompose in the presence of air, moringa oil is comparable to olive oil. Cold-pressed moringa oil has higher viscosity and acidity than oil extracted using *n*-hexane. The increased viscosity is due to water bound in the oil during extraction, while the higher acidity results from the water added during seed milling prior to cold pressing (Leone et al. 2016). Its lower iodine value compared with olive oil indicates less unsaturation, while its saponification value is similar, suggesting a comparable fatty acid composition.

The oil contains approximately 21% saturated fatty acids (SFAs), with palmitic acid (C16:0) as the dominant component. It is rich in monounsaturated fatty acids (MUFAs) (77%), primarily oleic acid (C18:1, 74%), while its polyunsaturated fatty acid (PUFA) content is low (about 1%). Table 13.3 compares the fatty acid composition of moringa oil (Leone et al. 2016) with that of olive oil (Revelou et al. 2021). With a high MUFA-to-SFA ratio, similar to olive oil, *M. oleifera* oil could serve as a viable substitute in regions where *M. oleifera* is cultivated. However, due to its low PUFA content, additional dietary sources of polyunsaturated fatty acids may be necessary.

TABLE 13.3
Fatty Acid Composition of Moringa Oil and Olive Oil

Fatty acid	Moringa oil	Olive oil
Palmitic acid (C16:0)	6.25	12.5
Stearic acid (C18:0)	4.97	2.49
Oleic acid (C18:1, Δ^9)	73.6	71.8
Linoleic acid (C18:2, $\Delta^{9,12}$)	0.76	10.1
Arachidic acid (C20:0)	3.23	0.43
Behenic acid (C22:0)	6.02	0.13
$\sum$SFAs	21.2	15.6
$\sum$MUFAs	76.7	72.9
$\sum$PUFAs	1.18	12.0
MUFA/PUFA	65.0	7.43
C18:1/C18:2	96.8	7.98

Composition (values in g/100 g oil): SFA, saturated fatty acid; MUFA, monounsaturated fatty acid; PUFA, polyunsaturated fatty acid.

M. oleifera oil's sterol composition is distinct from olive oil, with β-sitosterol, stigmasterol, campesterol and Δ^5-avenasterol comprising 92% of the sterols. These sterols may offer cholesterol-lowering and antidiabetic benefits, though the cardiovascular impact of plant sterols remains uncertain.

The oil is rich in tocopherols (T), including α-T (13.2 mg/100 g, range 9.8–22.7 mg/100 g), γ-T (6.4 mg/100 g, range 2.7–7.2 mg/100 g) and δ-T (8.1 mg/100 g, range 4.8–21.7 mg/100 g), contributing to its antioxidant properties (Leone et al. 2016). These levels are higher than those in other plant oils, such as olive oil (α-T: 9.0 mg, γ-T: 0.5 mg, δ-T: 0.04 mg/100 g), sunflower oil (α-T: 56.4 mg, γ-T: 0.4 mg, δ-T: 0.09 mg/kg), peanut oil (α-T: 14.1 mg, γ-T: 13.1 mg, δ-T: 0.92 mg/100 g) and soybean oil (α-T: 17.9 mg, γ-T: 60.4 mg, δ-T: 37.1 mg/100 g) (Belitz, Grosch and Schieberle 2004). Antioxidant activity generally increases from α to δ, and *M. oleifera* oil with a relatively high δ-T concentration is expected to have improved stability during processing and storage.

Pramanik and Islam (1998) found that the chemical composition of *M. oleifera* flowers differs significantly between their mature and premature states. Mature flowers contain free neutral sugars (mannose and glucose in a 1:5 ratio), proteins (13.5%), ascorbic acid (9.8%) and carbohydrate materials, but no polysaccharide was isolated. The extract also lacked D-galactose and D-glucuronic acid, with a higher ascorbic acid content compared with premature flowers. In contrast, premature flowers contain free sugars (mannose, galactose and glucose) along with a polysaccharide composed of glucose, galactose and glucuronic acid. The polysaccharide was isolated and characterised, revealing a repeating unit structure with specific linkages between the sugars. The presence of galactose and glucuronic acid in the polysaccharide of premature flowers and their absence in mature flowers, coupled with higher ascorbic acid in mature flowers, suggests enzymatic conversion of sugars to ascorbic acid during maturation.

Prabakaran et al. (2018) reported that the roots of *M. oleifera* are rich in polyphenolic compounds and exhibit significant antioxidant and antibacterial properties. The highest total phenolic content and total flavonoid content (TFC) were observed in the ethyl acetate extracts. Notable phenolic compounds identified in the roots include myricetin (1100 mg/kg), quercetin (685 mg/kg), kaempferol (24 mg/kg), biochanin A (45 mg/kg) and gentisic acid (85 mg/kg). The antioxidant activity of the root extracts was particularly high in the ethyl acetate extract, which showed substantial activity in DPPH, reductive potential and total antioxidant capacity assays.

13.3 EXTRACTION AND PROCESSING TECHNOLOGY

Pop, Kerezsi and Ciont (2022) reviewed various methods for extracting and encapsulating bioactive compounds from *M. oleifera*. Extraction techniques are categorised into conventional and non-conventional methods. Conventional solvent extraction, one of the simplest and earliest techniques, uses solvents such as water, ethanol, methanol, ethyl acetate and chloroform to extract bioactive compounds. The efficiency of this method depends on factors like solvent type, temperature, solid-to-solvent ratio and extraction time. Another conventional method is hot water extraction, which is particularly suitable for thermostable compounds. By immersing *M. oleifera* material in hot water (50–100°C), various polar compounds are extracted.

Non-conventional extraction methods, such as ultrasound-assisted extraction (USAE), use ultrasonic waves to disrupt cell walls, facilitating bioactive release, improving solvent penetration and enhancing mass transfer. Similarly, microwave-assisted extraction (MWAE) and enzymatic extraction offer promising approaches to improving extraction efficiency, reducing processing time and enhancing extract quality. Solid-state fermentation (SSF), which involves cultivating microorganisms like *Aspergillus niger* on a solid substrate, enhances compound extraction through enzymatic action, increasing protein content while reducing indigestible components.

The review by Bhadange, Carpenter and Saharan (2024) highlights several modern extraction methods designed to efficiently isolate bioactive compounds while preserving their bioactivity. These techniques offer significant advantages over traditional methods, such as maceration, distillation and Soxhlet extraction, by increasing efficiency, shortening extraction times, minimising solvent consumption and enhancing selectivity and preservation of delicate compounds. Additionally, they promote environmentally friendly practices by reducing dependence on harmful solvents.

MWAE rapidly heats solvents, causing cell rupture and facilitating the release of bioactive compounds, while USAE enhances mass transfer by disrupting cell structures. MWAE offers high yields, reduced extraction times and lower solvent consumption, but may be limited by overheating and inefficacy with non-polar solvents. Similarly, excessive ultrasonic power can lead to sample degradation. Recent innovations focus on hybrid approaches, such as combining MWAE with ultrasound, to achieve synergistic effects and further enhance extraction performance.

Subcritical solvent-assisted extraction operates at temperatures and pressures below the solvent's critical point, enhancing solvation while allowing the use of environmentally friendly alternatives to conventional solvents. In contrast, supercritical fluid extraction (SCFE) utilises a supercritical fluid, typically supercritical CO_2, to extract bioactives with high selectivity and no residual organic solvents. While SCFE is both efficient and environmentally friendly, it requires costly equipment and precise process control. Recent research focuses on optimising the SCFE of *M. oleifera* through the use of cosolvents and machine learning techniques.

Solid-phase microextraction (SPME) is a solvent-free technique that utilises a solid-phase fibre coated with a sorbent material to extract bioactives, offering high sensitivity and versatility. Although SPME is rapid and effective, its applicability is limited by sample volume and reduced efficiency for certain compounds. Therefore, this chapter will focus on alternative extraction methods.

13.3.1 Ultrasound- and Microwave-Assisted Extraction

Thangaiah et al. (2024) compared four solid-liquid extraction methods – maceration, Soxhlet, USAE, and MWAE – to optimise the recovery of bioactive compounds and minerals from *M. oleifera* leaves using a 70% ethanol and 30% water (v/v) mixture as the solvent. After extraction, the recovered extracts were concentrated and freeze-dried, and extraction yield was expressed as the mass of freeze-dried extract relative to the initial dry leaf mass. The study found that USAE and MWAE significantly outperformed conventional methods in recovering bioactive compounds. USAE extracts also showed the highest TPC (148.86 mg gallic acid equivalents (GAE)/g freeze-dried extract) and total flavonoid content (TFC; 23.18 mg quercetin equivalents

(QE)/g freeze-dried extract), while MWAE gave similarly high values. The improved recovery observed in both methods was attributed to cavitation in USAE and microwave–water molecule interactions in MWAE.

Antioxidant activity, measured by the DPPH free-radical scavenging assay, was highest in USAE extracts (86.25%). These extracts also showed the lowest IC50 value (the concentration required for 50% scavenging), indicating strong antioxidant capacity. In addition to phytochemicals, USAE showed higher reported protein response and vitamin C content than the conventional methods. Relative to Soxhlet extraction, the reported protein value was approximately 22% higher, while vitamin C content was about 73% higher. Furthermore, USAE and MWAE extracts had higher levels of essential minerals such as Fe, Zn, Ca, Mg, K and P, as confirmed by inductively coupled plasma optical emission spectroscopy (ICP-OES) analysis, due to greater cell disruption.

Morphological analysis using scanning electron microscopy revealed extensive structural damage in USAE- and MWAE-treated leaves, which correlated with the enhanced release of intracellular compounds. High-performance thin-layer chromatography (HPTLC) identified USAE extracts as having the highest quercetin content (674.70 μg/mL), consistent with enhanced recovery of quercetin, a key antioxidant flavonoid. Given the improved extraction efficiency, the study suggests that these advanced methods have strong potential for applications in nutraceuticals and biofortified foods. However, a challenge identified was the high hygroscopicity of the extracts, which could affect storage and stability.

Pereira et al. (2021) investigated the extraction of antioxidants from *M. oleifera* leaves using USAE and ultra-homogenisation-assisted extraction (HAE), optimising parameters such as solvent composition and solid-to-solvent ratio through response surface methodology. Both methods effectively extracted antioxidants, with optimal conditions being 4 g of sample per 200 mL of a 1:1 ethanol-water solvent. Phytochemical analysis using ultra-high-performance liquid chromatography-quadrupole time-of-flight mass spectrometry/mass spectrometry (UHPLC-QTOF-MS/MS) identified 30 compounds, including quercetin, rutin and kaempferol. The study found that USAE required a shorter extraction time (five minutes) than HAE (ten minutes), while HAE equipment may be more accessible in some laboratory settings. The TPC ranged from 1.06 to 2.75 g GAE/100 g dry leaf for HAE and from 1.06 to 2.66 g GAE/100 g dry leaf for USAE, with HAE showing slightly higher efficiency, likely due to its high shear rates that enhance metabolite release. The TFC ranged from 0.86 to 2.52 g QE/100 g dry leaf for HAE and from 0.58 to 2.33 g QE/100 g dry leaf for USAE, with higher values obtained when using a 1:1 ethanol-water solvent mixture and lower sample mass. These results support previous findings that solvent polarity is crucial in optimising the extraction of phenolic compounds, confirming that hydroalcoholic mixtures outperform pure solvents in antioxidant recovery.

13.3.2 Enzyme-Assisted Extraction

The extraction of oil from *M. oleifera* seeds has gained significant interest due to its high oleic acid content and similarity to olive oil, making it a valuable alternative in the food industry. Traditional mechanical extraction methods yield only 20–40% of the oil, while solvent extraction, though more efficient at around 80%, may raise safety and environmental considerations associated with organic solvents. Enzymatic extraction has emerged as a potentially safer and more sustainable alternative, enhancing oil recovery through enzymatic disruption of cell-wall components under mild conditions while reducing reliance on organic solvents.

Abdulkarim et al. (2006) investigated the use of four enzymes: Neutrase (neutral protease), Termamyl (α-amylase), Pectinex (pectinase) and Celluclast (cellulase), individually and in combination, to optimise oil recovery. Neutrase was found to be the most effective individual enzyme, yielding ≈72% oil recovery, while a combination of all four enzymes resulted in the highest recovery at 74%. The enzymatically extracted oil exhibited similar iodine value and melting point to solvent-extracted oil; however, the latter had significantly higher free fatty acid content and colour intensity, necessitating further refining. In contrast, solvent-extracted oil showed lower viscosity than enzymatically extracted oil. The enzymatic process may eliminate the need for degumming, reduce refining costs, and improve oil recovery compared with conventional aqueous extraction methods. Compared with solvent extraction, enzymatic extraction offers a more environmentally friendly alternative without compromising oil quality, making it a promising approach for large-scale oil recovery.

Abdulkarim et al. (2005) compared the physico-chemical properties of *M. oleifera* seed oil extracted using solvent extraction with petroleum ether and aqueous enzymatic extraction with 2% Neutrase 0.8L. Neutrase was identified as the most effective enzyme, yielding the highest oil recovery. Fatty acid analysis showed that both oils were rich in oleic acid (67.9% for solvent-extracted and 70.0% for enzyme-extracted), along with significant amounts of palmitic, stearic and behenic acids. Both oils were highly unsaturated and liquid at room temperature, with a pale-yellow colour and a peanut-oil-like flavour. The enzyme-extracted oil exhibited better quality attributes, including slightly higher oleic acid and iodine values, together with lower free fatty acid and unsaponifiable matter (non-fatty acid components), while the overall fatty acid composition remained broadly similar between the two extraction methods.

Mat Yusoff et al. (2016) explored aqueous enzymatic extraction as a sustainable alternative to solvent-based methods for extracting *M. oleifera* oil. Oil recovery was significantly influenced by pH, moisture/kernel ratio and shaking speed, with optimised conditions achieving up to 80% recovery while minimising emulsion formation. The study highlighted the necessity of protease and cellulase for effective cell wall breakdown, emphasising enzymatic synergy. However, reusing the aqueous phase resulted in zero recovery and increased the formation of stable emulsions, which hinder oil separation and highlight the complexity of enzyme interactions. Further research is needed to enhance de-emulsification strategies and improve overall process efficiency.

13.3.3 Solid-State Fermentation for Bioactive Enrichment

SSF with *A. niger* has been explored as a strategy to enhance bioactive compound content in moringa leaf flour (MLF). Feitosa et al. (2020) demonstrated that SSF significantly increased polyphenolic compounds, with TPC rising by about 136% and TFC by about 783%, depending on the initial moisture level and extraction solvent. The greatest enhancement was observed with 80% ethanol and 80% acetone (both 80:20 solvent–water mixtures) as extraction solvents, which have significantly lower polarity than distilled water.

Antioxidant activity, measured via the ferric reduction antioxidant power (FRAP) assay, was notably improved, with the 80% acetone extract from fermented MLF showing the highest activity (277.2 μmol Trolox/g MLF). Trolox, a water-soluble antioxidant used as a standard in the FRAP assay, is a vitamin E analogue that helps quantify antioxidant capacity. Additionally, solid-state fermentation improved antimicrobial properties, as extracts from fermented MLF inhibited *Bacillus subtilis*, whereas unfermented extracts had no effect.

HPLC analysis identified gallic acid as the major compound in all extracts, while 14 new bioactive compounds were released or synthesised post-fermentation. The 80% acetone extract from fermented MLF (50% initial moisture) contained the highest number of novel compounds, including epigallocatechin, epigallocatechin gallate, quercetin, kaempferol-3-glycoside, caffeic acid, taxifolin and *p*-coumaric acid, which were absent in unfermented extracts. These findings highlight SSF as an effective method to obtain *Moringa* extracts enriched with bioactive compounds, demonstrating enhanced antioxidant and antimicrobial properties with potential applications in the food, cosmetic and pharmaceutical industries.

13.3.4 Subcritical Solvent-Assisted Extraction

Bhutada et al. (2016) investigated Soxhlet solvent-assisted extraction of oil from *M. oleifera* seeds, examining the effects of extraction temperature, solvent type, and chloroform–methanol solvent ratio on oil yield. The highest yield was achieved with a chloroform–methanol ratio of 3:1 at 100°C. The use of mixed solvents improved extraction efficiency compared with individual solvents, likely because methanol facilitated the extraction of more polar minor constituents such as pigments, antioxidants, and tocopherols, while chloroform enhanced lipid solubility. Gas chromatography showed that the extracted oil was rich in oleic acid (73.5%), with appreciable amounts of stearic, arachidic, and behenic acids. Fourier-transform infrared spectroscopy and proton nuclear magnetic resonance confirmed characteristic triglyceride functional groups and indicated no substantial oxidative degradation, while thermogravimetric analysis showed major thermal decomposition between approximately 300 and 428°C, demonstrating good thermal stability. The extracted oil remained liquid and stable at room temperature, confirming that Soxhlet solvent extraction is an effective method for recovering moringa seed oil.

Camacho et al. (2017) evaluated two methods for extracting oil from *M. oleifera* seeds: pressurised fluid extraction and ethanol extraction. Pressurised extraction was conducted at 30°C and 8 MPa using propane at a flow rate of 1 mL/min for 60–90 minutes, resulting in a 94% reduction in oil content. Ethanol extraction involved soaking shelled seeds in 1 M NaCl, blending and extracting with 95% ethanol (5% w/v of seeds to ethanol) for 30 minutes, followed by centrifugation and drying, reducing the oil content by 76%. The fat/oil content of the seeds was measured as approximately 29.5% for the initial seed powder, 1.87% after pressurised extraction and 7.07% after extraction with ethanol.

Oladipo and Betiku (2019) investigated the solvent extraction of *M. oleifera* seed oil using *n*-hexane, ethyl acetate and ethanol. The study assessed the effects of extraction time (one to three hours), solid-to-solvent ratio (50–200 g/L) and solvent type on oil yield. Optimal conditions included a three-hour extraction time, a 50 g/L solid–solvent ratio and *n*-hexane as the solvent, yielding about 42% oil, followed by ethyl acetate (14%) and ethanol (8.2%). This finding aligns with Mani, Jaya and Vadivambal (2007), who determined *n*-hexane to be the most suitable solvent for oil extraction due to its effectiveness, low cost and commercial viability, despite acetone reducing extraction time and being effective with larger particle sizes. The study also revealed that the physical and chemical properties of the oil varied depending on the solvent used. The acid values were 8.25, 3.40 and 2.88 mg KOH/g for ethanol, ethyl acetate and *n*-hexane, respectively, indicating a higher concentration of organic acids in the more polar ethanol extract.

13.3.5 Green Solvent-Assisted Extraction

Jaglan et al. (2024) investigated the extraction of bioactive compounds from *M. oleifera* flowers using deep eutectic solvents (DES) as an eco-friendly alternative to conventional and advanced extraction methods. Deep eutectic solvents are formed by mixing a hydrogen bond acceptor, such as choline chloride or L-proline, with a hydrogen bond donor, such as glycerol, malic acid, lactic acid or glucose, in specific molar ratios. The study optimised the extraction conditions by evaluating different DES compositions and varying temperatures (50 to 80°C). The researchers found that DES containing proline and glycine exhibited the highest antioxidant activity, with a peak value of about 75% at 70°C. The study also analysed the lipase inhibitory activity of the extracts, demonstrating significant inhibition comparable to Orlistat, a medication used for weight loss that works by inhibiting lipase, thereby reducing fat absorption. The extracts showed an IC_{50} value of 62 μg/mL, lower than Orlistat's (74 μg/mL), suggesting potential applications in obesity management. Furthermore, the glucose uptake assay indicated that the extracts could influence glucose metabolism, highlighting their potential for diabetes-related applications. The findings confirmed that DES were effective solvents for extracting bioactive compounds from *M. oleifera* flowers, yielding extracts with high antioxidant and lipase inhibitory activities. The study reinforced DES as a promising green extraction technology, offering an efficient and sustainable alternative to conventional solvents for recovering bioactive compounds with potential health benefits.

Lakka et al. (2020) developed a green extraction process for polyphenols from *M. oleifera* leaves using a DES composed of glycerol and nicotinamide. *M. oleifera* leaves were freeze-dried, ground and mixed with a 70% w/v aqueous DES at a defined liquid-to-solid ratio (20–100 mL/g). The mixture was stirred at 50°C, with optimisation of temperature, stirring speed and extraction duration. Ultrasonication at ambient temperature for 30 minutes was used as a pretreatment in some experiments. After extraction, the mixture was centrifuged to separate the solid residue from the liquid extract. Response surface methodology optimised the extraction conditions, yielding 83 mg GAE/g dry mass after 30 minutes of ultrasonic pretreatment, which was not significantly different from the yield without pretreatment. LC-MS analysis showed no significant impact on the polyphenolic profile, with major compounds including neochlorogenic acid, chlorogenic acid, kaempferol and quercetin glycosides.

The extraction kinetics were studied by measuring mg GAE/g dry mass over time at temperatures ranging from 40 to 80°C. Data were fitted to an empirical kinetic model, and the second-order extraction rate constant was determined. The rate constant showed an Arrhenius dependence on temperature, from which the activation energy of the extraction process was calculated. Ultrasonication pretreatment increased the activation energy for polyphenol extraction from 11.24 kJ/mol (without pretreatment) to 34.02 kJ/mol. This was attributed to the removal of easily extractable polyphenols during pretreatment, leaving more difficult-to-access compounds that required more energy to extract.

Wei et al. (2023) investigated the efficient extraction of flavonoids from *M. oleifera* leaves using DES and evaluated their antioxidant, antimicrobial and antitumour activities. The study employed a DES composed of choline chloride (ChCl) and glycine (Gly) for flavonoid extraction, followed by purification using macroporous resin (D101). The resin achieved an adsorption rate of about 35% and a desorption rate of about 88%, indicating its effectiveness in capturing and recovering the flavonoid extract. HPLC-MS/MS analysis identified ten flavonoids in the extract, including vitexin, osajin, quercetin 3-β-D glucoside, quercetin, nictoflorin, cynaroside, cyanidin, taxifolin, kaempferol-G-M and fisetin, with osajin being reported for the first time in *M. oleifera* leaves. The extracted compounds exhibited strong antioxidant activity, surpassing vitamin C in ABTS radical scavenging (IC_{50} value of about 9.5 μg/mL). Additionally, the extract demonstrated antibacterial and antitumour potential.

Okuda et al. (1999) optimised protein extraction from *M. oleifera* seeds using water and NaCl. They found the optimal NaCl concentration for extraction was 1.0 mol/L, which enhanced protein solubility through the 'salting-in' effect, while higher concentrations (3.0 mol/L) caused 'salting-out'. Other salts tested showed no significant differences, indicating that ionic strength is the key factor. Dialysis experiments confirmed NaCl's role in maintaining protein solubility and activity. Oladoja and Pan (2015) described an extraction process using a 5% (w/v) seed cake suspension in 1 M NaCl, followed by agitation, filtration and ammonium sulphate precipitation to isolate active components. The crude extract had a protein content of 2.1 g/L, while the precipitated extract contained 7.3 g/L, indicating effective protein concentration.

13.3.6 Supercritical Fluid Extraction

Chen et al. (2022) optimised the extraction of *M. oleifera* seed oil using supercritical CO_2 (SC-CO_2; critical temperature, $T_c \approx 31.1$°C; critical pressure, $P_c \approx 7.38$ MPa) to maximise yield and assess its antioxidant activity. Using response surface methodology and Box–Behnken design, the study examined the effects of extraction time (one to three hours), temperature (40–50°C) and pressure (40–50 MPa). The results revealed that extraction time had the most significant impact on yield, followed by pressure and temperature. The optimal extraction conditions were 45°C, 2.5 hours and 50

MPa, yielding about 38.5% oil. GC-MS analysis showed that the oil was rich in oleic acid (≈85%), along with other fatty acids and sterols. The oil exhibited potent antioxidant properties, outperforming commercial peanut oil in DPPH, hydroxyl and superoxide radical-scavenging abilities. These findings support the use of moringa seed oil in functional foods, nutraceuticals and cosmetics, due to its high yield, health benefits and antioxidant activity.

Nguyen et al. (2011) investigated SC-CO_2 extraction of oil from *M. oleifera* seeds and evaluated ethanol-assisted extraction (SC-CO_2–EtOH) as a means to enhance yield. Ethanol (10% or 15% w/w), expressed relative to the weight of 500 mL of CO_2 under operating conditions, was added directly to the extractor along with ground kernels. The substrate was soaked under processing conditions for approximately 45 minutes before SC-CO_2 extraction commenced, which continued for up to seven hours. Key extraction parameters, including pressure (15–30 MPa), temperature (35–60 °C) and particle size (0.16–1.12 mm), were optimised using a pilot-scale system. The study also examined the effect of raw material configuration by comparing randomly packed and multi-layered kernels. Fatty acid profiles of oils extracted via SC-CO_2, SC-CO_2–EtOH and *n*-hexane (Soxhlet) were similar, with oleic acid as the predominant component. Oil yield increased by approximately 27% with multi-layered packing and by 10% with 10% ethanol pretreatment. Yields from SC-CO_2–EtOH and *n*-hexane were nearly identical (37.8% vs. 37.5%). Using response surface methodology, optimal extraction conditions were predicted at 28.97 MPa, 44.30°C and 0.54 mm particle size, resulting in a theoretical oil yield of 37.85%.

Zhao and Zhang (2014) investigated the solubility of *M. oleifera* oil in SC-CO_2 across a temperature range of 60–100 °C and a pressure range of 20–50 MPa. The primary aim was to obtain experimental solubility data and to correlate this data using thermodynamic and density-based models. The oil solubility was found to be between 0.64 and 12.68 g/kg of CO_2 under the investigated conditions. The results indicated that the solubility increased with increasing pressure at a constant temperature. However, the effect of temperature on solubility was more complex: at lower pressures (20 and 30 MPa), solubility decreased slightly with temperature, while at higher pressures (40 and 50 MPa), it increased. This phenomenon was attributed to the competing effects of reducing solvent density and increasing solute vapour pressure with rising temperature. The study determined that a CO_2 flow rate of 3.68 g/min was sufficient to ensure saturation of the oil in the supercritical fluid.

The fatty acid composition of moringa seed oil extracted at 40 MPa and 60°C was analysed using gas chromatography. Oleic acid was found to be the predominant fatty acid, constituting about 67% of the total, along with other unsaturated and saturated fatty acids. The study explored how fatty acid composition influences the solubility of moringa oil in supercritical CO_2 by comparing its solubility with that of other vegetable oils, including sunflower, corn, rosehip, borage, lunaria and olive husk oils, under various pressure and temperature conditions. Moringa oil's solubility was similar to lunaria oil but differed from oils like sunflower and rosehip. The study identified that oleic acid (67%) and *cis*-vaccenic acid (6%) played a significant role in solubility. Oils with shorter-chain fatty acids were generally more soluble, and the presence of unsaturated fatty acids like oleic acid contributed to the oil's solubility. Despite having similar oleic acid content, moringa oil's solubility differed from olive husk oil, suggesting the influence of other fatty acids like *cis*-vaccenic acid. Oleic acid is Δ^9-octadecenoic acid, with a double bond at the ninth carbon, while *cis*-vaccenic acid is Δ^{11}-octadecenoic acid, with a double bond at the 11th carbon. The study showed that fatty acid composition, including chain length and unsaturation, is a crucial factor in determining the solubility of oils in SC-CO_2, with specific fatty acids influencing solubility even in oils with similar major components.

To model the experimental solubility data – measuring the solubility of *M. oleifera* oil in SC-CO_2 at temperatures of 333 K, 353 K and 373 K, and pressures of 20 MPa, 30 MPa, 40 MPa and 50 MPa – four models were employed: the Peng–Robinson equation of state and three density-based models: Chrastil, del Valle and Aguilera, and Adachi and Lu. While all four models showed good correlation with the experimental data, the del Valle and Aguilera model provided the best fit. The Peng–Robinson equation was used to explain the observed solubility trends with varying temperature and pressure thermodynamically. Additionally, the study introduced a new method for estimating the binary interaction parameters for the Peng–Robinson equation, noting their linear variation with temperature.

Ruttarattanamongkol et al. (2014) conducted a pilot-scale investigation of SC-CO_2 extraction for oil recovery from *M. oleifera* seeds, comparing its efficiency and product quality with conventional extraction methods. The study assessed extraction yield, physicochemical properties and the composition of fatty acids, tocopherols and sterols in the extracted oils. The effects of pressure (15–35 MPa) and temperature (25–35°C) on oil yield were examined. It was found that pressure had a more pronounced influence on extraction efficiency than temperature within the tested range. The highest accumulated yield, approximately 83%, was achieved at 35 MPa and 35°C. The fatty acid profile of the SC-CO_2 extracted moringa seed oil was dominated by oleic acid, which accounted for over 73% of the total fatty acids. Additionally, the oil contained a significant amount of behenic acid (C22:0), a saturated fatty acid valued in the cosmetics industry. The high oleic acid content contributes to the oil's oxidative stability, reinforcing its potential as a healthy cooking oil, while the presence of behenic acid highlights its applications in food and personal care formulations. The study also analysed tocopherol and sterol compositions, comparing these minor bioactive components with those of conventionally extracted oils. The results further demonstrated the effectiveness of SC-CO_2 in preserving oil quality while eliminating the need for organic solvents.

The three-step sequential extraction method developed by Rodríguez-Pérez et al. (2016) efficiently recovered bioactive compounds from *M. oleifera* leaves using supercritical fluids. Each step targeted compounds of different polarities: SC-CO_2 was used in the first step to extract non-polar compounds like fatty acids and alkanes, optimised at 15 MPa and 50°C for 60 minutes. The second step employed CO_2-expanded ethanol (CXE) to extract intermediate polarity compounds, such as glycosylated flavonoids, at 7 MPa and 50°C for 160 minutes. The final step, pressurised hot water extraction (PHWE), recovered highly polar compounds like phenolic acids and enhanced antioxidant activity, using 125°C and 7 MPa for 60 minutes. This approach resulted in distinct fractions with varied compositions, maximising the extraction of bioactive compounds. The extracts were characterised using GC-MS and HPLC coupled to electrospray ionisation quadrupole time-of-flight mass spectrometry (HPLC-ESI-QTOF-MS). The PHWE fraction exhibited the highest antioxidant activity, as determined by the ABTS assay, and was also richer in total phenolic content. In contrast, the CXE fraction contained a higher concentration of flavonoids.

13.3.7 Microencapsulation

Pop, Kerezsi and Ciont (2022) highlighted encapsulation techniques for their role in protecting *M. oleifera* bioactives from degradation and improving their stability, controlled release and bioavailability. Microencapsulation, a method of entrapping bioactive compounds within a protective matrix, is commonly achieved through techniques such as spray drying, electrospinning, ionic gelation and liposomal encapsulation. Spray drying involves spraying a bioactive solution into a hot gas stream to form dry microparticles (Castro-López et al. 2020), while electrospinning uses an electric field to create fine fibres. Ionic gelation forms a gel matrix using polyelectrolytes and multivalent ions, and liposomal encapsulation traps bioactives in spherical lipid vesicles. Nanoencapsulation, which uses nanocarriers such as polymeric nanoparticles, nanoliposomes and nanoemulsions, offers enhanced solubility, bioavailability and targeted delivery. Emulsification is another technique that disperses bioactives within oil or water phases, stabilised by emulsifiers. The choice of encapsulation method depends on the bioactive compound's properties, the desired release profile and the intended application. The encapsulating materials, such as polymers, lipids and carbohydrates, are crucial for ensuring the protection and efficient release of the bioactives. Encapsulation not only enhances stability but also masks undesirable flavours and improves the overall efficacy of *M. oleifera* extracts. Castro-López et al. (2020) demonstrated that spray-dried microencapsulation of *M. oleifera* extract improved storage stability, modulated gastrointestinal release, and preserved antioxidant activity.

To enhance bioaccessibility, Wei et al. (2023) explored microencapsulation of the flavonoids extracted from *M. oleifera* leaves. The wall materials used to create these microcapsules were gelatin and soybean protein isolate (SPI), which were modified by xylose through the Maillard reaction. The Maillard reaction occurs between amino acids and reducing sugars; in this case, between the proteins in soybean protein isolate and xylose. This modification was intended to improve the encapsulation efficiency and bioaccessibility of the flavonoids. Microcapsules prepared with modified SPI exhibited more uniform and spherical morphology compared with those prepared with unmodified SPI. In vitro digestion studies showed that microencapsulation protected the flavonoid extract in simulated oral and gastric fluids, reducing premature losses and preserving more flavonoids for the intestinal phase. As a result, the modified microcapsules produced a 7.4-fold higher measurable flavonoid content in the simulated intestinal phase than the non-encapsulated extract.

13.4 PROPERTIES AND APPLICATIONS

13.4.1 Nutrition

M. oleifera can be used as an important source of human, animal and plant nutrition. It is valued for its nutritional content, providing vitamins, minerals, proteins and essential amino acids. The leaves and seeds are rich in vitamins A and C, calcium, potassium, iron and various phytochemicals, making it a beneficial dietary addition.

Makkar and Becker (1997) reported that both *M. oleifera* leaves and the protein-rich residue remaining after seed processing could be useful ingredients for animal feed. The leaves contained about 26% crude protein, of which around 64% was degraded in the rumen after 24 h, while about one-third was estimated to remain available for digestion in the intestine. Most of the leaf protein was true protein (87%), and its essential amino acid profile was comparable to soybeans. The processed seed residue contained about 70% crude protein, with 69% of this protein estimated to remain available for intestinal digestion. These findings suggest that moringa leaves may serve as a forage protein source, while the seed residue could partially replace conventional oilseed meals and provide a valuable source of sulphur-containing amino acids and other nutrients for livestock diets (Falowo et al. 2018).

M. oleifera leaf extracts contain zeatin, a plant hormone, and have been shown to enhance plant growth and boost the crop yields of wheat, maize and rice (Leone et al. 2015; Ashfaq, Basra and Ashfaq 2012).

13.4.2 Water Treatment

M. oleifera seed proteins have emerged as a promising, sustainable alternative to conventional aluminium salts for water treatment applications. Their biodegradability and nontoxicity make them particularly attractive for addressing water contamination challenges. The active ingredients in the *M. oleifera* seed are cationic proteins with a molecular weight of 13 kDa and an isoelectric point between 10 and 11 (Ndabigengesere and Narasiah 1998). The amino acid sequences are also available as determined by Gassenschmidt et al. (1995) and Freire et al. (2015). The MO2 and *Mo*-CBP_3 proteins are rich in arginine (and glutamine) residues and promote flocculation through a charge-patch mechanism.

Extensive research has demonstrated their efficacy in treating a wide range of polluted water sources, including cyanobacteria-laden water containing *Microcystis aeruginosa*, a primary driver of harmful algal blooms (Orimolade et al. 2025; Alhaithloul et al. 2024; Camacho et al. 2017). Additionally, they have shown effectiveness in removing helminth parasite eggs from irrigation water (Sengupta et al. 2012), dyes from textile industry wastewater (Patel and Vashi 2012), pollutants from coffee fermentation wastewater (Garde et al. 2017), melanoidin pigments from distillery molasses spent wash (Prasad 2009) and suspended solids from various water sources (Boulaadjoul et al. 2018; Kane et al. 2016; Madrona et al. 2012). Desta and Bote (2021) reported that *M. oleifera* seed powder can achieve up to ~99% turbidity removal, ~90–98% color removal, and ~66% COD removal at an optimal dosage of ~0.8 g/L, with performance influenced by pH and response surface methodology enabling efficient optimization of treatment conditions.

Orimolade et al. (2025) systematically compared the flocculation dynamics of *M. aeruginosa* algae using alum and aqueous *M. oleifera* defatted seed extract as flocculants. Light scattering and fractal analysis were applied to evaluate floc growth, structural evolution and shear stability across a range of algal cell densities (200 to 800 μg/L chlorophyll *a*) and flocculant concentrations. Alum induced rapid diffusion-limited flocculation, initially forming lower fractal dimension flocs, which densified over time through structural rearrangement. However, alum flocs were more susceptible to shear-induced breakage, limiting their size and stability. In contrast, the *M. oleifera* seed extract produced larger, more stable flocs with fractal dimension values stabilising around 2.0, likely due to patch charge interactions from its cationic proteins. Higher *M. oleifera* seed protein concentrations further enhanced floc size and shear resistance. Both flocculants were effective in treating 800 μg/L chlorophyll *a* suspensions, but the *M. oleifera* seed protein demonstrated superior performance in terms of floc size and resilience at similar mass concentrations.

The aqueous extracts of defatted *M. oleifera* seeds were characterised using UV-vis absorption, light scattering and surface tension measurements. The seed protein extract in 1 M NaCl had an extinction coefficient of 0.74 (g BSA equivalent/L)$^{-1}$ cm^{-1} at 280 nm. The protein reduced surface tension to 57.4 mN/m at 10 g/L, and dynamic light scattering revealed large protein aggregates.

Morgan et al. (2020) investigated the chemical composition of *M. oleifera* root powder solution and its effectiveness in reducing *Escherichia coli* in contaminated water. The study highlighted the potential of moringa roots, which are less explored than the seeds, as an alternative water treatment option in regions with limited access to clean water. The root powder contained essential macronutrients like potassium, phosphorus, sulphur, magnesium, sodium and calcium, with no harmful heavy metals. The study showed that moringa root powder reduced *E. coli* concentrations by up to 87% at 600 mg/L for higher initial *E. coli* levels, with significant reductions at lower concentrations (250 and 450 mg/L).

13.4.3 Antioxidant and Antimicrobial Properties

Vats and Gupta (2017) analysed the phytochemical profile and antioxidant activity of hydroethanolic extracts from various parts of *M. oleifera*. Leaves exhibited the highest total phenolics (9.58 mg/g), β-carotene (14.10 mg/g) and lycopene (2.60 mg/g), while flowers contained the highest flavonoids (3.5 mg/g) and ascorbic acid (7.10 mg/g), and bark had the greatest anthocyanin content (52.80 mg/g). Anthocyanins are known to exhibit antidiabetic and lipid-lowering effects. Antioxidant assays showed that leaves had the strongest nitric oxide scavenging activity (IC_{50} = 120 μg/mL) and hydroxyl radical-scavenging activity (IC_{50} = 178 μg/mL), whereas flowers had the highest 2,2-diphenyl-1-picrylhydrazyl (DPPH) radical-scavenging potential (IC_{50} = 405 μg/mL). In these assays, lower IC_{50} values indicate stronger antioxidant activity. GC-MS analysis identified major bioactive compounds such as epiglobulol (bark), phytol (leaves) and β-sitosterol (flowers).

Kooltheat et al. (2014) demonstrated the antioxidant and anti-inflammatory potential of *M. oleifera* leaf extracts. An ethyl acetate fraction prepared from fresh leaves showed high phenolic content and strong antioxidant activity. In human macrophages, pre-treatment with the extract significantly suppressed lipopolysaccharide- and cigarette smoke extract-induced production of the pro-inflammatory cytokines TNF, IL-6, and IL-8 at both protein and mRNA levels. The extract also downregulated *RelA*, a key regulator of NF-κB inflammatory signaling. These findings suggest that *M. oleifera* leaf extracts may help reduce cigarette smoke-induced inflammatory responses relevant to lung damage.

Lee et al. (2017) investigated the antibiofilm properties of *M. oleifera* extracts obtained through SC-CO_2 extraction. The extracts from leaves and seeds effectively inhibited biofilm formation by

Staphylococcus aureus, with key bioactive components being unsaturated fatty acids, including palmitoleic acid (C16:1, Δ^9), oleic acid (C18:1, Δ^9), linoleic acid (C18:2, $\Delta^{9,12}$) and gondoic acid (C20:1, Δ^{11}). Gondoic acid, abundant in the leaf extract, demonstrated significant antibiofilm and anti-haemolytic effects. Moyo, Masika and Muchenje (2012) reported that the acetone extract of *M. oleifera* leaves, at a concentration of 5 mg/mL, exhibited bactericidal activity against *E. coli* and *Micrococcus kristinae*. Notably, *M. kristinae* was the most susceptible, with its growth inhibited at 0.5 mg/mL. The extract was also bacteriostatic against *S. aureus*, *Enterobacter cloacae* and *Proteus vulgaris*.

Semen cryopreservation damages sperm due to oxidative stress. Carrera-Chávez et al. (2020) evaluated the effect of *M. oleifera* seed extract on ram sperm quality after freezing and thawing. The extract improved antioxidant activity, sperm viability and progressive motility, particularly at a concentration of 0.5 mg/mL. The 0.5 mg/mL group showed significantly better sperm membrane integrity and less damage compared with the control group. The extract's antioxidant compounds, such as polyphenols and flavonoids, likely contributed to these improvements. The study suggests that *M. oleifera* extract could be a potential alternative to conventional antioxidants for semen cryopreservation.

Sreelatha and Padma (2009) examined the antioxidant activity and total phenolic content of *M. oleifera* leaves at two maturity stages: mature and tender. Both leaf extracts demonstrated strong antioxidant activity, crucial for protecting against oxidative stress-related diseases. The study focused on assays for radical scavenging and lipid peroxidation inhibition, as lipid peroxidation can damage cell membranes, particularly in the liver, which plays a key role in lipid metabolism. The results showed that mature leaves had higher phenolic content and stronger antioxidant activity than tender leaves. Both extracts also offered protection against DNA damage.

Fortification of processed foods with *M. oleifera* has been shown to improve nutritional value, organoleptic properties, oxidative stability and shelf life (Falowo et al. 2018). The study by Ramachandran et al. (2017) investigated the effectiveness of *M. oleifera* seed and leaf extracts as natural preservatives for sugarcane juice, offering a potential alternative to chemical preservatives like sodium benzoate. *M. oleifera* extracts were prepared by drying, grinding and extracting the seeds and leaves with ethanol and chloroform. The antibacterial activity of the extracts was tested against 32 microorganisms using the agar well-diffusion method. The most effective extract combination, *M. oleifera* seed extract, was added at 5% and 10% to both fresh and pasteurised sugarcane juice samples, which were stored at 2°C, 6°C and 30°C for ten days. The study found that pasteurised sugarcane juice with 10% *M. oleifera* seed extract, stored at 2°C, showed the best shelf life and sensory properties, outperforming chemical preservatives.

Prabakaran et al. (2018) investigated the polyphenol composition and antimicrobial activity of solvent extracts from various parts of *M. oleifera* (leaves, roots, flowers, barks and seeds). The leaves exhibited the highest TPC and TFC, with myricetin and quercetin as dominant flavonols, while the roots had elevated levels of gentisic acid and biochanin A. Leaf extracts, particularly methanol and ethyl acetate extracts, showed the highest antioxidant activity and strong antibacterial effects, notably inhibiting *Pseudomonas aeruginosa* and *Erwinia carotovora*. Similarly, El-Sherbiny et al. (2024) found that *M. oleifera* leaf extract demonstrated potent antibacterial activity against standard and foodborne bacterial strains and significant antioxidant potential due to its high phenolic and flavonoid content, and thus, holds promising potential for applications in both the food and medical industries.

M. oleifera leaf extracts, particularly hydroalcoholic extracts from leaves harvested in February, demonstrated strong antioxidant, photoprotective, anti-inflammatory and antimicrobial properties, making them promising candidates for cosmetic and pharmaceutical applications (Baldisserotto et al. 2023). February extracts showed the highest polyphenol content and antioxidant activity. Their photostabilising effect was evaluated against the commercial UVA filter avobenzone, known for its photo-instability due to keto-enolic tautomerism upon UV exposure. Formulations with 5% avobenzone and 5% moringa extracts enhanced photostability compared with those with 5% octocrylene, a conventional stabiliser. Ferulic acid, the primary active in February extracts, was also tested at equivalent concentrations (0.19% for hydroalcoholic and 0.09% for methanolic extract and infusion). The extracts outperformed ferulic acid alone, suggesting a synergistic effect. Additionally, the February hydroalcoholic extract reduced pro-inflammatory mediator production in human keratinocyte cells and exhibited antimicrobial activity against skin-related microorganisms with low cytotoxicity. These findings underscore the importance of optimal harvesting and extraction methods to maximise the beneficial properties of *M. oleifera* extracts.

13.4.4 Medical

Some early medicinal applications of *M. oleifera* are mentioned in the *Sūtra-sthān* of the *Āyurveda* treatise *Caraka Saṃhitā* (Sharma and Dash 2009), where it is noted as an ingredient in ointments for treating skin diseases. Moringa oil is also mentioned in the context of oleation therapy. The review by Pareek et al. (2023) provides a comprehensive overview of the traditional uses, phytochemical composition, pharmacological activities and safety aspects of *M. oleifera*. It highlights the plant's therapeutic potential, including its antimicrobial, anti-inflammatory, anticancer, hepatoprotective, cardiovascular and antidiabetic effects, supported by preclinical studies.

Gupta et al. (2012) investigated the antidiabetic and antioxidant effects of methanol extracts of *M. oleifera* pods in streptozotocin-induced diabetic albino rats. Administering two doses (150 and 300 mg/kg) for 21 days significantly reduced serum glucose and nitric oxide levels, increased insulin, protein and antioxidant enzymes (superoxide dismutase, glutathione and catalase) and decreased lipid peroxidation in pancreatic tissue. Histological analysis showed a reversal of β-cell degeneration. Phytochemical analysis identified quercetin and kaempferol as active constituents. These findings suggest that *M. oleifera* pods have potential as a therapeutic agent for diabetes management.

Akinduti et al. (2022) examined the antimicrobial activity of *M. oleifera* leaf extracts against multidrug-resistant *S. aureus*, a common cause of abscesses. The extracts showed significant antimicrobial effects, with key phytochemicals such as alkaloids, flavonoids and saponins. *M. oleifera* demonstrated an IC_{50} of 4.80 mg/mL and an IC_{90} of 9.6 mg/mL, suggesting its potential for treating skin infections such as abscesses. While not the most potent extract, *M. oleifera* offers a safe, non-toxic alternative to synthetic drugs and could be developed into herbal formulations for wound care.

Saad El-Din et al. (2023) investigated the effects of *M. oleifera* leaf extract, alone or in combination with praziquantel (PZQ), on liver and spleen damage caused by *Schistosoma mansoni* infection in mice. Treatment with the leaf extract significantly reduced worm count, egg load and hepatic granulomas, improved liver function and alleviated histopathological damage in both the liver and spleen. Immunohistochemical staining revealed reduced NF-κB expression, suggesting decreased inflammation. The study concluded that *M. oleifera*, due to its phenolic compounds like caffeic acid, effectively mitigated liver and spleen damage, with enhanced effects when combined with PZQ.

Al-Asmari et al. (2015) investigated the anticancer effects of *M. oleifera* leaf, bark, and seed extracts on breast (MDA-MB-231) and colorectal (HCT-8) cancer cells. Leaf and bark extracts significantly reduced cell survival, colony formation, and motility, whereas seed extracts showed little or no significant activity. Apoptosis assays revealed a marked increase in apoptotic cells in both cell lines, particularly in MDA-MB-231 cells, while cell-cycle analysis showed G2/M phase arrest, indicating inhibition of mitotic progression. GC–MS analysis identified several bioactive compounds, including eugenol and isopropyl isothiocyanate, which may have contributed to the observed anticancer effects. These findings suggest that *M. oleifera* leaves and bark possess promising anticancer potential and warrant further investigation.

13.5 FUTURE PROSPECTS

M. oleifera holds significant potential for future development due to its ease of cultivation, rapid growth and resilience in regions where other crops may fail. Its reliable yield supports sustainable large-scale biomass sourcing. Extracting its valuable phytochemicals at a commercial scale requires adopting efficient and eco-friendly methods. While traditional solvent extraction is common, supercritical fluid extraction offers a green alternative without toxic solvents. Additionally, the use of green solvents, such as deep eutectic solvents, presents a promising approach; however, large-scale separation processes must be developed to ensure the cost-effective recovery of extracted compounds. Optimising extraction techniques, including ultrasound-assisted and microwave-assisted methods, can further enhance yields and efficiency.

Ensuring the stability and controlled release of *M. oleifera*'s bioactive compounds is crucial for applications in pharmaceuticals, nutraceuticals and cosmetics. Addressing challenges such as the hygroscopic nature of leaf extracts through encapsulation and other stabilisation techniques will maximise efficacy.

In addition, *M. oleifera* seed proteins show immense promise as sustainable water treatment solutions. These bio-flocculant compounds effectively clarify water and remove contaminants, particularly benefiting communities in developing regions. Expanding large-scale extraction and commercialisation or establishing local extraction facilities will enhance accessibility and impact. Advancements in these areas will unlock the full potential of *M. oleifera* as a valuable resource for health, industry and environmental applications.

ABBREVIATIONS

ABTS – 2,2′-Azino-bis(3-ethylbenzothiazoline-6-sulfonic acid);
ADF – Acid detergent fiber;
ADIP – Acid detergent insoluble protein;
ADL – Acid detergent lignin;
BSA – Bovine serum albumin;
CP – Crude protein;
CXE – CO_2-expanded ethanol;
DE – Diosgenin equivalents;
DES – Deep eutectic solvent(s);
DPPH – 2,2-Diphenyl-1-picrylhydrazyl;
EtOH – Ethanol;
FRAP – Ferric reducing antioxidant power;
GAE – Gallic acid equivalents;
GC-MS – Gas chromatography–mass spectrometry;
HAE – Homogenizer-assisted extraction;
HPLC – High-performance liquid chromatography;
HPLC-ESI-QTOF-MS – High-performance liquid chromatography–electrospray ionization quadrupole time-of-flight mass spectrometry;
HPTLC – High-performance thin-layer chromatography;
ICP-OES – Inductively coupled plasma optical emission spectroscopy;
IC_{50} – Half-maximal inhibitory concentration;
IV – Iodine value;
LC-MS – Liquid chromatography–mass spectrometry;
MLF – Moringa leaf flour;
MO2 – Purified protein fraction from *Moringa oleifera* seeds;
Mo-CBP_3 – *Moringa oleifera* chitin-binding protein 3;
MPa – Megapascal;

MUFA – Monounsaturated fatty acid;
MWAE – Microwave-assisted extraction;
NDF – Neutral detergent fiber;
NF-κB – Nuclear factor kappa-light-chain-enhancer of activated B cells
PHWE – Pressurized hot water extraction;
PUFA – Polyunsaturated fatty acid;
QE – Quercetin equivalents;
SC-CO_2 – Supercritical carbon dioxide;
SCFE – Supercritical fluid extraction;
SFA – Saturated fatty acid;
SPI – Soybean protein isolate;
SPME – Solid-phase microextraction;
SSF – Solid-state fermentation;
SV – Saponification value;
TAE – Tannic acid equivalents;
TFC – Total flavonoid content;
TNF – Tumor necrosis factor;
TPC – Total phenolic content;
UHPLC-QTOF-MS/MS – Ultrahigh-performance liquid chromatography quadrupole time-of-flight tandem mass spectrometry;
USAE – Ultrasound-assisted extraction;
UV-vis – Ultraviolet-visible spectroscopy;
WHO – World Health Organization

REFERENCES

Abdulkarim, S. M., O. M. Lai, S. K. S. Muhammad, K. Long, and H. M. Ghazali. 2006. "Use of Enzymes to Enhance Oil Recovery During Aqueous Extraction of *Moringa oleifera* Seed Oil." *Journal of Food Lipids* 13 (2): 113–30. https://doi.org/10.1111/j.1745-4522.2006.00038.x.

Abdulkarim, S. M., K. Long, O. M. Lai, S. K. S. Muhammad, and H. M. Ghazali. 2005. "Some Physico-Chemical Properties of *Moringa oleifera* Seed Oil Extracted Using Solvent and Aqueous Enzymatic Methods." *Food Chemistry* 93 (2): 253–63. https://doi.org/10.1016/j.foodchem.2004.09.023.

Akinduti, P. A., V. Emoh-Robinson, H. F. Obamoh-Triumphant, Y. D. Obafemi, and T. T. Banjo. 2022. "Antibacterial Activities of Plant Leaf Extracts Against Multi-Antibiotic Resistant *Staphylococcus aureus* Associated with Skin and Soft Tissue Infections." *BMC Complementary Medicine and Therapies* 22 (1): 47. https://doi.org/10.1186/s12906-022-03527-y.

Al-Asmari, A. K., S. M. Albalawi, M. T. Athar, A. Q. Khan, H. Al-Shahrani, and M. Islam. 2015. "*Moringa oleifera* as an Anticancer Agent Against Breast and Colorectal Cancer Cell Lines." *PLoS One* 10 (8): e0135814. https://doi.org/10.1371/journal.pone.0135814.

Alhaithloul, Haifa A. S., Zakaria A. Mohamed, Abdullah A. Saber, Ibtisam Mohammed Alsudays, Mohamed A. Abdein, Mesfer M. Alqahtani, Noha G. AbuSetta, et al. 2024. "Performance Evaluation of *Moringa oleifera* Seeds Aqueous Extract for Removing *Microcystis Aeruginosa* and Microcystins from Municipal Treated-Water." *Frontiers in Bioengineering and Biotechnology* 11. https://doi.org/10.3389/fbioe.2023.1329431.

Ashfaq, M., Shahzad M. A. Basra, and Umair Ashfaq. 2012. "*Moringa*: A Miracle Plant for Agro-Forestry." *Journal of Agriculture and Social Sciences* 8 (3): 115–22.

Baldisserotto, A., R. Barbari, C. Tupini, R. Buzzi, E. Durini, I. Lampronti, S. Manfredini, BaldiniE. , and S. Vertuani. 2023. "Multifunctional Profiling of *Moringa oleifera* Leaf Extracts for Topical Application: A Comparative Study of Different Collection Time." *Antioxidants* 12 (2): 411. https://doi.org/10.3390/antiox12020411.

Belitz, H.-D., W. Grosch, and P. Schieberle. 2004. *Food Chemistry*. 3rd ed. Berlin, Germany: Springer. https://doi.org/10.1007/978-3-540-69934-7.

Bhadange, Yogesh A, Jitendra Carpenter, and Virendra Kumar Saharan. 2024. "A Comprehensive Review on Advanced Extraction Techniques for Retrieving Bioactive Components from Natural Sources." *ACS Omega* 9 (29): 31274–97. https://doi.org/10.1021/acsomega.4c02718.

Bhutada, Payal R., Ananda J. Jadhav, Dipak V. Pinjari, Parag R. Nemade, and Ratnesh D. Jain. 2016. "Solvent Assisted Extraction of Oil from *Moringa oleifera* Lam. Seeds." *Industrial Crops and Products* 82: 74–80. https://doi.org/10.1016/j.indcrop.2015.12.004.

Boukandoul, S., F. Zaidi, C. S. P. Santos, and S. Casal. 2023. "*Moringa oleifera* Oil Nutritional and Safety Impact on Deep-Fried Potatoes." *Foods* 12 (24): 4416. https://doi.org/10.3390/foods12244416.

Boulaadjoul, Soumia, Hassiba Zemmouri, Zoubida Bendjama, and Nadjib Drouiche. 2018. "A Novel Use of *Moringa oleifera* Seed Powder in Enhancing the Primary Treatment of Paper Mill Effluent." *Chemosphere* 206: 142–49.

Camacho, Franciele Pereira, Vânia Serrão Sousa, Rosângela Bergamasco, and Margarida Ribau Teixeira. 2017. "The Use of *Moringa oleifera* as a Natural Coagulant in Surface Water Treatment." *Chemical Engineering Journal* 313: 226–37.

Carrera-Chávez, J. M., E. E. Jiménez-Aguilar, T. P. Acosta-Pérez, J. A. Núñez-Gastélum, A. Quezada-Casasola, A. M. Escárcega-Ávila, and E. Orozco-Lucero. 2020. "Effect of *Moringa oleifera* Seed Extract on Antioxidant Activity and Sperm Characteristics in Cryopreserved Ram Semen." *Journal of Applied Animal Research* 48 (1): 114–20. https://doi.org/10.1080/09712119.2020.1741374.

Castro-López, Cecilia, Catarina Gonçalves, Janeth M. Ventura-Sobrevilla, Lorenzo M. Pastrana, Cristóbal N. Aguilar-González, and Guillermo C. G. Martínez-Ávila. 2020. "*Moringa oleifera*—Storage Stability, *in Vitro*-Simulated Digestion and Cytotoxicity Assessment of Microencapsulated Extract." *Processes* 8 (7): 770.

Chen, Xuexiang, Zhuobin Li, Sarah A Smith, Mingxiu Chen, Hanbin Liu, Jing Zhang, Lirong Tang, Jie Li, Qishan Liu, and Xian Wu. 2022. "Optimization of Supercritical CO_2 Extraction of *Moringa oleifera* Seed Oil Using Response Surface Methodological Approach and Its Antioxidant Activity." *Frontiers in Nutrition* 8: 829146. https://doi.org/10.3389/fnut.2021.829146.

Desta, Wendesen Mekonin, and Million Ebba Bote. 2021. "Wastewater Treatment Using a Natural Coagulant (*Moringa oleifera* Seeds): Optimization through Response Surface Methodology." *Heliyon* 7 (11): e08451. https://doi.org/10.1016/j.heliyon.2021.e08451

Dzuvor, Christian K. O., Sharadwata Pan, Charles Amanze, Prosper Amuzu, Charles Asakiya, and Francis Kubi. 2022. "Bioactive Components from *Moringa oleifera* Seeds: Production, Functionalities and Applications–a Critical Review." *Critical Reviews in Biotechnology* 42 (2): 271–93. https://doi.org/10.1080/07388551.2021.1931804.

El-Sherbiny, Gamal M., Amira J. Alluqmani, Islam A. Elsehemy, and Mohamed H. Kalaba. 2024. "Antibacterial, Antioxidant, Cytotoxicity, and Phytochemical Screening of *Moringa oleifera* Leaves." *Scientific Reports* 14 (1): 30485. https://doi.org/10.1038/s41598-024-80700-y.

Falowo, Andrew B., Felicitas E. Mukumbo, Idamokoro Emrobowansan M. , José M. Lorenzo, Anthony J. Afolayan, and Voster Muchenje. 2018. "Multi-Functional Application of *Moringa oleifera* Lam. In Nutrition and Animal Food Products: A Review." *Food Research International* 106: 317–34.

Feitosa, Paula Ribeiro Buarque, Tacila Rayane Jericó Santos, Nayjara Carvalho Gualberto, Narendra Narain, and Luciana Cristina Lins de Aquino Santana. 2020. "Solid-State Fermentation with *Aspergillus niger* for the Bio-Enrichment of Bioactive Compounds in *Moringa oleifera* (*Moringa*) Leaves." *Biocatalysis and Agricultural Biotechnology* 27: 101709. https://doi.org/10.1016/j.bcab.2020.101709.

Freire, Jose E. C., Ilka M. Vasconcelos, Frederico B. M. B. Moreno, Adelina B. Batista, Marina D. P. Lobo, Mirella L. Pereira, Joao P. M. S. Lima, Ricardo V. M. Almeida, Antonio J. S. Sousa, and Ana C. O. Monteiro-Moreira. 2015. "*Mo*-CBP_3, an Antifungal Chitin-Binding Protein from *Moringa oleifera* Seeds, Is a Member of the 2S Albumin Family." *PLoS One* 10 (3): e0119871. https://doi.org/10.1371/journal.pone.0119871.

Garde, William K., Steven G. Buchberger, David Wendell, and Margaret J. Kupferle. 2017. "Application of *Moringa oleifera* Seed Extract to Treat Coffee Fermentation Wastewater." *Journal of Hazardous Materials* 329: 102–9. https://doi.org/10.1016/j.jhazmat.2017.01.006.

Gassenschmidt, Ursula, Klaus D. Jany, Tauscher Bernhard, and Heinz Niebergall. 1995. "Isolation and Characterization of a Flocculating Protein from *Moringa oleifera* Lam." *Biochimica Et Biophysica Acta (BBA)-General Subjects* 1243 (3): 477–81.

Gopalakrishnan, Lakshmipriya, Kruthi Doriya, and Devarai Santhosh Kumar. 2016. "*Moringa oleifera*: A Review on Nutritive Importance and Its Medicinal Application." *Food Science and Human Wellness* 5 (2): 49–56. https://doi.org/10.1016/j.fshw.2016.04.001.

Gupta, Rajnish, Manas Mathur, Vijay K. Bajaj, Pawan Katariya, Sunita Yadav, Raka Kamal, and Radhey S. Gupta. 2012. "Evaluation of Antidiabetic and Antioxidant Activity of *Moringa oleifera* in Experimental Diabetes." *Journal of Diabetes* 4 (2): 164–71. https://doi.org/10.1111/j.1753-0407.2011.00173.x.

Jaglan, Poonam, Mukul Kumar, Deepika Kaushik, Ashwani Kumar, Dimitrios Argyropoulos, Fatih Oz, and Charalampos Proestos. 2024. "Optimization of the Extraction Process of *Moringa oleifera* Flower by Using Deep Eutectic Solvents (DES)." *Results in Chemistry* 7: 101445. https://doi.org/10.1016/j.rechem.2024.101445.

Kane, Cheikhou, Aïcha Bâ, Seïd Ali Malloum Mahamat, Nicolas Ayessou, Maryam Khadim Mbacké, and Codou Guèye Mar Diop. 2016. "Combination of Alum and Extracted *Moringa oleifera* Bioactive Molecules Powder for Municipal Wastewater Treatment." *International Journal of Biological and Chemical Sciences* 10 (4): 1918–29. https://doi.org/10.4314/ijbcs.v10i4.39.

Kooltheat, Nateelak, Rungnapa Pankla Sranujit, Pilaipark Chumark, Pachuen Potup, Nongnit Laytragoon-Lewin, and Kanchana Usuwanthim. 2014. "An Ethyl Acetate Fraction of *Moringa oleifera* Lam. Inhibits Human Macrophage Cytokine Production Induced by Cigarette Smoke." *Nutrients* 6 (2): 697–710. https://doi.org/10.3390/nu6020697.

Lakka, Achillia, Spyros Grigorakis, Olga Kaltsa, Ioanna Karageorgou, Georgia Batra, Eleni Bozinou, Stavros Lalas, and Dimitris P. Makris. 2020. "The Effect of Ultrasonication Pretreatment on the Production of Polyphenol-Enriched Extracts from *Moringa oleifera* L. (Drumstick Tree) Using a Novel Bio-Based Deep Eutectic Solvent." *Applied Sciences* 10 (1): 220. https://doi.org/10.3390/app10010220.

Lee, Jin-Hyung, Yong-Guy Kim, Jae Gyu Park, and Jintae Lee. 2017. "Supercritical Fluid Extracts of *Moringa oleifera* and Their Unsaturated Fatty Acid Components Inhibit Biofilm Formation by *Staphylococcus Aureus*." *Food Control* 80: 74–82. https://doi.org/10.1016/j.foodcont.2017.04.035.

Leone, Alessandro, Antonella Spada, Anna Battezzati, Jessica Schiraldi, Andrea Aristil, and Bruno Bertoli. 2015. "Cultivation, Genetic, Ethnopharmacology, Phytochemistry and Pharmacology of *Moringa oleifera* Leaves: An Overview." *International Journal of Molecular Sciences* 16 (6): 12791–835. https://doi.org/10.3390/ijms160612791.

Leone, Alessandro, Alberto Spada, Alberto Battezzati, Alberto Schiraldi, Junior Aristil, and Simona Bertoli. 2016. "*Moringa oleifera* Seeds and Oil: Characteristics and Uses for Human Health." *International Journal of Molecular Sciences* 17 (12): 2141. https://doi.org/10.3390/ijms17122141.

Madrona, GrasieleScaramal, Ivanise Guilherme Branco, Vanessa JurcaSeolin, Benício de Abreu Alves Filho, Marcia Regina Fagundes-Klen, and Rosangela Bergamasco. 2012. "Evaluation of Extracts of *Moringa oleifera* Lam. Seeds Obtained with NaCl and Their Effects on Water Treatment." *Acta Scientiarum. Technology* 34 (3): 289–93.

Makkar, H. P. S., and K. Becker. 1997. "Nutrients and Antiquality Factors in Different Morphological Parts of the *Moringa oleifera* Tree." *The Journal of Agricultural Science* 128 (3): 311–22. https://doi.org/10.1017/S0021859697004292.

Mani, S., S. Jaya, and R. Vadivambal. 2007. "Optimization of Solvent Extraction of *Moringa* (*Moringa oleifera*) Seed Kernel Oil Using Response Surface Methodology." *Food and Bioproducts Processing* 85 (4): 328–35. https://doi.org/10.1205/fbp07075.

Mat Yusoff, Masni, Michael H. Gordon, Onyinye Ezeh, and Keshavan Niranjan. 2016. "Aqueous Enzymatic Extraction of *Moringa oleifera* Oil." *Food Chemistry* 211: 400–8. https://doi.org/10.1016/j.foodchem.2016.05.050.

Morgan, CR, C Opio, and S Migabo. 2020. "Chemical Composition of *Moringa* (*Moringa oleifera*) Root Powder Solution and Effects of Moringa Root Powder on *e. Coli* Growth in Contaminated Water." *South African Journal of Botany* 129: 243–48. https://doi.org/10.1016/j.sajb.2019.07.020.

Moyo, Busani, Patrick Julius Masika, and VosterMuchenje. 2012. "Antimicrobial Activities of *Moringa oleifera* Lam. Leaf Extracts." *African Journal of Biotechnology* 11 (11): 2797–802. https://doi.org/10.5897/AJB10.686.

Ndabigengesere, Anselme, and K Subba Narasiah. 1998. "Quality of Water Treated by Coagulation Using *Moringa oleifera* Seeds." *Water Research* 32 (3): 781–91.

Nguyen, Hoang N., Pag-asa D. Gaspillo, Julius B. Maridable, Roberto M. Malaluan, Hirofumi Hinode, Chris Salim, and Ha K. P. Huynh. 2011. "Extraction of Oil from *Moringa oleifera* Kernels Using Supercritical Carbon Dioxide with Ethanol for Pretreatment: Optimization of the Extraction Process." *Chemical Engineering and Processing: Process Intensification* 50 (11): 1207–13. https://doi.org/10.1016/j.cep.2011.08.006.

Okuda, Tetsuji, Aloysius U Baes, WataruNishijima, and Mitsumasa Okada. 1999. "Improvement of Extraction Method of Coagulation Active Components from *Moringa oleifera* Seed." *Water Research* 33 (15): 3373–78.

Oladipo, Babatunde, and Eriola Betiku. 2019. "Process Optimization of Solvent Extraction of Seed Oil from *Moringa oleifera*: An Appraisal of Quantitative and Qualitative Process Variables on Oil Quality Using d-Optimal Design." *Biocatalysis and Agricultural Biotechnology* 20: 101187. https://doi.org/10.1016/j.bcab.2019.101187.

Oladoja, NurudeenAbiola, and Gang Pan. 2015. "Modification of Local Soil/Sand with *Moringa oleifera* Extracts for Effective Removal of Cyanobacterial Blooms." *Sustainable Chemistry and Pharmacy* 2: 37–43. https://doi.org/10.1016/j.scp.2015.08.003.

Orimolade, Temitope, Ngoc-Tram Le, Lyle Trimble, Bandaru Ramarao, and Sitaraman Krishnan. 2025. "Light Scattering Study of Algal Floc Growth and Structure: Alum Vs. Polymeric Plant-Derived Flocculant." *Soft Matter* 21: 561–72. https://doi.org/10.1039/D4SM00837E.

Pareek, Ashutosh, Malvika Pant, Madan Mohan Gupta, Pushpa Kashania, Yashumati Ratan, Vivek Jain, Aaushi Pareek, and Anil A Chuturgoon. 2023. "*Moringa oleifera*: An Updated Comprehensive Review of Its Pharmacological Activities, Ethnomedicinal, Phytopharmaceutical Formulation, Clinical, Phytochemical, and Toxicological Aspects." *International Journal of Molecular Sciences* 24 (3): 2098. https://doi.org/10.3390/ijms24032098.

Patel, Himanshu, and R. T. Vashi. 2012. "Removal of Congo Red Dye from Its Aqueous Solution Using Natural Coagulants." *Journal of Saudi Chemical Society* 16 (2): 131–36. https://doi.org/10.1016/j.jscs.2010.12.003.

Pereira, Juliana M. G., Franciele Leila G. Viell, Patricia C. de Lima, Evandro Silva, Eduardo J. Pilau, CorrêaRúbia C. G. , Evandro Bona, and Angélica M. S. Vieira. 2021. "Optimization of the Extraction of Antioxidants from *Moringa* Leaves: A Comparative Study Between Ultrasound- and Ultra-Homogenizer-Assisted Extractions." *Journal of Food Processing and Preservation* 45 (6): e15512. https://doi.org/10.1111/jfpp.15512.

Pop, OanaLelia, Andreea Diana Kerezsi, and Călina Ciont. 2022. "A Comprehensive Review of *Moringa oleifera* Bioactive Compounds—Cytotoxicity Evaluation and Their Encapsulation." *Foods* 11 (23): 3787.

Prabakaran, Mayakrishnan, Seung-Hyun Kim, AsokanSasireka, Murugesan Chandrasekaran, and Ill-Min Chung. 2018. "Polyphenol Composition and Antimicrobial Activity of Various Solvent Extracts from Different Plant Parts of *Moringa oleifera*." *Food Bioscience* 26: 23–29. https://doi.org/10.1016/j.fbio.2018.09.003.

Pramanik, Amar, and Syed Sirajul Islam. 1998. "Chemical Investigation of Aqueous Extract of the Mature and Premature Flowers of *Moringa oleifera* (Sajina) and Structural Studies of a Polysaccharide Isolated from Its Premature Flowers." *Indian Journal of Chemistry* 37B: 676–82.

Prasad, R. Krishna. 2009. "Color Removal from Distillery Spent Wash Through Coagulation Using *Moringa oleifera* Seeds: Use of Optimum Response Surface Methodology." *Journal of Hazardous Materials* 165 (1): 804–11. https://doi.org/10.1016/j.jhazmat.2008.10.068.

Ramachandran, C., K. V. Peter, and P. K. Gopalakrishnan. 1980. "Drumstick (*Moringa oleifera*): A Multipurpose Indian Vegetable." *Economic Botany* 34 (3): 276–83. http://www.jstor.org/stable/4254186.

Ramachandran, C., R. Sudha Rani, K. Lavanya, S. Nivetha, and A. Usha. 2017. "Optimization of Shelf Stability of Sugarcane Juice with Natural Preservatives." *Journal of Food Processing and Preservation* 41 (1): e12868. https://doi.org/10.1111/jfpp.12868.

Revelou, Panagiota-Kyriaki, Marinos Xagoraris, Athanasia Alexandropoulou, Charalabos D. Kanakis, George K. Papadopoulos, Christos S. Pappas, and Petros A. Tarantilis. 2021. "Chemometric Study of Fatty Acid Composition of Virgin Olive Oil from Four Widespread Greek Cultivars." *Molecules* 26 (14): 4151. https://doi.org/10.3390/molecules26144151.

Rodríguez-Pérez, C., J. A. Mendiola, R. Quirantes-Piné, E. Ibáñez, and A. Segura-Carretero. 2016. "Green Downstream Processing Using Supercritical Carbon Dioxide, CO_2-Expanded Ethanol and Pressurized Hot Water Extractions for Recovering Bioactive Compounds from *Moringa oleifera* Leaves." *The Journal of Supercritical Fluids* 116: 90–100. https://doi.org/10.1016/j.supflu.2016.05.009.

Ruttarattanamongkol, Khanitta, Susanne Siebenhandl-Ehn, Matthias Schreiner, and Angelika M. Petrasch. 2014. "Pilot-Scale Supercritical Carbon Dioxide Extraction, Physico-Chemical Properties and Profile Characterization of *Moringa oleifera* Seed Oil in Comparison with Conventional Extraction Methods." *Industrial Crops and Products* 58: 68–77. https://doi.org/10.1016/j.indcrop.2014.03.020.

Saa, Romuald Willy, Edith NigFombang, Elie Baudelaire Ndjantou, and Nicolas YanouNjintang. 2019. "Treatments and Uses of Moringa Oleifera Seeds in Human Nutrition: A Review." *Food Science & Nutrition* 7 (6): 1911–19. https://doi.org/10.1002/fsn3.1057.

Saad El-Din, Marwa I., Heba N. Gad EL-Hak, Mahi A. Ghobashy, and Ranwa A. Elrayess. 2023. "Parasitological and Histopathological Studies to the Effect of Aqueous Extract of *Moringa oleifera* Lam. Leaves Combined with Praziquantel Therapy in Modulating the Liver and Spleen Damage Induced by Schistosoma Mansoni to Male Mice." *Environmental Science and Pollution Research* 30 (6): 15548–60. https://doi.org/10.1007/s11356-022-23098-2.

Saini, Ramesh Kumar, Iyyakkannu Sivanesan, and Young-Soo Keum. 2016. "Phytochemicals of *Moringa oleifera*: A Review of Their Nutritional, Therapeutic and Industrial Significance." *3 Biotech* 6:1–14. https://doi.org/10.1007/s13205-016-0526-3.

Sengupta, Mita E., Bernard Keraita, Annette Olsen, Osei K. Boateng, Stig M. Thamsborg, Guðný R. Pálsdóttir, and Anders Dalsgaard. 2012. "Use of *Moringa oleifera* Seed Extracts to Reduce Helminth Egg Numbers and Turbidity in Irrigation Water." *Water Research* 46 (11): 3646–56. https://doi.org/10.1016/j.watres.2012.04.011.

Sharma, Ram Karan, and Vaidya Bhagwan Dash. 2009. *Caraka Saṃhitā: Text with English Translation & Critical Exposition Based on Cakrapāṇi Datta's Āyurveda Dīpikā*. Varanasi: Chowkhamba Sanskrit Series.

Sreelatha, S., and P. R. Padma. 2009. "Antioxidant Activity and Total Phenolic Content of *Moringa oleifera* Leaves in Two Stages of Maturity." *Plant Foods for Human Nutrition* 64: 303–11. https://doi.org/10.1007/s11130-009-0141-0.

Thangaiah, Arumugam, Sandeep Gunalan, Vijayalatha Kulandaivelu Rathnasamy, Rajasekar Aruliah, Mohammad S. AlSalhi, SandhanasamyDevanesan, Rajaram Rajamohan, and Tabarak Malik. 2024. "Optimization of Ultrasound-Assisted Phytomolecules Extraction from *Moringa* Leaves (*Moringa oleifera* Lam.) Using Response Surface Methodology." *Cogent Food & Agriculture* 10 (1): 2309834. https://doi.org/10.1080/23311932.2024.2309834.

Vats, Sharad, and Tanya Gupta. 2017. "Evaluation of Bioactive Compounds and Antioxidant Potential of Hydroethanolic Extract of *Moringa oleifera* Lam. From Rajasthan, India." *Physiology and Molecular Biology of Plants* 23 (1): 239–48. https://doi.org/10.1007/s12298-016-0407-6.

Wei, Ping, Yue Zhang, Yao-Ying Wang, Jin-Feng Dong, Zi-Hao Lin, Wu Li, Lei Liu, et al. 2023. "Efficient Extraction and Excellent Activity of Flavonoid from *Moringa oleifera* Leaves and Its Microencapsulation." *LWT* 184: 115021. https://doi.org/10.1016/j.lwt.2023.115021.

WHO. 2007. *Protein and Amino Acid Requirements in Human Nutrition: Report of a Joint FAO/WHO/UNU Expert Consultation*. Vol. 935. WHO Technical Report Series. Geneva, Switzerland: World Health Organization. https://iris.who.int/bitstream/handle/10665/43411/WHO_TRS_935_eng.pdf.

Zhao, Suwei, and Dongke Zhang. 2014. "An Experimental Investigation into the Solubility of *Moringa oleifera* Oil in Supercritical Carbon Dioxide." *Journal of Food Engineering* 138: 1–10. https://doi.org/10.1016/j.jfoodeng.2014.03.031.

14 *Moringa oleifera* Quality Control Parameters

Shiv Shankar Shukla, Manisha Verma, Ravindra Kumar Pandey and Bina Gidwani

14.1 INTRODUCTION

Native to the Indian subcontinent, *Moringa oleifera* is a fast-growing, drought-resistant tree that is also referred to as the horseradish tree, drumstick tree, or just moringa. It belongs to the Moringaceae family and is commonly grown in tropical and subtropical regions of Asia, Africa and Latin America. Respected for its nutritional and medicinal properties, *M. oleifera* is often referred to as a 'miracle tree'. In traditional herbal medicine, the plant's edible parts – leaves, pods, seeds, flowers and roots – are used. The leaves are particularly rich in calcium, potassium, protein and the vitamins A, C and E, making them a helpful supplement in the fight against malnutrition (Seifu and Teketay 2020). In addition to its nutritional value, *M. oleifera* has antibacterial, anti-inflammatory and antioxidant properties. In addition to being a natural pesticide, it can be used to enhance soil and purify water. Because of its ability to flourish in unfavourable soil conditions and resilience to harsh climates, it is a valuable and sustainable resource in areas where environmental degradation and food insecurity are prevalent. All things considered, *M. oleifera* is a highly versatile plant with significant benefits for human health, the environment and agriculture. *M. oleifera* is an effective treatment for malnutrition. Moringa is rich in nutrients due to the many essential phytochemicals present in its leaves, pods and seeds. Moringa is actually said to have 25 times more iron than spinach, 17 times more calcium than milk, nine times more protein than yoghurt, 15 times more potassium than bananas, seven times more vitamin C than oranges and ten times more vitamin A than carrots (Gopalakrishnan et al. 2016).

14.2 THE IMPORTANCE OF *M. OLEIFERA*

M. oleifera holds immense importance due to its wide range of nutritional, medicinal and environmental applications. Rich in essential nutrients such as vitamins, minerals, antioxidants and amino acids, it plays a significant role in the fight against malnutrition, especially in developing countries. The leaves are often eaten in powdered form or added to foods to boost their nutritional content. In both traditional and modern medicine, moringa is used to treat conditions such as diabetes, high blood pressure, inflammation and digestive problems because of its anti-inflammatory, antimicrobial and antioxidant properties. Apart from their health benefits, moringa seeds have the ability to purify water by removing impurities and bacteria, making them beneficial in areas with scarce water resources. By being used in agriculture as a natural fertiliser and animal feed, it promotes sustainable farming. Its ability to withstand severe weather conditions and unfavourable soil conditions highlights its significance in reducing food insecurity and advancing environmental preservation. All things considered, *M. oleifera* is a multifunctional tree that greatly enhances sustainability, economic growth and health (Boopathi et al. 2021). The moringa tree has many medicinal uses, including prevention and treatment. Its bark, sap, roots, leaves, seeds, oil and flowers are used in traditional medicine in many countries for gastric problems, catarrh, cancer, gastric ulcers, skin

DOI: 10.1201/9781003617860-16

disorders, lowering blood sugar, increasing bone density, neurological disorders, diabetes, fatigue, increased lactation, hayfever, impotence, oedema, cramps, haemorrhoids, headaches and sore gums, as well as to strengthen the eyes, brain, liver, gall, digestive, respiratory and immune systems and as a blood builder and cleanser. The leaves were traditionally applied as a poultice to the abdomen to rid the body of intestinal worms. A leaf infusion is used as an eye wash to treat conjunctivitis (Anwar et al. 2007; Mahajan et al. 2007).

14.3 PLANT PROFILE

14.3.1 Botanical Description

The small to medium-sized, deciduous tree *M. oleifera* grows quickly and usually reaches a height of 5–10 m. It features a straight or slightly twisted trunk, frequently with drooping branches, and soft, whitish-grey bark. The tree is well-suited to arid and semi-arid climates because of its deep taproot system. The compound, tripinnate (feather-like) leaves have a maximum length of 60 cm. They have several tiny, oval-shaped leaflets grouped in opposing pairs along a central stem, and they are bright green in colour. These leaflets have complete margins and are thin and smooth.

The fragrant, creamy-white flowers of *M. oleifera* measure 1–1.5 cm in diameter. The flowers are bisexual, zygomorphic (bilaterally symmetrical) and pollinated by insects, especially bees.

The fruit is a long, thin, triangular pod that can reach a length of 45 cm. It is commonly referred to as a drumstick. It has round-to-oval, dark-brown seeds with wings. The seeds have a hard shell and can germinate and produce oil. *M. oleifera* is known for its quick growth, adaptability and variety of morphological traits, all of which support its many applications in agriculture, medicine and nutrition (Abubkar et al. 2021; Boopathi et al. 2021).

14.3.2 Taxonomy and Nomenclature

M. oleifera is a member of the class Eudicots, phylum Angiosperms (flowering plants) and kingdom Plantae. It belongs to the Brassicales order, which also contains other economically significant plants such as capers, mustard and cabbage. *M. oleifera* is the most extensively grown and researched of the 13 species in the genus *Moringa*, which makes up the family Moringaceae. Originally from the Himalayan foothills of northwest India, this quick-growing, drought-resistant tree has since spread throughout the world's tropical and subtropical zones. The Latin word 'oleifera' (oil-producing) and the Tamil word 'murungai' (twisted pod) were combined to create its scientific name, *Moringa oleifera*, which refers to its edible seed oil. Taxonomically, it is closely related to other *Moringa* species, such as *M. stenopetala* (African moringa), but *M. oleifera* remains the most nutritionally and commercially significant. Its classification reflects its unique biological and agricultural importance as a versatile plant used for food, medicine and environmental benefits (Mallenakuppe et al. 2015; Sharmin et al. 2019). The taxonomical classification of *Moringa* is listed in Table 14.1.

14.3.3 Distribution and Habitat

In northwest India, specifically in areas such as Punjab, Himachal Pradesh and Uttar Pradesh, *M. oleifera* is indigenous to the foothills of the Himalayas. Africa, Southeast Asia, Latin America, the Caribbean and the Pacific Islands are just a few of the tropical and subtropical regions where it has been introduced and naturalised because of its exceptional adaptability. The tree is a valuable crop in areas with limited water resources because it can withstand poor soils and protracted droughts and thrive in arid and semi-arid climates. Although it can withstand partial shade, it thrives in well-drained sandy or loamy soils with a pH between 6.0 and 7.0. It also prefers full sunlight. *M. oleifera* is susceptible to frost and wet conditions, even though it can tolerate temperatures as high as 48°C (118°F). Its ability to grow in low-fertility soils and resist pests has contributed to its

TABLE 14.1
Taxonomic Classification of *M. oleifera*

S. no	Taxonomic rank	Classification
1.	Kingdom	Plantae
2.	Subkingdom	Tracheobionta (Vascular plants)
3.	Division	Magnoliophyta (Angiosperms)
4.	Class	Magnoliopsida (Dicotyledons)
5.	Order	Brassicales
6.	Family	Moringaceae
7.	Genus	*Moringa*
8.	Species	*Moringa oleifera* Lam.

widespread cultivation in Africa (Nigeria, Ethiopia, Kenya), the Philippines, Indonesia, Central America (Nicaragua, Haiti) and even parts of the southern United States (Florida, Texas). In many of these regions, it is grown not only for its nutritional and medicinal benefits but also for soil stabilisation, agroforestry and livestock feed, demonstrating its ecological and agricultural versatility (Bania et al. 2023; Popoola et al. 2013). Although *M. oleifera* can thrive in regions with 250–2600 mm of annual rainfall, it needs 700–2200 mm of rainfall per year to grow to its full potential. *M. oleifera* is a hardy crop that thrives in semi-arid and arid areas, as evidenced by its minimum annual rainfall requirements of 250 mm (Zaku et al. 2015).

14.3.4 Morphological Characteristics

14.3.4.1 Leaves

M. oleifera has compound, tripinnate (feather-like) leaves that alternately grow along the stem. Each large leaf, which is made up of many small, smooth-bordered, oval to obovate-shaped leaflets, can reach a maximum length of 60 cm. The leaflets have a soft, delicate texture and a bright-to-dark green upper surface with a lighter shade on the underside. These leaves are a valuable dietary supplement in tropical and subtropical regions where the tree thrives because they are rich in essential vitamins (A, C and B complex), minerals (calcium, iron and potassium) and antioxidants. This is in addition to their morphological qualities. Moringa is known as a 'miracle tree' for its ability to prevent malnutrition due to its rapid growth and high nutritional value (Zhigila et al. 2015).

14.3.4.2 Flowers

The delicate creamy-white to pale yellow flowers of *M. oleifera* are bisexual and zygomorphic, or bilaterally symmetrical. They develop along the stems in loose, branching clusters called axillary panicles, which measure 1–1.5 cm in diameter. Because of their gentle, sweet scent, these blooms draw pollinators, especially bees and other insects, which helps with cross-pollination. Along with its nutritional and therapeutic advantages, the tree's agricultural and environmental value is increased by the flowers' aesthetic appeal, delicate fragrance and ecological function in assisting pollinators. The tree's quick reproductive cycle in tropical climates is highlighted by the short lifespan, which is usually one to two days (Ruiz Hernandez et al. 2021).

14.3.4.3 Seeds and Pods

Often referred to as 'drumsticks', *M. oleifera*'s pods (fruits) are long, thin and triangular in cross-section, usually ranging in length from 20 to 45 cm. They have a smooth, slightly ridged texture and are bright green when immature; when they mature, they become woody and brown. Round to oval seeds (about 1 cm in diameter) with a hard, dark brown to black outer coat and characteristic papery white wings that help with wind dispersal are found inside the pods. Because of their coagulant properties, these seeds are used in traditional water purification, are rich in edible oil (of a quality comparable to olive oil) and are increasingly being investigated as a sustainable source for biodiesel production. Moringa's reputation as a multipurpose 'miracle tree' is further enhanced by its high nutritional and industrial value (Islam et al. 2020).

14.3.4.4 Bark and Roots

M. oleifera trees have grey to whitish bark that is smooth when young but rough and corky as they get older. When cut, this moderately thick substance releases a gummy resin that has long been utilised in adhesives and basic medicine. Usually pale brown to off-white in colour, the root system consists of a thick, fleshy taproot with spreading lateral branches. The tree is known as a 'horseradish tree' because of the strong, horseradish-like smell that its damaged roots release. In addition to being drought-resistant and anchoring, the roots contain bioactive compounds with antibacterial properties; however, their use in medicine should be done with caution as high dosages may be toxic (Mobarak et al. 2017).

14.3.5 *M. oleifera* Collection and Cultivation

M. oleifera is a fast-growing, drought-resistant tree that thrives in tropical and subtropical climates. It can be propagated either through seeds or stem cuttings, with seed propagation being more common for large-scale cultivation. Moringa prefers well-drained sandy or loamy soil with a neutral to slightly acidic pH and requires full sunlight for optimal growth. The seeds are usually sown directly into the ground or in nursery beds and transplanted when they are about 30–40 cm in height. Usually, plants are 1–2 m apart. Once established, the plant requires little water, making it suitable for arid areas. Although moringa thrives in poor soil, the soil can be improved by adding compost or organic manure. The tree can be harvested several times a year and starts to bloom six to eight months after planting. While pods are gathered when they are soft enough to use as vegetables or allowed to mature in order to extract the seed oil, leaves are hand-picked or cut with stems. Frequent pruning improves leaf yield and promotes bushier growth. To maintain the nutritional value of the harvested parts, proper drying and storage methods are crucial (Gandji et al. 2018). Various species of *M. oleifera* are shown in Table 14.2.

TABLE 14.2
Various Species of *M. oleifera*

S. no	Species Name	Common Name	Native Region
1	*Moringa oleifera*	Drumstick tree, horseradish tree	Indian subcontinent
2	*Moringa stenopetala*	African moringa	East Africa (Kenya, Ethiopia)
3	*Moringa peregrina*	Wild moringa	Arabian Peninsula, NE Africa
4	*Moringa concanensis*	Konkan moringa	Western India
5	*Moringa drouhardii*	Bottle tree	Madagascar
6	*Moringa hildebrandtii*	Hildebrandt's moringa	Madagascar (extinct in the wild)
7	*Moringa borziana*	–	Somalia
8	*Moringa arborea*	–	Kenya, Somalia
9	*Moringa longituba*	–	Northeastern Africa
10	*Moringa pygmaea*	–	Somalia
11	*Moringa ovalifolia*	Ghost tree	Namibia, Angola
12	*Moringa rivae*	–	Kenya, Ethiopia
13	*Moringa ruspoliana*	–	East Africa (Kenya, Ethiopia, Somalia)

TABLE 14.3
Traditional Remedies of Various Plant Parts of *M. oleifera*

Part used	Traditional remedy
Leaves	Fever, inflammation, digestive problems, skin infections and anaemia.
Seeds	Antimicrobial, purgative and used to treat urinary tract issues.
Bark	Snake bites, joint pain and digestive tonic.
Flowers	Used to help nursing moms produce more milk.
Root	Used as a cardiac tonic (in very small amounts because higher doses are toxic) and to treat epilepsy and asthma.

14.4 TRADITIONAL AND ETHNOBOTANICAL USES OF *M. OLEIFERA*

Moringa, often referred to as the 'miracle tree', has been traditionally used for centuries in various cultures for its nutritional and medicinal properties. Moringa leaves, pods and seeds have been used in Ayurvedic medicine to treat conditions such as infections, inflammation and digestive issues. The leaves are commonly consumed as a nutrient-rich food, either cooked or dried and powdered, to help combat malnutrition due to their high vitamin and mineral content. The roots and bark of moringa are used in African traditional medicine to treat diabetes, reduce fever and ease pain, while the oil extracted from the seeds is applied topically for skin and hair care issues. Many tropical regions also include moringa in their daily diets as a natural energy boost and immune system booster. Its numerous applications in traditional practices underscore its importance as a plant that promotes health and well-being. Traditional remedies of various plant parts of *M. Oleifera* are shown in Table 14.3 (Mishra et al. 2011; Senthilkumar et al. 2018).

14.5 PHYTOCHEMICAL SCREENING OF *M. OLEIFERA*

M. oleifera's phytochemical screening identifies a wide range of bioactive substances that give it its many therapeutic benefits. Numerous significant primary and secondary metabolites have been identified in various plant parts, including leaves, seeds, bark, roots and pods. The phytochemical constituents of *M. oleifera* are shown in Table 14.4 (Bagheri et al. 2020).

14.6 NUTRITIONAL PROFILE, MICROSCOPY AND MACROSCOPY OF *M. OLEIFERA*

14.6.1 Nutritional Profile

Moringa is a nutritional powerhouse, packed with an impressive array of vitamins, minerals and bioactive compounds. In particular, the leaves are extremely rich in essential nutrients, including higher concentrations of vitamin C (more abundant than oranges) and vitamin A (β-carotene), which support immune and visual function. Compared with common sources like milk and spinach, they also contain higher concentrations of calcium, which is vital for bone health, and iron, which helps prevent anaemia. As a good source of protein that contains all nine essential amino acids, moringa is also good for muscle growth and repair. Additionally rich in potassium, magnesium and dietary fibre, the leaves support healthy digestion, nerve function and the heart. In addition to essential nutrients, moringa has anti-inflammatory and beneficial fatty acid compounds, as well as potent antioxidants that fight oxidative stress, such as quercetin and chlorogenic acid. Moringa is a very powerful supplement for strengthening immunity, preventing malnutrition and improving general health because of its rich nutritional profile (Castillo et al. 2017; Tekle et al. 2015).

TABLE 14.4
Phytochemical Constituents of *M. Oleifera*

Phytochemical group	Inference	Biological significance
Alkaloids	Present	Analgesic, antimicrobial
Flavonoids	Abundant	Antioxidant, anti-inflammatory
Tannins	Present	Astringent, antimicrobial
Saponins	Present	Immunostimulant, cholesterol-lowering
Terpenoids	Present	Anticancer, antimicrobial
Phenolic compounds	High concentration	Antioxidant, cardioprotective
Glycosides	Present	Cardiotonic, anti-inflammatory
Steroids	Detected	Hormonal regulation, anti-inflammatory
Vitamins (A, C, E, B)	Abundant in leaves	Nutritional, antioxidant
Minerals (Ca, Fe, Zn)	High concentration	Hematopoietic, bone health
Isothiocyanates	Present	Anticancer, antidiabetic

14.6.2 Microscopic Properties of *M. oleifera*

Microscopic examination of *M. oleifera* reveals distinct anatomical features in its leaves, seeds and roots, essential for identification and quality control. The leaf displays a single-layered epidermis with anisocytic stomata, non-glandular trichomes and a mesophyll differentiated into palisade and spongy parenchyma, facilitating photosynthesis and gas exchange. The seeds exhibit a multilayered, lignified seed coat, an oil- and protein-rich endosperm and a well-developed embryo with starch-containing cotyledons. The root and bark feature cork cells in the outer layer, a cortex with sclereids, well-organised phloem and xylem and calcium oxalate crystals – often as raphides or prisms – scattered in the parenchyma. These microscopic characteristics not only aid in authentication but also reflect the plant's structural adaptations and bioactive storage, supporting its pharmacological applications. Physicochemical properties of *M. oleifera* are shown in Table 14.5 (Fatmawati et al. 2021).

14.7 PHARMACOLOGICAL ACTIVITIES

The plant's rich composition of bioactive compounds, such as flavonoids, polyphenols, glucosinolates and alkaloids, gives it a wide range of pharmacological activities, including strong antioxidant properties that help scavenge free radicals and lower oxidative stress, which is linked to chronic diseases such as diabetes and cancer; anti-inflammatory properties that inhibit pro-inflammatory cytokines and enzymes like COX-2, making it beneficial for conditions like arthritis and cardiovascular

TABLE 14.5
Physicochemical Properties of *M. Oleifera*

Physicochemical parameters	Acceptable range
Moisture content (LOD)	Not more than 5–10%
Total ash	Not more than 10%
Acid-insoluble ash	Not more than 1–2%
Water-soluble ash	2–4%
Extractive values	Water-soluble: 20–30%; alcohol-soluble: 10–20%
pH of 1% solution	~6–7

disease; antidiabetic effects due to its ability to improve insulin sensitivity and reduce hepatic glucose production; and its antimicrobial activity against bacteria, fungi and viruses supports its historical use in the treatment of infections. Moringa also has hypolipaemic qualities that help lower cholesterol and hepatoprotective effects that protect the liver from toxins. Its anticancer potential is also supported by research, with compounds such as niazimicin exhibiting chemopreventive effects. These various pharmacological effects highlight the therapeutic benefits of moringa in both conventional and alternative medicine.

14.7.1 Antioxidant Activity of *M. oleifera*

M. oleifera has strong antioxidant activity that helps combat oxidative stress and prevents chronic diseases because of its high polyphenol, flavonoid and vitamin content. It supports its role as a nutraceutical by lowering lipid peroxidation and DNA damage and boosting cellular antioxidant defences; however, more clinical validation is needed (Santos et al. 2012).

14.7.2 Antimicrobial Activity of *M. oleifera*

M. oleifera demonstrates broad-spectrum antimicrobial activity due to bioactive compounds like isothiocyanates and flavonoids, effectively inhibiting bacteria, fungi and viruses. Its mechanism includes disrupting microbial membranes and enzymes, supporting its traditional use, though clinical validation is still required (Moyo et al. 2012).

14.7.3 Anti-Inflammatory Activity of *M. oleifera*

M. oleifera demonstrates significant anti-inflammatory effects by suppressing mediators like TNF-α, interleukins and COX-2 through its flavonoids and isothiocyanates. Supported by both lab and animal studies, it shows promise as a natural treatment for inflammatory conditions like arthritis and asthma (Koheil et al. 2011).

14.7.4 Antidiabetic Activity of *M. oleifera*

M. oleifera shows strong antidiabetic potential by improving insulin sensitivity, enhancing secretion and reducing glucose absorption, thanks to its rich phytochemicals such as isothiocyanates and flavonoids. Additionally, it lowers blood sugar and HbA1c levels, protects pancreatic β-cells and supports its use as a natural adjunct in diabetes management (Fatoumata et al. 2020).

14.7.5 Antihypertensive Activity of *M. oleifera*

M. oleifera exhibits strong antihypertensive effects due to compounds such as nitrile and isothiocyanate glycosides that relax blood vessels and improve circulation. Its potassium content and antioxidant and diuretic properties further support blood pressure regulation, making it a promising natural remedy for hypertension (Ma et al. 2021).

14.7.6 Anticancer Activity of *M. oleifera*

M. oleifera exhibits notable anticancer potential through bioactive compounds like niazimicin and quercetin, which induce apoptosis and inhibit tumour growth. Its antioxidant and anti-inflammatory effects support its use against cancers such as breast, lung, colon and prostate, though clinical validation is still required (Vasanth et al. 2014).

14.7.7 Antianaemic Activity of *M. oleifera*

M. oleifera helps combat anaemia due to its high iron, vitamin C and folate content, which support red blood cell production and iron absorption. Regular intake improves haemoglobin levels and alleviates anaemia symptoms, making it a natural remedy, especially for at-risk groups like children and pregnant women (Manjegowda et al. 2024).

14.7.8 Anthelmintic Activity of *M. oleifera*

M. oleifera shows strong anthelmintic activity due to compounds like tannins, saponins and flavonoids that disrupt parasitic worms' metabolism and structure. Traditionally used for deworming, it offers a natural, affordable alternative to synthetic drugs, though more clinical research is needed to confirm its efficacy and dosing (Rastogi et al. 2009).

14.7.9 Skin Protection Activity of *M. oleifera*

M. oleifera supports skin health through its antioxidants, essential fatty acids and anti-inflammatory and antimicrobial compounds. Used in both traditional and modern skincare, it helps combat ageing, acne and irritation while promoting hydration and healing, making it ideal for all skin types (Athikomkulchai et al. 2020).

14.8 ANALYSIS OF *M. OLEIFERA* BY SPECTROSCOPIC METHODS

14.8.1 Analysis of *M. oleifera* by UV-Vis Spectrophotometry

Various studies using UV-vis spectrophotometry have highlighted the rich phytochemical profile and biological potential of *M. oleifera*. Protein analysis in moringa fruit revealed concentrations ranging from 2.4883 to 2.7477 mg/50g, while seed extracts showed significant levels of alkaloids (916.87 µg/g), saponins (6.367%), tannins (3724.5 µg/g) and flavonoids (0.255%). The phytochemical composition and antioxidant activity of moringa leaves were found to vary with age and extraction solvent, with notable contents of flavonoids, chlorophylls, carotenoids and tocopherols. Additionally, phenolic content (11.85 mg GAE/g) and flavonoid levels (measured as catechin equivalents) were linked to antimicrobial and antioxidant activities, supporting the plant's medicinal potential (Das et al. 2022; Fahal et al. 2018; Nobossé et al. 2018; Younis et al. 2022). Analysis by the UV-vis spectroscopic method of *M. Oleifera* is shown in Table 14.6.

14.8.2 Analysis of *M. oleifera* by NMR

Advanced spectroscopic techniques have been effectively employed to characterise the bioactive compounds in *M. oleifera*; ^{13}C NMR analysis of the seed oil revealed the presence of only monounsaturated fatty acids – oleic, eicosenoic (ω-9) and vaccenic (ω-7) – with accurate identification of their positional arrangement on the glycerol backbone. In the methanolic leaf extract, five novel kaempferol and kaempferide glycosides, along with known flavonoids such as rutin, gallic acid and quercetin 3-O-β-glucoside, were isolated. A tissue-specific metabolic profiling study using 1D-1H NMR identified about 30 metabolites across leaf and stem tissues, with distinct compounds such as tyrosine and adenosine in leaves and glutamate and tryptophan in stems, contributing to a comprehensive metabolite database. Additionally, saponins were isolated from pods and characterised using TLC, HPLC, IR and 1H NMR, confirming their structural identity through key functional groups, reinforcing the phytochemical richness and pharmacological potential of *M. oleifera* (Mahmud et al. 2014; Manguro et al. 2007; Sharma et al. 2013; Vlahov et al. 2002). Analysis by the NMR method of *M. Oleifera* is shown in Table 14.6.

TABLE 14.6
Analysis by Spectroscopy

S. no	Study title	Key findings	Reference
Analysis by the UV-vis spectroscopic method			
1.	Qualitative and quantitative phytochemical analysis of *Moringa oleifera* (Lam) Pods Analysed compounds: proteins	Protein content measured in three samples: A = 2.614 mg/50g, B = 2.4883 mg/50g and C = 2.7477 mg/50g. UV-vis spectrophotometry effectively quantified protein levels in moringa fruit.	Fahal et al. (2018)
2.	Effects of age and extraction solvent on phytochemical content and antioxidant activity of fresh *Moringa oleifera* L. leaves Analysed compounds: alkaloids, flavonoids, saponins, tannins	Quantitative results: alkaloids = 916.87 μg/g, flavonoids = 0.255%, saponins = 6.367%, tannins = 3,724.5 μg/g. The study confirmed the presence and concentrations of key secondary metabolites in moringa seeds.	Nobossé et al. (2018)
3.	Investigation on the phenolic content in *Moringa oleifera* and its antimicrobial activity Analysed compounds: flavonoids, chlorophylls, carotenoids, tocopherols	Flavonoid content measured at 510 nm; chlorophylls at 645 and 663 nm; carotenoids at 450 nm; and tocopherols at 270 nm. The study demonstrated that the extraction solvent and leaf age significantly affect phytochemical content and antioxidant activity.	Das et al. (2022)
4.	Phytochemical and antioxidant screening of *Moringa oleifera* for its utilization in the management of hepatic injury Analysed compounds: flavonoids	Total flavonoid content expressed as catechin equivalents. UV-vis spectrophotometry at 510 nm effectively quantified flavonoid levels, supporting the plant's antioxidant potential.	Younis et al. (2022)
Analysis by nuclear magnetic resonance (NMR)			
5.	^{13}C NMR characterization of triacylglycerols of *Moringa oleifera* seed oil: An 'oleic-vaccenic acid' oil	The composition of *M. oleifera* seed oil triacylglycerols was analysed using 13C NMR spectroscopy, revealing only mono-unsaturated fatty acids: oleic, eicosenoic (ω-9) and vaccenic (ω-7). The method enabled precise identification and positional assignment (1,3- and 2-positions) of these acyl chains on the glycerol backbone.	Vlahov et al. (2002)
6.	Phenolics of *Moringa oleifera* leaves	From the methanolic extract of *M. oleifera* leaves, five novel kaempferol and kaempferide glycosides, along with benzoic acid and benzaldehyde glycosides, were isolated and characterised using spectroscopic methods. Known flavonoids and phenolic compounds such as kaempferol, rutin, gallic acid and quercetin 3-O-β-glucoside were also identified.	Manguro et al. (2007)
7.	Tissue-specific metabolic profile study of *Moringa oleifera* L. using nuclear magnetic resonance spectroscopy	This study used 1D proton NMR to profile metabolites in *M. oleifera* leaf and stem tissues, identifying around 30 metabolites with distinct tissue-specific distributions. While 22 metabolites were common, others, such as tyrosine and adenosine, were leaf-specific, and glutamate and tryptophan were stem-specific, linking to 36 biochemical pathways and supporting the creation of a tissue-specific metabolite database.	Mahmud et al. (2014)

(*Continued*)

TABLE 14.6 (CONTINUED)
Analysis by Spectroscopy

S. no	Study title	Key findings	Reference
8.	Isolation and characterization of saponins from *Moringa oleifera* (Moringaeceae) pods	Saponin was successfully isolated from *M. oleifera* pods using thin-layer chromatography (TLC) and high-performance liquid chromatography (HPLC) techniques. Characterisation via IR and ^{1}H NMR confirmed the presence of hydroxyl, ester, ether, alkyne and aromatic functional groups, validating the compound's saponin nature.	Sharma et al. (2013)
Analysis by Fourier transform infrared (FTIR) spectroscopy			
9.	Extraction and quantification of *Moringa oleifera* leaf powder extracts by HPLC and FTIR	High concentrations of antioxidants such as quercetin, gallic acid and chlorogenic acid were found in this study's extraction and characterisation of bioactive compounds from *M. oleifera* leaf powder using HPLC and FTIR. The presence of various functional groups supports its potential as a dietary supplement for combating oxidative stress and related chronic diseases.	Khalid et al. (2023)
10.	FTIR analysis of phenolic extracts from *Moringa oleifera* leaves	*M. oleifera* leaves, rich in phenolic compounds with nutritional and therapeutic benefits, were analysed using FTIR to identify key functional groups. This non-destructive method complements traditional techniques used for quantifying total phenol content, offering an alternative approach for plant analysis.	Cruz Espinoza et al. (2012)
11.	Rapid authentification of *Moringa oleifera* powder: Detection and quantification of adulteration using ATR-FTIR spectroscopy and chemometrics	This study used ATR-FTIR spectroscopy combined with chemometric models (PLS-DA and OPLS-DA) to detect and quantify henna adulteration in *M. oleifera* leaf powder. The models achieved 100% classification accuracy and high predictive reliability, demonstrating a robust method for quality assurance and ensuring product integrity.	El Orche et al. (2025)
12.	Evaluation of phytochemical constituents of methanol extract of *Moringa oleifera* Lam. whole leaf by gas chromatography-mass spectrometry and Fourier transform infrared spectroscopy analysis	Methanol extracts of *M. oleifera* leaves were analysed by FTIR and gas chromatography-mass spectrometry (GC-MS), revealing functional groups indicative of alcohols, esters, aldehydes and phenols. GC-MS identified 20 compounds, with major bioactive constituents, such as N,N'-pentamethylenebis[s-3-aminopropyl thiosulfuric acid and 2-myristynoyl pantetheine, supporting the plant's traditional medicinal uses.	Okechukwu et al. (2021)

14.8.3 Analysis of *M. oleifera* by FTIR

Fourier transform infrared (FTIR) spectroscopy, along with complementary techniques such as HPLC, GC-MS and chemometric modelling, has been effectively used to analyse and authenticate *M. oleifera* leaf extracts. High concentrations of antioxidants like quercetin, gallic acid and chlorogenic acid were detected by HPLC and FTIR and functional groups confirming their function in preventing oxidative stress were also found. The presence of important functional groups associated with phenolic compounds was verified by FTIR analysis alone, providing a non-destructive substitute for nutritional and medicinal assessment. Additionally, ATR-FTIR spectroscopy strengthened quality control by enabling 100% accurate detection of adulterants such as henna in moringa powder when paired with chemometric tools like PLS-DA and OPLS-DA. Furthermore, FTIR and GC-MS analyses of methanol extracts revealed 20 phytoconstituents and a variety of functional groups, including important bioactives that support the plant's traditional medicinal uses (Cruz Espinoza et al. 2012; El Orche et al. 2025; Khalid et al. 2023; Okechukwu et al. 2021). Analysis by the FTIR method of *M. Oleifera* is shown in Table 14.6.

14.9 ANALYSIS OF *M. OLEIFERA* BY CHROMATOGRAPHIC METHODS

14.9.1 Analysis of *M. oleifera* by HPLC

High-performance liquid chromatography (HPLC) has been extensively used for the quality control and standardisation of *M. oleifera* by enabling precise quantification of key bioactive compounds. Validated methods have been developed to simultaneously analyse antioxidant compounds such as cryptochlorogenic acid, isoquercetin and astragalin, ensuring consistency in pharmaceutical formulations. Microscale HPLC techniques have also enabled accurate measurement of nutritional components such as α- and γ-tocopherol in leaves, flowers and beans, aiding dietary evaluations. Furthermore, reverse-phase HPLC effectively quantified bioactive nitrile glycosides – niaziridin and niazirin – highlighting their differential distribution in plant parts. Additionally, HPLC methods targeting isothiocyanates and astragalin have been validated for their precision and reliability in evaluating anti-inflammatory and antioxidant activity, promoting their application in cosmetics and health product development. Analysis by the HPLC method of *M. Oleifera* is shown in Table 14.7 (Engsuwan et al. 2017; Sánchez Machado et al. 2006; Shanker et al. 2007; Vongsak et al. 2014).

14.9.2 Analysis of *M. oleifera* by HPTLC

High-performance thin-layer chromatography (HPTLC) has proven to be a powerful tool for analysing and ensuring the quality of *M. oleifera* extracts. Several studies have developed and validated HPTLC methods for the simultaneous estimation of key phytochemicals such as gallic acid, quercetin, rutin, chlorogenic acid, caffeic acid, p-coumaric acid and kaempferol across various plant parts. These methods demonstrated high accuracy, precision and clear separation of compounds, with results indicating that different extraction solvents (ethanol and water) yield varied concentrations of bioactives. HPTLC fingerprinting also helped establish specific retention factor (Rf) values for identification and standardisation purposes. Overall, these validated methods offer cost-effective, efficient and reliable approaches for quality control, phytochemical profiling and authentication of *M. oleifera* in herbal and pharmaceutical preparations. Analysis by the HPTLC method of *M. Oleifera* is shown in Table 14.7 (Alam et al. 2020; Elangovan et al. 2015; Niranjan et al. 2017).

14.9.3 Analysis of *M. oleifera* by TLC

Thin-layer chromatography (TLC) and TLC-densitometric methods have been effectively employed for the analysis and quality control of *M. oleifera* extracts. A preliminary study on the

TABLE 14.7
Analysis by Chromatography

S. no	Method	Findings	Reference
Analysis by HPLC			
1.	Simultaneous HPLC quantitative analysis of active compounds in leaves of *Moringa oleifera* Lam.	A validated HPLC method was developed for the simultaneous quantification of the antioxidant compounds cryptochlorogenic acid, isoquercetin and astragalin in *M. oleifera* leaves, showing high precision and recovery. The method enables effective standardisation of leaf extracts and pharmaceutical products from different regions.	Vongsak et al. (2014)
2.	High-performance liquid chromatography method to measure α- and γ-tocopherol in leaves, flowers and fresh beans from *Moringa oleifera*	A microscale HPLC method with optimised saponification and extraction was developed to quantify α- and γ-tocopherol in *M. oleifera* leaves, flowers and beans. The method showed good precision and was applied in nutritional studies of edible plants from northwest México.	Sanchez Machado et al. (2006)
3.	Determination of bioactive nitrile glycoside(s) in drumstick (*Moringa oleifera*) by reverse phase HPLC	A precise HPLC method was developed to quantify the bioactive nitrile glycosides niaziridin and niazirin in *M. oleifera* leaves and pods, with niaziridin higher in pods and niazirin higher in leaves; both compounds were absent in bark. This robust method supports quality assurance of pharmacologically active extracts.	Shanker et al. (2007)
4.	HPLC methods for quality control of *Moringa oleifera* extract using isothiocyanates and astragalin as bioactive markers	This study developed and validated HPLC methods to quantify isothiocyanates and astragalin – key anti-inflammatory and antioxidant compounds – in *M. oleifera* extracts from various plant parts. The methods demonstrated good accuracy and precision, supporting their use in quality control for cosmetic and health product development.	Engsuwan et al. (2017)
Analysis by high-performance thin-layer chromatography (HPTLC)			
5.	Preliminary phytochemical screening and HPTLC fingerprinting profile of leaf extracts of *Moringa oleifera* and *Phyllanthus emblica*	This study used HPTLC to establish physical constants and fingerprint profiles of *Phyllanthus emblica* and *M. oleifera* leaf extracts. The extracts showed distinct phytochemical profiles with specific Rf values, identifying key components – most notably at Rf 0.99 in *M. oleifera* and 0.95 in *P. emblica*. These fingerprints serve as reliable markers for plant identification and quality control.	Elangovan et al. (2015)
6.	A high performance thin layer chromatographic method using a design of experiment approach for estimation of phytochemicals in extracts of *Moringa oleifera* leaves	A robust HPTLC method, optimised using design of experiments (DoE), was developed for the simultaneous estimation of gallic acid, quercetin and rutin in *M. oleifera* leaf extracts. The method showed high accuracy, precision and effective separation, with ethanolic extracts richer in quercetin and rutin and aqueous extracts containing more gallic acid. This validated technique offers a cost-effective and efficient tool for phytochemical analysis in herbal products.	Thomas et al. (2020)

(*Continued*)

TABLE 14.7 (CONTINUED)
Analysis by Chromatography

S. no	Method	Findings	Reference
7.	Simultaneous quantification of six phenolic compounds in various parts of *Moringa oleifera* Lam. using high-performance thin-layer chromatography	This study developed a simple and sensitive HPTLC method for the simultaneous quantification of six phenolic compounds in various parts of *M. oleifera*. The method effectively detected gallic acid, caffeic acid, p-coumaric acid, chlorogenic acid, quercetin and kaempferol, with high precision and recovery. Results showed varying phenolic content across plant parts, highlighting *M. oleifera* as a valuable source of nutraceuticals for industrial applications.	Niranjan et al. (2017)
8.	Simultaneous identification of rutin, chlorogenic acid and gallic acid in *Moringa oleifera* by densitometric high performance thin layer chromatography method	A precise and reliable HPTLC method was developed for the simultaneous estimation of rutin, chlorogenic acid and gallic acid in different parts of *M. oleifera*. The method showed sharp, well-resolved peaks with excellent linearity and accuracy, making it suitable for quantifying flavonoids and for quality control of bioactive compounds in medicinal plants.	Alam et al. (2020)
Analysis by TLC			
9.	Preliminary phytochemical investigation and thin layer chromatography profiling of sequential extracts of *Moringa oleifera* pods	The study focused on preliminary phytochemical screening and TLC profiling of *M. oleifera* pods. It revealed the presence of phenolics, triterpenoids, cardiac glycosides, steroids, alkaloids and saponins across sequential extracts. TLC analysis using solvents of varying polarity confirmed *M. oleifera* pods as a rich source of natural antioxidants.	Sharma et al. (2013a; 2013b)
10.	Simultaneous determination of crypto-chlorogenic acid, isoquercetin, and astragalin contents in *Moringa oleifera* leaf extracts by TLC-densitometric method	A validated TLC-densitometric method was developed for the simultaneous quantification of cryptochlorogenic acid, isoquercetin and astragalin in 70% ethanolic extracts of *M. oleifera* leaves from 12 locations. The method showed high accuracy (98.28–99.65%), precision and linearity ($r>0.9961$), making it reliable and cost-effective for quality control of *M. oleifera* leaf extracts. Leaves from Chiang Mai province had the highest levels of all three compounds.	Vongsak et al. (2013)
Analysis by LC-MS			
11.	Determination of flavonoids by LC/MS and anti-inflammatory activity in *Moringa oleifera*	*M. oleifera* leaves from sub-Saharan Africa were analysed and found rich in flavonoids such as quercetin and kaempferol glucosides, with total flavonoid content ranging from 0.18% to 1.64%. Acid hydrolysis enabled accurate quantification, and the plant showed notable anti-inflammatory activity, though its flavonoid malonates were thermally unstable.	Coppin et al. (2013)
12.	LC-MS/MS-based chemical profiling of water extracts of *Moringa oleifera* leaves and pharmacokinetics of their major constituents in rat plasma	*M. oleifera* leaves were analysed using advanced mass spectrometry, identifying 42 bioactive compounds including flavonoids and phenolic acids. Pharmacokinetic studies revealed rapid absorption, with vicenin-2 showing high bioavailability and kaempferol-3-O-rutinoside exhibiting a longer half-life, supporting their potential for clinical applications.	Wang et al. (2024)

(*Continued*)

TABLE 14.7 (CONTINUED)
Analysis by Chromatography

S. no	Method	Findings	Reference
13.	Anti-inflammatory properties and potential bioactive components from *Moringa oleifera* leaves revealed by affinity ultrafiltration LC–MS and molecular docking	This study demonstrated the anti-inflammatory potential of *M. oleifera* leaf extracts (MOLEs) by inhibiting COX-2 and reducing pro-inflammatory cytokines (IL-1β, IL-6, PGE2, TNF-α) in vitro. Fifteen active phytochemicals were identified via UF-LC/MS, with quercetin and kaempferol glycosides – especially acetylated quercetin derivatives – showing strong COX-2 binding through molecular docking. The findings support *M. oleifera*'s potential as a natural anti-inflammatory agent for functional food and healthcare applications.	Chen, G. et al. (2021)
14.	*Moringa oleifera*: Study of phenolics and glucosinolates by mass spectrometry	This study assessed the bioactive compound profile in various parts of *M. oleifera* seedlings, focusing on phenolics and glucosinolates known for their antioxidant and anticarcinogenic properties. Using FI-ESI-MS with chemometric analysis and LC-MS/MS, researchers successfully identified and quantified these compounds, highlighting moringa's potential in preventing chronic degenerative diseases and supporting detoxification.	Maldini et al. (2014)
15.	Ultrasound-microwave assisted extraction of proteins from *Moringa oleifera* leaves: Comparative optimization study and LC-MS analysis of the protein concentrate	This study optimised the extraction of high-quality protein from *M. oleifera* leaves, achieving higher yield with ultrasound-microwave assisted extraction (UMAE) compared with traditional methods (82.07 mg/g vs. 68.99 mg/g). LC-MS and bioinformatics confirmed the nutritional and functional potential of the extracted proteins, supporting their use as nutritional supplements and food ingredients.	Cheng, F. et al. (2021)
16.	Effect of extraction conditions on the bioactive compounds from *Moringa oleifera* (*PKM 1*) seeds and their identification using LC–MS	This study optimised the extraction of bioactive compounds from *M. oleifera* seeds using a central composite design, with ideal conditions being 49.8% ethanol, 80°C, 45 min and 0.62 mm particle size. LC-MS analysis identified 12 bioactive compounds, including luteolin-6-C-glucoside and hesperidin, known for their diverse biological and nutraceutical benefits.	Premi et al. (2017)
Analysis by GC-MS			
17.	Multiresidue analysis of pesticides in moringa pods by GC-MS/MS and LC-MS/MS	This study developed and validated a multiresidue method using GC-MS/MS and LC-MS/MS for detecting over 380 pesticides in moringa pods. The method showed excellent linearity (r^2>0.99), precision (RSD<20%) and recovery (70–120%), with effective sample cleanup and reduced matrix effects, meeting SANTE/12682/2019 guidelines.	Dutta et al. (2020)
18.	LC/MS, GC/MS screening and in vivo anti-inflammatory activity of Malaysian *Moringa oleifera* Lam leaf extracts and fractions against carrageenan-induced paw oedema in rats	This study evaluated the anti-inflammatory activity of three *M. oleifera* leaf extracts using a carrageenan-induced paw oedema model in rats. The 95% ethanol extract and its dichloromethane fraction (250 mg/kg) showed the highest activity. LC/MS and GC/MS analyses identified 26 bioactive compounds, including flavonoids and phenol glucosides. Further studies, especially in humans, are recommended to confirm mechanisms and active constituents.	Mahdi et al. (2017)

pods revealed the presence of key phytochemicals such as phenolics, triterpenoids, cardiac glycosides, steroids, alkaloids and saponins, highlighting their antioxidant potential. Complementarily, a validated TLC-densitometric method was developed for the simultaneous quantification of cryptochlorogenic acid, isoquercetin and astragalin in 70% ethanolic leaf extracts from various locations. This method demonstrated excellent accuracy (98.28–99.65%), precision and linearity (r>0.9961), establishing it as a reliable and cost-effective tool for the standardisation and quality assessment of *M. oleifera* leaf products, with Chiang Mai samples showing the highest bioactive content. Analysis by the TLC method of *M. Oleifera* is shown in Table 14.7 (Sharma et al. 2013b; Vongsak et al. 2013).

14.9.4 Analysis of *M. oleifera* by LC-MS

M. oleifera, a highly valuable tropical tree, is renowned for its nutritional and medicinal benefits, though its potential remains underutilised. A study focusing on moringa leaves from sub-Saharan Africa utilised HPLC–UV–MS to identify 12 flavonoids, with quercetin and kaempferol glucosides and glucoside malonates as the primary components. To ensure accurate quantification, acid hydrolysis was employed during extraction, converting flavonoid conjugates into aglycones, which simplified measurement. The validated method showed high recovery rates (92.6–107.5%), confirming its reliability. Analysis of moringa samples from Ghana, Senegal and Zambia revealed total flavonoid content ranging from 0.18% to 1.64% (g/dry weight), varying by geographical origin and variety. The study also highlighted the anti-inflammatory properties of flavonoid-rich varieties. Additionally, thermal stability tests demonstrated that flavonoid malonates, key constituents in fresh leaves, degrade easily under heat, breaking down into glycosides. This finding is crucial for processing methods, as excessive heat may alter the bioactive profile. Another study using UPLC-Q-Exactive Orbitrap-MS identified 42 compounds in moringa leaves, including phenolic acids, flavonoids and isothiocyanates, with pharmacokinetic studies showing rapid absorption (within 0.5 hours) and varying bioavailability among compounds. Further research explored moringa's anti-inflammatory mechanisms, linking quercetin and kaempferol glycosides to COX-2 inhibition and cytokine suppression. Additionally, optimisation studies on protein and bioactive compound extraction from moringa leaves and seeds demonstrated enhanced yields using advanced techniques such as ultrasound-microwave-assisted extraction. These findings collectively underscore moringa's potential as a functional food and natural remedy, though careful processing is required to preserve its bioactive integrity. Analysis by the LC-MS method of *M. oleifera* is shown in Table 14.7 (Cheng, F. et al. 2021; Chen, G. et al. 2021; Coppin et al. 2013; Maldini et al. 2014; Premi et al. 2017; Wang et al. 2024).

14.9.5 Analysis of *M. oleifera* by GC-MS

This study developed and validated a multi-residue method for analysing multi-class pesticides in moringa pods using GC-MS/MS and LC-MS/MS. The sample preparation involved acetonitrile extraction followed by dispersive solid-phase extraction (d-SPE) cleanup using primary secondary amine (PSA), graphitised carbon black (GCB) and C18 sorbents. The LC-MS/MS portion was directly analysed, while the GC-MS/MS portion required reconstitution in ethyl acetate to reduce matrix effects. The method was validated per SANTE/12682/2019 guidelines, covering 180 pesticides by GC-MS/MS and 203 by LC-MS/MS, demonstrating good linearity (r^2>0.99), precision (RSD<20%) and accuracy (70–120% recoveries). Matrix-matched calibration effectively corrected recovery variations. In a related investigation, the anti-inflammatory activity of moringa leaf extracts was assessed using a carrageenan-induced paw oedema model in rats. The 95% ethanol extract (250 mg/kg) and its dichloromethane fraction exhibited the strongest anti-inflammatory effects. LC-MS and GC-MS analysis of the active fraction identified 18 and eight compounds, respectively, including flavonoids, phenol glucosides, amino acids and vitamins. These findings suggest that moringa's anti-inflammatory properties may be attributed to these bioactive compounds. However, further research, including human trials, is needed to identify the exact active phytochemicals and their mechanism of action. Analysis by the GC-MS method of *M. Oleifera* is shown in Table 14.7 (Dutta et al. 2020; Mahdi et al. 2017).

14.10 CONCLUSION

The comprehensive analytical profiling of *M. oleifera* confirms its rich phytochemical diversity and significant therapeutic potential. Various advanced analytical techniques such as UV-vis spectrophotometry, HPTLC, HPLC, FTIR, NMR and GC-MS have effectively identified and quantified a wide range of bioactive compounds, including flavonoids (quercetin, rutin, kaempferol), phenolic acids (gallic acid, chlorogenic acid), isothiocyanates, nitrile glycosides, tocopherols and saponins. These methods demonstrated high precision, accuracy, sensitivity and reliability, making them suitable for the standardisation and quality control of *M. oleifera*-based formulations. The analytical data not only validate the plant's traditional uses but also support its application in pharmaceutical, nutraceutical and food industries. Overall, the robust analytical characterisation affirms *M. oleifera* as a valuable medicinal and nutritional plant with well-defined chemical markers essential for ensuring product safety, efficacy and consistency.

REFERENCES

Akter, T., Rahman, M.A., Moni, A., Apu MAI, Fariha A, & Hannan M.A., (2021). Prospects for protective potential of *Moringa oleifera* against kidney diseases. Plants. https//doi.org/10.3390/plants10122818.

Alam, P., Alam, P., Sharaf-Eldin, M. A., & Alqarni, M. H. (2020). Simultaneous identification of rutin, chlorogenic acid and gallic acid in Moringa oleifera by densitometric high-performance thin-layer chromatography method. *JPC–Journal of Planar Chromatography–Modern TLC*, *33*, 27–32.

Anwar, F., Latif, S., Ashraf, M., & Gilani, A. H. (2007). Moringa oleifera: A food plant with multiple medicinal uses. *Phytotherapy Research: An International Journal Devoted to Pharmacological and Toxicological Evaluation of Natural Product Derivatives*, *21*(1), 17–25.

Athikomkulchai, S., Tunit, P., Tadtong, S., Jantrawut, P., Sommano, S. R., & Chittasupho, C. (2020). Moringa oleifera seed oil formulation physical stability and chemical constituents for enhancing skin hydration and antioxidant activity. *Cosmetics*, *8*(1), 2.

Bagheri, G., Martorell, M., Ramírez-Alarcón, K., Salehi, B., & Sharifi-Rad, J. (2020). Phytochemical screening of Moringa oleifera leaf extracts and their antimicrobial activities. *Cellular and Molecular Biology*, *66*(1), 20–26.

Bania, J. K., Deka, J. R., Hazarika, A., Das, A. K., Nath, A. J., & Sileshi, G. W. (2023). Modelling habitat suitability for Moringa oleifera and Moringa stenopetala under current and future climate change scenarios. *Scientific Reports*, *13*(1), 20221.

Boopathi, N. M., & Abubakar, B. Y. (2021). Botanical descriptions of Moringa spp. *The Moringa Genome*, 11–20.

Boopathi, N. M., & Raveendran, M. (2021). Moringa and its importance. *The Moringa Genome*, 1–9.

Boopathi, N. M., Raveendran, M., & Kole, C. (2021). *The Moringa genome*. Springer, Cham, Switzerland.

Castillo-Lopez, R. I., Leon-Felix, J., Angulo-Escalante, M. A., Gutierrez-Dorado, R., Muy-Rangel, M. D., & Heredia, J. B. (2017). Nutritional and phenolic characterization of Moringa oleifera leaves grown in Sinaloa, Mexico. *Pakistan Journal of Botany*, *49*(1), 161–168.

Chen, G., Xu, Y., Wu, J., Li, N., & Guo, M. (2021). Anti-inflammatory properties and potential bioactive components from Moringa oleifera leaves revealed by affinity ultrafiltration LC–MS and molecular docking. *ACS Food Science & Technology*, *1*(10), 1953–1962.

Cheng, F., Shu, G., Chen, L., Dai, C., Wan, H., Chen, H., & Dong, X. (2021). Ultrasound-microwave assisted extraction of proteins from Moringa oleifera leaves: Comparative optimization study and LC-MS analysis of the protein concentrate. *Journal of Food Processing and Preservation*, *45*(6), e15547.

Coppin, J. P., Xu, Y., Chen, H., Pan, M. H., Ho, C. T., Juliani, R., ... Wu, Q. (2013). Determination of flavonoids by LC/MS and anti-inflammatory activity in Moringa oleifera. *Journal of Functional Foods*, *5*(4), 1892–1899.

Cruz-Espinoza, J. E., Orduña-Díaz, A., Rosales-Perez, M., Zaca-Morán, O., Delgado-Macuil, R., Gayou, V. L., & Rojas-López, M. (2012). FTIR analysis of phenolic extracts from Moringa oleifera leaves. *Journal of Biometrics & Biostatistics, 11*, 2802.

Das, S. K., Dharan, B. J., Pavitra, P. V., Das, S., Behera, S. P., Veilumuthu, P., & Christopher, J. G. (2022). Investigation on the phenolic content in Moringa oleifera and its antimicrobial activity. *Indian Journal of Agricultural Research, 56*(3), 255–261.

Dutta, A., Hingmire, S., & Banerjee, K. (2020). Multiresidue analysis of pesticides in moringa pods by GC-MS/MS and LC-MS/MS. *Journal of AOAC International, 103*(6), 1486–1497.

El Orche, A., El Karbane, M., Ait El Alia, O., Bouchafra, H., Zarayby, L., & Bouatia, M. (2025). Rapid authentication of Moringa oleifera powder: Detection and quantification of adulteration using ATR-FTIR spectroscopy and chemometrics. *Food Analytical Methods, 18(6)*, 898–910.

Elangovan, N. M., Dhanarajan, M. S., & Elangovan, I. (2015). Preliminary phytochemical screening and HPTLC fingerprinting profile of leaf extracts of Moringa oleifera and Phyllanthus emblica. *International Research Journal of Pharmaceutical and Biosciences, 2*, 32–40.

Engsuwan, J., Waranuch, N., Limpeanchob, N., & Ingkaninan, K. (2017). HPLC methods for quality control of Moringa oleifera extract using isothiocyanates and astragalin as bioactive markers. *ScienceAsia, 43*(3), 168–176. https://doi.org/10.2306/scienceasia1513-1874.2017.43.168.

Fahal, E. M., Rani, B. M. A., Aklakur, M. D., Chanu, T. I., & Saharan, N. (2018). Qualitative and quantitative phytochemical analysis of Moringa oleifera (Lam) Pods. *International Journal of Current Microbiology and Applied Sciences, 7*(5), 657–665.

Fatmawati, A., Sucianingsih, D., Kurniawati, R., & Abdurrahman, M. (2021). Microscopic identification and determination of total flavonoid content of moringa leaves extract and ethyl acetate fraction (Moringa oleifera L.). *Indonesian Journal of Pharmaceutical Science and Technology, 1*, 66–74.

Fatoumata, B. A., Mohamet, S. E. N. E., Sambou, J. K., MbackÃ, M., & El HadjiMakhtar, B. A. (2020). Antidiabetic properties of Moringa oleifera: A review of the literature. *Journal of Diabetes and Endocrinology, 11*(1), 18–29.

Gandji, K., Salako, V. K., Fandohan, A. B., Assogbadjo, A. E., & GlèlèKakaï, R. L. (2018). Factors determining the use and cultivation of Moringa oleifera Lam. in the Republic of Benin. *Economic Botany, 72*, 332–345.

Gopalakrishnan, L., Doriya, K., & Kumar, D. S. (2016). Moringa oleifera: A review on nutritive importance and its medicinal application. *Food Science and Human Wellness, 5*(2), 49–56.

Islam, M. A., Sheikh, A., Waterman, C., & Hosenuzzaman, M. D. (2020). Morphology, pod yield and nutritional quality of two cultivars of Moringa (Moringa oleifera) in Bangladesh. *Indian Journal of Science and Technology, 13*(36), 3725–35.

Khalid, S., Arshad, M., Mahmood, S., Siddique, F., Roobab, U., Ranjha, M. M. A. N., & Lorenzo, J. M. (2023). Extraction and quantification of Moringa oleifera leaf powder extracts by HPLC and FTIR. *Food Analytical Methods, 16*(4), 787–797.

Koheil, M. A., Hussein, M. A., Othman, S. M., & El-Haddad, A. (2011). Anti-inflammatory and antioxidant activities of Moringa peregrina seeds. *Free Radicals and Antioxidants, 1*(2), 49–61.

Ma, K., Wang, Y., Wang, M., Wang, Z., Wang, X., Ju, X., & He, R. (2021). Antihypertensive activity of the ACE–renin inhibitory peptide derived from Moringa oleifera protein. *Food & Function, 12*(19), 8994–9006.

Mahajan, S. G., Mali, R. G., & Mehta, A. A. (2007). Effect of Moringa oleifera Lam. seed extract on toluene diisocyanate-induced immune-mediated inflammatory responses in rats. *Journal of Immunotoxicology, 4*(2), 85–96.

Mahdi, H. J., Khan, N. A., Mahmud, R., Asmawi, M. Z., Vikneswaran, A., & Murugaiyah, L. (2017). LC/MS, GC/MS screening and in vivo anti-inflammatory activity of Malaysian Moringa oleifera Lam leaf extracts and fractions against carrageenan-induced paw oedema in rats. *Journal of Innovative Pharmaceutical and Biological Sciences, 4*(3), 48–54.

Mahmud, I., Chowdhury, K., & Boroujerdi, A. (2014). Tissue-specific metabolic profile study of Moringa oleifera L. using nuclear magnetic resonance spectroscopy. *Plant Tissue Culture & Biotechnology/ Bangladesh Association for Plant Tissue Culture & Biotechnology; BAPTC&B, 24*(1), 77.

Maldini, M., Maksoud, S. A., Natella, F., Montoro, P., Petretto, G. L., Foddai, M., ... Pintore, G. (2014). Moringa oleifera: study of phenolics and glucosinolates by mass spectrometry. *Journal of Mass Spectrometry, 49*(9), 900–910.

Mallenakuppe, R., Homabalegowda, H., Gouri, M. D., Basavaraju, P. S., & Chandrashekharaiah, U. B. (2015). History, taxonomy and propagation of Moringa oleifera–a review. *Crops, 3*(3.28), 3–15.

Manguro, L. O. A., & Lemmen, P. (2007). Phenolics of Moringa oleifera leaves. *Natural Product Research*, *21*(1), 56–68.

Manjegowda, N., Kumar, M. R., Raghunathanaidu, B. D., Babu, A., Sridhara, K., Ramesh, B., & Ahmed, S. S. Anti-anemic potential of Moringa oleifera flower extract against phenylhydrazineinduced anemia in rats. *ACTA Pharmaceutica Sciencia*, *62*(2), 448–56. https://doi.org/10.23893/1307-2080.APS6228..

Mishra, G., Singh, P., Verma, R., Kumar, S., Srivastav, S., Jha, K. K., & Khosa, R. L. (2011). Traditional uses, phytochemistry and pharmacological properties of Moringa oleifera plant: An overview. *Der Pharmacia Lettre*, *3*(2), 141–164.

Mobarak, A. A., Shaltout, K., Ali, H. I., Baraka, D., & Aly, S. (2017). Morphological VariabilityAmong Moringa oleifera (Lam.) Populations in Egypt. *Egyptian Journal of Botany*, *57*(1), 241–257.

Moyo, B., Masika, P. J., & Muchenje, V. (2012). Antimicrobial activities of Moringa oleifera Lam leaf extracts. *African Journal of Biotechnology*, *11*(11), 2797–2802.

Niranjan, A., Ngpoore, N. K., Anis, N., Kumar, A., Lehri, A., Shirke, P. A., & Tewari, S. K. (2017). Simultaneous quantification of six phenolic compounds in various parts of Moringa oleifera Lam. using high-performance thin-layer chromatography. *JPC-Journal of Planar Chromatography-Modern TLC*, *30*(6), 502–509.

Nobossé, P., Fombang, E. N., & Mbofung, C. M. (2018). Effects of age and extraction solvent on phytochemical content and antioxidant activity of fresh Moringa oleifera L. leaves. *Food Science & Nutrition*, *6*(8), 2188–2198.

Okechukwu, V. U., Eze, S. O., Omokpariola, D. O., & Okereke, J. C. (2021). Evaluation of phytochemical constituents of Methanol extract of Moringa oleifera Lam. whole leaf by gas chromatography-mass spectrometry and fourier transform infrared spectroscopy analysis. *World News of Natural Sciences*, *37*, 18–30.

Popoola, J. O., & Obembe, O. O. (2013). Local knowledge, use pattern and geographical distribution of Moringa oleifera Lam.(Moringaceae) in Nigeria. *Journal of Ethnopharmacology*, *150*(2), 682–691.

Premi, M., & Sharma, H. K. (2017). Effect of extraction conditions on the bioactive compounds from Moringa oleifera (PKM 1) seeds and their identification using LC–MS. *Journal of Food Measurement and Characterization*, *11*, 213–225.

Rastogi, T., Bhutda, V., Moon, K., Aswar, P. B., & Khadabadi, S. S. (2009). Comparative studies on anthelmintic activity of Moringa oleifera and Vitex negundo. *Asian Journal of Research in Chemistry*, *2*(2), 181–182.

Ruiz Hernández, R., Pérez-Vázquez, A., García Pérez, E., Landeros Sánchez, C., Morales Trejo, F., & Soto Hernández, R. M. (2021). Morphological characterization of Moringa oleifera accessions from the South-Southeast of Mexico. *Revistamexicana de cienciasagrícolas*, *12*(7), 1209–1222.

Sánchez-Machado, D. I., Lopez-Cervantes, J., & Vázquez, N. R. (2006). High-performance liquid chromatography method to measure α-and γ-tocopherol in leaves, flowers and fresh beans from Moringa oleifera. *Journal of Chromatography A*, *1105*(1–2), 111–114.

Santos, A. F., Argolo, A. C., Paiva, P. M., & Coelho, L. C. (2012). Antioxidant activity of Moringa oleifera tissue extracts. *Phytotherapy Research*, *26*(9), 1366–1370.

Seifu, E., & Teketay, D. (2020). Introduction and expansion of *Moringa oleifera* Lam. in Botswana: Current status and potential for commercialization. *South African Journal of Botany*, *129*, 471–479.

Senthilkumar, A., Karuvantevida, N., Rastrelli, L., Kurup, S. S., & Cheruth, A. J. (2018). Traditional uses, pharmacological efficacy, and phytochemistry of Moringa peregrina (Forssk.) Fiori.—a review. *Frontiers in Pharmacology*, *9*, 465.

Shanker, K., Gupta, M. M., Srivastava, S. K., Bawankule, D. U., Pal, A., & Khanuja, S. P. (2007). Determination of bioactive nitrile glycoside (s) in drumstick (Moringa oleifera) by reverse phase HPLC. *Food Chemistry*, *105*(1), 376–382.

Sharma, V., Paliwal, R., Janmeda, P., & Sharma, S. (2013). Phytochemical analysis and evaluation of antioxidant and free radical scavenging activity of *Moringa oleifera* Lam. pods. J Pharm. *6(2)*, 334-40. https//doi.org/10.1016/j.jopr.2013.01.011.

Sharma, V., & Paliwal, R. (2013a). Isolation and characterization of saponins from Moringa oleifera (moringaceae) pods. *International Journal of Pharmacy and Pharmaceutical Sciences*, *5*(1), 179–183.

Sharma, V., & Paliwal, R. (2013b). Preliminary phytochemical investigation and thin layer chromatography profiling of sequential extracts of Moringa oleifera pods. *International Journal of Green Pharmacy (IJGP)*, *7*(1).

Sharmin, F., Sarker, M. S. K., Sarker, N. R., Huque, K. S., Hossain, S. M. J., & Basher, M. K. (2019). Taxonomical identification, biomass production and nutrient composition of Moringa sp. as a fodder crop. *Bangladesh Journal of Livestock Research*, *26*(1–2), 61–72.

Tekle, A., Belay, A., Kelem, K., Wodajo, B., & Tesfaye, Y. (2015). Nutritional profile of Moringa stenopetala species samples collected from different places in Ethiopia. *European Journal of Nutrition & Food Safety*,*5*(5), 1100–1101.https://doi.org/10.9734/ejnfs/2015/21263

Thomas, A., Kanakdhar, A., Shirsat, A., Deshkar, S., & Kothapalli, L. (2020). A high performance thin layer chromatographic method using a design of experiment approach for estimation of phytochemicals in extracts of Moringa oleifera leaves. *Turkish Journal of Pharmaceutical Sciences*, *17*(2), 148.

Vasanth, K., Ilango, K., MohanKumar, R., Agrawal, A., & Dubey, G. P. (2014). Anticancer activity of Moringa oleifera mediated silver nanoparticles on human cervical carcinoma cells by apoptosis induction. *Colloids and Surfaces B: Biointerfaces*, *117*, 354–359.

Vlahov, G., Chepkwony, P. K., & Ndalut, P. K. (2002). 13C NMR characterization of triacylglycerols of Moringa oleifera seed oil: an "oleic-vaccenic acid" oil. *Journal of Agricultural and Food Chemistry*, *50*(5), 970–975.

Vongsak, B., Sithisarn, P., & Gritsanapan, W. (2013). Simultaneous determination of crypto-chlorogenic acid, isoquercetin, and astragalin contents in moringa oleifera leaf extracts by TLC-densitometric method. *Evidence-Based Complementary and Alternative Medicine*, *2013*(1), 917609.

Vongsak, B., Sithisarn, P., & Gritsanapan, W. (2014). Simultaneous HPLC quantitative analysis of active compounds in leaves of Moringa oleifera Lam. *Journal of Chromatographic Science*, *52*(7), 641–645.

Wang, J., Du, Y., Jiang, L., Li, J., Yu, B., Ren, C., ... He, B. (2024). LC-MS/MS-based chemical profiling of water extracts of Moringa oleifera leaves and pharmacokinetics of their major constituents in rat plasma. *Food Chemistry: X*, *23*, 101585.

Younis, N., Khan, M. I., Zahoor, T., & Faisal, M. N. (2022). Phytochemical and antioxidant screening of Moringa oleifera for its utilization in the management of hepatic injury. *Frontiers in Nutrition*, *9*, 1078896.

Zaku, S. G., Emmanuel, S., Tukur, A. A., & Kabir, A. (2015). Moringa oleifera: An underutilized tree in Nigeria with amazing versatility: A review. *African Journal of Food Science*, *9*(9), 456–461.

Zhigila, D. A., Mohammed, S., Oladele, F. A., & Sawa, F. B. (2015). Numerical analyses of leaf and fruit external morphology in Moringa oleifera Lam. *JurnalTeknologi (Sciences & Engineering)*, *77*(13), 75–82. https://doi.org/10.11113/jt.v77.6030.

15 *Moringa oleifera* Seed Oils *Extraction Methods, Quality Parameters and Biological Activities*

Halida Rahmania, Gunawan Indrayanto, Yugo Susanto and Abdul Rohman

15.1 INTRODUCTION

Moringa oleifera Lam, known as the horseradish tree or drumstick tree, is commonly called the 'miracle tree' because it has functional food and nutraceutical properties. Therefore, this plant has been used to improve human health (Leone et al., 2016). In Indonesia, *M. oleifera* is known as *Kelor* and has been used for thousands of years by numerous civilisations, including Indian, Greek and Egyptian, to prevent and treat diseases (Amin et al., 2024). The seed of *M. oleifera* is reported to be used for hair care, skincare, the treatment of eye disease, fever, gastritis, gout and ulcers. The plant is also used as a traditional herbal medicine in the Ayurvedic, Unani and Siddha systems of medicine (Matic et al., 2018). The extracts, fractions and phytochemicals of *M. oleifera* have been reported to exert pharmacological activities, including working as a purgative to alleviate constipation, lower cholesterol and reduce body weight (Li et al., 2024). *M. oleifera* possesses the ability to mitigate malnutrition as a result of its abundant reservoirs of proteins, vitamins, minerals and carbohydrates. In addition to its nutritional constituents, *M. oleifera* is the only source of a diverse array of non-nutritive molecules often known as phytochemicals. Plants that possess abundant natural compounds are exploited in various applications, both directly and indirectly, to promote the well-being of both human populations and domestic animals (Raja et al., 2016).

In recent years, there has been a significant body of research on isolating oil from the seeds of *M. oleifera*. The primary emphasis of these investigations was directed towards the extraction and assessment of the benefits of *M. oleifera* seed oil (MOSO) (Irnawati et al., 2024). MOSO is known as an edible oil with a high price in the fats and oils industry due to valuable bioactive components having pharmacological activities that are beneficial to human health, including phenolic contents and unsaturated fatty acids such as oleic acid (C18:1) and linoleic acid (C18:2) (Nadeem and Imran, 2016). MOSO is recognised in skincare; therefore, it is not surprising that MOSO is formulated in cosmetic preparations (Athikomkulchai et al., 2021).

In the Indonesian fats and oils market, MOSO is sold at US$15–20, while common vegetable oils such as corn, soybean and palm are sold at prices of US$2–4; therefore, MOSO can be a target for adulteration with these vegetable oils to gain financial profit. Irnawati et al. (2024) have successfully authenticated MOSO against adulteration with soybean oil. From this perspective, the determination of physical properties such as density and viscosity, as well as chemical properties such as iodine and saponification values, of MOSO is very important for characterising it, as any discrepancies from the respective standards could indicate that MOSO has been adulterated.

DOI: 10.1201/9781003617860-17

15.2 EXTRACTION OF *M. OLEIFERA* SEED OILS

Before the extraction process, *Moringa oleifera* (MO) seeds underwent several pre-treatments. These included high-pressure processing (HPP), ultrasound treatment and microwave treatment. Each treatment was applied to enhance extraction efficiency and improve oil yield.

15.2.1 HPP

HPP is a food preservation technique that uses high-pressure pascalisation or cold pasteurisation, usually using water as a medium. This method has been applied as a pre-treatment to MO kernels, prior to MO oil extraction with aqueous enzymatic extraction (AEE). HPP is intended to promote protein hydrolysis for higher free oil recovery (Latif et al., 2011; Abdulkarim et al., 2005), as the MO kernel cells predominantly contain protein (Mat Yusoff et al., 2016). In the study by Mat Yusoff et al. (2017), it was reported that HPP treatment (50–250 MPa, 20–60°C, 10–60 minutes) significantly enhanced free oil recovery while minimising emulsion layer thickness, especially when applied to ground-sieved kernels rather than whole kernels. Remarkably, pre-treating with HPP at 50 MPa and 60°C for 35 minutes yielded around 73% (w/w) free oil recovery and resulted in a thinner emulsion layer compared with relying solely on AEE. Similarly, findings by Sun et al. (2019) demonstrated that HPP treatment enhanced oil extraction efficiency by 9.03% compared with traditional methods, further emphasising its potential as a superior technique for oil recovery optimisation.

15.2.2 Ultrasound Treatment

Ultrasonic pre-treatment has proven highly effective for enhancing oil extraction from seeds. This technique operates by generating cavitation bubbles through the application of ultrasonic waves in the solvent. The collapse of these bubbles induces localised pressure and temperature fluctuations, thereby facilitating the mass transfer of solutes into the solvent (J. Liu et al., 2014). A study by Thirugnanasambandham (2018) reported that ultrasonic pre-treatment of MO seed oil extraction under optimal conditions (temperature of 35°C and extraction time of 20 minutes) yielded a maximum oil content of 59%. Similarly, another study observed a 1.24% increase in oil yield compared with samples that did not undergo pre-treatment (Kayanan and Sagum, 2021).

The physicochemical properties of MO oil were largely unaffected by ultrasonic pre-treatment. Studies have shown no significant differences in fatty acid composition, acylglycerol profiles or thermal properties between pre-treated and untreated oils (Zhong et al., 2018). Moreover, this technique has been shown to enhance the content of bioactive compounds in the oil. For instance, ultrasonic pre-treatment significantly increased the total phytosterol content from 72.69 mg/kg in untreated samples to 369.32 mg/kg in pre-treated oil (Kayanan and Sagum, 2021). This pre-treatment method has been associated with improved oil stability over time. Ultrasonic pre-treatment also reduces extraction time while maintaining high oil recovery rates. Reported recovery rates range from 91% to 94%, demonstrating its superiority in efficiency and effectiveness over conventional extraction methods (Zhong et al., 2018).

15.2.3 Microwave Treatment

Microwave treatment uses microwave energy to improve the efficiency and quality of oil extraction processes. This method generates heat by causing polar molecules, such as water, to rapidly oscillate, producing volumetric heating that is more efficient than conventional techniques due to its direct interaction with the treated material (Azadmard-Damirchi et al., 2011). The process disrupts cell walls and membranes, releasing oil more effectively. This disruption results from rapid heating and pressure changes within the cells, leading to increased extraction rates and yields (Leone et al., 2015).

In MO seed oil extraction, microwave pre-treatment has been shown to significantly enhance oil yields. For instance, pre-treatment using 100 W of microwave power for 30 seconds achieved a maximum oil yield of 35.28% w/w when paired with supercritical CO_2 (SC-CO_2) extraction (Da Porto et al., 2016). Similarly, Ahangarnejad et al. (2021) reported an optimal extraction efficiency of 56.55% with 380.92 W of microwave power for 3.02 minutes. de Souza et al. (2024) demonstrated that microwave pre-treatment at 175 W and 50°C reduced the Soxhlet extraction time from eight hours to just one hour, significantly improving efficiency. Regarding oil composition, microwave pre-treatment did not alter the fatty acid composition, acylglycerol profile or thermal properties of MO oil. The MO microwave pre-treatment oil also improved the bioactive components, including phytosterols and antioxidants. It prolonged its stability, making it a highly effective method for producing high-quality oils with an extended shelf life.

In order to ensure a high-quality MOSO, highly effective extraction methods are therefore required. It is widely accepted that organic solvents are conventionally used for oil extractions from seeds (Gharsallah et al., 2021). Some extraction methods have been developed and proposed to obtain MOSO with desired qualities. Some experimental design techniques were also introduced to achieve the optimum conditions during MOSO extractions. The extraction methods of MOSO included mechanical pressing, solvent extraction, ultrasound-assisted extraction (UAE), AEE and SC-CO_2 extraction.

15.3 MECHANICAL PRESSING

Mechanical pressing is a widely utilised method for extracting oil from oleaginous seeds without the need for solvents. This technique is favoured for its safety, environmental sustainability and ability to preserve valuable natural components in the extracted oils (Savoire et al., 2013; Vovk et al., 2023). Several factors influence the oil yield in this method. Pressure is a critical factor; while higher pressure generally increases yield, it must be optimised beyond a certain point to avoid reaching a plateau where additional pressure has minimal effect (Andasuryani et al., 2020). Temperature and screw rotation speed are equally important, as they significantly affect the efficiency and yield of the extraction process. Additionally, moisture content plays a vital role, with higher moisture levels often leading to improved yields (Souza et al., 2024).

Mechanical pressing is commonly used to extract MO oil from seeds. For example, Ayu et al. (2024) reported that using MO seeds with an 8% moisture content and maintaining a pressing temperature of 100°C yielded 27.3%–28.8%. The extracted MO oil also exhibited acceptable quality parameters, including peroxide value, acidity and saponification value, indicating good stability and overall quality (Ayu et al., 2024; Pereira, Galvaõ et al., 2016). Moreover, oil obtained through mechanical pressing is rich in unsaturated fatty acids, particularly oleic acid, comprising approximately 73.60% to 77.8% of the total composition (Cervera-Chiner et al., 2024; Eman and Muhamad, 2016). This composition closely resembles that of olive oil, making it ideal for applications in the food, cosmetics and pharmaceutical industries.

Despite its advantages, the oil yield from mechanical pressing is generally lower compared with other extraction methods. Studies have shown that mechanical pressing achieves oil yields of approximately 11.4% to 13.0% from MO seeds, whereas solvent extraction can yield up to 41.9% (Moneim and Al-Anbari, 2024; Pereira, De Brito Neto et al., 2016). Nevertheless, the absence of solvents in mechanical pressing enhances its safety and eco-friendliness (Ngcobo and Mbuyisa, 2024; Vovk et al., 2023)

15.4 SOLVENT EXTRACTION

Solvent extraction is a widely employed technique for oil extraction due to its high efficiency and capacity to produce high-quality oil suitable for industrial and commercial purposes. This process uses various solvents to extract oil from the seeds, which is then separated and collected (Ewunie et al., 2021). Among the various oil extraction methods, solvent extraction is particularly prevalent for MOSO.

Several studies optimised MOSO extraction using various solvents. Oladipo and Betiku (2019) demonstrated that using *n*-hexane as the solvent achieved an impressive oil yield of 42.0±0.5 wt%, with optimal conditions set at a three-hour extraction time and a solid-to-solvent ratio of 0.050 g/mL. Similar findings were also reported by Silva et al. (2016) that extraction of MO seed using *n*-hexane yielded 42.6%, while propane also showed promise with yields ranging from 32.8% to 42.1%. Ethanol, recognised as a less toxic alternative to *n*-hexane, has also been optimised for MO oil extraction (Arguelles-Peña et al., 2021). Both ethanol and *n*-hexane extractions produced similar fatty acid profiles, primarily comprising oleic, palmitic and stearic acids. Furthermore, Souza et al. (2024) also reported that while Soxhlet extraction with ethanol yielded a lower oil output (26.0%) than with *n*-hexane (38.9%), incorporating microwave pre-treatment increased ethanol-based yields to 42.0%, surpassing *n*-hexane's 35.9%. However, solvent extraction raises significant environmental concerns. The volatility and toxicity of solvents such as *n*-hexane pose notable health and ecological risks (Ahmad et al., 2019). The process emits volatile organic compounds (VOCs), contributing to air pollution and potential respiratory issues (Xiong et al., 2021). Research is still ongoing to develop greener solvents and improve the sustainability of the extraction process.

15.5 AQUEOUS ENZYMATIC EXTRACTION

AEE is a novel, environmentally sustainable approach for extracting oil from various oil-bearing materials. This method offers multiple advantages, including low energy consumption, product safety, mild reaction conditions and simultaneous separation of oil and protein. By utilising water and specific enzymes to degrade cell walls and release oil, AEE presents a cleaner alternative to traditional extraction techniques. Its green credentials stem from avoiding harmful organic solvents, thereby reducing environmental pollution and minimising health risks (Benkirane et al., 2022).

AEE has also been applied to the extraction of MO oil. Under optimised conditions – pH 4.5, a moisture-to-kernel ratio of 8:1, shaking speed of 300 strokes/min at 40°C for one hour – AEE has demonstrated oil recovery rates of approximately 70% (Mat Yusoff et al., 2017). Additionally, enzymes such as Viscozyme L, Celluclast 1.5L, Nutrase 0.8L and Alcalase 2.4LFG have been evaluated for their effectiveness in MOSO extraction. Studies indicate that using a ternary enzyme mixture can enhance extraction yields, although the results remain slightly lower than those of *n*-hexane extraction (Mehanni et al., 2014). Nevertheless, oil obtained via AEE exhibits superior oxidative stability and higher tocopherol retention compared with solvent-extracted oil (Mat Yusoff et al., 2020).

One of the primary challenges of AEE is emulsion formation, which can hinder oil separation. HPP pre-treatment has been shown to reduce emulsion thickness, thereby improving MO oil recovery (Mat Yusoff et al., 2017). Additionally, Zhang et al. (2022) addressed emulsion issues in Jiangyong pomelo seed oil by employing calcium chloride-assisted demulsification, a technique that could potentially be adapted for MO oil extraction. Continued research and optimisation aim to improve the efficiency and applicability of AEE for various oil-bearing materials, particularly for MO oil.

15.6 SUPERCRITICAL CO_2 EXTRACTION

SC-CO_2 extraction is an advanced and efficient technique for extracting oil from MO seeds. This method utilises the unique properties of CO_2 in its supercritical state, where it exhibits high diffusivity and low viscosity, enabling the effective extraction of bioactive compounds (Casas et al., 2009). Key parameters influencing the SC-CO_2 extraction process include pressure, temperature, particle size, CO_2 flow rate and extraction time (Smith et al., 2022; Ruttarattanamongkol and Petrasch, 2016). SC-CO_2 extraction is widely regarded as an environmentally sustainable method, as it eliminates the need for toxic organic solvents, reducing both environmental and health hazards (Ruttarattanamongkol et al., 2014).

Ruttarattanamongkol and Petrasch (2016) demonstrated that SC-CO_2 extraction at 35 MPa and 30°C achieved a maximum oil yield of 75.27% on a dry-weight basis. When compared with oil extracted using the Soxhlet method, SC-CO_2-extracted oil showed reduced oxidative stability, as evidenced by its weaker DPPH (1,1-diphenyl-2-picrylhydrazyl) scavenging activity and lower oxidative stability index (OSI). However, SC-CO_2 extraction is reported to be more selective than conventional methods, producing purer extracts with fewer unwanted compounds (Zhao and Zhang, 2013).

Further optimisation of the SC-CO2 process was explored by Chen et al. (2022), who achieved a maximum oil yield of 38.54% under conditions of 45°C, 50 MPa, 2.5 hours extraction time and a CO_2 flow rate of 240 L/h. Analysis of the extracted oil revealed a high content of unsaturated fatty acids, predominantly oleic acid, which comprises approximately 70–76% of the total fatty acid composition. Other fatty acids present include octadecanoic acid, palmitic acid, stearic acid and eicosanoic acid (Kumar et al., 2024).

15.7 PHYSICOCHEMICAL PROPERTIES OF *M. OLEIFERA* SEED OILS

Edible oils, including MOSO, were characterised using several physicochemical properties to assess their quality and authenticity. Acid value, peroxide value, anisidine value and some oxidation products are used as parameters for edible oil quality, while saponification value, iodine value, fatty acid composition and triacylglycerol compositions can be used as tools for the authenticity of edible oils since these parameters are characteristics of the corresponding studied oils (Taerian Esfahani et al., 2024). Irnawati et al. reported that MOSO extracted from *M. oleifera* seeds using the maceration technique, applying hexane as the extracting solvent obtained from different regions in Indonesia, has acid values (mg KOH/g MOSO) of 0.32±0.01, iodine values (g I_2/100 g MOSO) of 78.30±4.24, saponification values (mg KOH/g MOSO) of 182.50±5.95 and peroxide values (meq O_2/kg MOSO) of 1.12±0.02. Fatty acid compositions can be considered fingerprinting tools for the characterisation of edible fats and oils, since no two fats or oils have the same fatty acid composition in terms of fatty acid types and levels (van Ruth et al., 2010). Table 15.1 shows fatty acid compositions of MOSO obtained from different regions in Indonesia. The dominant fatty acids of MOSO were oleic acid and palmitic acid, accounting for 68.31–76.89% and 6.17–8.27%, respectively.

Cervera-Chiner et al. (2024) evaluated the physicochemical properties and fatty acid composition of extracted MOSO using an automatic screw press extractor at temperatures of 70, 100, 130, 160, 190 and 220°C. In this study, acid values ranged from 3 to 3.6 mg KOH/g MOSO, iodine values from 124 to 168 g I_2/100 g MOSO, saponification values from 187 to 219 mg KOH/g MOSO and peroxide values of 0.19±0.12 meqO2/kg MOSO. In addition, the specific extinction coefficients for UV absorption of MOSO at 232 nm (K_{232}) and 270 nm (K_{270}) were 270–300 and 10.3–23.46, respectively. The analysed MOSO contained unsaturated fatty acids, with oleic acid (approximately 77.8%) and linolenic acid (3.4%), as well as behenic acid (7%), palmitic acid (6%), stearic acid (5%) and arachidic acid (0.2%).

The physical and chemical properties of MOSO extracted from *M. oleifera* seeds subjected to irradiation using radioactive ^{60}Co at dose levels of 0.0 (control, unirradiated), 5 kGy, 10 kGy and 15 kGy were investigated. The density of MOSO subjected to irradiation was not statistically different (in the range of 0.9013–0.9210 g/cm^3), while the iodine value was reduced from 76.50 to 55.21 g I_2/100 g. In addition, the acid value and saponification value were increased in all irradiated samples. The contents of total phenolics and total flavonoids were also studied, in which MOSO irradiated at 10 kGy had the highest phenolics and flavonoids, accounting for 82.21 mg gallic acid equivalent/100 g MOSO and 93.28 mg quercetin equivalent/100 g MOSO, respectively, compared with unirradiated MOSO, accounting for 50.29 mg gallic acid equivalent/100 g and 81.49 mg/100 g quercetin equivalent/100 g MOSO (Aly et al., 2016).

TABLE 15.1
Fatty Acid Compositions of MOSO Obtained from Several Regions in Indonesia

M. oleifera seed oils	Fatty acid composition (%)													
	C12:0	C14:0	C16:0	C16:1	C17:0	C18:0	C18:1	C18:2	C18:3	C20:0	C20:1	C22:0	C22:1	C24:0
Wakatobi	0.011±0.000^{bc}	0.064±0.004^{f}	7.942±0.016^{b}	1.422±0.020^{de}	0.031±0.053^{ab}	5.424±0.041^{de}	70.801±1.408^{ef}	0.772±0.007^{cd}	0.056±0.000^{g}	3.451±0.019^{bc}	2.534±0.065^{bcd}	6.232±0.105^{cd}	nd	0.837±0.032^{a}
West Muna	0.007±0.000^{bc}	0.113±0.005^{bc}	7.459±0.054^{cd}	1.508±0.015^{d}	0.065±0.004^{ab}	3.726±0.116^{g}	73.930±1.383^{bcde}	0.927±0.012^{a}	0.145±0.005^{a}	2.481±0.040^{fg}	2.476±0.046^{cd}	5.392±0.076^{f}	nd	0.889±0.053^{a}
South Buton	nd	0.094±0.006^{de}	6.124±0.046^{h}	1.411±0.010^{de}	0.000±0.000^{b}	3.683±0.044^{g}	75.637±1.450^{abc}	0.676±0.008^{fg}	0.072±0.000^{f}	2.436±0.034^{g}	2.121±0.085^{e}	6.358±0.128^{bc}	nd	0.688±0.033^{abc}
Kendari	nd	0.127±0.006^{b}	7.368±0.157^{cd}	1.891±0.083^{a}	0.025±0.043^{ab}	6.304±0.077^{ab}	72.569±1.400^{cdef}	0.785±0.009^{bcd}	0.145±0.000^{a}	3.399±0.071^{bc}	1.812±0.048^{f}	4.349±0.123^{g}	nd	0.658±0.022^{abc}
East Kolaka	nd	0.053±0.000^{f}	7.173±0.150^{de}	1.171±0.020^{gh}	0.000±0.000^{b}	3.547±0.071^{g}	75.757±1.014^{abc}	0.746±0.010^{de}	0.058±0.000^{g}	2.305±0.033^{g}	2.355±0.080^{d}	5.079±0.141^{f}	nd	0.766±0.298^{ab}
Buton	nd	0.113±0.005^{bc}	7.715±0.052^{bc}	1.068±0.013^{hi}	0.024±0.028^{ab}	6.012±0.130^{bc}	71.990±1.158^{def}	0.779±0.010^{cd}	0.133±0.000^{b}	3.455±0.044^{bc}	2.414±0.241^{d}	5.468±0.061^{ef}	nd	0.859±0.034^{a}
North Buton	0.225±0.015^{a}	0.632±0.016^{a}	8.570±0.289^{a}	0.701±0.019^{j}	0.000±0.000^{b}	4.774±0.096^{f}	72.009±0.514^{def}	0.773±0.031^{cd}	0.085±0.000^{e}	2.673±0.031^{ef}	1.825±0.014^{f}	5.080±0.060^{f}	nd	0.196±0.178^{c}
Konawe	nd	0.129±0.010^{b}	7.139±0.081^{de}	1.045±0.013^{i}	0.000±0.000^{b}	5.668±0.127^{cd}	72.434±0.290^{cdef}	0.816±0.014^{bc}	0.050±0.000^{g}	3.293±0.070^{cd}	2.714±0.000^{b}	6.328±0.228^{bc}	nd	0.709±0.043^{abc}
Bau-Bau	nd	0.058±0.000^{f}	6.834±0.105^{ef}	1.460±0.053^{de}	0.000±0.000^{b}	5.240±0.092e	77.860±0.736^{a}	0.613±0.014^{h}	0.000±0.000^{h}	2.755±0.075^{e}	1.325±0.051^{g}	4.450±0.275^{g}	0.000±0.000^{f}	0.228±0.395^{bc}
Kolaka	0.014±0.000^{b}	0.114±0.000^{bc}	6.750±0.101^{efg}	1.794±0.032^{ab}	0.083±0.004^{a}	6.525±0.099^{a}	69.344±0.931^{f}	0.943±0.025^{a}	0.137±0.000^{b}	4.433±0.059^{a}	3.321±0.956^{a}	5.864±0.097^{de}	0.107±0.006^{a}	1.184±0.031^{a}
Central Buton	nd	0.054±0.000^{f}	7.375±0.259^{cd}	1.353±0.045^{ef}	0.000±0.000^{b}	4.546±0.124^{f}	75.129±0.547^{abcd}	0.607±0.012^{h}	0.000±0.000^{h}	2.745±0.137^{e}	2.136±0.015^{e}	5.068±0.133^{f}	0.000±0.000^{f}	0.705±0.019^{abc}
Muna	nd	0.104±0.000^{cd}	6.353±0.040^{gh}	1.248±0.016^{fg}	0.043±0.042^{ab}	5.348±0.123^{de}	76.167±1.024^{ab}	0.761±0.020^{de}	0.101±0.000^{d}	3.158±0.050^{d}	2.469±0.146^{cd}	6.107±0.240^{cd}	0.051±0.000^{d}	0.762±0.010^{ab}
South Konawe	0.008±0.000^{bc}	0.123±0.005^{b}	8.115±0.132^{b}	1.367±0.026^{e}	0.024±0.042^{ab}	3.811±0.077^{g}	77.305±1.139^{a}	0.714±0.014^{ef}	0.098±0.000^{d}	2.504±0.080^{fg}	2.634±0.009^{bc}	6.678±0.148^{b}	0.058±0.000^{c}	0.725±0.019^{abc}
Pati	nd	0.083±0.000^{e}	6.599±0.160^{fg}	1.736±0.021^{bc}	0.060±0.001^{ab}	5.581±0.086^{de}	71.166±1.224^{ef}	0.654±0.024^{gh}	0.112±0.000^{c}	3.564±0.126^{b}	2.501±0.061^{cd}	7.394±0.065^{a}	0.089±0.000^{b}	0.748±0.07^{3ab}
Bone	0.008±0.000^{bc}	0.120±0.000^{bc}	6.601±0.163^{fg}	1.627±0.067^{c}	0.080±0.001^{a}	6.631±0.306^{a}	69.385±1.489^{f}	0.835±0.033^{b}	0.151±0.009^{a}	4.405±0.045^{a}	2.433±0.042^{d}	6.419±0.091^{bc}	0.029±0.000^{e}	0.751±0.548^{ab}

Means with different lowercase letters within a column in each origin are significantly different ($p<0.05$) after applying a one-way ANOVA and Tukey's post hoc test. Values are mean±SD of triplicate; nd, not detected with a limit of detection of 0.001%. Taken from Irnawati et al. (2024) with permission.

15.8 PHARMACOLOGICAL ACTIVITIES OF *M. OLEIFERA* SEED OILS

In 2016, Leone et al. highlighted the value of MO seeds in food and non-food applications, attributing this to their rich content of healthy monounsaturated fats, sterols, tocopherols and sulphur-rich proteins (Leone et al., 2016). While MO seeds and their oil are often used in traditional medicine, their effects on humans have not been scientifically studied until now. Recent studies have shown that MOSO exhibits many useful pharmacological effects.

Díaz-Domínguez et al. (2022) discussed the application of MOSO for treating various diseases, including its antimicrobial, antifungal, antioxidant, antitubercular and anticancer activities, citing its purported biological properties. However, many of the claims made in cited publications lack support from robust evidence, such as placebo-controlled, randomised clinical trials. Furthermore, these studies often lack detailed descriptions of validation methods and may not have been published in high-impact, peer-reviewed journals. Gharsallah et al. (2022) reported that MOSO exhibited antioxidant and anti-inflammatory activity. The composition of fatty acids, sterols and phenolic compounds was also described. The therapeutic effects of bioactive compounds of MOSO, namely oleic acids, β-sitosterol and phenolic compounds, were also summarised and discussed. Some phenolic compounds have demonstrated potential for use in cancer treatment, as well as in the management of diabetes and skin disorders. Shahbaz et al. (2024) reviewed the pharmacological potential of moringa seeds, including MOSO. It suggests that MOSO may reduce the risk of cardiovascular diseases by inhibiting malondialdehyde formation and modulating the activity of cardiac antioxidant enzymes, such as superoxide dismutase and glutathione peroxidase. Unfortunately, these reviews did not evaluate the validation methods employed by the cited publications, and the accurate chemical specification of the MOSO used was not reported, so the integrity of the reported data cannot be proved, raising concerns about the reliability of the cited publications (Gharsallah et al., 2022).

This section will provide a summary and discussion of MOSO's bioactivity and pharmacological properties, as reported in original articles from 2019 to 2025. Famurewa et al. (2019) reported that commercial MOSO helps to restore the balance of oxidation and reduction (redox) in the body and influences a specific chain of cellular events (the iNOS/NF-kB/caspase-3 signalling pathway). These actions likely contribute to the oil's ability to protect against kidney damage caused by the anticancer drug 5-fluorouracil in rats by reducing oxidative stress, inflammation and cell death. Cretella et al. (2020) investigated the use of commercial MOSO for treating skin diseases in female Swiss mice characterised by keratinocyte hyperproliferation. Oleic acid (OA), the main component of MOSO, is also effective in acute inflammation. MOSO and OA depend on glucocorticoid receptor activation for anti-inflammatory effects, but do not exhibit the same adverse effects seen in topical treatment with dexamethasone. Das et al. (2024) demonstrated that self-extracted MOSO exhibits cytotoxic and antiproliferative effects against oral cancer cell lines (CAL27 and SCC15). This research suggests that MOSO may be a promising anticancer agent for oral cancer treatment. GCxGC TOF-MS analysis was employed to identify the major constituents of MOSO. Saka et al. reported that self-extracted MOSO exhibits hepatoprotective and nephroprotective properties. While the chemical characterisation of MOSO was not detailed, the study revealed that MOSO effectively mitigated the toxic effects of dichlorvos exposure in male Wistar rats, with significant improvements in key hepatic and renal biomarkers being observed. According to Zouboulis et al. (2023), the anti-inflammatory OA found in MOSO likely contributes to its low levels of pro-inflammatory cytokine secretion and its ability to induce cell death. MOSO appears to possess several desirable properties for cultured human sebocytes SZ95 cell lines, including a high concentration of the anti-inflammatory fatty acid OA. It induces cell proliferation and lipogenic patterns similar to OA and promotes lipogenesis with a low inhibition of pro-inflammatory cytokine secretion. The composition (%) of the main fatty acid in MOSO was reported, but unfortunately, the exact origin of MOSO is not specified. Oliveira de Sousa et al. (2024) reported that although the fixed oil extracted by pressured *n*-propane from moringa seeds did not exhibit antibacterial activity against the tested bacterial strains, it demonstrated synergistic effects by reducing the minimum inhibitory concentrations of the antibiotics from the aminoglycoside, β-lactam and fluoroquinolone classes, for both sensitive and multidrug-resistant bacterial strains. The fatty acid composition of MOSO was determined using GC-FID.

Guedes et al. (2022) found that protein and fixed oil isolated or extracted from MO seeds exhibited the ability to inhibit the growth of *Trichophyton tonsurans* when tested separately. Interestingly, combining these extracts resulted in a synergistic effect, meaning their combined antifungal activity was stronger than the sum of their individual effects. This enhanced activity was observed both independently and when combined with some commercially available antifungal drugs. These findings suggest that these extracts from MO seeds have considerable potential for treating dermatophytosis caused by this fungus. The identities of fatty acid methyl esters in MOSO were determined by GC-MS, while quantification was carried out using GC-FID. Utilising DDY strain mice obtained from Biofarma Laboratory in Indonesia, Arozal et al. (2024) demonstrated that MOSO and aqueous leaf extracts of MO (MOE) possess neuroprotective properties. Notably, MOSO exhibited a more pronounced ability to inhibit brain senescence compared with MOE. Their findings revealed that administering MOE or MOSO led to an increased percentage of alternation and recognition of novel objects. Furthermore, both extracts prevented the upregulation of p16 and p21 expression and ameliorated senescent cell-associated β-galactosidase (SA-β-gal) activity within the brain. Edeogu et al. (2020) found that MOSO effectively reduced serum creatinine and urea levels in gentamicin (GM)-induced kidney injury. Additionally, MOSO significantly reduced oxidative stress by lowering renal malondialdehyde levels, enhancing the activity of antioxidant enzymes, such as superoxide dismutase, catalase and glutathione peroxidase, and increasing reduced glutathione levels. Moreover, MOSO restored renal levels of inflammatory cytokines (IL-1β, IL-6, TNF-α) and nitric oxide by downregulating the expression of NF-κB, iNOS and caspase-3. These biochemical and antioxidant improvements were accompanied by a significant reduction in histopathological kidney damage.

Elmalawany et al. (2024) reported that both MOSO and praziquantel effectively decreased the number of parasite eggs of *Schistosoma mansoni* in mice, thereby reducing the size of liver lesions (granulomas) and improving the overall health of the liver tissue. MOSO and/or praziquantel significantly reduced egg numbers, granuloma size and liver damage. MOSO specifically improved liver function by reducing hepatic enzyme activity, increasing protein levels, enhancing antioxidant activity and decreasing oxidative stress markers. Additionally, MOSO reduced pro-inflammatory cytokine levels. De Oliveira et al. (2023) reported that MOSO extracted with pressurised *n*-propane can potentially inhibit biofilms of different *Staphylococcus aureus* strains, thus contributing to microbiological safety during food processing. Tunit et al. (2022) reported that emulgels containing *Perilla frutescens* seed oil, MOSO and a mix of the two seed oils do not induce cytotoxicity to peripheral blood mononuclear cells nor induce skin irritation and allergic reactions in healthy volunteers. Mottaghi et al. (2024) reported that the use of solid lipid nanoparticles (SLNs) as a nano-carrier for *M. oleifera* oil extract (MOOE) and *Cinnamomum zeylanicum* oil extract could kill *Toxoplasma gondii* tachyzoites with low cytotoxicity. This suggests the effectiveness of these nano-emulsions, along with chemical agents, in treating *Toxoplasma*. The origin and specification of the oil extracts were not reported. Aldayel et al. (2023) reported that commercial MOSO significantly ameliorated Ehrlich ascites carcinoma (EAC) in Swiss female albino mice by increasing body weight and reducing EAC cell viability. The major compounds identified in MOSO by HPLC were cinnamic acid, ellagic acid, quercetin, gallic acid, vanillin and hesperidin. The molecular docking study highlighted the virtual binding of the identified compounds within GSH and SOD proteins, with quercetin exhibiting a particularly promising binding affinity and a good interactive binding mode with the key amino acids. Ventura et al. (2021) described commercial MOSO and oleic acid as effective agents for wound healing, particularly in difficult-to-treat conditions such as diabetes and immunosuppression. No chemical characterisation of the MOSO was reported. Sakr et al. (2021) underscore the harmful effects of tributyltin on brain health and highlighted the potential of commercial MOSO as a protective agent, warranting further exploration in both preclinical and clinical settings. Liu et al. (2020) reported that self-extracted MOSO modulates sleep architecture by activating GABAergic systems, suggesting its potential as a natural sleep aid. The sterol and fatty acid composition of MOSO was determined.

The original articles above demonstrated that MOSO exhibits numerous beneficial pharmacological effects for humans. However, several shortcomings were identified. First, no validation methods for the bioassays were described or discussed. Second, several articles failed to report the chemical characteristics of the MOSO used in their bioactivity studies. In some cases, a commercial MOSO of unspecified origin was used in the experiments. These omissions raise concerns about the integrity of the cited research data, as the results may not be reproducible by other researchers. To ensure data integrity, it is crucial to: (1) perform a complete chemical characterisation of MOSO, (2) authenticate all biological resources and (3) validate all methodologies before routine application.

REFERENCES

Abdulkarim, S. M., Lai, O. M., Long, K., & Ghazali, H. M. (2005). Aqueous extraction of moringa oleifera seed oil. *Journal of Food Lipids*, *60*(13), 113–130.

Ahangarnejad, N., Khoshtaghaza, M. H., Ghobadian, B., & Sahari, M. A. (2021). Optimization of extraction from moringa peregrina seed oil by microwave pretreatment. *Journal of Food Science and Technology*, *18*(120), 229–239. https://doi.org/10.52547/fsct.18.120.18

Ahmad, S. I., Yunus, N. A., Hashim, H., & Rashid, R. (2019). Inherent safety assessment of solvent alternatives for palm oil recovery. *Chemical Engineering Transactions*, *72*, 49–54. https://doi.org/10.3303/CET1972009

Aldayel, T. S., Gad El Hak, H. N., Nafie, M. S., Saad, R., Abdelrazek, H. M. A., & Kilany, O. E. (2023). Evaluation of antioxidant, anti-inflammatory, anticancer activities and molecular docking of Moringa oleifera seed oil extract against experimental model of Ehrlich ascites carcinoma in Swiss female albino mice. *BMC Complementary Medicine and Therapies*, *23*(1), 1–15. https://doi.org/10.1186/s12906-023-04279-z

Aly, A. A., Maraei, R. W., & Ali, H. G. M. (2016). Fatty acids profile and chemical composition of Egyptian moringa oleifera seed oils. *Journal of the American Oil Chemists' Society*, *93*(3), 397–404. https://doi.org/10.1007/s11746-015-2781-6

Amin, M. F., Ariwibowo, T., Putri, S. A., & Kurnia, D. (2024). Moringa oleifera: A review of the pharmacology, chemical constituents, and application for dental health. *Pharmaceuticals*, *17*(1). https://doi.org/10.3390/ph17010142

Andasuryani, Ifmalinda, Derosya, V., &Syukri, D. (2020). The effect of moisture on the yield of sweet passion fruit (Passiflora lingularis Juss cv. Gumanti) seed oil from Indonesia by mechanical extraction. *IOP Conference Series: Earth and Environmental Science*, *515*(1). https://doi.org/10.1088/1755-1315/515/1/012021

Arguelles-Peña, K., Olguín-Rojas, J. A., Acosta-Osorio, A. A., Carrera, C., Barbero, G. F., García-Alvarado, M. Á., & Rodríguez-Jimenes, G. D. C. (2021). An evaluation of the equilibrium properties in hexane and ethanol extractive systems for Moringa oleifera seeds and fatty acid profiles of the extracts. *Separations*, *8*(11), 217. https://doi.org/10.3390/separations8110217

Arozal, W., Safutra, M. S., Barinda, A. J., Hardi, H., Dwita, N. C., & Lee, H. J. (2024). Comparative neuroprotective effects of moringa oleifera seed oil and aqueous extract on cognitive functions on a high-fat, high-fructose diet mice: Focus on senescence markers. *Scientific World Journal*, *2024*(2024), 1–9. https://doi.org/10.1155/2024/8034401

Athikomkulchai, S., Tunit, P., Tadtong, S., Jantrawut, P., Sommano, S. R., & Chittasupho, C. (2021). Moringa oleifera seed oil formulation physical stability and chemical constituents for enhancing skin hydration and antioxidant activity. *Cosmetics*, *8*(1), 1–18. https://doi.org/10.3390/COSMETICS8010002

Ayu, L. R., Aliwarga, L., & Adisasmito, S. (2024). Mechanical and Solvent Extraction of Moringa Oleifera Seeds for Vegetable Oil. *Key Engineering Materials*, *989*, 99–107. https://doi.org/10.4028/P-BYR9LQ

Azadmard-Damirchi, S., Alirezalu, K., & Achachlouei, B. F. (2011). Microwave pretreatment of seeds to extract high quality vegetable oil. *World Academy of Science, Engineering and Technology*, *81*(9), 513–516.

Benkirane, C., Allay, A., Ben Moumen, A., Belhaj, K., Melhaoui, R., Mansouri, F., Abid, M., Serghini-Caid, H., & Elamrani, A. (2022). Aqueous enzymatic extraction of safflower (Carthamus tinctorius L.) seed oil: A preliminary study. *E3S Web of Conferences*, *337*, 1–9. https://doi.org/10.1051/e3sconf/202233704010

Casas, L., Mantell, C., Rodríguez, M., López, E., & De La Ossa, E. M. (2009). Industrial design of multifunctional supercritical extraction plant for agro-food raw materials. *Chemical Engineering Transactions, 17,* 1585–1590. https://doi.org/10.3303/CET0917265

Cervera-Chiner, L., Pageo, S., Juan-Borrás, M., García-Mares, F. J., Castelló, M. L., &Ortolá, M. D. (2024). Fatty acid profile and physicochemical properties of Moringa oleifera seed oil extracted at different temperatures. *Foods, 13*(17), 2733. https://doi.org/10.3390/FOODS13172733

Chen, X., Li, Z., Smith, S. A., Chen, M., Liu, H., Zhang, J., Tang, L., Li, J., Liu, Q., & Wu, X. (2022). Optimization of supercritical CO2 extraction of Moringa oleifera seed oil using response surface methodological approach and its antioxidant activity. *Frontiers in Nutrition, 8,* 829146. https://doi.org/10.3389/FNUT.2021.829146/BIBTEX

Cretella, A. B. M., Soley, B. da S., Pawloski, P. L., Ruziska, R. M., Scharf, D. R., Ascari, J., Cabrini, D. A., & Otuki, M. F. (2020). Expanding the anti-inflammatory potential of Moringa oleifera: Topical effect of seed oil on skin inflammation and hyperproliferation. *Journal of Ethnopharmacology, 254,* 112708. https://doi.org/10.1016/j.jep.2020.112708

Da Porto, C., Decorti, D., &Natolino, A. (2016). Microwave pretreatment of Moringa oleifera seed: Effect on oil obtained by pilot-scale supercritical carbon dioxide extraction and Soxhlet apparatus. *Journal of Supercritical Fluids, 107,* 38–43. https://doi.org/10.1016/j.supflu.2015.08.006

Das, M., Mohanty, J. N., Bhuyan, S. K., & Bhuyan, R. (2024). Anti-cancer activity of Moringa oleifera Lam. seed oil on oral cancer. *Beni-Suef University Journal of Basic and Applied Sciences, 13*(1), 25. https://doi.org/10.1186/s43088-024-00475-z

de Sousa, E. O., Brito, A. dos S., Feitoza, D. M. A., Ribeiro, A. de A., Tavares, V. H. N. D., Costa, M. G. M., Nobre, C. de A., Oliveira-Tintino, C. D. de M., Borges, J. A. de O., de Araújo, G. A., Moura, T. F., de Lima, J. P. M., Florencio, S. G. L., Coutinho, H. D. M., de Menezes, I. R. A., & Tintino, S. R. (2024). Chemical analysis of Moringa oleifera (Moringaceae) seed oil and potentiation of antibiotic activity against standard and multidrug-resistant bacterial strains. *Food Chemistry Advances, 5,* 100765. https://doi.org/10.1016/j.focha.2024.100765

Díaz-Domínguez, Y., Piloto-Rodríguez, R., Fernández-Santan, E., Rondón-Macias, M., & Tabio-García, D. (2022). Chapter 5 - Biological activities of Moringa seeds oil. In A. A. Mariod (Ed.), *Multiple biological activities of unconventional seed oils* (1st ed., pp. 55–75). Academic Press Inc. https://doi.org/10.1016/j.sajb.2019.06.017

Edeogu, C. O., Kalu, M. E., Famurewa, A. C., Asogwa, N. T., Onyeji, G. N., & Ikpemo, K. O. (2020). Nephroprotective effect of Moringa Oleifera seed oil on gentamicin-induced nephrotoxicity in rats: Biochemical evaluation of antioxidant, anti-inflammatory, and antiapoptotic pathways. *Journal of the American College of Nutrition, 39*(4), 307–315. https://doi.org/10.1080/07315724.2019.1649218

Elmalawany, A. M., Osman, G. Y., Mohamed, A. H., Khalaf, F. M., & Yassien, R. I. (2024). Schistosomicidal effects of moringa oleifera seed oil extract on schistosoma mansoni-Infected Mice. *Parasite Immumology, 46*(11), e13070. https://doi.org/10.1111/pim.13070

Eman, N. A., & Muhamad, K. N. S. (2016). Comparison of Moringa Oleifera seeds oil characterization produced chemically and mechanically. *IOP Conference Series: Earth and Environmental Science, 36*(1), 012063. https://doi.org/10.1088/1755-1315/36/1/012063

Ewunie, G. A., Morken, J., Lekang, O. I., &Yigezu, Z. D. (2021). Factors affecting the potential of Jatropha curcas for sustainable biodiesel production: A critical review. *Renewable and Sustainable Energy Reviews, 137,* 110500. https://doi.org/10.1016/j.rser.2020.110500

Famurewa, A. C., Asogwa, N. T., Aja, P. M., Akunna, G. G., Awoke, J. N., Ekeleme-Egedigwe, C. A., Maduagwuna, E. K., Folawiyo, A. M., Besong, E. E., Ekpono, E. U., & Nwoha, P. A. (2019). Moringa oleifera seed oil modulates redox imbalance and iNOS/NF-κB/caspase-3 signaling pathway to exert antioxidant, anti-inflammatory and antiapoptotic mechanisms against anticancer drug 5-fluorouracil-induced nephrotoxicity in rats. *South African Journal of Botany, 127,* 96–103. https://doi.org/10.1016/j.sajb.2019.08.038

Gharsallah, K., Rezig, L., & Mahfoudhi, N. (2022, September). Beneficial effects of Moringa oleifera seed oil bioactive compounds. In *Handbook of research on advanced phytochemicals and plant-based drug discovery* (pp. 268–291). https://doi.org/10.4018/978-1-6684-5129-8.ch013

Gharsallah, K., Rezig, L., Msaada, K., Chalh, A., & Soltani, T. (2021). Chemical compositionand profile characterization of moringa oleifera seed oil. *South African Journal of Botany, 137,* 475–482. https://doi.org/10.1016/j.sajb.2020.11.014

Guedes, C. C. da S., Buonafina-Paz, M. D. S., Rocha, S. K. L., Coelho, L. C. B. B., Navarro, D. M. do A. F., Neves, R. P., Napoleão, T. H., de Oliveira, A. P. S., da Silva, P. M., & Paiva, P. M. G. (2022). Antimycotic potential of protein preparation and fixed oil obtained from Moringa oleifera seeds against Trichophyton tonsurans. *South African Journal of Botany, 150*, 443–450. https://doi.org/10.1016/j.sajb.2022.08.023

Irnawati, I., Rohman, A., Ahmad, N., Azrina, A., Windarsih, A., & Asmah, R. (2024). Case studies in chemical and environmental engineering Moringa oleifera seed oils : Physico-chemical characterization and its authentication using FTIR spectroscopy and chemometrics. *Case Studies in Chemical and Environmental Engineering, 10*, 100994. https://doi.org/10.1016/j.cscee.2024.100994

Kayanan, B. U. R., & Sagum, R. S. (2021). Microwave and ultrasound pretreatment of moringa oleifera lam. Seeds: Effects on oil expression, oil quality, and bioactive component. *Journal of Oleo Science, 70*(7), 875–884. https://doi.org/10.5650/jos.ess20357

Kumar, S., Rai, A., & Prasad, K. (2024). Optimization and multifactor assessments of supercritical Moringa oleifera seed oil extraction. *Journal of Food Measurement and Characterization, 19*(3), 1586–1602. https://doi.org/10.1007/s11694-024-03045-0

Latif, S., Anwar, F., Hussain, A. I., & Shahid, M. (2011). Aqueous enzymatic process for oil and protein extraction from Moringa oleifera seed. *European Journal of Lipid Science and Technology, 113*(8), 1012–1018. https://doi.org/10.1002/ejlt.201000525

Leone, A., Spada, A., Battezzati, A., Schiraldi, A., Aristil, J., & Bertoli, S. (2016). Moringa oleifera seeds and oil: Characteristics and uses for human health. *International Journal of Molecular Sciences, 17*(12), 1–14. https://doi.org/10.3390/ijms17122141

Leone, A., Tamborrino, A., Zagaria, R., Sabella, E., & Romaniello, R. (2015). Plant innovation in the olive oil extraction process: A comparison of efficiency and energy consumption between microwave treatment and traditional malaxation of olive pastes. *Journal of Food Engineering, 146*, 44–52. https://doi.org/10.1016/j.jfoodeng.2014.08.017

Li, Z., Li, X., Tang, S., Gao, Q., Li, C., Chen, P., Yue, X., Fu, R., Huang, X., Zhang, Y., Yang, H., & Yang, B. (2024). Moringa oleifera Lam. Leaf improves constipation of rats induced by low-fiber-diet : A proteomics study. *Journal of Ethnopharmacology, 318*, 116936. https://doi.org/10.1016/j.jep.2023.116936

Liu, J., Chen, P., He, J., Deng, L., Wang, L., Lei, J., & Rong, L. (2014). Extraction of oil from Jatropha curcas seeds by subcritical fluid extraction. *Industrial Crops and Products, 62*, 235–241. https://doi.org/10.1016/j.indcrop.2014.08.039

Liu, W. L., Wu, B. F., Shang, J. H., Zhao, Y. L., & Huang, A. X. (2020). Moringa oleifera lam seed oil augments pentobarbital-induced sleeping behaviors in mice via GABAergic systems. *Journal of Agricultural and Food Chemistry, 68*(10), 3149–3162. https://doi.org/10.1021/acs.jafc.0c00037

Mat Yusoff, M., Gordon, M. H., Ezeh, O., & Niranjan, K. (2016). Aqueous enzymatic extraction of Moringa oleifera oil. *Food Chemistry, 211*, 400–408. https://doi.org/10.1016/j.foodchem.2016.05.050

Mat Yusoff, M., Gordon, M. H., Ezeh, O., & Niranjan, K. (2017). High pressure pre-treatment of Moringa oleifera seed kernels prior to aqueous enzymatic oil extraction. *Innovative Food Science and Emerging Technologies, 39*, 129–136. https://doi.org/10.1016/j.ifset.2016.11.014

Mat Yusoff, M., Niranjan, K., Mason, O. A., & Gordon, M. H. (2020). Oxidative properties of Moringa oleifera kernel oil from different extraction methods during storage. *Journal of the Science of Food and Agriculture, 100*(4), 1588–1597. https://doi.org/10.1002/jsfa.10167

Matic, I., Guidi, A., Kenzo, M., Mattei, M., & Galgani, A. (2018). Investigation of medicinal adietary plants review traditionally supplements: Onmoringa used oleiferaas. *Journal of Public Health in Africa, 9*(3), 191–199. https://doi.org/10.4081/jphia.2018.841

Mehanni, A. E., El-Reffaei, W. H., Akdowa, E. P., & Bouba, A. A. (2014). Enzyme aqueous extraction of Moringa oleifera and Canola seed oils and their effect on physiochemical characteristics and stability. *Rivista Italiana Delle Sostanze Grasse, 91*(3), 187–197.

Moneim, S. A., & Al-Anbari, I. H. A. (2024). Effect of moringa (moringa oleifera lam.) Seed oil extraction methods on its physicochemical properties. *Sabrao Journal of Breeding and Genetics, 56*(5), 2143–2151. https://doi.org/10.54910/SABRAO2024.56.5.37

Mottaghi, M., Karami, P., Hesari, Z., Nemati, S., Mohammad Rahimi, H., &Mirjalali, H. (2024). Evaluation of anti-Toxoplasma effects of solid lipid nanoparticles carrying Cinnamon zeylanicum and Moringa oleifera oil extracts. *BMC Complementary Medicine and Therapies, 24*(1), 375. https://doi.org/10.1186/s12906-024-04677-x

Nadeem, M., & Imran, M. (2016). Promising features of Moringa oleifera oil: Recent updates and perspectives. *Lipids in Health and Disease, 15*(1), 1–8. https://doi.org/10.1186/s12944-016-0379-0

Ngcobo, B. L., & Mbuyisa, S. (2024). Recent developments in environmentally friendly techniques for extracting bioactive compounds from moringa plant parts. *Acta Horticulturae*, *1394*, 93–98. https://doi.org/10.17660/ACTAHORTIC.2024.1394.12

Oladipo, B., & Betiku, E. (2019). Process optimization of solvent extraction of seed oil from Moringa oleifera: An appraisal of quantitative and qualitative process variables on oil quality using D-optimal design. *Biocatalysis and Agricultural Biotechnology*, *20*, 101187. https://doi.org/10.1016/j.bcab.2019.101187

Oliveira, A. M. de, Anjos Szczerepa, M. M. dos, BronharoTognim, M. C., Abreu Filho, B. A. de, Cardozo-Filho, L., Gomes, R. G., & Bergamasco, R. (2023). Moringa oleifera seed oil extracted by pressurized n-propane and its effect against Staphylococcus aureus biofilms. *Environmental Technology*, *44*(8), 1083–1098. https://doi.org/10.1080/09593330.2021.1994653

Pereira, F. S. G., De Brito Neto, E. X., Wei, S., Galvão, C. C., De Lima, V. F., Da Silva, V. L., & De Lima Filho, N. M. (2016). Production of methyl biodiesel using purified oil of Moringa oleífera lamarck. *Revista Virtual de Química*, *8*(3), 873–888. https://doi.org/10.5935/1984-6835.20160063

Pereira, F. S. G., Galvaõ, C. C., De Lima, V. F., Da Rocha, M. F. A., Schuler, A. R. P., Da Silva, V. L., & De Lima Filho, N. M. (2016). The versatility of the Moringa oleifera oil in sustainable applications. *OCL*, *23*(6), A601. https://doi.org/10.1051/OCL/2016027

Raja, W., Bera, K., & Ray, B. (2016). Polysaccharides from Moringa oleifera gum: structural elements, interaction with beta-blactoglobulin and antioxidative activity. *RSC Advances*, *6*, 75699–75706. https://doi.org/10.1039/C6RA13279K

Ruttarattanamongkol, K., & Petrasch, A. (2016). Oxidative susceptibility and thermal properties of Moringa Oliefera seed oil obtained by pilot-scale subcritical and supercritical carbon dioxide extraction. *Journal of Food Process Engineering*, *39*(3), 226–236. https://doi.org/10.1111/jfpe.12213

Ruttarattanamongkol, K., Siebenhandl-Ehn, S., Schreiner, M., & Petrasch, A. M. (2014). Pilot-scale supercritical carbon dioxide extraction, physico-chemical properties and profile characterization of Moringa oleifera seed oil in comparison with conventional extraction methods. *Industrial Crops and Products*, *58*, 68–77. https://doi.org/10.1016/j.indcrop.2014.03.020

Sakr, S., Rashad, W. A., & Abaza, M. T. (2021). The ameliorative effect of Moringa oleifera oil on tributyltin-induced brain toxicity in albino rats. *Environmental Toxicology*, *36*(10), 2025–2039. https://doi.org/10.1002/tox.23320

Savoire, R., Lanoisellé, J. L., &Vorobiev, E. (2013). Mechanical continuous oil expression from oilseeds: A review. *Food and Bioprocess Technology*, *6*(1), 1–16. https://doi.org/10.1007/S11947-012-0947-X/FIGURES/8

Shahbaz, M., Naeem, H., Batool, M., Imran, M., Hussain, M., Mujtaba, A., Alsagaby, S. A., Al Abdulmonem, W., El-Ghorab, A. H., Ghoneim, M. M., Shaker, M. E., Abdelgawad, M. A., & Al Jbawi, E. (2024). Antioxidant, anticancer, and anti-inflammatory potential of Moringa seed and Moringa seed oil: A comprehensive approach. *Food Science and Nutrition*, *June*, 6157–6173. https://doi.org/10.1002/fsn3.4312

Silva, M. O., Camacho, F. P., Ferreira-Pinto, L., Giufrida, W. M., Vieira, A. M. S., Visentaine, J. V., Vedoy, D. R. L., & Cardozo-Filho, L. (2016). Extraction and phase behaviour of Moringa oleifera seed oil using compressed propane. *Canadian Journal of Chemical Engineering*, *94*(11), 2195–2201. https://doi.org/10.1002/cjce.22614

Smith, S., Chen, X., Li, Z., Smith, S., Chen, M., Liu, H., & Wu, X. (2022). Optimization of supercritical CO2 extraction of Moringa oleifera seed oil using response surface methodological approach and its antioxidant activity. *Current Developments in Nutrition*, *6*, 535. https://doi.org/10.1093/cdn/nzac077.038

Souza, D. E. S., Melo, J. J. C. de, Santos, F. F. dos, Vasconcelos, A. L. dos S., Jesus, A. dos S. de, Freitas, L. dos S., Souza, R. L. de, & Soares, C. M. F. (2024). Microwave-Assisted vs. Conventional extraction of Moringa oleifera seed oil: Process optimization and efficiency comparison. *Foods*, *13*(19), 3141. https://doi.org/10.3390/FOODS13193141

Sun, Y., Qin, X., Zhong, J., & X, L. (2019). Effects of different pretreatments on extracting Moringa Oleifera seed oil and relevant mechanisms. *Food and Fermentation Industries*, *45*, 179–190. https://doi.org/10.13995/J.CNKI.11-1802/TS.019592

Taerian Esfahani, S., Zamindar, N., Esmaeili, Y., & Sharifian, S. (2024). Effect of initial quality of oil and thermal processing on oxidation indexes in canned tuna. *Applied Food Research*, *4*(2), 100553. https://doi.org/10.1016/j.afres.2024.100553

Thirugnanasambandham, K. (2018). Ultrasound-assisted extraction of oil from Moringa oleifera Lam. seed using various solvents. *Energy Sources, Part A: Recovery, Utilization and Environmental Effects*, *40*(3), 343–350. https://doi.org/10.1080/15567036.2017.1416708

Tunit, P., Chittasupho, C., Sriyakul, K., Tungsuruthai, P., Chakkavittumrong, P., Na-Bangchang, K., & Kietinun, S. (2022). Emulgels containing perilla frutescens seed oil, Moringa oleifera seed oil, and mixed seed oil: Microemulsion and safety assessment. *Polymers, 14*(12), 2348. https://doi.org/10.3390/polym14122348

van Ruth, S. M., Villegas, B., Akkermans, W., Rozijn, M., van der Kamp, H., & Koot, A. (2010). Prediction of the identity of fats and oils by their fatty acid, triacylglycerol and volatile compositions using PLS-DA. *Food Chemistry, 118*(4), 948–955. https://doi.org/10.1016/j.foodchem.2008.10.047

Ventura, A. C. S. S. B., de Paula, T., Gonçalves, J. P., Soley, B. da S., Cretella, A. B. M., Otuki, M. F., & Cabrini, D. A. (2021). The oil from Moringa oleifera seeds accelerates chronic skin wound healing. *Phytomedicine Plus, 1*(3), 1–11. https://doi.org/10.1016/j.phyplu.2021.100099

Vovk, H., Karnpakdee, K., Ludwig, R., & Nosenko, T. (2023). Enzymatic Pretreatment of Plant Cells for Oil Extraction. *Food Technology and Biotechnology, 61*(2), 160. https://doi.org/10.17113/FTB.61.02.23.7896

Xiong, Y., Zhou, J., Xing, Z., & Du, K. (2021). Cancer risk assessment for exposure to hazardous volatile organic compounds in Calgary, Canada. *Chemosphere, 272,* 129650. https://doi.org/10.1016/j.chemosphere.2021.129650

Zhang, J., Liu, Q., Wang, D., Xiao, X., & Jiang, L. (2022). Technology of aqueous enzymatic extraction for jiangyong pomelo seed oil with calcium chloride demulsification. *Journal of the Chinese Cereals and Oils Association, 37,* 101–107.

Zhao, S., & Zhang, D. (2013). Supercritical fluid extraction and characterisation of Moringa oleifera leaves oil. *Separation and Purification Technology, 118*, 497–502. https://doi.org/10.1016/j.seppur.2013.07.046

Zhong, J., Wang, Y., Yang, R., Liu, X., Yang, Q., & Qin, X. (2018). The application of ultrasound and microwave to increase oil extraction from Moringa oleifera seeds. *Industrial Crops and Products, 120,* 1–10. https://doi.org/10.1016/j.indcrop.2018.04.028

Zouboulis, C. C., Hossini, A. M., Hou, X., Wang, C., Weylandt, K. H., & Pietzner, A. (2023). Effects of Moringa oleifera seed oil on cultured human sebocytes in vitro and comparison with other oil types. *International Journal of Molecular Sciences, 24*(12), 10332. https://doi.org/10.3390/ijms241210332

16 *Moringa oleifera* in Food Technology as a Dietary Superfood

Jahidul Hassan, Mijanur Rahman Rajib, Liakat Ali, Syada Nizer Sultana, Saiful Islam, Sadia Tasnim, Kamrun Sanjana Badhon, Zubayer and Yukio Ozaki

16.1 INTRODUCTION

The emergence of diets with 'superfoods' is a direct consequence of the increasing health consciousness among financially secure, elderly populations and the inclination to increase economic productivity through self-optimisation. 'A nutrient-dense food regarded as particularly advantageous for health and wellness' is the definition of a superfood in the *Oxford English Dictionary* (2026). Loyer (2016) mentioned superfoods as a growing category of health foods known for their exceptional nutrition, medicinal value, traditional use and natural authenticity in modern times.

Industrialised nations, such as the United States and those in Europe, have become substantial consumers of fitness products that are classified as 'superfoods'. According to Ahuja and Mamtani (2019), international commerce for *Moringa oleifera* (MO) is currently estimated at US$5.8 billion, with a projected doubling in the next five years. This has established MO as a prominent emerging superfood in industrialised nations (Graeff-Hönninger & Khajehei, 2019). As demonstrated by its pricing, marketing strategies and consumer expectations, MO is unquestionably classified as a superfood. MO has been used for over 4000 years in traditional medicine and nutrition in humans and livestock. Diverse therapeutic applications of the MO tree are delineated in Ayurveda (Schrott & Ammon, 2012). MO is frequently touted as a superfood due to its high concentration of essential minerals, vitamins, β-carotene, amino acids and diversified phenolic compounds, as well as phytohormones such as cytokinin and zeatin (Matic et al., 2018). MO is claimed to contain more Fe than spinach, higher levels of carotenoids than carrots, more ascorbic acid than citrus fruits, greater Ca content than milk, higher potassium levels than bananas and higher protein levels than yoghurt (Gopalakrishnan et al., 2016). Despite this, nutrients are frequently absent in individuals from the least developed countries. Additionally, there is no justification for the assumption that the average consumer in industrialised nations, who lives in a state of prosperity, would rely on a nutrient to ensure an adequate intake of proteins or vitamins. Therefore, there is a significant opportunity to convert MO into a variety of process products and integrate it into food technology to further investigate its potential sustainably.

In the interim, MO may be elevated to the status of a superfood due to specific components. The presence of specific active chemicals is indicated by the use of different morphological parts of the MO tree to treat conditions such as malaria, arthritis and hypertension, as well as boosting innate immunity (Leone et al., 2015). The discovery of several glucosinolate types, which vary in structure and concentration across distinct MO species, demonstrates cytoprotective effects across mammalian cell lines (Fahey et al., 2018). The European market is currently oversaturated with a multitude of MO products, which are frequently found on store shelves. Tea blends, desiccated and

DOI: 10.1201/9781003617860-18

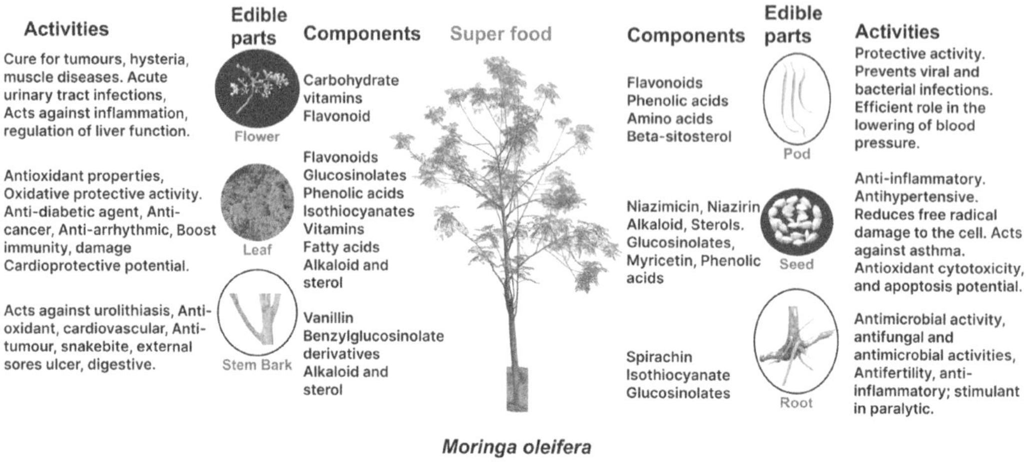

FIGURE 16.1 Different Parts of MO and Its Potential as a Superfood.

powdered leaves, smoothies, pills and raw MO fruits are among the available products. It has been widely grown in warm and temperate regions across countries due to its substantial potential as a nutraceutical and therapeutic diet (Lin et al., 2019; Xu et al., 2019). MO leaves contain flavonoids, kaempferol, alkaloids, tannins and saponins (Rocchetti et al., 2020).

The health benefits of MO have been extensively studied (Kou et al., 2018). It is cultivated for its nutrient-rich, edible plant parts, which serve as both a medicinal resource and a dietary component (Saucedo-Pompa et al., 2018) (Figure 16.1). MO leaves are the most commonly used part, rich in vitamin C, vitamin E and β-carotene (Suresh et al., 2020). Extensive research highlights their potential as a functional food with therapeutic properties, including free radical scavenging, inflammation modulation, blood sugar regulation, tumour suppression, support for heart health, liver protection and respiratory relief, validated through in vitro and in vivo studies (Fotio et al., 2020). Additionally, they have shown efficacy in the treatment of neurodegenerative disorders (Zeng et al., 2019). In addition to the leaves, other components of MO have been identified as effective in contributing to human nutrition by containing essential phytochemicals. Consequently, it is advantageous to conduct a thorough examination of MO.

This chapter summarises the composition of phenolic compounds, their traditional applications, physiological functions, biochemical pathway targeting and potential applications as health-promoting and skincare agents. The objective is to promote new research and uncover novel therapeutic applications of MO, and to improve understanding of MO phytochemistry.

16.2 GEOGRAPHICAL DISTRIBUTION AND ECOLOGICAL REQUIREMENTS

MO is an aggressive, perennial tropical deciduous plant that is an affiliate of the Moringaceae family. Originally from India, it has been widely grown across Asia, Africa, parts of Europe and Central America's tropical and subtropical areas. This tree species may survive internationally in virtually all tropical and subtropical environments. However, it is considered to be indigenous to the Indian subcontinent (Pareek et al., 2023). It is widespread in Asia, Africa, the Caribbean, Latin America, the Pacific Islands, Florida, Madagascar, Central America, Cuba, the Philippines, Ethiopia and Nigeria (Trigo et al., 2020). Nowadays, MO is broadly grown in Paraguay, Brazil, South Florida, Mexico, Peru and practically every other continent in the equatorial regions of Asia, Central America and South America. Requiring little in the way of environmental conditions, this tree grows quickly and may be reproduced both vegetatively and by seed (Klimek-Szczykutowicz et al., 2024). MO grows best in warm, mostly dry tropical regions, as it is a drought-tolerant crop that needs 250–3000 mm of rainfall per year. Sand, loamy and clay soils are among the soil types in which MO can flourish. It does, however, like soils that drain well and have a pH between 6.0 and 7.5. It grows mostly from 0 to 1400 m above sea level. Nevertheless, its expansion has been documented at an elevation of 2000 m. It can endure a broad spectrum of weather extremes and infertile soils, including high heat, droughts and mild frosts (James & Zikankuba, 2017). Although it can withstand 48°C for short periods, its optimal temperature limit is 25 to 35 °C (Trigo et al., 2020). It may acclimate to areas affected by changing climatic conditions, such as the Mediterranean Basin, due to its significant resilience to elevated temperatures.

16.3 THE EDIBLE PARTS OF MO

MO is regarded as one of the most versatile plants globally, with all its parts used to treat various ailments, including antimicrobial, blood pressure, cholesterol and blood sugar, muscle-relaxing, ulcer-healing and inflammation-reducing conditions. Each part of the plant possesses valuable functional and nutraceutical properties, making it a diverse biomaterial for food and related uses. MO contains various acids, phenols and antioxidants across its different parts. Due to its high nutritional value, every part of the tree serves both dietary and commercial purposes. The leaves, seeds, bark, roots, sap and flowers are widely used in medications, while the leaves and immature pods are consumed as vegetables. Its medicinal qualities also make it a potential alternative to coriander leaves for enhancing flavour (Gandji et al., 2018).

16.4 THE MEDICINAL IMPORTANCE OF THE EDIBLE COMPONENTS OF MO

The MO tree is often called the 'miracle tree' because every part of it has been used for various benefits. Different parts of the plant are utilised for nutrition and therapeutic purposes.

16.4.1 Leaves

MO leaves are exceptionally malleable and rich in nutrients, rendering them beneficial for several applications. Bioactive compounds and phenols in green and dried leaves also contain higher amounts of sugars (38.1%) and amino acids (22%). Additionally, they possess essential oils that inhibit microbial activity (Marrufo et al., 2013). MO leaf powder is incorporated into the diets of undernourished individuals, including infants and expectant mothers. The diverse uses of MO leaves include a vegetable that can be cooked like spinach, a dried powder used as a condiment, a medicine to regulate blood sugar levels and a nutritious herbal tea to boost the immune system.

16.4.2 Immature Seed Pods

MO pods are highly nutrient-dense, offering a rich blend of phytochemicals, structural and non-structural carbohydrates, proteins, lipids and essential minerals. Additionally, they are a valuable source of key fatty acids, including oleic, linoleic, palmitic and linolenic acids. These pods possess anti-helminthic substances and are employed in the treatment of liver and spleen infections and alleviate articular pains. The pods also guard against bacterial activity, cancer, oxidative stress, antidiabetic effects, irritation, liver injury, hepatic fibrosis, hypercholesterolaemia and other harmful conditions (Shah et al., 2022). The immature pods are consumed in various parts of Asia and Africa, often cooked in soups, stews or curries.

16.4.3 Mature Seeds

MO seeds are a reputable supplier of antioxidant activity, lipids, proteins and minerals such as calcium, magnesium and zinc, with 30 inherent antioxidants. The high levels of protein, minerals and unsaturated fatty acids, primarily oleic acid, make them valuable raw materials for culinary applications (Milla et al., 2021). The seeds of MO have the potential to cure hypercholesterolaemia by decreasing the circulation levels of low-density lipoprotein, thanks to the unsaturated fatty acids they contain. Also, they are used as a purgative and to treat hysteria, tumours and scurvy. According to recent studies, MO seed powder increases the nutritional content of a variety of foods, including bread, biscuits, cakes, snacks and cereal gruels (Rabie et al., 2020). The seed powder is commonly used as a natural flocculant to elucidate turbid water. Powdered seeds can be used to purify honey without boiling and purify sugarcane juice. Meanwhile, the seed kernel is valued for its medicinal properties, particularly in treating bronchial asthma.

16.4.4 Oil Pressed from Seeds

MO seeds are exceptionally rich in oil, with oleic acid as its predominant fatty acid, similar to olive oil (Gharby et al., 2021). This oil holds significant potential as a promising commercial source of edible oil for the food industry. Additionally, MO oil contains other key fatty acids found in olive oil (Nadeem & Imran, 2016). When cold-pressed, MO seeds also yield oil, which is highly prized in the beauty industry due to its moisturising and rejuvenating benefits. Since the earliest times, MO oil has been used for the production of beauty products and scents, as lubricants in machinery and, lately, for biodiesel manufacture.

16.4.5 Flowers

Flowers have many uses, including in food, medicine and as a natural preservative. Flowers have an aromatic, fresh scent and a mild, subtly sweet flavour, with younger flowers offering the best texture and taste. When cooked, they take on a flavour reminiscent of asparagus and mushrooms. In Ayurveda, these flowers are used in teas and tonics to alleviate cold symptoms, cleanse the body, regulate blood sugar and enhance insulin sensitivity.

16.4.6 Roots

MO roots exhibit elevated quantities of antinutrients in comparison with other plant components, thereby restricting their suitability for consumption. Roots are characterised by increased levels of tannins and oxalates, which do not serve as beneficial sources of nutrition. MO root is eaten as a spice. Furthermore, they contain elevated amounts of carbohydrates, arginine, lysine, vitamin C and sodium (Vargas-Sánchez et al., 2019).

16.4.7 Bark

MO bark holds both curative and preventative potential. Its extract has demonstrated antifungal properties, while the juice from the bark and stem exhibits antibacterial activity against *Staphylococcus aureus*. The bark is used to treat respiratory tract infections, eye problems, tumours and ulcers. MO bark is used to treat scurvy and diarrhoea. The root bark juice is applied to the ears to relieve earaches and is also used as a pain reliever in dental cavities (Qi et al., 2019). MO bark can be consumed in various ways, including mixing it into salad dressings, incorporating it into baking, adding it to smoothies, sprinkling it over soups, steeping it as a tea or stirring it into hot water with a touch of lemon and honey.

16.4.8 Trunk

Gum extracted from a cut tree trunk has been used in calico printing, as a medicinal ingredient and as a mild-flavoured condiment.

16.5 THE NUTRITIONAL VALUES OF DIFFERENT PARTS OF MO

The MO tree has been dubbed the 'wonder tree' owing to its numerous nutritional applications and its capacity to alleviate a wide spectrum of afflictions. Nutrients are stored in all sections of the plant (Table 16.1). This tree has been used to fight undernourishment, especially among vulnerable people. The leaves of the MO plant have a great deal of minerals. Obese people can still eat the leaves because they are not calorie-dense. Proteins, minerals and all nine vital amino acids are abundant in its leaves. Pods, which contain around 46.78% fibre and 20.66% protein, are highly regarded for their ability to alleviate digestive issues and cancer of the colon (Tiwari et al., 2023). MO also has significant levels of β-carotene, pyridoxine, nicotinic acid and vitamins C, D and E. There is little doubt that consuming MO powder in situations of imminent hunger provides substantial health benefits due to its eminent dietary properties.

16.5.1 Leaves

MO is regarded as a superfood because it is abundant in nutrients. Among all components of this species, the leaves are predominantly utilised for medicinal and culinary applications. There are a lot of vitamins in the leaves. The amount of vitamin A found in 100 g of fresh MO leaves might vary between 11,000 and 23,000 IU, as reported by Leone et al. (2015). The vitamin C content of the leaves is 200 mg/100 g, which is greater than that of oranges. Furthermore, the herb can function as a potent antioxidant due to the high concentration of vitamin E in its leaves, which is equivalent to that found in nuts at around 9.0 mg/100 g, primarily in the form of α-tocopherol. B vitamins, including niacin, riboflavin and thiamin,e are also plentiful in the plant. Thiamine levels in leaves range from 0.06 to 0.6 mg/100 g, whereas riboflavin levels are between 0.05 and 0.17 mg/100 g. The niacin content varies from 0.8 to 0.82 mg per 100 g. These quantities are similar to those found in a variety of vegetables and fruits (Leone et al., 2015).

A number of minerals are found in fresh MO leaves, including iron, copper, sodium, potassium, magnesium, phosphorus, selenium and zinc. In addition, MO fresh leaves provide a lot of protein and many important amino acids. MO contains at least 18 amino acids, including at least eight vital ones. The leaves of MO contain fibre and fatty acids, making them positive for fighting hunger (Falowo et al., 2018). Furthermore, citing Falowo et al. (2018), the nutrients listed above have higher levels per standard serving than those of other edible plants. They went on to say that the MO plant contains fatty acids, but most are unsaturated, and saturated fatty acids are quite rare. Compared with other parts of MO, the leaves have the highest nutrient concentration, particularly protein (19–29%). Their remarkably high calcium and high iron content make them suitable for treating anaemia. The potassium content is three times that of a banana. All nutrients, with the exception of vitamin C, are more abundant in powdered MO leaves than in fresh leaves. An intriguing aspect of this is that dried leaves may be preserved, ensuring their use all year round (Arias Sabín, 2014). In several impoverished societies, they frequently serve as the sole basis of accompanying nutrients. Moreover, their low levels of lipids, carbs and phosphate render them one of the superior plant-based diets.

16.5.2 Seeds

MO seeds are known for their high levels of oil and fatty acids. According to Falowo et al. (2018), for every 100 g of *M. oleifera* seeds, you can expect to find 9.98–51.80 g of crude protein, 17.26–20.00 g of crude fibre, 3.36–18.00 g of carbohydrates, 38.67–43.60 g of fat and 3.60–5.00 g of ash. MO seeds support the treatment of hyperthyroidism, chronic disorders, arthritis, rheumatism, gout, cramps, epilepsy and sexually transmitted infections and possess antibacterial and anti-inflammatory properties. They comprise oleic acid (Ben oil), the antibiotic pterygospermin and fatty acids, along with phytochemicals such as tannins, saponins, phenolics, phytate, flavonoids, terpenoids

TABLE 16.1
Phytochemical Constituents and Medicinal Uses of MO Parts

Plant part	Group of compounds	Phytoconstituents	Biological activity	References
Leaves	Flavonoids	Isorhamnetin, kaempferol, quercetin, myricetin, rutin	Antioxidant properties, oxidative protective activity and antidiabetic agent	Klimek-Szczykutowicz et al. (2024)
	Glucosinolates	4-O-(α-L-rhamnopyranosyloxy)-benzylglucosinolate	Antioxidant activity	Ahmed et al. (2016)
	Phenolic acids	Chlorogenic acid, ellagic acid, ferulic acid, gallic acid, caffeic acid, o-coumaric acid, syringic acid	Antioxidant and antimicrobial activity, prevent infections, anticancer, anti-inflammatory and boosts athletic performance	Pareek et al. (2023)
	Isothiocyanates	4-[4′-O-acetylalpha-L-rhamnosyloxy) benzyl] isothiocyanate	Anticancer properties	Ma et al. (2020)
	Vitamins	Vitamins A, B1, B2, B3, C and E, β-carotene, α-tocopherol	Improve immune function and cure bronchitis, piles, sore throats, scurvy, catarrh and eye infections	Ma et al. (2020)
	Fatty acids	Arachidic, behenic, caprylic, henicosanoic, lauric, ligoceric, α-linolenic, γ-linolenic, linolenic, margaric, myristic, oleic, palmitic, stearic, tricosanoic and vaccenic acid	Pharyngitis, cholesterol-lowering action, haemorrhoids, anti-inflammatory, antihypertensive, anti-proliferative, antioxidant, anxiolytic, diuretic and antihyperthyroidism	Trigo et al. (2020)
	Amino acids	Almost all amino acids	Cholesterol-lowering action, pharyngitis, diuretics, antioxidants, antiproliferatives, haemorrhoids and antihyperthyroidism	Klimek-Szczykutowicz et al. (2024); Trigo et al. (2020)
	Alkaloids and sterols	Niaziminin, N, vincosamide, α-L-rhamnopyranosyl	Inhibition of virus activation and in vivo and in vitro cardioprotective potential	Panda et al. (2013)
	Benzyl isothiocyanate	Niazirin, niazirinin, niaziminin, niazimicin A and B	Anticonvulsant, antioxidant, antihypertensive, antibacterial and anticancerous	Milla et al. (2021); Srivastava et al. (2023)
Flowers	Vitamins	Vitamin C	A stimulant to cure tumours, hysteria, muscle diseases and inflammation, and lower serum cholesterol	Vergara-Jimenez et al. (2017)
	Carbohydrates	D-glucose, D-mannose	Treatment of genetic defects and urinary tract infections	Maurya et al. (2014)
	4-(α-L-rhamnopyranosyloxy) benzylglucosinolate	Isoquercitin, kaemopherol, kaempferitin	Acts against inflammation	Miller et al. (2018)
	Flavanoid	Quercetin	Have impacts on the regulation of liver function	Lalas and Tsaknis (2002)

Pods	Flavonoids	Kaempferol, quercetin, rutin	Oxidative damage protective activity	Klimek-Szczykutowicz et al. (2024)
	Phenolic acids	Chlorogenic acid, ellagic acid, ferulic acid, gallic acid and syringic acid	Prevents viral and bacterial infections	Klimek-Szczykutowicz et al. (2024)
	Amino acids	All amino acids except asparagine	Treatment of joint discomfort, diarrhoea and spleen and liver issues	Mashamaite et al. (2024)
	4-(4-O-acetylalpha-L-Rhamnosyloxy) benzyl isothiocyanate	Isothiocyanate nitrites, β-sitosterol	Inflammation, antihelmintics and lowering of blood pressure	Srivastava et al. (2023)
Seeds	Phenolic acids	Chlorogenic acid, ellagic acid, ferulic acid, gallic acid and caffeic acid	Antioxidant and antimicrobial activity, anticancer, antithrombotic and anti-inflammatory	Klimek-Szczykutowicz et al. (2024)
	Flavonoids	Kaempferol, quercetin, rutin, myricetin	Cardioprotective and diabetes protective	Leone et al. (2015); Rue et al. (2018)
	Glucosinolates	4-O-(α-L-rhamnopyranosyloxy)-benzylglucosinolate	Anti-inflammatory, pain-relieving, antihypertensive	Borgonovo et al. (2020)
	Fatty acids	Arachidic acid, behenic acid, eicosenoic acid, erucic acid, α-linolenic acid, oleic acid, palmitic acid and stearic acid	Reduces blood pressure and free radical damage, membrane localisation of the enzyme, trypanocidal and antileukemic effect	Athikomkulchai et al. (2020); Sari et al. (2020)
	Sterols	β-sitosterol, stigmastanol	Anti-inflammatory	Liao et al. (2018)
	Amino acids	All amino acids	Antitumor, antispasmodic, rheumatism, anthelmintic, arthritis and ulcers	Dalei et al. (2016)
	Alkaloid	Pterygospermin	Antibiotic properties, antioxidant cytotoxicity and apoptosis potential	Maiyo et al. (2016)
	Niazimicin	Moringine, niazimicin, niazirin	Acts against asthma	Saucedo-Pompa et al. (2018)
Stem, bark	Glucosinolates	4-O-(α-L-rhamnopyranosyloxy)-benzylglucosinolate (glucomoringin), benzylglucosinolate derivatives	Dental caries, hysteria, antitumour, snakebites, ulcers, digestive, headache, birth control, colitis and acts against urolithiasis	Chhikara et al. (2021); Arora and Arora (2021)
	Alkaloid and sterol	β-Sitosterol-3-O-β-D-galactopyranoside	Antioxidant, cardiovascular and immunomodulatory	Liao et al. (2018)
	Vanillin	Vanillin, β-carotene	Acts against inflammation.	Balakumbahan et al. (2020)
Roots	Glucosinolates	Benzylglucosinolate,4-O-(α-L-rhamnopyranosyloxy)-benzyl glucosinolate (glucomoringin)	Anti-inflammatory, stimulant in paralytic afflictions, cardiac for treatment of rheumatism and for lower back pain	Prajapati et al. (2022)
	Alkaloids and sterols	Aurantiamide acetate and 1,3- dibenzyl urea	Anti-inflammatory and antinociceptive agents	Arora and Arora (2021)
	Spirachin	Moringine, moringinine, spirachin, p-cymene	Antifertility	Aziz et al. (2021)
	Isothiocyanate	Pterygospermin, 4-L-rhamnosyloxy-benzyl isothiocyanate	Showed antimicrobial activity	Trigo et al. (2020)

and lectins. In addition to these, seeds contain lipids, fibre, proteins, minerals, vitamins and amino acids. Its anti-inflammatory properties are attributed to the presence of flavonoids. Antimicrobial properties are attributed to the antibiotic pterygospermin (Alhassan et al., 2022). The additional phytochemicals are useful for treating a variety of illnesses. Seeds contain a substantial amount of cystine and methionine, comparable to that found in eggs and milk. In some regions, they are also consumed as roasted nuts (James & Zikankuba, 2017).

16.5.3 Pods (Fruits)

By the end of the second year, the plant has developed pods and blooms with lobed edges. Harvesting the pods at their youngest, most delicate, most verdant stages is ideal, as these immature pods are rich in nutrients and essential amino acids. Roughly 46.78% of the weight of an immature pod is fibre, while roughly 20.66% is protein. The proportion of these amino acids in pods is 30% (Leone et al., 2015). Pods also have a low lipid content and a decent quantity of unsaturated and essential fatty acids, mainly oleic acid (Benítez, 2012). Pods can be eaten by both animals and humans. Recent research demonstrated that consuming one bowl of sliced fresh pods, which weighs 100 g, provides 157% of the recommended daily intake of vitamin C for adults (Vijayanchali, 2024). Preparing immature pods as peas or green beans or eating them raw is said to give them an asparagus flavour; frying mature pods gives them a peanut flavour. Ben oil, an edible oil, accounts for 38–40% of their production. This oil is odourless, transparent and sweet; therefore, it will probably not change the flavour (Liu, 2018). Its nutritional profile is similar to that of olive oil, and its anti-inflammatory properties help reduce discomfort associated with inflammatory disorders (Gopalakrishnan et al., 2016).

16.5.4 Flowers

MO blossoms are processed into ready-to-eat snacks, tea and nutritional supplements because they contain several nutrients, such as proteins, calcium, potassium, α and γ tocopherol (antioxidants) and fats that are polyunsaturated. The taste of fried MO blossoms resembles that of mushrooms (Liu et al., 2018). Because of the phytoconstituents found in the flowers, Divya et al. (2019) state that they contain anticancer, antioxidant, anti-inflammatory and antimicrobial properties.

16.5.5 Roots and Bark

MO's roots and bark are rich in amino acids, ascorbic acid and carbohydrates; however, they are deficient in B vitamins. Its roots and bark have antiulcer properties, but they also contain a lot of antinutrients that limit food intake by preventing the body from absorbing essential nutrients. Certain root extracts contain analgesics called moringin and moringinine, which may contribute to their efficacy against lumbago when taken at the right dosages. The underground roots, seeds, bark and even leaves all pose health risks if consumed in excess (Canett-Romero et al., 2014).

16.6 MO PRODUCTS AVAILABLE ON THE MARKET

As more people become aware of MO, a variety of products have emerged to satisfy customer demands. The health benefits of MO are predicted to cause a sharp increase in demand worldwide. The global market for MO has expanded recently. The largest markets for MO-based products are in Asia-Pacific, Europe and North America, and the growing recognition of the advantages of plant-based diets is a major factor propelling market growth. The MO products market is projected to expand at a cumulative annual growth rate of 9.20%, with its value expected to reach approximately £11.2 billion by 2032.

16.6.1 Moringa Powder

One of the most widely used items made from MO foliage is MO powder. Usually, the foliage is dehydrated and crushed into a fine powder. Vitamins C, A, E and numerous B vitamins (including B1, B2, B3 and B6) are all abundant in MO powder. Antioxidants like β-carotene, quercetin and chlorogenic acid are abundant in MO powder (Pareek et al., 2023). MO powder is renowned as a superfood for its high antioxidant content, and it is frequently used in teas, soups, shakes, smoothies and health supplements. Demand for MO powder has increased dramatically as a result of the growing popularity of plant-based diets, especially among natural supplement users and health-conscious people. Organic MO powder, which is available in health food shops and online marketplaces, is one of the most popular items.

16.6.2 Moringa Capsules, Tablets and Gum

MO is also available as tablets and capsules to meet the growing demand for nutritional supplements. These products, which do not require heating or preparation, contain powdered MO leaves or extracts, providing a quick approach to incorporating the plant's health benefits into daily activities. MO tablets are frequently used to boost energy levels, aid in digestion and support immunological function. MO-based capsules, which are promoted for their anti-inflammatory, antioxidant and cholesterol-lowering qualities, have become increasingly popular in the supplement sector (Gopalakrishnan et al., 2016). MO gum is an excellent natural source of pharmaceutical ingredients. A number of studies established the suitability of MO gum as a sustained and mucoadhesive polymer, disintegrant and binder (Srivastava et al., 2023).

16.6.3 Moringa Oil

Nearly all of the oil may be recovered by solvent extraction, usually *n*-hexane, whereas cold-press processing produces a lower yield. In reality, cold pressing typically extracts an average of 69% of the total oil contained in seeds. Rural residents boil dehusked seeds in water to extract the edible oil, then collect the oil from the water's surface. Palmitic acid accounts for the majority of the saturated fatty acid content (21.18%), followed closely by behenic, stearic and arachidic acids. Due to its high behenic acid concentration, the oil is commercially known as 'Ben' or 'Behen' oil. MO seed oil has also been shown to contain trace amounts of cerotic, lignoceric, myristic, margaric and caprylic acids. On average, 76.73% of the oil is composed of monounsaturated fatty acids (Leone et al., 2015). MO oil is widely utilised in the beauty and cosmetics sectors because of its strong antioxidant content.

16.6.4 Moringa Tea

Dehydrated MO leaves are infused with warm water to make MO tea. It is a well-liked beverage among people seeking natural detoxification and health advantages such as better digestion, healthier skin and more energy. This product comes in loose-leaf and tea-bag versions and is advertised mostly as a health beverage. With a subtle, earthy flavour, MO tea can be drunk by itself or combined with other herbal teas. Tea made from MO leaves contains terpenoids, flavonoids, alkaloids, saponins, tannins and polyphenolic substances (Rahmawati et al., 2023).

16.6.5 Moringa Snacks

Due to its popularity, snack items enhanced with MO have also been developed. Usually, adding various components of MO to products such as bread, pastries, snacks and drinks enhances the food value of these commodities (Trigo et al., 2023). MO leaf powder-fortified date energy bars have good nutritional content and might be utilised as a nutritious, high-energy substitute for athletes and children who are severely malnourished (Eid et al., 2025). Bread dough is mostly fortified with seed flour and powdered dry MO leaves. Seed flour and dry MO leaf powder are mainly used to revive bread dough. Bread enriched with MO has higher levels of protein and fibre (Trigo et al., 2023).

16.6.6 Moringa Leaf Extracts and Concentrates

MO leaf extract is available in both powdered and concentrated liquid forms. MO Leaf extract is a great source of phenolic and flavonoid chemicals. Additionally, it scavenges free radicals, demonstrating high antioxidant activity (Nizioł-Łukaszewska et al., 2020). It is most commonly used in wellness and health products that support a healthy digestive system, increase immunity and offer a natural energy source.

16.6.7 Moisturiser and Skin Care Cream

Applying a 3% MO leaf cream topically over three winter months considerably enhanced skin hydration compared with a local cream. Additionally, it seemed to reduce the visual appearance of wrinkles on the cheeks in young adult males, suggesting that MO may offer both moisturising and anti-ageing benefits.

16.7 THE WORLDWIDE MO UTILISATION PATTERN

MO is popular for its many applications and capacity to thrive in arid and wet climates (Gandji et al., 2018). On top of its beauty, the MO tree is commercially beneficial since practically every part may be eaten (Table 16.2). Owing to its plethora of active constituents, it exhibits numerous pharmacological properties (Gao et al., 2022). The different components of MO provide high-value health-promoting nutrition, making them useful in multiple industries, including medicine and water purification (Trigo et al., 2020). Consequently, MO has always garnered a lot of interest owing to its reputation as a tree of miracles. The primary applications of MO encompass human nutrition, alley cropping, animal fodder, biogas, cleaning artefact (crushed leaves), toxin absorbent, fence, compost, foliar nutrient and green manure (Leone et al., 2015). The species is primarily used for gum extraction, honey production, juice clarification, medicinal purposes, aesthetic enhancement, biopesticide application (to mitigate damping-off in seedlings), rope (from bark), hide tanning and water purification (Leone et al., 2015).

MO is a cost-effective and trustworthy solution for ensuring appropriate nutrition and for the treatment and prevention of several ailments (Paliwal et al., 2011). The versatility and high-yielding potential of MO attract considerable attention, as it can help alleviate nutritional deficiencies in areas where they are most prevalent (Leone et al., 2015). Its farming is limited in underdeveloped nations, where several portions of the plant are used (Gandji et al., 2018). MO leaf powder is the most marketable product from an industrial standpoint. It is a multivitamin supplement that contains significant concentrations of amino acids, minerals, natural iron and calcium (Trigo et al., 2020). In addition to its many medicinal uses, this plant has practical applications, including nutritional supplements, water purification, biofuel, gum and honey. Owing to its fragrant, attractive blossoms and dense canopy, it is utilised as an ornamental tree in several parts of the world. This tree serves as a fence and supplier of pulpwood, making it compatible with agroforestry systems.

16.8 CHALLENGES OF MO AS A SUPERFOOD IN FOOD TECHNOLOGY

Many classic, unorthodox and neglected food crops are disappearing from the natural environment due to urbanisation. It is essential to motivate communities to persist in cultivating these crops to ensure the continuation of their advantageous qualities. This encompasses promoting a

TABLE 16.2
Utilisation Status of MO-Derived Ingredients for Multiple Purposes Worldwide

Purpose of use	Plant part used	Purpose of use	Available form	Country or area	References
Traditional and modern medicine	Leaf, seed oil, bark, leaf paste	Nutritional supplement, skin and body pain, treating infections, wound healing, antimicrobial and to prevent diabetes	Capsule, powder, tea, oil and bark extract	India, Bangladesh, Europe, Latin America, Nigeria, Ghana, Tanzania and Ethiopia	Doglikuu et al. (2021)
Cosmetics	Seed oil, leaf extract	Skin, hair and body care and anti-ageing products	Oil, lotion, cream, soap, moisturisers and serums	India, United States, Europe, China, Japan and Ghana	Pareek et al. (2023); Peñalver et al. (2022)
Food	Green pods, leaves, young flowers, roots, seed oil	Vegetables, salad, spices, cooking oil to boost immunity and digestion, sources of nutrients	Fresh green pods, raw flower, oil and leaf powder	Asian countries, Ethiopia, Ghana, Tanzania and Indonesia	El Bilali et al. (2024)
Animal feed	Seed cake, leaf powder, a combination of leaf and other plant parts	Improve nutrition, enhance growth, boost immunity, improve milk and meat production	Seed cake, leaf powder, forage and pellets	India, Africa, South Asia, Latin America and China	Da Silva Junior et al. (2024); Song et al. (2024)
Water purifier	Seed cake, seed powder	Remove impurities and bacteria	Seed extracted powder	Africa, South Asia, Brazil and Peru	Addich et al. (2024)
Biofertiliser	Seed cake, leaf powder	Enhance crop growth and productivity, increase tolerance to abiotic stress	Seed cake, liquid extract and foliar spray	India, Latin America, Bangladesh and Africa	Mashamaite et al. (2022); Leone et al. (2015)
Biofuels	Seed oil, leaves, stems, branches	Renewable energy, eco-friendly	Biodiesel, bioethanol and charcoal	Ghana, Kenya, India, Indonesia and the Philippines	Takase et al. (2022); Trigo et al. (2020)
Fodder	Leaves, green pods and branches, seed residues	Cattle and poultry feeding	Fresh processed hay and seed cake	India, Bangladesh and Africa	Su and Chen (2020)
Gum	Bark	Thickening, emulsifier, stabilising agents	Jelly-like gum	India, Pakistan, Bangladesh and Sri Lanka,	Badwaik et al. (2020)
Ornamental	Whole plant	Flower fragrance, attractive appearance	Flower and trees	Africa, Thailand, the Philippines, India and Indonesia	Pereira et al. (2018)

more nutritious diet for Indigenous communities that have historically relied on these crops, thus safeguarding the intricate relationship between humanity and the natural world. Furthermore, the cultivation of these crops fosters species diversification, thereby offering a broader gene pool. In various Asian and African nations, particularly those classified as underdeveloped or developing, a significant number of individuals experience hidden hunger and confront the challenges of malnutrition. In contemporary discourse, the designation 'superfood' has captured widespread interest, frequently linked to marketing tactics aimed at elevating the profile of exotic fruits and vegetables renowned for their nutraceutical attributes. The classification of superfoods encompasses a variety of compounds, such as vitamins and diverse phytochemicals. These compounds exhibit a series of beneficial actions, including antioxidant, anti-inflammatory, prebiotic, antimicrobial and hypocholesterolaemic assets, which are significant in mitigating the risk of serious diseases. Nevertheless, these foods tend to be more expensive and are often out of reach for most of the population. In this context, MO can be regarded as a category of superfoods, thereby providing sustenance to the undernourished and impoverished populations of developing nations in pursuit of the objective of eradicating hunger.

While the phytochemicals found in MO offer numerous advantageous applications, it is essential to address the potential challenges related to the presence of harmful substances (Padayachee & Baijnath, 2020). The detrimental constituents encompass alkaloids and various phytotoxins, which may lead to nerve paralysis and other adverse effects when ingested in substantial amounts (Tshabalala et al., 2019). Phenols such as tannin, pectinesterase, moringinine, moringine and oestrogene represent a selection of these botanical compounds (Elsayed et al., 2020). Considering that elevated concentrations of these botanical compounds reside in the root and bark of the plant, research has demonstrated that the leaf represents the plant's most secure dietary element (Martínez et al., 2019). The 4-(-lramnopyranosyloxy)-benzyl glucosinolate, a potential mutagen, suggests a certain level of genotoxicity in roasted MO seeds (Asare et al., 2012). Research has demonstrated that benzyl glucosinolates confer various fitness advantages associated with urgency response, irritation, antioxidant activity and antimicrobial properties. Nevertheless, the elevated levels present in MO seeds can lead to various adverse outcomes, including anaemia, goitre, gastrointestinal disorders and growth obstacles, along with renal and hepatic distress (Bischoff, 2021).

In light of the continuous investigation into its safety, there remains a lack of empirical evidence to substantiate claims of the lethal and noxious effects of MO and its derivatives on both humans and livestock (De Barros et al., 2022). The external characteristics and unpleasant flavour of MO-enhanced products continue to pose challenges, affecting consumer acceptance (Olusanya et al., 2020). Consequently, it is anticipated that extensive research will be undertaken, addressing all pertinent issues related to the phytochemicals and antinutritional profiling of MO, to establish it as a functional superfood for future nutrition security.

16.9 FUTURE TRENDS OF MO AS A DIETARY SUPERFOOD IN FOOD TECHNOLOGY

The maintenance of its antioxidant capabilities and the utilisation of food complements in cases of nutritional inadequacy are emerging trends in incorporating MO into bread products. This condition suggests that these substances may be utilised as nutraceuticals based on dietary assessment. Research into protein-based molecules with bioactive effects on humankind, including antiproteolytic, procoagulant, antidiabetic, antihypertensive and anti-inflammatory properties, facilitates the expansion of innovative, purposeful foods and nutritive complements. This plant is highly cost-effective and extensively used due to its numerous advantages. The present industry exhibits a substantial requirement for organic medications, dietary supplements, fitness products, pharmaceuticals and cosmetics. Most dietary supplements have not been rigorously assessed. Moreover, numerous widely consumed food types possess ethnopharmacological relevance; however, they lack clinical evaluation.

16.10 CONCLUSION

The focus of this chapter was to establish the scientific rationale for the utilisation of MO in various food products. The chapter targeted the therapeutic applications of a nutritive plant that holds prominence across various medical disciplines. The results indicate that MO is an excellent source of both macronutrients and micronutrients. Consequently, it may be further investigated to support the development of a nutritious diet that could help treat various disorders and thereby enhance quality of life. Conversely, recent studies have sought to develop gluten-free foods that possess unique nutritional attributes by substituting a majority or the entirety of wheat flour with MO. The prevailing trend regarding MO indicates that it serves as an exceptional supplement for various ailments, contributing to nutrient-rich and nutraceutical diets, while also enhancing the nutritional attributes of fortified products. Consequently, future research should focus on developing and investigating MO-enriched fortified functional foods for human health.

REFERENCES

Addich, M., Fatni, A., Bouzrour, S., El Baraka, N., Ousaa, A., Albrimi, Y.A., & Laknifli, A. (2024). Optimization of coagulation-flocculation in urban water treatment using *Moringa oleifera*. *Desalination and Water Treatment*, *320*, 100707. https://doi.org/10.1016/j.dwt.2024.100707

Ahmed, K.S., Banik, R., Hossain, M.H., & Jahan, I.A. (2016). Vitamin C (L-ascorbic acid) content in different parts of *Moringa oleifera* grown in Bangladesh. *American Chemical Science Journal*, *11*(1), 1–6. https://doi.org/10.9734/ACSJ/2016/21119

Ahuja, K., & Mamtani, K. (2019). *Phycocyanin market analysis 2019–2025 growth statistics report.* Available online: https://www.gminsights.com/industry-analysis/phycocyanin-market (accessed on 31 March 2021).

Alhassan, Y.J., Sanchi, I.D., Dorh, L.E., & Sunday, J.A. (2022). Review of the nutritive, medicinal and general economic potentials of *Moringa oleifera*. *Cross Current International Journal of Agriculture and Veterinary Sciences*, *4*(1), 1–8. https://doi.org/10.36344/ccijavs.2022.v04i01.001

Arias Sabín, C. (2014). *Estudio de las posibles zonas de introducción de la Moringa oleífera Lam. en la Península Ibérica, Islas Baleares e Islas Canarias*. Madrid, febrero.

Arora, S., & Arora, S. (2021). Nutritional significance and therapeutic potential of *Moringa oleifera*: The wonder plant. *Journal of Food Biochemistry, 45*(10), e13933. https://doi.org/10.1111/jfbc.13933

Asare, G.A., Gyan, B., Bugyei, K., Adjei, S., Mahama, R., Addo, P., Otu-Nyarko, L., Wiredu, E.K., & Nyarko, A. (2012). Toxicity potentials of the nutraceutical *Moringa oleifera* at supra-supplementation levels. *Journal of Ethnopharmacology*, *139*(1), 265–272. https://doi.org/10.1016/j.jep.2011.11.009.

Athikomkulchai, S., Tunit, P., Tadtong, S., Jantrawut, P., Sommano, S.R., & Chittasupho, C. (2020). *Moringa oleifera* seed oil formulation physical stability and chemical constituents for enhancing skin hydration and antioxidant activity. *Cosmetics, 8*(1), 2. https://doi.org/10.3390/cosmetics8010002

Aziz, I., Saqib, Q.N., Younus, M., Ali, E., Qadir, M.I., Siddique, F.A., Wazir, M.A., Shaheer, M.T., Adnan, Q., Shirazi, J.H., & Rehman, K.U. (2021). Physicochemical, antidiarrheal and antidiabetic potential of super food (*Moringa oleifera* Lam.). *Pakistan Journal of Pharmaceutical Sciences*, 34(2 Suppl): 723–9. doi.org/10.36721/PJPS.2021.34.2.SUP.723-729.1

Badwaik, H.R., Al Hoque, A., Kumari, L., Sakure, K., Baghel, M., & Giri, T.K. (2020). Moringa gum and its modified form as a potential green polymer used in biomedical field. *Carbohydrate Polymers*, *249*, 116893. https://doi.org/10.1016/j.carbpol.2020.116893

Balakumbahan, R., Sathiyamurthy, V.A., & Janavi, G.J. (2020). Moringa leaf–A super food. *Biotica Research Today*, *2*(6), 438–440.

Benítez, W.M. (2012). Aprovechamiento poscosecha de la moringa (*Moringa oleífera*). *Revista iberoamericana de tecnología postcosecha*, *13*(2), 171–174.

Bischoff, K. L. (2021). Glucosinolates. In *Nutraceuticals* (pp. 903–909). Academic Press. https://doi.org/10.1016/B978-0-12-821038-3.00053-7

Borgonovo, G., De Petrocellis, L., Schiano Moriello, A., Bertoli, S., Leone, A., Battezzati, A., Mazzini, S., & Bassoli, A. (2020). Moringin, a stable isothiocyanate from *Moringa oleifera*, activates the somatosensory and pain receptor TRPA1 channel in vitro. *Molecules*, *25*(4), 976. https://doi.org/10.3390/molecules25040976

Canett-Romero, R., Arvayo-Mata, K.L., & Ruvalcaba-Garfias, N.V. (2014). Aspectos tóxicos más relevantes de *Moringa oleifera* y sus posibles daños. *Biotecnia*, *16*(2), 36–43. https://doi.org/10.18633/bt.v16i2.45

Chhikara, N., Kaur, A., Mann, S., Garg, M. K., Sofi, S. A., & Panghal, A. (2021). Bioactive compounds, associated health benefits and safety considerations of *Moringa oleifera* L.: An updated review. *Nutrition & Food Science, 51*(2), 255–277. https://doi.org/10.1108/NFS-03-2020-0087

Da Silva Junior, R.V., Rabello, C.B.V., Ludke, M.D.C.M.M., Lopes, C.D.C., de Medeiros Ventura, W.R.L., Soares, E.D.S.R., Guedes Paiva, P.M., Napoleão, T.H., Ribeiro, A.G., Nascimento, J.C.D.S., & Souza, L.F.A.D. (2024). Performance and quality of eggs of laying hens fed with *Moringa oleifera* leaf flour. *PLoS One*, *19*(12), e0314905. https://doi.org/10.1371/journal.pone.0314905

Dalei, J., Rao, V.M., Sahoo, D., Rukmini, M., & Ray, R. (2016). Review on nutritional and pharmacological potencies of *Moringa oleifera. European Journal of Pharmaceutical and Medical Research*, *3*(1), 150–155.

De Barros, M.C., Silva, A.G.B., dos Santos Souza, T.G., Chagas, C.A., Machado, J.C.B., Ferreira, M.R.A., Soares, L.A.L., Xavier, V.L., de Araújo, L.C.C., de Oliveira Borba, E.F., & da Silva, T.G. (2022). Evaluation of acute toxicity, 28-day repeated dose toxicity, and genotoxicity of *Moringa oleifera* leaves infusion and powder. *Journal of Ethnopharmacology*, *296*, 115504. https://doi.org/10.1016/j.jep.2022.115504.

Divya, K.G., Rubeena, M., Andalil, R., Erni, B., Sundaramoorthy, B., Thatipelli, S., Kumar, K.N.S., & Shakila, R. (2019). Identity profile of *Moringa oleifera* Lam. flower. *Identity*, *4*(4), 90–99.

Doglikuu, B. I. D., Abubakari, A., Yaseri, M., Shakibazadeh, E., Djazayery, A., & Mirzaei, K. (2021). The potential role of plantains, moringa, plantain-moringa combined diets, and other plant-based dietary patterns in controlling glycaemia among T2DM persons, a hospital based cross sectional survey in Ghana. *Journal of Diabetes & Metabolic Disorders*, *20*, 1529–1536. https://doi.org/10.1007/s40200-021-00896-y

Eid, W.A., Azab, D.E.S.H., & Negm, S.H. (2025). Characterization of a novel date energy bar fortified with *Moringa oleifera* leaves powder. *Journal of Future Foods*, *5*(3), 266–275. https://doi.org/10.1016/j.jfutfo.2024.07.006

El Bilali, H., Dan Guimbo, I., Nanema, R.K., Falalou, H., Kiebre, Z., Rokka, V.M., Tietiambou, S.R.F., Nanema, J., Dambo, L., Grazioli, F., & Naino Jika, A.K. (2024). Research on Moringa (*Moringa oleifera* Lam.) in Africa. *Plants*, *13*(12), 1613. https://doi.org/10.3390/plants13121613

Elsayed, E.A., Farooq, M., Sharaf-Eldin, M.A., El-Enshasy, H.A., & Wadaan, M. (2020). Evaluation of developmental toxicity and anti-angiogenic potential of essential oils from *Moringa oleifera* and *Moringa peregrina* seeds in zebrafish (*Danio rerio*) model. *South African Journal of Botany*, *129*, 229–237. https://doi.org/10.1016/j. sajb.2019.07.022.

Fahey, J.W., Olson, M.E., Stephenson, K.K., Wade, K.L., Chodur, G.M., Odee, D., Nouman, W., Massiah, M., Alt, J., Egner, P.A., & Hubbard, W.C. (2018). The diversity of chemoprotective glucosinolates in Moringaceae (*Moringa* spp.). *Scientific Reports*, 8(1), 7994. https://doi.org/10.1038/s41598-018-26058-4

Falowo, A.B., Mukumbo, F.E., Idamokoro, E.M., Lorenzo, J.M., Afolayan, A.J., & Muchenje, V. (2018). Multi-functional application of *Moringa oleifera* Lam. in nutrition and animal food products: A review. *Food Research International*, *106*, 317–334. https://doi.org/10.1016/j.foodres.2017.12.079

Fotio, A.L., Nguepi, M.S.D., Tonfack, L.B., Temdie, R.J.G., & Nguelefack, T.B. (2020). Acetaminophen induces liver injury and depletes glutathione in mice brain: Prevention by *Moringa oleifera* extract. *South African Journal of Botany*, *129*, 317–323. https://doi.org/10.1016/j.sajb.2019.08.037

Gandji, K., Chadare, F.J., Idohou, R., Salako, V.K., Assogbadjo, A.E., & Kakaï, R.G. (2018). Status and utilisation of *Moringa oleifera* Lam: A review. *African Crop Science Journal*, *26*(1), 137–156. https://doi.org/10.4314/acsj.v26i1.10

Gao, Q., Wei, Z., Liu, Y., Wang, F., Zhang, S., Serrano, C., Li, L., & Sun, B. (2022). Characterization, large-scale HSCCC separation and neuroprotective effects of polyphenols from *Moringa oleifera* leaves. *Molecules*, *27*(3), 678. https://doi.org/10.3390/molecules27030678

Gharby, S., Hajib, A., Ibourki, M., Nounah, I., Moudden, H.E., Elibrahimi, M., & Harhar, H. (2021). Induced changes in olive oil subjected to various chemical refining steps: A comparative study of quality indices, fatty acids, bioactive minor components, and oxidation stability kinetic parameters. *Chemical Data Collections*, *33*, 100702. https://doi.org/10.1016/j.cdc.2021.100702.

Gopalakrishnan, L., Doriya, K., & Kumar, D.S. (2016). *Moringa oleifera*: A review on nutritive importance and its medicinal application. *Food Science and Human Wellness, 5*(2), 49–56. https://doi.org/10.1016/j.fshw.2016.04.001

Graeff-Hönninger, S. & Khajehei, F. 2019. The demand for superfoods: Consumers' desire, production viability and bio-intelligent transition. In *Food tech transitions: Reconnecting agri-food, technology and society* (pp. 81–94). Cham: Springer International Publishing. https://doi.org/10.1007/978-3-030-21059-5_5

James, A., & Zikankuba, V. (2017). *Moringa oleifera* a potential tree for nutrition security in sub-Sahara Africa. *American Journal of Research Communication*, *5*(4), 1–14.

Klimek-Szczykutowicz, M., Gaweł-Bęben, K., Rutka, A., Blicharska, E., Tatarczak-Michalewska, M., Kulik-Siarek, K., ... Szopa, A. (2024). *Moringa oleifera* (drumstick tree)—nutraceutical, cosmetological and medicinal importance: A review. *Frontiers in Pharmacology*, *15*, 1288382. https://doi.org/10.3389/fphar.2024.1288382

Kou, X., Li, B., Olayanju, J.B., Drake, J.M., & Chen, N. (2018). Nutraceutical or pharmacological potential of *Moringa oleifera* Lam. *Nutrients, 10*(3), 343. https://doi.org/10.3390/nu10030343

Lalas, S., & Tsaknis, J. (2002). Extraction and identification of natural antioxidant from the seeds of the *Moringa oleifera* tree variety of Malawi. *Journal of the American Oil Chemists' Society, 79*, 677–683. https://doi.org/10.1007/s11746-002-0542-2

Leone, A., Spada, A., Battezzati, A., Schiraldi, A., Aristil, J., & Bertoli, S. (2015). Cultivation, genetic, ethnopharmacology, phytochemistry and pharmacology of *Moringa oleifera* leaves: An overview. *International Journal of Molecular Sciences*, *16*(6), 12791–12835. https://doi.org/10.3390/ijms160612791

Liao, P.C., Lai, M.H., Hsu, K.P., Kuo, Y.H., Chen, J., Tsai, M.C., Li, C.X., Yin, X.J., Jeyashoke, N., & Chao, L.K.P. (2018). Identification of β-sitosterol as in vitro anti-inflammatory constituent in *Moringa oleifera. Journal of Agricultural and Food Chemistry*, *66*(41), 10748–10759. https://doi.org/10.1021/acs.jafc.8b04555

Lin, H., Zhu, H., Tan, J., Wang, H., Wang, Z., Li, P., Zhao, C., & Liu, J. (2019). Comparative analysis of chemical constituents of *Moringa oleifera* leaves from China and India by ultra-performance liquid chromatography coupled with quadrupole-time-of-flight mass spectrometry. *Molecules*, *24*(5), 942. https://doi.org/10.3390/molecules24050942

Liu, Y., Wang, X.Y., Wei, X.M., Gao, Z.T., & Han, J.P. (2018). Values, properties and utility of different parts of *Moringa oleifera*: An overview. *Chinese Herbal Medicines*, *10*(4), 371–378. https://doi.org/10.1016/j.chmed.2018.09.002

Loyer, J. (2016). Communicating superfoods: A case study of Maca packaging. In *Proceedings of the Oxford symposium on food and cookery 2015. Oxford Symposium.* 236–246..

Ma, Z.F., Ahmad, J., Zhang, H., Khan, I., & Muhammad, S. (2020). Evaluation of phytochemical and medicinal properties of Moringa (*Moringa oleifera*) as a potential functional food. *South African Journal of Botany, 129*, 40–46. https://doi.org/10.1016/j.sajb.2018.12.002

Maiyo, F.C., Moodley, R., & Singh, M. (2016). Cytotoxicity, antioxidant and apoptosis studies of quercetin-3-O glucoside and 4-(β-D-glucopyranosyl-1→ 4-α-L-rhamnopyranosyloxy)-benzyl isothiocyanate from *Moringa oleifera. Anti-Cancer Agents in Medicinal Chemistry (Formerly Current Medicinal Chemistry-Anti-Cancer Agents)*, *16*(5), 648–656.

Marrufo, T., Nazzaro, F., Mancini, E., Fratianni, F., Coppola, R., De Martino, L., Agostinho, A.B., & De Feo, V. (2013). Chemical composition and biological activity of the essential oil from leaves of *Moringa oleifera* Lam. cultivated in Mozambique. *Molecules, 18*(9), 10989–11000. https://doi.org/10.3390/molecules180910989.

Martínez, L., Bastida, P., Castillo, J., Ros, G., & Nieto, G. (2019). Green alternatives to synthetic antioxidants, antimicrobials, nitrates, and nitrites in clean label Spanish chorizo. *Antioxidants*, *8*(6), 184. https://doi.org/10.3390/antiox8060184

Mashamaite, C.V., Ngcobo, B.L., Manyevere, A., Bertling, I., & Fawole, O.A. (2022). Assessing the usefulness of *Moringa oleifera* leaf extract as a biostimulant to supplement synthetic fertilizers: A review. *Plants*, *11*(17), 2214. https://doi.org/10.3390/plants11172214

Mashamaite, C.V., Ramatsitsi, M.N., & Manyevere, A. (2024). *Moringa oleifera* Lam.: A versatile climate-smart plant for nutritional security and therapeutic usage in semi-arid regions. *Journal of Agriculture and Food Research*, p.101217. https://doi.org/10.1016/j.jafr.2024.101217

Matic, I., Guidi, A., Kenzo, M., Mattei, M., & Galgani, A. (2018). Investigation of medicinal plants traditionally used as dietary supplements: A review on *Moringa oleifera. Journal of Public Health in Africa*, *9*(3), 841. https://doi.org/10.4081/jphia.2018.841

Maurya, S.K., & Singh, A.K. (2014). Clinical efficacy of *Moringa oleifera* Lam. stems bark in urinary tract infections. *International Scholarly Research Notices, 2014*(1), 906843. https://doi.org/10.1155/2014/906843

Milla, P.G., Peñalver, R., & Nieto, G. (2021). Health benefits of uses and applications of *Moringa oleifera* in bakery products. *Plants, 10*(2), 318. https://doi.org/10.3390/plants10020318

Miller, G.J., Embalabala, R., Bennett, C., Buck, J., Russell, B., Holder, E., Blanchard, H., & Henry, L. (2018). *Moringa oleifera*, miracle tree and superfood: Antibacterial evidence and nutritional benefits.

Nadeem, M., & Imran, M. (2016). Promising features of *Moringa oleifera* oil: recent updates and perspectives. *Lipids in Health and Disease, 15*, 1–8. https://doi.org/10.1186/s12944-016-0379-0.

Nizioł-Łukaszewska, Z., Furman-Toczek, D., Bujak, T., Wasilewski, T., & Hordyjewicz-Baran, Z. (2020). *Moringa oleifera* L. extracts as bioactive ingredients that increase safety of body wash cosmetics. *Dermatology Research and Practice, 2020*(1), 8197902. https://doi.org/10.1155/2020/8197902

Olusanya, R.N., Kolanisi, U., Van Onselen, A., Ngobese, N.Z., & Siwela, M. (2020). Nutritional composition and consumer acceptability of *Moringa oleifera* leaf powder (MOLP)-supplemented mahewu. *South African Journal of Botany, 129*, 175–180. https://doi.org/ 10.1016/j.sajb.2019.04.022.

Padayachee, B., & Baijnath, H.J.S. (2020). An updated comprehensive review of the medicinal, phytochemical and pharmacological properties of *Moringa oleifera. South African Journal of Botany, 129*, 304–316. https://doi.org/10.1016/j.sajb.2019.08.021.

Paliwal, R., Sharma, V., & Pracheta, P. (2011). A Review on horse radish tree (*Moringa oleifera*): A multipurpose tree with high economic and commercial importance. *Asian Journal of Biotechnology, 3*(4), 317–28. https://doi.org/10.3923/ajbkr.2011.317.328

Panda, S., Kar, A., Sharma, P., & Sharma, A. (2013). Cardioprotective potential of N, α-l-rhamnopyranosyl vincosamide, an indole alkaloid, isolated from the leaves of *Moringa oleifera* in isoproterenol induced cardiotoxic rats: In vivo and in vitro studies. *Bioorganic & Medicinal Chemistry Letters, 23*(4), 959–962. https://doi.org/10.1016/j.bmcl.2012.12.060

Pareek, A., Pant, M., Gupta, M.M., Kashania, P., Ratan, Y., Jain, V., Pareek, A., & Chuturgoon, A.A. (2023). *Moringa oleifera*: An updated comprehensive review of its pharmacological activities, ethnomedicinal, phytopharmaceutical formulation, clinical, phytochemical, and toxicological aspects. *International Journal of Molecular Sciences, 24*(3), 2098. https://doi.org/10.3390/ijms24032098

Peñalver, R., Martínez-Zamora, L., Lorenzo, J.M., Ros, G., & Nieto, G. (2022). Nutritional and antioxidant properties of *Moringa oleifera* leaves in functional foods. *Foods, 11*(8), 1107. https://doi.org/10.3390/foods11081107

Pereira, F.S.G., de Sobral, A.D., da Silva, A.M.R.B., & da Rocha, M.A.G. (2018). *Moringa oleifera*: a promising agricultural crop and of social inclusion for Brazil and semi-arid regions for the production of energetic biomass (biodiesel and briquettes). *OCL, 25*(1), D106. https://doi.org/10.1051/ocl/2017047

Prajapati, C., Ankola, M., Upadhyay, T.K., Sharangi, A.B., Alabdallah, N.M., Al-Saeed, F.A., Muzammil, K., & Saeed, M. (2022). *Moringa oleifera*: Miracle plant with a plethora of medicinal, therapeutic, and economic importance. *Horticulturae, 8*(6), 492. https://doi.org/10.3390/horticulturae8060492

Qi, L., Zhou, Y., Li, W., Zheng, M., Zhong, R., Jin, X., & Lin, Y. (2019). Effect of *Moringa oleifera* stem extract on hydrogen peroxide-induced opacity of cultured mouse lens. *BMC Complementary and Alternative Medicine, 19*, 1–9. https://doi.org/10.1186/s12906-019-2555-z

Rabie, M.M., Ibrahim, F.Y., Youssif, M.R.G., & Ezz El-Ragal, N.M. (2020). Effect of *Moringa oleifera* leaves and seeds powder supplementation on quality characteristics of cookies. *Journal of Food and Dairy Sciences, 11*(2), 65–73. https://dx.doi.org/10.21608/jfds.2020.78888

Rahmawati, I., Anggraeni, S.D., Bernika, S.O., & Julianti, A.I. (2023). Phenolic content and antioxidant activity of *Moringa oleifera* leaf infusions and tea. In *AIP Conference Proceedings* (Vol. 2572, No. 1). AIP Publishing. https://doi.org/10.1063/5.0118400

Rocchetti, G., Pagnossa, J.P., Blasi, F., Cossignani, L., Piccoli, R.H., Zengin, G., Montesano, D., Cocconcelli, P.S., & Lucini, L. (2020). Phenolic profiling and in vitro bioactivity of *Moringa oleifera* leaves as affected by different extraction solvents. *Food Research International, 127*, 108712. https://doi.org/10.1016/j.foodres.2019.108712

Rue, E.A., Rush, M.D., & van Breemen, R.B. (2018). Procyanidins: A comprehensive review encompassing structure elucidation via mass spectrometry. *Phytochemistry Reviews, 17*, 1–16. https://doi.org/10.1007/s11101-017-9507-3

Sari, K., Sirajuddin, S., Maddepungeng, M., Hadju, V., Saleh, A., Tanziha, I., & Hastuti, H. (2020). *Moringa oleifera* intake during pregnancy and breastfeeding toward docosahexaenoic acid and arachidonic acid levels in breast milk. *Open Access Macedonian Journal of Medical Sciences, 8*(B), 757–761. https://doi.org/10.3889/oamjms.2020.4614

Saucedo-Pompa, S., Torres-Castillo, J.A., Castro-López, C., Rojas, R., Sánchez-Alejo, E.J., Ngangyo-Heya, M., & Martínez-Ávila, G.C.G. (2018). Moringa plants: Bioactive compounds and promising applications in food products. *Food Research International*, *111*, 438–450. https://doi.org/10.1016/j.foodres.2018.05.062

Schrott, E., & Ammon, H. P. T. (2012). *Heilpflanzen der ayurvedischen und der westlichen Medizin*. Springer Berlin Heidelberg.

Shah, L., Baxi, N., Shah, A., & Akhani, T.K. (2022). An overview of the health benefits of drumstick pods. *Asian Pacific Journal of Health Sciences*, *9*(3), 10–4. https://doi.org/10.21276/apjhs.2022.9.3.03

Song, Q., Zou, Z., Chen, X., Ai, G., Xiong, P., Song, W., Liu, G., Zheng, A., & Chen, J. (2024). Effect of *Moringa oleifera* leaf powder supplementation on growth performance, digestive enzyme activity, meat quality, and cecum microbiota of Ningdu yellow chickens. *Agriculture*, *14*(9), 1523. https://doi.org/10.3390/agriculture14091523

Srivastava, S., Pandey, V.K., Dash, K.K., Dayal, D., Wal, P., Debnath, B., Singh, R., & Dar, A.H. (2023). Dynamic bioactive properties of nutritional superfood *Moringa oleifera*: A comprehensive review. *Journal of Agriculture and Food Research*, *14*, 100860. https://doi.org/10.1016/j.jafr.2023.100860

Su, B., & Chen, X. (2020). Current status and potential of *Moringa oleifera* leaf as an alternative protein source for animal feeds. *Frontiers in Veterinary Science*, *7*, 53. https://doi.org/10.3389/fvets.2020.00053

Super food (2026). Oxford English Dictionary. Available at: https://www.oed.com (Accessed: 17 April 2026)

Suresh, S., Chhipa, A.S., Gupta, M., Lalotra, S., Sisodia, S.S., Baksi, R., & Nivsarkar, M. (2020). Phytochemical analysis and pharmacological evaluation of methanolic leaf extract of *Moringa oleifera* Lam. in ovalbumin induced allergic asthma. *South African Journal of Botany*, *130*, 484–493. https://doi.org/10.1016/j.sajb.2020.01.046

Takase, M., Essandoh, P.K., Asare, R.K., & Nazir, K.H. (2022). The value chain of *Moringa oleifera* plant and the process of producing its biodiesel in Ghana. *The Scientific World Journal*, *2022*(1), 1827514. https://doi.org/10.1155/2022/1827514

Tiwari, A. K., Bano, S., Jaiswal, S., & Tiwari, A. (2023). A comprehensive review on *Moringa oleifera*: The green superfood for wellness. *Journal of Emerging Technologies and Innovative Research*, *10*(10), 271-272.

Trigo, C., Castelló, M.L., & Ortolá, M.D. (2023). Potentiality of *Moringa oleifera* as a nutritive ingredient in different food matrices. *Plant Foods for Human Nutrition*, *78*(1), 25–37. https://doi.org/10.1007/s11130-022-01023-9

Trigo, C., Castello, M.L., Ortola, M.D., Garcia-Mares, F.J., & Desamparados Soriano, M. (2020). *Moringa oleifera*: An unknown crop in developed countries with great potential for industry and adapted to climate change. *Foods*, *10*(1), 31. https://doi.org/10.3390/foods10010031

Tshabalala, T., Ncube, B., Madala, N.E., Nyakudya, T.T., Moyo, H.P., Sibanda, M., & Ndhlala, A.R. (2019). Scribbling the cat: A case of the "miracle" plant, *Moringa oleifera*. *Plants*, *8*(11), 510. https://doi.org/10.3390/plants8110510

Vargas-Sánchez, K., Garay-Jaramillo, E., & González-Reyes, R.E. (2019). Effects of *Moringa oleifera* on glycaemia and insulin levels: A review of animal and human studies. *Nutrients*, *11*(12), 2907. https://doi.org/10.3390/nu11122907

Vergara-Jimenez, M., Almatrafi, M.M., & Fernandez, M.L. (2017). Bioactive components in *Moringa oleifera* leaves protect against chronic disease. *Antioxidants*, *6*(4), 91. https://doi.org/10.3390/antiox6040091

Vijayanchali, S.S. (2024). The role of *Moringa oleifera* pods, seeds, bark and root as functional food. *World Journal of Advanced Research and Reviews*, *21*(3), 1937–1942. https://doi.org/10.30574/wjarr.2024.21.3.0933

Xu, Y.B., Chen, G.L., & Guo, M.Q. (2019). Antioxidant and anti-inflammatory activities of the crude extracts of *Moringa oleifera* from Kenya and their correlations with flavonoids. *Antioxidants*, *8*(8), 296. https://doi.org/10.3390/antiox8080296

Zeng, K., Li, Y., Yang, W., Ge, Y., Xu, L., Ren, T., Zhang, H., Zhuo, R., Peng, L., Chen, C., & Zhou, Y. (2019). *Moringa oleifera* seed extract protects against brain damage in both the acute and delayed stages of ischemic stroke. *Experimental Gerontology*, *122*, 99–108. https://doi.org/10.1016/j.exger.2019.04.014

Part 3

Formulation Development

17 *Moringa oleifera* as a Nutraceutical

Nabeeha Koya, Pavan Ramrao Chavan, Pravin Kumar, Abhishek Ambawatiya, Saqib Hameed, Priya Bisht, Shalini Choudhary and Nitesh Kumar

17.1 INTRODUCTION

Moringa oleifera, a monogenous species, falls within the family Moringaceae, native to Afghanistan, Bangladesh, northern India, Pakistan and so on. Currently, it is harvested across subtropical and tropical regions, including Asia, South America, Australia, the Middle East and Africa (Bhattacharya et al., 2018). It is generally referred to as 'drumstick tree', 'miracle tree', 'horseradish tree' or 'sahjan'. These trees can bear drought and grow up to 12 m in height. Moringa's versatility, nutritional worth and quick growth make it a valuable crop. The plant is of high value because nearly every part of it serves a function and provides nutrients, including vitamins, minerals and other elements. The World Health Organization (WHO) has also promoted the use of moringa to treat malnutrition. The pods, seeds, leaves, bark, roots and flowers are all utilised in the diet (Pareek et al., 2023).

M. oleifera leaves are a powerhouse of minerals such as iron, zinc, calcium, magnesium, potassium, copper and manganese. Apart from minerals, the leaves also contain vitamins (vitamin C), phenolic compounds, β-carotene, a great source of proteins and amino acids (lysine, tryptophan, cysteine and methionine). Due to their natural antioxidant content, including ascorbic acid, phenolics and carotenoids, they can extend the shelf life of fat-containing foods (Anwar et al., 2007; Masitlha et al., 2024). Despite being low in fat and carbohydrates, *M. oleifera* leaves are considered a complete dietary supplement since they provide a rich supply of protein and vital amino acids (Milla et al., 2021; Patil et al., 2022).

Malnutrition has been treated using flowers, especially in newborns and nursing mothers. Vitamin A, which is found in moringa flowers, can strengthen immunity and safeguard eye health. The blossoms also include potassium, iron, amino acids and calcium to build bones and vitamin C to alleviate inflammation (Pareek et al., 2023). To lessen cold symptoms and cleanse the body, moringa blossoms are steeped in teas and used in tonics in traditional treatments worldwide, particularly in Ayurveda. These flowers are abundant in antioxidants such as quercetin and chlorogenic acid, which may help decrease blood pressure. The flowers have anti-inflammatory properties that may help reduce systemic inflammation, which is associated with several ailments, such as digestive issues, arthritis and cardiovascular disease (Milla et al., 2021; Prabsattroo et al., 2015).

The medicinal benefits of moringa tree roots have been utilised since ancient times in Egypt, Greece and early Rome (Hamza & Azmach, 2017). The roots contain a higher quantity of nutrients than any other part of the tree. Because of their antibacterial qualities, moringa roots cure many ailments, including cancer, inflammation, skin problems, gastritis, asthma, digestive issues and sexual performance. Even women in India's most rural areas utilise it as a long-term method of birth control (Abd-Rabou et al., 2017).

M. oleifera seeds contain a significant quantity of vitamins A and B1. In addition, they include micronutrients, minerals and bioactive substances such as sterols, flavonoids, phytates, saponins and trypsin inhibitors. The seed's lipid content ranges from 13% to 46%, which might make it an

DOI: 10.1201/9781003617860-20

oilseed. The seeds are therefore presented as an excellent source of fibre, protein and lipids (Saa et al., 2019). Seed extracts are used as a water purification agent and have antibacterial action. Numerous studies have shown that moringa seeds can lower blood pressure, blood sugar, inflammation and oxidative stress (Islam et al., 2021).

17.2 NUTRITIONAL COMPOSITION (QUANTITATIVE)

17.2.1 Macronutrients

Macronutrients, including proteins, carbohydrates and lipids, are essential in large quantities to supply energy, support hormone production, synthesise vital molecules and regulate metabolic processes (Espinosa-Salas & Gonzalez-Arias, 2025). While *M. oleifera* leaves contain carbohydrates and fats, they are particularly notable for being an excellent source of protein and vital amino acids (Patil et al., 2022). Based on dry weight (DW) analysis, moringa leaves reveal the presence of crude protein (10.74–30.29%), carbohydrates (13.41–63.11%), fats (6.50–20%), crude fibre (7.09–35%) and mineral matter (7.64–10.71%) (Piyush Kashyap et al., 2022).

Proteins are essential macronutrients that serve as fundamental components of the human body's structure and function, playing a crucial role in physiological growth and development. They are composed of long amino acid sequences, including non-essential and essential types, linked by peptide bonds. While humans can produce non-essential amino acids endogenously, vital amino acids must be supplied through the diet, with familiar sources including fish, red meat, eggs and poultry. For vegetarians, however, consuming a complete spectrum of essential amino acids can be difficult, as many plant-based proteins are deficient in one or more of them.

M. oleifera leaves present an effective solution to this challenge. They are protein-rich and contain a comprehensive range of essential amino acids, making moringa leaf powder a valuable plant-based protein source for vegetarians. Moreover, the leaves have health-promoting phytochemicals, including sterols, tannins, flavonoids, alkaloids, saponins and terpenoids, which further contribute to their nutritional benefits (Islam et al., 2021).

17.2.2 Micronutrients

Micronutrients are required in trace amounts to support various biochemical processes, including gene transcription, enzymatic activities and defence against oxidative stress. These nutrients are broadly classified into two categories: vitamins and minerals, each playing distinct roles in maintaining human health (Espinosa-Salas & Gonzalez-Arias, 2025). In addition to macronutrients, the animal body relies on essential micronutrients that either act as carriers or facilitate the metabolic breakdown of macronutrients. Among these, vitamins are vital for energy metabolism. Vitamin deficiencies can result in well-known health conditions such as beriberi, rickets and scurvy.

M. oleifera is recognised as a rich supplier of vital vitamins, such as vitamins A (β-carotene), B (nicotinic acid, folic acid, pyridoxine), C, D and E. Regarding mineral content, moringa also provides essential elements necessary for physiological development. Notably, the calcium content of dried moringa leaf powder has been reported to be up to 17 times higher than that of milk (Islam et al., 2021). Table 17.1 shows the overview of macronutrient and micronutrient constituents in *M. oleifera* leaves and seeds, and Table 17.2 shows the nutritional composition of fresh and dried *M. oleifera* leaves in comparison with familiar food sources (all nutrient values per 100 g of material).

17.3 LIPIDS

Lipids are a rich source of essential fatty acids, energy and fat-soluble vitamins, as well as carriers of flavour and various bioactive compounds that support numerous physiological functions (Nadeem & Imran, 2016). Chemically, lipids are fatty, waxy, or oily substances that are soluble in

TABLE 17.1
Overview of Macronutrient and Micronutrient Constituents in *M. oleifera* Leaves and Seeds

Nutrient	Fresh leaves	Dried leaves	Seeds
Energy	92 kcal	329 kcal	–
Macronutrients			
Protein	6.7 g	29.4 g	35.97±0.19 g
Carbohydrates	12.5 g	41.2 g	8.67±0.12 g
Fat	1.7 g	5.2 g	38.67±0.03 g
Dietary fibre	0.9 g	12.5 g	2.87±0.03 g
Micronutrients			
Vitamin A	6.8 mg	18.9 mg	–
Vitamin B1 – thiamine	0.06 mg	2.02 mg	0.05 mg
Vitamin B2 – riboflavin	0.05 mg	21.3 mg	0.06 mg
Vitamin B3 – niacin	0.8 mg	7.6 mg	0.2 mg
Vitamin C	220 mg	15.8 mg	4.5±0.17 mg
Vitamin E	448 mg	10.8 mg	751.67±4.41 mg
Minerals			
Calcium	440 mg	2185 mg	45 mg
Phosphorus	70 mg	252 mg	75 mg
Magnesium	42 mg	448 mg	635±8.66 mg
Copper	0.07 mg	0.49 mg	5.20±0.15 mg
Potassium	259 mg	1236 mg	–
Sulphur	–	–	0.05 mg
Iron	0.85 mg	25.6 mg	–

(Gopalakrishnan et al., 2016).

TABLE 17.2
Nutritional Composition of Fresh and Dried *M. oleifera* Leaves in Comparison with Familiar Food Sources

Nutrients	Typical source	Moringa (fresh leaves)	Moringa (dried leaves)
Vitamin A (mg)	Carrots, 1.8	6.8	18.9
Potassium (mg)	Banana, 88	259	1236
Calcium (mg)	Milk, 120	440	2185
Vitamin C (mg)	Orange, 30	220	15.8
Protein (g)	Yoghurt, 3.1	6.7	29.4

Note: All nutrient values per 100 g of material. (Devkota et al., 2020; Gopalakrishnan et al., 2016).

organic solvents but insoluble in polar solvents like water. They include triglycerides (fats and oils), phospholipids, waxes and steroids (Ahmed et al., 2023).

Moringa oleifera oil is particularly abundant in oleic acid, closely resembling the composition of olive oil, and it also includes other beneficial fatty acids. Owing to this composition, it holds substantial potential as a commercial edible oil within the food industry. Historically, *Moringa oleifera* oil has found applications in formulating perfumes, personal care items, machine lubricants, and, more

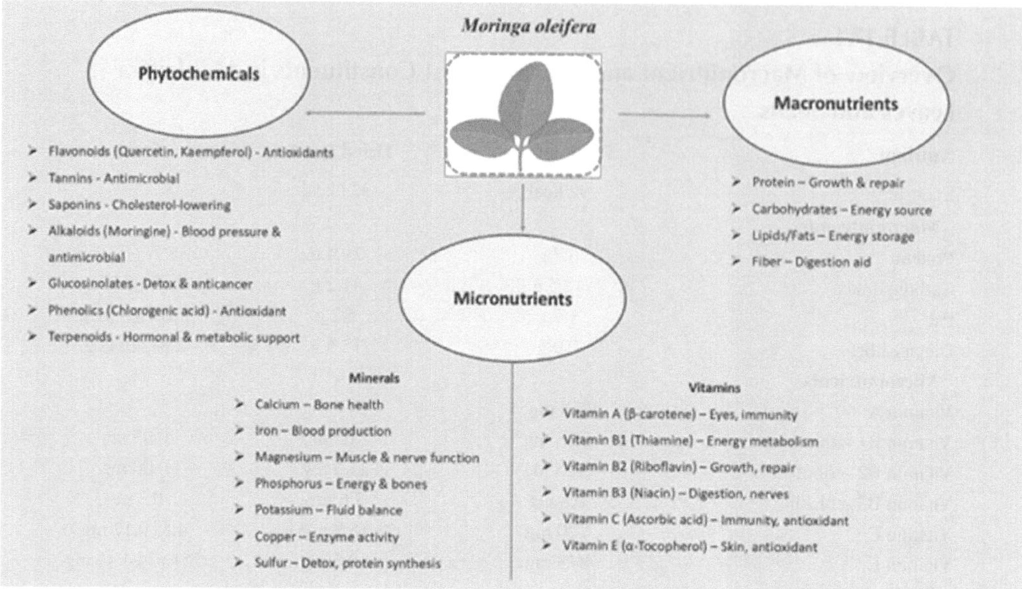

FIGURE 17.1 Infographic on *M. oleifera* Nutrient Density.

recently, as a biofuel source. The primary triacylglycerol in moringa oil is triolein, which accounts for approximately 36.7% of its content. Moreover, the oil demonstrates superior thermal stability during frying compared to canola, soybean, and palm olein oils. Its mild, peanut-like aroma further adds to its appeal. Beyond culinary uses, Moringa oil is widely used as a protective and hair care commodity.

(Owon et al., 2021)

17.4 KEY BIOACTIVE COMPOUNDS AND MECHANISMS

An overview of the key nutritional and phytochemical constituents of *M. oleifera* is given in Figure 17.1. Aside from its beneficial nutritional content, *M. oleifera* provides a remarkable range of medicinal advantages. Various parts of this plant contain substantial amounts of protein, vitamins, amino acids, β-carotene and different phenolics, along with a profile of essential minerals (Anwar et al., 2007). The leaves are packed with nutrients, including proteins, β-carotene, fat- and water-soluble vitamins (A, B, C and E), comprising riboflavin, niacin, folic acid and pyridoxine, along with amino acids, minerals and phenolic substances (Moyo et al., 2012). The approach to complex mechanisms of action and the characterisation of biologically active compounds found within *M. oleifera* play an essential role in managing various conditions and provide optimistic information for further approaches to medicine (Chis et al., 2023). Table 17.3 shows multiple bioactive compounds and therapeutic properties of the plant parts.

17.4.1 Polyphenols and Flavonoids

M. oleifera holds many bioactive phytochemicals, including flavonoids, terpenoids, tannins, anthocyanins and proanthocyanidins. Flavonoids and anthocyanins are widely recognised for their potent anti-inflammatory and antioxidant properties (G. M. El-Sherbiny et al., 2024). Terpenoids have been shown to possess considerable anticancer and antibacterial properties, whereas tannins exhibit anti-inflammatory and antimicrobial resistance. Proanthocyanidins are valued for their notable antioxidant capacity and role in cardiovascular health. Moringa's synergistic effects contribute significant

TABLE 17.3
Bioactive Compounds and Therapeutic Properties of *M. oleifera* Plant Parts

Plant part	Key bioactive compounds	Health benefits and traditional uses	References
Roots	Alkaloids, glycosides, saponins, terpenoids	To treat ear or tooth pain, ulcers and inflammation	Anwar et al., 2007; Gopalakrishnan et al., 2016; Pareek et al., 2023
Bark	Tannins, alkaloids, saponins	Used for wounds, skin infections, fever and as an abortifacient	Anwar et al., 2007; Pareek et al., 2023
Leaves	Flavonoids (quercetin, kaempferol), vitamins, alkaloids, phenolics, isothiocyanates, saponins	Used for wounds, pain, ulcers, liver and heart disease, inflammation, insomnia and malnutrition	Ajantha et al., 2020; Anwar et al., 2007; Chis et al., 2023; Pareek et al., 2023; Selasa et al., 2022
Stem, drumstick	Vitamins, minerals, amino acids, phenolics	To treat hepatitis and joint pain; it is a nutritious vegetable	Anwar et al., 2007; Pareek et al., 2023
Flower	Flavonoids, vitamins	Less commonly used; some nutritional and medicinal applications	Anwar et al., 2007; Gopalakrishnan et al., 2016; Pareek et al., 2023b
Seed	Isothiocyanates, proteins, fatty acids, sterols	Used as a laxative, for tumours, prostate/bladder problems, arthritis and water purification	Anwar et al., 2007; Pareek et al., 2023; Waterman et al., 2014; Xie et al., 2022
Gum	Polysaccharides, proteins	Treats fever, used to induce abortion in traditional medicine	Pareek et al., 2023

antioxidant, antigenotoxic (DNA-protective) and immunostimulatory (immune-enhancing) potential, emphasising its therapeutic potential for improving health and preventing disease (Abd El-Hack et al., 2022; G. M. El-Sherbiny et al., 2024). Phenolic compounds and flavonoids play a significant role in inhibiting lipid peroxidation and enhancing oxidative stability, thereby preventing protein degradation and supporting various other biological functions. *M. oleifera* leaf extracts measured total phenolic content of 121.60 mg/100 g (aqueous extract) and 130.57 mg/100 g (alcoholic extract). Correspondingly, the flavonoid content measured 2.66 mg/100 g and 3.68 mg/100 g, respectively (Jassim & Ali, 2024).

The major flavonoids identified include quercetin, kaempferol, myricetin, isorhamnetin, rutin, apigenin and luteolin (Tao et al., 2025). These compounds are mainly present as glycosides, such as glucosides, malonyl glucosides and rutinosides. While the highest concentrations are typically present in the leaves, other portions of the plant, including the seeds, also contain appreciable amounts of these bioactive flavonoids (Vergara-Jimenez et al., 2017). The phenolic content of alcoholic extracts from *M. oleifera* flowers, leaves and seeds was measured at 114.49 mg/100 g, 170.07 mg/100 g and 22.43 mg/100 g, respectively. Although the seed extract exhibited the lowest phenolic content of the three, it still had significant quantities of bioactive compounds, indicating its potential for use as a natural preservative in food preservation and functional food formulations (do Nascimento et al., 2017). Table 17.4 shows the role and mechanisms of various polyphenols/flavonoids in *M. oleifera*.

17.4.2 Glucosinolates and Isothiocyanates

In domesticated varieties of *M. oleifera*, the predominant glucosinolate is glucomoringin, whereas wild types often contain higher concentrations of another glucosinolate known as glucosinalbin

TABLE 17.4
Role and Mechanisms of Various Polyphenols/Flavonoids in *M. oleifera*

Compounds	Role	Mechanisms
Quercetin	Induces apoptosis and inhibits proliferation and cytokine production. Furthermore, it inhibits pro-oxidant enzymes and scavenges ROS	Anticancer, anti-inflammatory and antioxidant properties
Kaempferol	Inhibits proliferation and pro-oxidant enzymes and induces apoptosis	Anticancer and antioxidant properties
Chlorogenic acid	Enhances insulin sensitivity and reduces glucose absorption	Antidiabetic properties
Isothiocyanates	Inhibit cytokine production and suppress the NF-κB pathway	Anti-inflammatory action
Gallic acid, ferulic acid	Scavenge ROS, inhibit pro-oxidant enzymes and chelate metals	Antioxidant properties

(Chodur et al., 2018). Among the edible parts analysed, the boiled seeds exhibited the highest glucosinolate levels, while the leaves also contributed significantly. Glucomoringin undergoes enzymatic hydrolysis, primarily catalysed by myrosinase, which leads to the formation of its principal isothiocyanate derivative known as moringin (4-α-L-rhamnopyranosyloxybenzyl isothiocyanate), also referred to as MIC-1 (Malabed & Noel, 2013). In addition to moringin, *M. oleifera* produces acetylated isothiocyanate derivatives, such as 4-O-acetylrhamnosyloxybenzyl isothiocyanate. Together, these compounds comprise more than 95% of the plant's total isothiocyanate content. Optimised extracts of *M. oleifera* can contain up to 1.66% isothiocyanates by weight, and these compounds show high stability, mainly due to the presence of the rhamnose sugar group (Waterman et al., 2014; Xie et al., 2021). Glucosinolates and their isothiocyanate derivatives have been shown to suppress the secretion of pro-inflammatory cytokines, such as tumour necrosis factor-alpha (TNF-α) and interleukin-1 beta (IL-1β), by inhibiting crucial inflammatory signalling pathways such as NF-κB. Isothiocyanates derived from *M. oleifera* enhance the antioxidant defence responses of cells, primarily through the Nrf2 pathway, thereby enhancing the expression of detoxification and cytoprotective enzymes. In addition, these compounds exhibit antimicrobial activity by disrupting bacterial cell membranes and impairing crucial metabolic functions (Chis et al., 2023; Malabed & Noel, 2013; Xie et al., 2021).

17.4.3 Alkaloids and Saponins

The concentrations of alkaloids and saponins in *M. oleifera* leaves vary depending on the methods of extraction and geographical origin. Reported levels of saponin and alkaloids range from 0.6 to 8.1% and 0.3 to 3.1%, respectively. These variations are influenced by the type of extraction solvent used, such as aqueous, ethanolic or methanolic solutions, as well as the specific plant part analysed and the environmental conditions of the region of origin (Ajantha et al., 2020; Bisong et al., 2019; Selasa et al., 2022). For instance, *M. oleifera* samples collected from Nigeria contained approximately 0.45% alkaloids and 0.84% saponins, equivalent to 446.67 mg/100 g and 844.17 mg/100 g, respectively. Alkaloids usually provide antimicrobial and antioxidant properties, whereas saponins have anticancer, cholesterol-lowering and antimicrobial benefits (Khalid et al., 2023; Pop et al., 2022).

17.4.4 Vitamins

M. oleifera leaves are an excellent source of vital vitamins, particularly vitamins A (β-carotene), C and E, making them a significant nutritional and therapeutic resource. The leaves are especially rich in provitamin A, with β-carotene levels reported between 89 and 112 mg per 100 g (Pareek et al., 2023). This concentration is significantly higher than in carrots, providing up to ten times more vitamin A. This vitamin is vital for eye health, immune system support, regulating cell differentiation and facilitating embryonic development (Qadir et al., 2022; Vergara-Jimenez et al., 2017). The vitamin C content in *M. oleifera* leaves changes according to the plant's developmental stage, with mature leaves generally exhibiting higher concentrations. Reported vitamin C levels in the leaves range from 143.14 to 278.50 mg/100 g of dry weight, with some studies documenting levels exceeding 200 mg/100 g (Vergara-Jimenez et al., 2017). These amounts are higher than those observed in citrus fruits such as oranges, indicating *M. oleifera* as a rich source of vitamin C. It is essential for maintaining immune function, promoting collagen synthesis and as a cofactor for several enzymatic reactions involved in antioxidant defence and metabolic regulation (Qadir et al., 2022; Sultana, 2020). The leaves contain vitamin E at levels comparable to those found in nuts and contribute to the inhibition of abnormal cell proliferation and enhance antioxidant defence (Sultana, 2020; Vergara-Jimenez et al., 2017). In addition to other vitamins, they are also a source of B complex vitamins, including thiamine (B1), riboflavin (B2), niacin (B3), vitamin B6 and folate (B9), which are vital for energy metabolism, nervous system health and the synthesis of neurotransmitters, supporting mood regulation and brain function (Gopalakrishnan et al., 2016).

17.4.5 Minerals

M. oleifera has a mineral content between 7.6% and 10.7% when dry weight is taken into consideration (P. Kashyap et al., 2022). A significant quantity of minerals, such as potassium, iron, calcium, manganese, magnesium and copper, as well as phenols, are found in the leaves. Among the various minerals, calcium is primarily needed for human development and growth. The calcium content of the leaves of *M. oleifera* and its powder is 1000 mg and 4000 mg, respectively. *M. oleifera* powder contains more iron than spinach and is a significant supplier of iron (17.5 mg/100 g DW). Its leaves are recommended as a possible alternative to iron pills for the treatment of anaemia, as they contain 28 mg of iron. Furthermore, the synthesis of RNA and DNA and sperm cell development depend on dietary zinc, an important mineral. Moringa leaves contain sufficient zinc (25.5–31.3 mg/kg) to meet daily dietary needs (Arora & Arora, 2021).

17.4.6 Fatty Acids and Essential Oils

In 2017, Kuete published a report on the phytochemical content of the volatile components of the ethanolic extract of *M. oleifera* leaves. Forty-four volatile chemicals were identified, containing major components such as pentacosane (17.4%), hexacosane (11.2%), (E)-phytol (7.7%) and 1-[2,3,6-trimethylphenyl]-2-butanone (3.4%). The oil of *M. oleifera* seeds is of superior quality. About 35% to 40% of the substance is seed oil. Its major fatty acid contents are oleic acid (65%), linoleic acid (16%), palmitic acid (12.31%), stearic acid (5.1%) and palmitoleic acid (2.1%) (Basuny & Al-Marzouq, 2016; Galts et al., 2019; Kuete, 2017).

17.4.7 Proteins and Amino Acids

Although *M. oleifera* leaves are lower in fat and carbohydrates, they are an excellent provider of complete, high-value protein comparable to milk and eggs (Paikra et al., 2017). The seeds of moringa contain roughly 52% protein. Additionally, the plant's leaves and pods contain vital amino acids. They are suggested as an excellent dietary supplement. *M. oleifera* contains several essential amino acids, including arginine, histidine, phenylalanine, leucine, isoleucine, lysine, threonine, methionine, tryptophan and valine. It also provides non-essential (endogenously synthesised) amino acids such as aspartic acid, alanine, glutamic acid, tyrosine, cysteine, proline, serine, glycine and hydroxyproline (Marta Klimek-Szczykutowicz et al., 2024).

17.4.8 Fibre (Prebiotic Effects)

The fruit, leaves and pods of *M. oleifera* are important sources of dietary fibre. According to clinical research, moringa pods enhance nutrient absorption and digestive health, which benefits digestion and gut health (Godinez-Ov et al., 2016; Abd El-Hack et al., 2018). The leaves of *M. oleifera* contain around 16–20% fibre (Teixeira et al., 2014). By eliminating potential carcinogens from the body, fibres aid in cleansing the digestive tract and help prevent the absorption of excessive cholesterol (Shil, 2021).

17.5 CLINICALLY VALIDATED HEALTH BENEFITS

17.5.1 Antidiabetic

Diabetes is a metabolic disorder with serious consequences that primarily presents as chronic hyperglycaemia due to impairment in insulin secretion (Frak et al., 2022). Increased thirst and appetite and frequent micturition are signs of high blood sugar. Diabetes-related hyperglycaemia is followed by abnormalities in protein, carbohydrate and lipid metabolism (Eller-Vainicher et al., 2020). Diabetes can cause a number of problems if left untreated, including heart disease, diabetic ketoacidosis, hyperosmolar hyperglycaemia, ulcers of the foot, stroke, CKD, eye damage and even death (Adnette, 2019).

Type I and type II diabetes can both be managed by moringa. Rutin, quercetin-3-glycoside and kaempferol glycosides are examples of polyphenols that lower glucose levels. It was discovered that moringa's alkaloids, terpenoids, flavonoids, glycosides and carotenoids exhibited antidiabetic properties. Moreover, polyphenols such as kaempferol glycosides, quercetin-3-glycoside and rutin reduced the levels of glucose (Al-Malki & El Rabey, 2015). According to a study conducted on rats given STZ, which is an extract of *M. oleifera* (aqueous) leaves, it is capable of treating both types of diabetes (Khan et al., 2017). Researchers explored how moringa leaf extract affected STZ-treated diabetic rats and discovered that the plant leaves significantly increased water consumption, weight loss, fasting plasma insulin and blood glucose. Additionally, it increases plasma albumin in rats and decreases fasting plasma levels of aspartate and alanine aminotransaminases (Chhikara et al., 2020). Similar to pioglitazone, a common antidiabetic medication, *M. oleifera* causes overexpression of GLUT 4, which is important in insulin resistance reversal (Adedapo et al., 2020).

17.5.2 Cardiovascular Protection

Up to 18% of people worldwide suffer from long-term inflammatory conditions such as psoriasis, systemic lupus erythematosus, rheumatoid arthritis and HIV infection. Compared with the general population, they are more likely to develop cardiovascular disorders linked to inflammation (Lazou et al., 2020; Sun & Tian, 2018). Since inflammation triggers the formation of atherosclerotic plaque, it contributes to atherosclerosis, heart failure and ischaemic heart disease. An important factor in the onset, progression and exacerbation of atherosclerosis has been identified as inflammation, even though many of these cardiovascular conditions are caused by common risk factors such as diabetes, high blood lipids, high blood pressure and smoking (Mehta et al., 2020). Patients with atherosclerosis have been found to have elevated serum levels of pro-inflammatory factors, which include IL-6, IL-12, TNF-α and inflammatory biomarkers such as homocysteine, YKL-40, fibrinogen, high-sensitivity C-reactive protein (hs-CRP) and C-reactive protein (CRP). It is well demonstrated that YKL-40 contributes significantly to the aetiology of cardiovascular disorders such as obesity and insulin resistance. Additionally, atherosclerotic plaques exhibit high levels of its expression (Deng et al., 2020). Significant antihypertensive or hypotensive effects have been reported from ethanolic extracts from *M. oleifera* leaves. From the *M. oleifera* ethyl acetate fraction, thiocarbamates have been discovered and separated, including 4-[(4′-O-acetyl-α-L-rhamnosyloxy) benzyl] isothiocyanate, 4-[(α-L-rhamnosyloxy)benzyl] isothiocyanate, niazinin A, niazicin A and niazirin. Any of these substances administered intravenously at doses of 3–10 mg/kg caused bradycardic and hypotensive effects in anaesthetised rats, which may have been caused by calcium antagonist effects (Abd Rani et al., 2018; Chis et al., 2023).

The alcoholic and aqueous extracts of seeds and pods contained phytochemicals such as thiocarbamate, glycosides, isothiocyanate, methyl p-hydroxybenzoate and β-sitosterol, which significantly lowered blood pressure (Faizi et al., 1998).

AChE activity was found to be decreased by extracts from moringa leaves and seeds, most likely due to their phenolic components (Adefegha et al., 2019). *M. oleifera*'s bioactive metabolites have vasodilatory and lipid-lowering effects that promote cardiovascular health. Two substances associated with improving lipid profiles and cardiovascular health are quercetin and β-sitosterol (Fatima et al., 2024).

17.5.3 Neuroprotection

Over 276 million people worldwide currently suffer from a severe neurological disorder, which lowers quality of life and is one of the primary causes of mortality (Galts et al., 2019). Pharmacological effects of moringa include diuretic, heart-stimulant, antiseptic, antiasthmatic and antispasmodic properties. As a result, scientists have begun using *M. oleifera* to treat neurological conditions; their results indicate that it may slow the progression of disorders like epilepsy, Alzheimer's disease and Parkinson's disease (Ghimire et al., 2021). Two bioactive compounds in moringa, i.e. flavonoids and isothiocyanates, have shown neuroprotective qualities. These metabolites may be helpful in preventing neurodegenerative disorders because of their neurotropic and antioxidant properties (Azlan et al., 2022).

Alzheimer's disease (AD) is a neurological disorder that results in β-amyloid buildup and tau protein phosphorylation, which leads to dementia. Because it improves blood flow to the brain, moringa is said to have neuroprotective properties. This is because cerebral ischaemia, resulting from insufficient blood circulation, triggers the generation of reactive oxygen species (ROS). By lowering ROS, moringa stimulates a neuroprotective antioxidant mechanism (Fahey, 2017). Continuous moringa supplementation improved spatial memory, reduced oxidative stress and escape latency and restored acetylcholine levels, all of which helped prevent the progression of AD (Onasanwo et al., 2021).

Parkinson's disease is a neurodegenerative disorder characterised by impaired movement, primarily caused by the progressive loss of dopamine-producing neurons in the substantia nigra pars compacta. Oxidative stress is recognised as a key driver of dopaminergic cell death (Soni & Kumar, 2022). Apoptosis, mitochondrial damage and ROS production are reported to be reduced by moringa extract.

Moringa extract is also used to treat Huntington's disease, epilepsy and other central nervous system disorders (Mundkar et al., 2022).

17.5.4 Antimicrobial Activity

The primary challenge of the modern era is antibiotic resistance. Inappropriate antibiotic use causes most pathogenic bacteria to develop a permanent resistance to antibiotics and chemotherapeutic drugs. New antimicrobial agents with various activities must be discovered. Many bioactive substances may be found in moringa leaves, and several of its extracts have demonstrated antibacterial, antifungal, antiviral and antiparasitic properties. However, many findings support the theory that the antibacterial properties of moringa leaves are due to chemical metabolites such as pterygospermin, moringine and benzyl isothiocyanate (Liu et al., 2013).

Alkaloids and glucosinolates found in *M. oleifera* have potent antimicrobial properties that work effectively against various diseases. The bioactive components of the plant exhibit inhibitory effects on bacteria, fungi and viruses, suggesting that they may find application in antimicrobial treatments (Adefegha et al., 2019).

Moringa leaves contain a large amount of glucosinolate, epicatechin, vicinin-like anthocyanins, phenols and catechins. Because they are all lipophilic, they could be the reason for the antibacterial action that the leaf extract has shown in several studies. Because phenolic compounds tend to interact with the cell membrane and other lipophilic cell components, their lipophilic nature increases their antimicrobial effectiveness. This interaction results in aggregations of cell contents and irreversible modifications to the cytoplasmic membrane, which inhibit channels and the enzyme's function (Sikkema et al., 1995).

By preventing bacterial quorum sensing, preserving membrane integrity and various other methods, several isothiocyanates exhibited potent antibacterial activity (Ganin et al., 2013). Additionally, *M. oleifera* was found to suppress *Staphylococcus aureus, Pseudomonas aeruginosa and Escherichia coli* (Rahman et al., 2009).

17.5.5 Anti-Inflammation Activity

The immune system's natural reaction to external stimuli, such as pathogens, is inflammation (Arulselvan et al., 2016). However, excessive and unchecked inflammation will also cause immunological disruption and immune cell apoptosis, which will lead to tumours and ongoing degenerative conditions (Cui et al., 2019). *M. oleifera* minimises inflammation in several ways: (a) its flavonoids, such as quercetin and kaempferol, block necessary enzymes (COX, LOX) that produce inflammatory agents; (b) its isothiocyanates affect pathways like NF-κB to decrease pro-inflammatory cytokines (TNF-α and IL-1β) and simultaneously elevate anti-inflammatory cytokines (IL-10); and (c) the antioxidants it contains, which include flavonoids and polyphenols, reduce oxidative stress, which further aids in inflammation control (Chis et al., 2023).

A 0.1 mL injection of 1% carrageenan in saline was used to inflame the left paw of rats to test the anti-inflammatory properties of *M. oleifera* root extracts. Depending on the concentration, *M. oleifera* extract decreased swelling by 35% to 45% (Ezeamuzie et al., 2008). Another study by Sara Avilés-Gaxiola et al. examined the in vitro anti-inflammatory properties of *M. oleifera* leaf protein hydrolysate (MOPH). They showed that MOPH dramatically decreased the production of nitric oxide (NO) after LPS stimulation, as determined by the Griess reaction, using RAW 264.7 macrophages. MOPH's potential as a natural anti-inflammatory agent was further supported by the study's confirmation that it was non-cytotoxic using the MTS assay (Avilés-Gaxiola et al., 2021). By inhibiting NO, IL-6 and TNF-α, the hexane extract of *M. oleifera* flowers, which is high in alkane hydrocarbons (heptacosane 46.2%), demonstrated notable anti-inflammatory properties (Fahmy et al., 2024). A substance called aurnatiamide acetate and 1,3-dibenzylurea that was extracted from *M. oleifera* roots prevented the production of TNF-α (Cuellar-Nunez et al., 2021). Active substances with anti-inflammatory qualities are tannins, phenols, alkaloids, flavonoids, carotenoids, β-sitosterol, vanillin and moringin (Bhattacharya et al., 2018). Although *M. oleifera* root extracts have encouraging anti-inflammatory effects in the carrageenan-induced rat paw model, the research does not provide comprehensive information on the precise mechanisms causing this decrease in swelling, raising concerns regarding the signalling pathways and inflammatory mediators at play. Correspondingly, although the clinical study on the potential of *M. oleifera* bark extract to treat UTIs shows promising initial results, the study's short duration and small sample size call for caution when making firm conclusions, highlighting the need for longer and more thorough research (M. Klimek-Szczykutowicz et al., 2024).

17.5.6 Hepatoprotective Effects

Leaf extracts from *M. oleifera* have been found to have hepatoprotective effects in recent studies. When Wistar rats were administered dimethyl 2,2-dichlorovinylphosphate (DDVP, also called dichlorvos), the effects of an ethanolic extract of *M. oleifera* seed oil on liver function were assessed. By controlling and raising plasma protein levels, *M. oleifera* seed oil decreased elevated levels of inflammatory and liver markers. Histopathological investigations showed that *M. oleifera* seed oil had a regenerative effect on liver damage brought on by dichlorvos (Adeoye et al., 2022).

M. oleifera (drumstick), a traditional Indian medicine remedy, was tested for its ability to protect rats' livers from hepatotoxicity induced by anti-TB drugs (INH, RMP and PZA) in a study conducted by L. Pari and N. Ashok Kumar. Lipid markers, bilirubin and liver enzyme levels (AST, ALT and ALP) were markedly improved by oral administration of the ethanolic leaf extract. Analysis of liver tissue confirmed these biochemical results. According to the study, *M. oleifera* extract may help people recover from drug-induced liver damage, as silymarin does (Pari & Kumar, 2002). Another study evaluated *M. oleifera*'s ability to protect rats' livers from cadmium toxicity. Cadmium treatment (200 ppm/kg), control and cadmium plus *M. oleifera* extract (500 mg/kg orally for 28 days) were the groups into which 18 Wistar rats were split. Increased lipid peroxidation, decreased antioxidant levels (SOD), elevated liver enzymes (AST, ALT and ALP) and cadmium accumulation in liver tissue were all consequences of cadmium exposure. Rats co-treated with *M. oleifera* exhibited a notable improvement in oxidative stress and enzyme markers, while the liver's cadmium levels stayed constant. These results imply that *M. oleifera* provides significant defence against liver damage brought on by cadmium (Toppo et al., 2015).

A study examined how *M. oleifera* leaf extract protected male Wistar rats' livers from damage caused by lead acetate. By enhancing antioxidant status, lowering inflammation (through the TNF-α and NF-κB pathways) and maintaining liver structure and function, moringa treatment dramatically reversed these effects (Abdel Fattah et al., 2020).

17.5.7 Anticancer

Studies have shown that constituents extracted from different parts of *M. oleifera* possess significant antitumour properties. In vitro and in vivo tests have shown that leaf compounds are effective against various cancer types (Phua et al., 2021). For instance, Swiss albino mice with Dalton's lymphoma ascites were used to test the leaf extract and fractions' in vivo anticancer activity (Krishnamurthy et al., 2015). Extracts from *M. oleifera* leaves have exhibited significant antitumour activity in vivo, particularly against Ehrlich solid tumours and Ehrlich ascites carcinoma models (Barhoi et al., 2021; Khalil et al., 2014). The traditional use of flowers in folk medicine for managing inflammation and tumours has also been well documented. Flower-based remedies often involve using aqueous, ethanolic or hydro-alcoholic extracts or preparing an infusion by steeping the flowers in hot water (Mehta et al., 2011). The anticancer effect of *M. oleifera* flower extract against human prostate cancer cells was mediated via the induction of apoptosis (Arulselvan et al., 2018).

According to Suphachai's study, the dichloromethane extract had higher anticancer and QR-inducing activity (CD = 91.36±1.26 μg/ml), while the methanol extract had more potent antioxidant activity. These findings corroborate the use of *M. oleifera* leaves in the development of anticancer and nutraceutical products by indicating that they have promising antioxidant, cytotoxic and chemopreventive potential (Suphachai, 2014).

17.5.8 Antifertility Activity

The hormonal and antifertility activity of an aqueous extract of *M. oleifera* roots was assessed in rats. In rats with ovariectomies, oral administration stimulated uterine tissue and increased uterine weight, indicating oestrogenic activity. However, when combined with it, the extract decreased the uterine changes caused by estradiol dipropionate (EDP), suggesting it may have anti-oestrogenic properties. Half of the test animals in a deciduoma test showed disruptions in decidual formation at the highest dose (600 mg/kg), suggesting antiprogesterone effects. At all doses, no decidual response was observed when given alone. According to these results, *M. oleifera* root extract may have antifertility effects via a variety of hormonal pathways (Shukla et al., 1988).

According to recent research on hot and cold extracts of *M. oleifera* leaves, consuming moringa before, during and after pregnancy may adversely affect the progression of the fetus by causing the uterine wall to contract rigorously (Attah et al., 2020).

17.5.9 Wound-Healing Properties

The researcher chose the most potent standardised extract using an in vitro test, and it was subsequently turned into a wound-healing film. The findings indicated that the aqueous extract of *M. oleifera* exhibited the most significant cell proliferation and migration activity among the extracts (Oguntibeju, 2019).

In diabetic animals, research investigated the impact of leaf extracts on the wound-healing process and showed upregulated vascular endothelial growth factor and downregulated inflammatory mediators in wound tissues, as well as notable antiproliferative effects on normal human dermal fibroblasts that are antimigratory (Al-Ghanayem et al., 2022).

The formulation of *M. oleifera* extract accelerated wound healing in infected wounds caused by *P. aeruginosa* and MRSA. The effect was more noticeable in the former than in the latter, suggesting that it might not be as effective in treating wounds infected with *P. aeruginosa*. According to earlier research, Gram-negative bacteria's innate resistance and cell wall structure allow them to withstand chemicals and natural substances. Gram-negative bacteria tend to exhibit greater resistance to antimicrobial agents compared with Gram-positive bacteria, primarily due to the presence of a protective lipopolysaccharide layer and the periplasmic space in their cell envelope (Al-Ghanayem et al., 2022; Koohsari et al., 2015).

17.5.10 Antioxidant

M. oleifera leaf extract's strong antioxidant qualities show a direct correlation with phenols, such as ferulic acid, which are involved in binding iron and copper, scavenging free radicals and protecting lipids from peroxidation (Gamal M. El-Sherbiny et al., 2024).

Potent antioxidants such as flavonoids, polyphenols and ascorbic acid, which help scavenge free radicals and lessen oxidative stress, are abundant in *M. oleifera* leaves. The body's defences against cellular damage are strengthened by these phytochemicals. Research has demonstrated that *M. oleifera* extract has potent anti-free radical properties. Its therapeutic effects, such as liver protection, are greatly influenced by its antioxidant potential (Younis et al., 2022).

Its leaves are edible in their entirety because the levels of antinutritional substances such as tannins, lecithins and protease inhibitors are negligible. They are also a key supplier of vitamins A and C and antioxidants, and they have a balanced profile of essential amino acids. We therefore question whether this plant will be a practical substitute for preventing food spoilage, as it has the nutritional profile to provide the micronutrients required in a healthy diet (Penalver et al., 2022).

17.6 NUTRACEUTICAL APPLICATIONS AND SAFETY

Foods and nutraceuticals derived from moringa have enormous potential to slow malnutrition in underdeveloped countries (Oyeyinka & Oyeyinka, 2018). According to some experts, moringa may be added to food products to improve the nutritional value by incorporating vitamins, minerals, vital amino acids and oils. Different parts of moringa can be used in foods (Milla et al., 2021).

17.6.1 Stiff Dough: 'Amala'

In Ghana and Nigeria, among other African countries, amala is a staple food. It is a stiff dough or starchy gel that is typically made with yam flour. Numerous researchers have been driven by amala's low nutritional content to fortify this staple with various fortifiers, including distillers' spent grain, soybean flour and *M. oleifera* leaf powder. Plantain flour amala fortified with 2.5% *M. oleifera* leaf powder (MOLP) has three times the protein content of the unfortified version, higher than the 1.6-fold increase seen in the yam-based counterpart. This difference is the result of variations in MOLP composition and the naturally higher protein content of plantain flour.

17.6.2 Bread

Across the globe, bread is a staple food. The quality of bread is influenced by a number of elements, such as the volume of the loaf, its colour and the texture of the crust and crumb. Its formulation's constituents determine its nutritional value. It has been observed that adding *M. oleifera* flowers, leaf powder and seeds to bread dough made with wheat flour alone or in conjunction with other flours increases the bread's nutritional value. In research, wheat flour bread fortified with 5% MOLP showed an approximate 54% increase in protein and a 56% rise in crude fibre content. Up to a 15% fortification of wheat flour with *M. oleifera* powdered seed in bread manufacturing has been shown to boost protein content by roughly 67% without appreciably changing the bread's sensory qualities. A product's acceptance is mostly determined by its sensory qualities, which include colour, taste and scent (Oyeyinka & Oyeyinka, 2018).

17.6.3 Dairy Products

Despite their many nutritional and healthful qualities, milk and its derived products are generally not thought to be a strong source of specific bioactive chemicals, such as polyphenols and antioxidants. As a result, health-conscious consumers are calling for an increasing number of innovative dairy products made with extracts or medicinal herbs. Various dairy products, such as labneh, cream cheese and soft white cheese, are formulated with 1% to 3% dried leaves of moringa or an alcoholic extract of the leaves. These products demonstrate improved nutritional profiles and acceptable sensory qualities. The content of certain amino acids, including alanine, glutamine and tyrosine, and antioxidant activity are also enhanced (Trigo et al., 2023).

17.6.4 Biscuits

MOLP is regarded as a potentially beneficial natural gluten-free (GF) component that can be incorporated into GF food products to increase their overall nutritional qualities. The process of making gluten-free biscuits involves adding more MOLP to a commercial GF flour mix. The results indicate that, even at the lowest MOLP-substitution level (5 g/100 g), the incorporation of MOLP helps to raise the samples' protein and dietary fibre content while lowering their overall starch content. Significant variations in colour, spread ratio and hardness values are also seen depending on the recipe's MOLP-substitution level. The starch pasting qualities of biscuits are altered by the addition of MOLP. Specifically, in MOLP-substituted biscuits, lower viscosity values following heating and cooling suggest decreased starch swelling power and retrogradation propensity. For GF biscuits, an MOLP-substitution level of 10% w/w offers a suitable balance between nutritional and physicochemical properties (Giuberti et al., 2020).

17.6.5 Herbal Tea

Since tea is currently the most popular beverage worldwide, it is considered a significant food product. Most people drink tea because of its appealing flavour and scent, as well as the special place it occupies in many cultures. Growing consumer knowledge of the health advantages has led to an increasing contemporary focus on tea. As a result, tea is part of the quickly growing wellness beverage sector. Herbal tea can be developed by incorporating different parts of *M. oleifera* in various proportions. Herbal tea with *M. oleifera* is prepared by selecting fresh, mature moringa leaves, cleaning them to remove dust and then drying them in a hot air oven at around 60°C until crisp. The dried leaves are ground into a fine powder and sieved for uniformity. Alongside moringa, other herbal ingredients, such as basil, lemongrass and tea leaves, are also processed similarly, cleaned, dried at controlled temperatures, powdered and sieved. These powders are then blended in specific proportions to create different formulations, typically with moringa leaf powder comprising 20–50% of the mix. The herbal blend is filled into filter paper pouches to make tea bags, each containing about 2 g of the mixture. These tea bags are sealed and labelled for easy use. One tea bag is steeped in hot water (around 95°C) for five to ten minutes to prepare the tea, allowing the bioactive compounds and flavours to infuse. The resulting beverage is rich in nutrients such as protein, calcium, iron and antioxidants, offering both health benefits and an aromatic, earthy taste (Shaizad et al., 2020).

17.6.6 Toothpaste

Toothpaste with moringa extract is prepared using the essential oil extracted from the root of *M. oleifera*. Fresh moringa roots are harvested, thoroughly washed, sun-dried and ground into a fine powder. This powder undergoes Soxhlet extraction with 95% ethanol to extract the essential oil, followed by concentration by rotary evaporation at 65°C, yielding a dark brown, peppermint-scented oil rich in antimicrobial phytochemicals such as flavonoids, alkaloids, phenolics and tannins. For toothpaste formulation, 0.5 ml of this concentrated oil is blended with 50 g of calcium carbonate as an abrasive, 20 g of sodium lauryl sulphate as a detergent, 0.1 g of methyl cellulose as a binder, 1 g of glycerine for sweetness and 50 g of water as the dispersion medium. These components are mixed in a stainless-steel mixer until a smooth paste is formed. The resulting toothpaste is white, homogeneous and has a pH of 8.0. It shows superior antimicrobial activity compared with commercial toothpaste, effectively inhibiting pathogens such as *Streptococcus mutans*, *Lactobacillus*, *S. aureus* and *Candida albicans*, making it a promising natural alternative for oral health (Ae & Cp, 2022).

REFERENCES

Abd-Rabou, A. A., Abdalla, A. M., Ali, N. A., & Zoheir, K. M. (2017). Moringa oleifera root induces cancer apoptosis more effectively than leave nanocomposites and its free counterpart. *Asian Pac J Cancer Prev*, *18*(8), 2141–2149. https://doi.org/10.22034/APJCP.2017.18.8.2141

Abd El-Hack, M. E., Alagawany, M., Elrys, A. S., Desoky, E.-S. M., Tolba, H. M. N., Elnahal, A. S. M., Elnesr, S. S., & Swelum, A. A. (2018). Effect of forage Moringa oleifera L. (moringa) on animal health and nutrition and its beneficial applications in soil, plants and water purification. *Agriculture*, *8*(9), 145. https://doi.org/10.3390/agriculture8090145

Abd El-Hack, M. E., Alqhtani, A. H., Swelum, A. A., El-Saadony, M. T., Salem, H. M., Babalghith, A. O., Taha, A. E., Ahmed, O., Abdo, M., & El-Tarabily, K. A. (2022). Pharmacological, nutritional and antimicrobial uses of Moringa oleifera Lam. leaves in poultry nutrition: an updated knowledge. *Poult Sci*, *101*(9), 102031. https://doi.org/10.1016/j.psj.2022.102031

Abd Rani, N. Z., Husain, K., & Kumolosasi, E. (2018). Moringa genus: A review of phytochemistry and pharmacology. *Front Pharmacol*, *9*, 108. https://doi.org/10.3389/fphar.2018.00108

Abdel Fattah, M. E., Sobhy, H. M., Reda, A., & Abdelrazek, H. M. A. (2020). Hepatoprotective effect of Moringa oleifera leaves aquatic extract against lead acetate-induced liver injury in male Wistar rats. *Environ Sci Pollut Res Int*, *27*(34), 43028–43043. https://doi.org/10.1007/s11356-020-10161-z

Adedapo, A. A., Ogunmiluyi, I. O., Falayi, O. O., Ogunpolu, B. S., Oyagbemi, A. A., Orishadipe, A., Omobowale, T. O., Yakubu, M. A., & Oguntibeju, O. O. (2020). The lyophilized aqueous leaf extract of Moringa oleifera blunts streptozocin-induced diabetes in rats through upregulation of GLUT 4 signaling pathway and anti-oxidant effect. *Sci Afr, 10*, e00619. https://doi.org/10.1016/j.sciaf.2020.e00619

Adefegha, S. A., Oboh, G., Iyoha, A. E., & Oyagbemi, A. A. (2019). Comparative effects of horseradish (Moringa oleifera) leaves and seeds on blood pressure and crucial enzymes relevant to hypertension in rat. *PharmaNutrition, 9*, 100152. https://doi.org/10.1016/j.phanu.2019.100152

Adeoye, S. W. A., Adeshina, O. S., Yusuf, M. G., & Omole, A. (2022). Hepatoprotective and renoprotective effect of Moringa oleifera seed oil on dichlorvos-induced toxicity in male wistar rats. *Niger J Physiol Sci, 37*(1), 119–126. https://doi.org/10.54548/njps.v37i1.15

Adnette, F. (2019). Diabetes mellitus: Classification, epidemiology, physiopathology, immunology, risk factors, prevention and nutrition. *Int J Adv Res, 7*(7), 855–863. https://doi.org/10.21474/ijar01/9433

Ae, A., & Cp, E. (2022). Production of herbal toothpaste using <em>Moringa </em>root essential oil extract. *Int J Adv Biochem Res, 6*(2), 49–51. https://doi.org/10.33545/26174693.2022.v6.i2a.134

Ahmed, S., Shah, P., & Ahmed, O. (2023). *Biochemistry, lipids*. StatPearls Publishing. https://pubmed.ncbi.nlm.nih.gov/30247827/

Ajantha, A., Kathirvelan, C., Purushothaman, M. R., & Visha, P. (2020). Studies on qualitative and quantitative phytochemical constituents of *Moringa oleifera* leaf meal. *Int J Curr Microbiol Appl Sci, 9*(6), 4195–4201. https://doi.org/10.20546/ijcmas.2020.906.491

Al-Ghanayem, A. A., Alhussaini, M. S., Asad, M., & Joseph, B. (2022). Moringa oleifera leaf extract promotes healing of infected wounds in diabetic rats: Evidence of antimicrobial, antioxidant and proliferative properties. *Pharmaceuticals (Basel), 15*(5), 528. https://doi.org/10.3390/ph15050528

Al-Malki, A. L., & El Rabey, H. A. (2015). The antidiabetic effect of low doses of Moringa oleifera Lam. seeds on streptozotocin induced diabetes and diabetic nephropathy in male rats. *Biomed Res Int, 2015*(1), 381040. https://doi.org/10.1155/2015/381040

Anwar, F., Latif, S., Ashraf, M., & Gilani, A. H. (2007). Moringa oleifera: A food plant with multiple medicinal uses. *Phytother Res, 21*(1), 17–25. https://doi.org/10.1002/ptr.2023

Arora, S., & Arora, S. (2021). Nutritional significance and therapeutic potential of Moringa oleifera: The wonder plant. *J Food Biochem, 45*(10), e13933. https://doi.org/10.1111/jfbc.13933

Arulselvan, P., Ju, J., Gothai, S., Hasanpourghadi, M., Nasser, A., Aziz Ibrahim, I., Shahzad, N., Pandurangan, A., Muniandy, K., & Kumar, S. (2018). Anticancer potential of Moringa oleifera flower extract in human prostate cancer PC-3 cells via induction of apoptosis and downregulation of AKT pathway. *Pharmacogn Mag, 14*(58). https://doi.org/10.4103/pm.pm_516_17

Arulselvan, P., Tan, W. S., Gothai, S., Muniandy, K., Fakurazi, S., Esa, N. M., Alarfaj, A. A., & Kumar, S. S. (2016). Anti-inflammatory potential of ethyl acetate fraction of *Moringa oleifera* in downregulating the NF-κB signaling pathway in lipopolysaccharide-stimulated macrophages. *Molecules (Basel, Switzerland), 21*(11), 1452. https://doi.org/10.3390/molecules21111452

Attah, A. F., Moody, J. O., Sonibare, M. A., Salahdeen, H. H., Akindele, O. O., Nnamani, P. O., Diyaolu, O. A., & Raji, Y. (2020). Aqueous extract of Moringa oleifera leaf used in Nigerian ethnomedicine alters conception and some pregnancy outcomes in Wistar rat. *S Afr J Bot, 129*, 255–262. https://doi.org/10.1016/j.sajb.2019.07.041

Avilés-Gaxiola, S., León-Félix, J., Jiménez-Nevárez, Y. B., Angulo-Escalante, M. A., Ramos-Payán, R., Colado-Velázquez, J., & Heredia, J. B. (2021). Antioxidant and anti-inflammatory properties of novel peptides from Moringa oleifera Lam. leaves. *S Afr J Bot, 141*, 466–473. https://doi.org/10.1016/j.sajb.2021.05.033

Azlan, U. K., Khairul Annuar, N. A., Mediani, A., Aizat, W. M., Damanhuri, H. A., Tong, X., Yanagisawa, D., Tooyama, I., Wan Ngah, W. Z., Jantan, I., & Hamezah, H. S. (2022). An insight into the neuroprotective and anti-neuroinflammatory effects and mechanisms of Moringa oleifera. *Front Pharmacol, 13*, 1035220. https://doi.org/10.3389/fphar.2022.1035220

Barhoi, D., Upadhaya, P., Barbhuiya, S. N., Giri, A., & Giri, S. (2021). Aqueous extract of moringa oleifera exhibit potential anticancer activity and can be used as a possible cancer therapeutic agent: A study involving in vitro and in Vivo approach. *J Am Coll Nutr, 40*(1), 70–85. https://doi.org/10.1080/07315724.2020.1735572

Basuny, A., & Al-Marzouq, M. (2016). Biochemical studies on Moringa oleifera seed oil. *MOJ Food Process Technol, 2*(2), 40–46. https://medcraveonline.com/MOJFPT/MOJFPT-02-00030.pdf

Bhattacharya, A., Tiwari, P., Sahu, P. K., & Kumar, S. (2018). A review of the phytochemical and pharmacological characteristics of Moringa oleifera. *J Pharm Bioallied Sci, 10*(4), 181–191. https://doi.org/10.4103/JPBS.JPBS_126_18

Bisong, S. A., Ajiwhen, I. O., Nku, C. O., & Uruakpa, K. C. (2019). Phytochemical and neurotoxicity evaluation of methanolic leaf-extract of Moringa oleifera. *J Phytopharmacol*, *8*(5), 210–215. https://doi.org/10.31254/phyto.2019.8501

Chhikara, N., Kaur, A., Mann, S., Garg, M. K., Sofi, S. A., & Panghal, A. (2020). Bioactive compounds, associated health benefits and safety considerations of Moringa oleifera L.: An updated review. *Nutr Food Sci*, *51*(2), 255–277. https://doi.org/10.1108/nfs-03-2020-0087

Chis, A., Noubissi, P. A., Pop, O. L., Muresan, C. I., Fokam Tagne, M. A., Kamgang, R., Fodor, A., Sitar-Taut, A. V., Cozma, A., Orasan, O. H., Heghes, S. C., Vulturar, R., & Suharoschi, R. (2023). Bioactive compounds in *Moringa oleifera*: Mechanisms of action, focus on their anti-inflammatory properties. *Plants (Basel)*, *13*(1), 20. https://doi.org/10.3390/plants13010020

Chodur, G. M., Olson, M. E., Wade, K. L., Stephenson, K. K., Nouman, W., Garima, & Fahey, J. W. (2018). Wild and domesticated Moringa oleifera differ in taste, glucosinolate composition, and antioxidant potential, but not myrosinase activity or protein content. *Sci Rep*, *8*(1), 7995. https://doi.org/10.1038/s41598-018-26059-3

Cuellar-Nunez, M. L., Gonzalez de Mejia, E., & Loarca-Pina, G. (2021). Moringa oleifera leaves alleviated inflammation through downregulation of IL-2, IL-6, and TNF-alpha in a colitis-associated colorectal cancer model. *Food Res Int*, *144*, 110318. https://doi.org/10.1016/j.foodres.2021.110318

Cui, C., Chen, S., Wang, X., Yuan, G., Jiang, F., Chen, X., & Wang, L. (2019). Characterization of Moringa oleifera roots polysaccharide MRP-1 with anti-inflammatory effect. *Int J Biol Macromol*, *132*, 844–851. https://doi.org/10.1016/j.ijbiomac.2019.03.210

Deng, Y., Li, G., Chang, D., & Su, X. (2020). YKL-40 as a novel biomarker in cardio-metabolic disorders and inflammatory diseases. *Clin Chim Acta*, *511*, 40–46. https://doi.org/10.1016/j.cca.2020.09.035

Devkota, S., Bhusal, K. K., & Yildiz, F. (2020). Moringa oleifera: A miracle multipurpose tree for agroforestry and climate change mitigation from the Himalayas – A review. *Cogent Food Agric*, *6*(1), 1805951. https://doi.org/10.1080/23311932.2020.1805951

do Nascimento, K., Reis, I. P., & Augusta, I. M. (2017). Total phenolic and antioxidant capacity of flower, leaf and seed of Moringa oleifera. *Nutr Res*, *1*(1), 1–9.

El-Sherbiny, G. M., Alluqmani, A. J., Elsehemy, I. A., & Kalaba, M. H. (2024). Antibacterial, antioxidant, cytotoxicity, and phytochemical screening of Moringa oleifera leaves. *Sci Rep*, *14*(1), 30485. https://doi.org/10.1038/s41598-024-80700-y

Eller-Vainicher, C., Cairoli, E., Grassi, G., Grassi, F., Catalano, A., Merlotti, D., Falchetti, A., Gaudio, A., Chiodini, I., & Gennari, L. (2020). Pathophysiology and management of type 2 diabetes mellitus bone fragility. *J Diabetes Res*, *2020*(1), 7608964. https://doi.org/10.1155/2020/7608964

Espinosa-Salas, S., & Gonzalez-Arias, M. (2025). Nutrition: Micronutrient intake, imbalances, and interventions. In *StatPearls*. StatPearls Publishing. https://www.ncbi.nlm.nih.gov/pubmed/37983324

Ezeamuzie, I. C., Ambakederemo, A. W., Shode, F. O., & Ekwebelem, S. C. (2008). Antiinflammatory effects of moringa oleifera root extract. *Int J Pharmacogn*, *34*(3), 207–212. https://doi.org/10.1076/phbi.34.3.207.13211

Fahey, J. W. (2017). Moringa oleifera: A review of the medicinal potential. *Acta Horticulturae* (1158), 209–224. https://doi.org/10.17660/ActaHortic.2017.1158.25

Fahmy, N. M., Fayez, S., Mohamed, R. W., Elissawy, A. M., Eldahshan, O. A., Zengin, G., & Singab, A. N. B. (2024). Moringa oleifera flowers: Insights into their aroma chemistry, anti-inflammatory, antioxidant, and enzyme inhibitory properties. *BMC Complement Med Ther*, *24*(1), 286. https://doi.org/10.1186/s12906-024-04579-y

Faizi, S., Siddiqui, B. S., Saleem, R., Aftab, K., Shaheen, F., & Gilani, A. H. (1998). Hypotensive constituents from the pods of Moringa oleifera. *Planta Med*, *64*(3), 225–228. https://doi.org/10.1055/s-2006-957414

Fatima, S., Usmani, M. A., & Srivastava, A. K. (2024). Bioactive metabolites of Moringa oleifera (Sahjan) with functional activities. *Res Pharm*, 25–37. https://doi.org/10.25081/rip.2024.v14.9060

Frak, W., Wojtasinska, A., Lisinska, W., Mlynarska, E., Franczyk, B., & Rysz, J. (2022). Pathophysiology of cardiovascular diseases: New insights into molecular mechanisms of atherosclerosis, arterial hypertension, and coronary artery disease. *Biomedicines*, *10*(8), 1938. https://doi.org/10.3390/biomedicines10081938

Galts, C. P. C., Bettio, L. E. B., Jewett, D. C., Yang, C. C., Brocardo, P. S., Rodrigues, A. L. S., Thacker, J. S., & Gil-Mohapel, J. (2019). Depression in neurodegenerative diseases: Common mechanisms and current treatment options. *Neurosci Biobehav Rev*, *102*, 56–84. https://doi.org/10.1016/j.neubiorev.2019.04.002

Ganin, H., Rayo, J., Amara, N., Levy, N., Krief, P., & Meijler, M. M. (2013). Sulforaphane and erucin, natural isothiocyanates from broccoli, inhibit bacterial quorum sensing. *Medchemcomm*, *4*(1), 175–179. https://doi.org/10.1039/c2md20196h

Ghimire, S., Subedi, L., Acharya, N., & Gaire, B. P. (2021). Moringa oleifera: A tree of life as a promising medicinal plant for neurodegenerative diseases. *J Agric Food Chem, 69*(48), 14358–14371. https://doi.org/10.1021/acs.jafc.1c04581

Giuberti, G., Bresciani, A., Cervini, M., Frustace, A., & Marti, A. (2020). Moringa oleifera L. leaf powder as ingredient in gluten-free biscuits: nutritional and physicochemical characteristics. *Eur Food Res Technol, 247*(3), 687–694. https://doi.org/10.1007/s00217-020-03656-z

Godinez-Ov, A., Guemes-Ver, N., & Acevedo-Sa, O. A. (2016). Nutritional and phytochemical composition of Moringa oleifera lam and its potential use as nutraceutical plant: A review. *Pak J Nutr, 15*(4), 397–405. https://doi.org/10.3923/pjn.2016.397.405

Gopalakrishnan, L., Doriya, K., & Kumar, D. S. (2016). Moringa oleifera: A review on nutritive importance and its medicinal application. *Food Sci Hum Well, 5*(2), 49–56. https://doi.org/10.1016/j.fshw.2016.04.001

Hamza, T. A., & Azmach, N. N. (2017). The miraculous moringa trees: From nutritional and medicinal point of views in tropical regions. *J Med Plants Stud, 5*(4), 151–162.

Islam, Z., Islam, S. M. R., Hossen, F., Mahtab-Ul-Islam, K., Hasan, M. R., & Karim, R. (2021). Moringa oleifera is a prominent source of nutrients with potential health benefits. *Int J Food Sci, 2021*(1), 6627265. https://doi.org/10.1155/2021/6627265

Jassim, A. A., & Ali, R. S. (2024). Study of phenolic compounds of moringa oleifera leaf extracts and their potential as antioxidants. *Pak J Life Soc Sci (PJLSS), 22*(1). https://doi.org/10.57239/pjlss-2024-22.1.00255

Kashyap, P., Kumar, S., Riar, C. S., Jindal, N., Baniwal, P., Guiné, R. P., Correia, P. M., Mehra, R., & Kumar, H. (2022). Recent advances in Drumstick (Moringa oleifera) leaves bioactive compounds: Composition, health benefits, bioaccessibility, and dietary applications. *Antioxidants, 11*(2), 402. https://doi.org/10.3390/antiox11020402.

Khalid, S., Arshad, M., Mahmood, S., Ahmed, W., Siddique, F., Khalid, W., Zarlasht, M., Asar, T. O., & Hassan, F. A. M. (2023). Nutritional and phytochemical screening of Moringa oleifera leaf powder in aqueous and ethanol extract. *Int J Food Prop, 26*(1), 2338–2348. https://doi.org/10.1080/10942912.2023.2246685

Khalil, W. K., Ghaly, I., Diab, K., & ELmakawy, A. (2014). Antitumor activity of Moringa oleifera leaf extract against Ehrlich solid tumor. *Int J Pharm, 4*(3), 68–82.

Khan, W., Parveen, R., Chester, K., Parveen, S., & Ahmad, S. (2017). Hypoglycemic potential of aqueous extract of moringa oleifera leaf and in Vivo GC-MS metabolomics. *Front Pharmacol, 8*, 577. https://doi.org/10.3389/fphar.2017.00577

Klimek-Szczykutowicz, M., Gawel-Beben, K., Rutka, A., Blicharska, E., Tatarczak-Michalewska, M., Kulik-Siarek, K., Kukula-Koch, W., Malinowska, M. A., & Szopa, A. (2024). Moringa oleifera (drumstick tree)-nutraceutical, cosmetological and medicinal importance: A review. *Front Pharmacol, 15*, 1288382. https://doi.org/10.3389/fphar.2024.1288382

Koohsari, H., Ghaemi, E. A., Sadegh Sheshpoli, M., Jahedi, M., & Zahiri, M. (2015). The investigation of antibacterial activity of selected native plants from North of Iran. *J Med Life, 8*(2), 38–42. https://www.ncbi.nlm.nih.gov/pubmed/28255395

Krishnamurthy, P. T., Vardarajalu, A., Wadhwani, A., & Patel, V. (2015). Identification and characterization of a potent anticancer fraction from the leaf extracts of Moringa oleifera L. *Indian J Exp Biol, 53*(2), 98–103. https://www.ncbi.nlm.nih.gov/pubmed/25757240

Kuete, V. (2017). Moringa oleifera. In Medicinal Spices and Vegetables from Africa (pp. 485–496). Academic Press. https://doi.org/10.1016/B978-0-12-809286-6.00022-4'

Lazou, A., Ikonomidis, I., Bartekova, M., Benedek, T., Makavos, G., Palioura, D., Cabrera Fuentes, H., & Andreadou, I. (2020). Chronic inflammatory diseases, myocardial function and cardioprotection. *Br J Pharmacol, 177*(23), 5357–5374. https://doi.org/10.1111/bph.14975

Liu, R., Zhang, H., Yuan, M., Zhou, J., Tu, Q., Liu, J. J., & Wang, J. (2013). Synthesis and biological evaluation of apigenin derivatives as antibacterial and antiproliferative agents. *Molecules (Basel, Switzerland), 18*(9), 11496–11511. https://doi.org/10.3390/molecules180911496

Malabed, R. S., & Noel, M. G. (2013). *Characterization of the glucosinolates and isothiocyanates in Malunggay (Moringa oleifera L.) extracts and determination of their myrosinase activity and anticancer properties*. Research Congress. Manila, Philippines: De La Salle University,

Masitlha, E. P., Seifu, E., & Teketay, D. (2024). Nutritional composition and mineral profile of leaves of Moringa oleifera provenances grown in Gaborone, Botswana. *Food Production, Processing and Nutrition, 6*(1), 3. https://doi.org/10.1186/s43014-023-00183-8

Mehta, J., Shukla, A., Bukhariya, V., & Charde, R. (2011). The magic remedy of moringa oliferia: An overview. *Int J Biomed Adv Res*, *2*(6), 215–227. https://doi.org/10.7439/ijbar.v2i6.35

Mehta, N. N., Dey, A. K., Maddineni, R., Kraus, W. E., & Huffman, K. M. (2020). GlycA measured by NMR spectroscopy is associated with disease activity and cardiovascular disease risk in chronic inflammatory diseases. *Am J Prev Cardiol*, *4*, 100120. https://doi.org/10.1016/j.ajpc.2020.100120

Milla, P. G., Penalver, R., & Nieto, G. (2021). Health benefits of uses and applications of Moringa oleifera in bakery products. *Plants (Basel)*, *10*(2), 318. https://doi.org/10.3390/plants10020318

Moyo, B., Oyedemi, S., Masika, P. J., & Muchenje, V. (2012). Polyphenolic content and antioxidant properties of Moringa oleifera leaf extracts and enzymatic activity of liver from goats supplemented with Moringa oleifera leaves/sunflower seed cake. *Meat Sci*, *91*(4), 441–447. https://doi.org/10.1016/j.meatsci.2012.02.029

Mundkar, M., Bijalwan, A., Soni, D., & Kumar, P. (2022). Neuroprotective potential of Moringa oleifera mediated by NF-kB/Nrf2/HO-1 signaling pathway: A review. *J Food Biochem*, *46*(12), e14451. https://doi.org/10.1111/jfbc.14451

Nadeem, M., & Imran, M. (2016). Promising features of Moringa oleifera oil: Recent updates and perspectives. *Lipids Health Dis*, *15*(1), 212. https://doi.org/10.1186/s12944-016-0379-0

Oguntibeju, O. O. (2019). Medicinal plants and their effects on diabetic wound healing. *Vet World*, *12*(5), 653–663. https://doi.org/10.14202/vetworld.2019.653–663

Onasanwo, S. A., Adamaigbo, V. O., Adebayo, O. G., & Eleazer, S. E. (2021). Moringa oleifera-supplemented diet protect against cortico-hippocampal neuronal degeneration in scopolamine-induced spatial memory deficit in mice: role of oxido-inflammatory and cholinergic neurotransmission pathway. *Metab Brain Dis*, *36*(8), 2445–2460. https://doi.org/10.1007/s11011-021-00855-9

Owon, M., Osman, M., Ibrahim, A., Salama, M. A., & Matthäus, B. (2021). Characterisation of different parts from Moringa oleifera regarding protein, lipid composition and extractable phenolic compounds. *Ocl*, *28*, 45. https://doi.org/10.1051/ocl/2021035

Oyeyinka, A. T., & Oyeyinka, S. A. (2018). Moringa oleifera as a food fortificant: Recent trends and prospects. *J Saudi Soc Agric Sci*, *17*(2), 127–136. https://doi.org/10.1016/j.jssas.2016.02.002

Paikra, B. K., Dhongade, H. K. J., & Gidwani, B. (2017). Phytochemistry and pharmacology of Moringa oleifera Lam. *J Pharmacopuncture*, *20*(3), 194–200. https://doi.org/10.3831/KPI.2017.20.022

Pareek, A., Pant, M., Gupta, M. M., Kashania, P., Ratan, Y., Jain, V., Pareek, A., & Chuturgoon, A. A. (2023). Moringa oleifera: An updated comprehensive review of its pharmacological activities, ethnomedicinal, phytopharmaceutical formulation, clinical, phytochemical, and toxicological aspects. *Int J Mol Sci*, *24*(3), 2098. https://doi.org/10.3390/ijms24032098

Pari, L., & Kumar, N. A. (2002). Hepatoprotective activity of Moringa oleifera on antitubercular drug-induced liver damage in rats. *J Med Food*, *5*(3), 171–177. https://doi.org/10.1089/10966200260398206

Patil, S. V., Mohite, B. V., Marathe, K. R., Salunkhe, N. S., Marathe, V., & Patil, V. S. (2022). Moringa tree, gift of nature: A review on nutritional and industrial potential. *Curr Pharmacol Rep*, *8*(4), 262–280. https://doi.org/10.1007/s40495-022-00288-7

Penalver, R., Martinez-Zamora, L., Lorenzo, J. M., Ros, G., & Nieto, G. (2022). Nutritional and antioxidant properties of moringa oleifera leaves in functional foods. *Foods*, *11*(8). https://doi.org/10.3390/foods11081107

Phua, C. Y. H., Teoh, Z. L., Goh, B. H., Yap, W. H., & Tang, Y. Q. (2021). Triangulating the pharmacological properties of thymoquinone in regulating reactive oxygen species, inflammation, and cancer: Therapeutic applications and mechanistic pathways. *Life Sci*, *287*, 120120. https://doi.org/10.1016/j.lfs.2021.120120

Pop, O. L., Kerezsi, A. D., & Ciont Nagy, C. (2022). A comprehensive review of moringa oleifera bioactive compounds-cytotoxicity evaluation and their encapsulation. *Foods*, *11*(23), 3787. https://doi.org/10.3390/foods11233787

Prabsattroo, T., Wattanathorn, J., Iamsaard, S., Somsapt, P., Sritragool, O., Thukhummee, W., & Muchimapura, S. (2015). Moringa oleifera extract enhances sexual performance in stressed rats. *J Zhejiang Univ Sci B*, *16*(3), 179–190. https://doi.org/10.1631/jzus.B1400197

Qadir, R., Anwar, F., Bashir, K., Tahir, M. H., Alhumade, H., & Mehmood, T. (2022). Variation in nutritional and antioxidant attributes of Moringa oleifera L. Leaves at different maturity stages. *Frontiers in Energy Research*, *10*, 888355. https://doi.org/10.3389/fenrg.2022.888355

Rahman, M. M., Sheikh, M. M. I., Sharmin, S. A., Islam, M. S., Rahman, M. A., Rahman, M. M., & Alam, M. (2009). Antibacterial activity of leaf juice and extracts of Moringa oleifera Lam. against some human pathogenic bacteria. *CMU J Nat Sci*, *8*(2), 219.

Saa, R. W., Fombang, E. N., Ndjantou, E. B., & Njintang, N. Y. (2019). Treatments and uses of Moringa oleifera seeds in human nutrition: A review. *Food Sci Nutr*, *7*(6), 1911–1919. https://doi.org/10.1002/fsn3.1057

Selasa, P., Waangsir, F. W. F., Rahmawati, E., & Kusmiyati, K. (2022). Alkaloids, flavonoids, tannins and saponins contents in Moringa Oleifera leaves. *Indonesian Journal of Global Health Research*, *4*(1), 139–144. https://doi.org/10.37287/ijghr.v4i1.832

Shaizad, S., Ajaj, S., Sadawarte, S. K., Machewad, G. M., Em, S., Sawate, A. R., & Singh, A. K. (2020). Studies on standardization, quality assessment of herbal tea with incorporated moringa olefera leaf powder. *International Journal of Chemical Studies*, *8*(6), 1379–1383. https://doi.org/10.22271/chemi.2020.v8.i6t.10951

Shil, S. (2021). Ethnomedicinal and nutraceutical potentialities of Moringa oleifera: A review. *J Phytopharmacol*, *10*(5), 415–420. https://doi.org/10.31254/phyto.2021.10522

Shukla, S., Mathur, R., & Prakash, A. O. (1988). Antifertility profile of the aqueous extract of Moringa oleifera roots. *J Ethnopharmacol*, *22*(1), 51–62. https://doi.org/10.1016/0378-8741(88)90230-9

Sikkema, J., de Bont, J. A., & Poolman, B. (1995). Mechanisms of membrane toxicity of hydrocarbons. *Microbiol Rev*, *59*(2), 201–222. https://doi.org/10.1128/mr.59.2.201-222.1995

Soni, D., & Kumar, P. (2022). GSK-3beta-mediated regulation of Nrf2/HO-1 signaling as a new therapeutic approach in the treatment of movement disorders. *Pharmacol Rep*, *74*(4), 557–569. https://doi.org/10.1007/s43440-022-00390-z

Sultana, S. (2020). Nutritional and functional properties of Moringa oleifera. *Metabol Open*, *8*, 100061. https://doi.org/10.1016/j.metop.2020.100061

Sun, H. H., & Tian, F. (2018). Inflammatory bowel disease and cardiovascular disease incidence and mortality: A meta-analysis. *Eur J Prev Cardiol*, *25*(15), 1623–1631. https://doi.org/10.1177/2047487318792952

Suphachai, C. (2014). Antioxidant and anticancer activities of Moringa oleifera leaves. *J Med Plants Res*, *8*(7), 318–325. https://doi.org/10.5897/jmpr2013.5353

Tao, L., Wang, Z., Yang, M., Kang, X., Dai, J., Zhao, C., Dai, T., Sheng, J., & Tian, Y. (2025). Phenolic compounds from Moringa oleifera Lam: Extraction, separation and identification, biological activities, and comprehensive applications. *Food Sci Hum Well*, *14*(3). https://doi.org/10.26599/fshw.2024.9250052

Teixeira, E. M., Carvalho, M. R., Neves, V. A., Silva, M. A., & Arantes-Pereira, L. (2014). Chemical characteristics and fractionation of proteins from Moringa oleifera Lam. leaves. *Food Chem*, *147*, 51–54. https://doi.org/10.1016/j.foodchem.2013.09.135

Toppo, R., Roy, B. K., Gora, R. H., Baxla, S. L., & Kumar, P. (2015). Hepatoprotective activity of Moringa oleifera against cadmium toxicity in rats. *Vet World*, *8*(4), 537–540. https://doi.org/10.14202/vetworld.2015.537-540

Trigo, C., Castello, M. L., & Ortola, M. D. (2023). Potentiality of Moringa oleifera as a nutritive ingredient in different food matrices. *Plant Foods Hum Nutr*, *78*(1), 25–37. https://doi.org/10.1007/s11130-022-01023-9

Vergara-Jimenez, M., Almatrafi, M. M., & Fernandez, M. L. (2017). Bioactive components in Moringa oleifera leaves protect against chronic disease. *Antioxidants (Basel)*, *6*(4), 91. https://doi.org/10.3390/antiox6040091

Waterman, C., Cheng, D. M., Rojas-Silva, P., Poulev, A., Dreifus, J., Lila, M. A., & Raskin, I. (2014). Stable, water extractable isothiocyanates from Moringa oleifera leaves attenuate inflammation in vitro. *Phytochemistry*, *103*, 114–122. https://doi.org/10.1016/j.phytochem.2014.03.028

Xie, J., Qian, Y. Y., Yang, Y., Peng, L. J., Mao, J. Y., Yang, M. R., Tian, Y., & Sheng, J. (2021). Isothiocyanate seeds inhibits the growth and migration of renal cancer cells by regulating the PTP1B-dependent Src/Ras/Raf/ERK signaling pathway. *Front Cell Dev Biol*, *9*, 790618. https://doi.org/10.3389/fcell.2021.790618

Xie, J., Qian, Y.-Y., Yang, Y., Peng, L.-J., Mao, J.-Y., Yang, M.-R., Tian, Y., & Sheng, J. (2022). Isothiocyanate from Moringa oleifera seeds inhibits the growth and migration of renal cancer cells by regulating the PTP1B-dependent Src/Ras/Raf/ERK signaling pathway. *Front Cell Dev Biol*, *9*, 790618.

Younis, N., Khan, M. I., Zahoor, T., & Faisal, M. N. (2022). Phytochemical and antioxidant screening of Moringa oleifera for its utilization in the management of hepatic injury. *Front Nutr*, *9*, 1078896. https://doi.org/10.3389/fnut.2022.1078896

18 *Moringa Oleifera* Drug Delivery Systems

Ushasi Das, Arindam Sarkar, Adarsha Bachhar, DhimanRoy and Sanchita Mandal

18.1 INTRODUCTION

The multifunctional *Moringa oleifera* (Lam.) 'drumstick tree', 'ben tree', 'radish tree' or 'horseradish tree', or 'malunggay' in some Asian countries, has been documented to cure several complications, including diabetes, cancer, oxidative stress and infections by microorganisms (Al-Ghanayem et al., 2022; Álvarez-Román et al., 2020). Moringa (the one that drops leaves) is found in the Southern Himalayas of Northern India. Because it is a valuable food source and can readily be cultivated throughout tropical and subtropical regions, it is cultivated in various countries. The plant is rich in a number of phytochemicals that are useful in diets for health, medicine, the environment and agriculture. Moringa, a plant with edible leaves, seeds, fruits, flowers, bark and roots used in food, traditional medicine and water cleaning, is a well-established food (Dhongade et al., 2017). The leaves are edible as a nutrient supplement to address malnutrition, as they are rich in vitamins, minerals, proteins and phytochemicals. Plant extracts have been reported for antioxidant, antibacterial, anti-inflammatory, antidiabetic and anticancer activities (Gopalakrishnan et al., 2016; Pareek et al., 2023). Often referred to as the 'tree of life' or the 'miracle tree', moringa has been used in Ayurvedic practice in India, where it is touted for its supposed healing and health benefits. It is rich in flavonoids, glucosinolates, chlorogenic acid and glucoside, the main bioactive compounds in the plant. Research supports its use in agriculture, medicine, nutrition and the environment (Abdel-Latif et al., 2022). Data from a handful of studies from China, Brazil, Nigeria and India in recent years have suggested that *M. oleifera* has been co-evolving with humans in many respects. A less expensive option, *M. oleifera* was widely used by farmers in pretreatment form as a biostimulant (Arif et al., 2023; Yuniati et al., 2022). Its effectiveness should be tested in a laboratory against diseases such as AIDS, coronavirus outbreaks and some cancers. In summary, *M. oleifera* is a potential phytomedicine and functional food, and several human chronic diseases might be treated by health practitioners as a substitute for treatment against a wide range of diseases (Meireles et al., 2020; Vergara-Jimenez et al., 2017).

Drug delivery systems are developed to minimise adverse effects, increase patient adherence and maximise the therapeutic efficacy of pharmaceuticals. By administering medications at a set pace for a preset duration, they minimise variations in drug concentrations and maintain ideal drug levels, while decreasing the number of doses, particularly in chronic diseases. Occasionally, delivery systems may be used to deliver therapeutics to specific clot sites by targeting binding reagents to specific proteins or endothelial markers within the clot, which will improve clot effectiveness and reduce exposure to native tissues. Bioavailability, when supported by drug delivery systems, can contribute to patient compliance. They also improve ease of administration and patient compliance by limiting side effects, simplifying routes of administration and masking the bad taste (Roy et al., 2026). High in vivo drug concentrations or drug release at a nontarget site, together with potential side effects due to these circumstances, can be prevented by controlling when and where the drug is released. In essence, drug release systems have been developed for specific applications of the

DOI: 10.1201/9781003617860-21

carrier, while considering that the drug release from the carrier should be maximal to the targeted lesion, enhancing efficacy and minimising side effects with improved patient compliance (Roy et al., 2025; Sarangi & Padhi, 2018).

18.2 PHYTOCHEMISTRY AND BIOACTIVE COMPONENTS

There is an antioxidant and antimicrobial effect in *M. oleifera* that displays a potential use in modern generations in drug delivery systems, as demonstrated in modern drug delivery carriers. There have been numerous phytochemical studies on this plant, which have disclosed a wide array of bioactive compounds supporting its use in folkloric herbal therapy. Different morphological forms of the plant – roots, bark, gum, leaves, seeds and immature pods – are employed in recipes to cure several diseases. Phytochemical investigations of these parts have been well documented, and their secondary metabolites have demonstrated antioxidant, antimicrobial, anticancer, anti-inflammatory and antiulcer effects. The isolated bioactive compounds and the pharmacological profile are presented in Table 18.1. Within the sizeable Moringaceae family, a wide consumption is reported for both vegetative and reproductive *M. oleifera* parts in food and medicines (Sánchez-Machado et al., 2010). The plant possesses great pharmacological potential, primarily on account of its wide range of medicinal ingredients. *M. oleifera* seeds are commonly used, even in some rural parts of continents, in the purification of water to floc out pathogenic bacteria of water-borne diseases. The seed aqueous extract is employed as a coagulant in water purification, and coagulation-flocculation sedimentation processes reduce impurities and microorganisms, rendering the cloudy water potable (Ndabigengesere & Subba Narasiah, 1998). Phenolic acids, found throughout the plant, offer cellular protection against a variety of chronic diseases, including neoplasia and cardiovascular disease. Recent investigations increasingly centre on the pharmacological impact of these antioxidant constituents on pathologies fostered by oxidative stress, emphasising their potential roles in cancer and cardiovascular aetiologies (Brandão et al., 2009). Furthermore, infusions of the flowers have been shown to attenuate the production of tumour necrosis factor-α, a cytokine upregulated by lipopolysaccharide stimulation and implicated in inflammatory and neoplastic contexts (Mahajan & Mehta, 2008).

TABLE 18.1
Bioactive Compounds with Their Pharmacological Profiles

Bioactive compounds	Pharmacological profile (activity)	References
Alkaloids	Cardioprotective	Vergara-Jimenez et al., 2017
Tannins	Anti-inflammatory, anticancer, antihepatotoxic, antiatherosclerosis	Vergara-Jimenez et al., 2017
Quercetin	Antidiabetic, antiatherosclerosis, hypolipidemic	Vergara-Jimenez et al., 2017
β-sitosterol	Decrease cholesterol absorption	Vergara-Jimenez et al., 2017
Isothiocyanate	Anti-inflammatory, anticancer, reduces insulin resistance	Vergara-Jimenez et al., 2017
Apigenin, myricetin, kaempferol	Antioxidant, anti-inflammatory	Pop et al., 2022
Glycosides	Antioxidant	Pop et al., 2022
Saponin	Antioxidant, anticancer	Vergara-Jimenez et al., 2017
Carotenoids	Antioxidant, acts as a provitamin	Pop et al., 2022
Vitamin A, C and E	Antioxidant, cellular protection, immune support	Ferreira et al., 2008; Vergara-Jimenez et al., 2017
Oleic acid, fatty acids	Cardiological health, membrane integrity	Ferreira et al., 2008

18.2.1 Major Phytoconstituents Relevant to Drug Delivery

M. oleifera is endowed with diverse classes of phytochemicals that exhibit significant pharmacological activities and recognised therapeutic potential (Putro et al., 2021). Among these are alkaloids, tannins, carotenoids, phenolic entities, glucosinolates, isothiocyanates, saponins, polyunsaturated fatty acids and bioactive proteins and peptides (Pop et al., 2022). Compounds such as apigenin, quercetin, luteolin, myricetin and kaempferol, along with phenolcarboxylic acids, provide antioxidant defence, while lutein, zeaxanthin and β-carotene serve as cellular protectants. In the foliage, alkaloids and isothiocyanates exert marked antifungal and antibacterial effects. Proteins and peptides are credited with a multifaceted pharmacological profile, including anticancer, hepatoprotective and antidiabetic actions, corroborating the plant's therapeutic breadth. The residual seed cake, generated post-oil extraction, is distinguished by elevated mineral concentration and a catalogue of 24 bioactive constituents, of which select components are conducive to the enhancement of soil fertility. Such collective bioactivities ascribe *M. oleifera* plant constituents a compelling promise as reservoirs of phytochemicals for the pharmaceutical sector. The gum exuded by the *M. oleifera* trunk and branches has been successfully isolated and comprehensively characterised; its rheological and chemical properties indicate utility as a novel polymer in creating formulations for sustained-release drug delivery (Kumar & Renuka, 2019).

18.2.2 Functional Properties

M. oleifera is known for both its nutritional richness and its therapeutic potential, and it demonstrates inhibitory effects on a range of both pathogenic bacteria and fungi. The plant serves as a prominent reservoir of bioactive metabolites, particularly flavonoids, polyphenols, alkaloids and terpenoids, which are preferentially distilled from the foliage, owing to their elevated concentrations and the capacity for rapid vegetative renewal. Among the polyphenolic constituents, flavonoids, phenolic acids and lignans predominate, mediating the antioxidant potential of the species; concurrently, carotenoids safeguard macromolecules from oxidative insult through radical-scavenging action. Protein isolates and computed peptide sequences reveal bioactivities encompassing anticarcinogenic, bactericidal, antioxidant, hepatoprotective and antidiabetic effects. Complementary secondary metabolites – such as alkaloids, glucosinolates, isothiocyanates, thiocyanates, sulphates and nitriles – foster both antifungal and antibacterial inhibition. Collectively, these phytochemicals, distributed throughout the plant, are consistently correlated with beneficial therapeutic outcomes (Gopalakrishnan et al., 2016).

18.3 *M. OLEIFERA* AS A PHARMACEUTICAL EXCIPIENT

M. oleifera is a widely cultivated tree known for its nutritious pods, flowers and seeds. Rheumatism, gout, arthritis and sexually transmitted diseases (STDs) have all been treated with the seeds in traditional medicine, and roasted seeds are also used to treat epilepsy and skin infections and as diuretics (Sailaja et al., 2016). Crude protein, fat, carbohydrates, methionine, cysteine, benzylglucosinolate, moringyne and triglycerides are all present in the seeds. The mucilage from the pods, called drumstick polysaccharide, is composed of galactose, dextrose, xylose and mineral salts. The seed extract has been employed as an eco-friendly clarifying agent for drinking water. The tree also shows great potential in drug delivery formulation (Kumar & Renuka, 2019). Natural polysaccharides from *M. oleifera* are commonly used as pharmaceutical excipients because they are readily available, biocompatible and biodegradable. Elevating their function as excipients, starch and protein-mucilage fractionates from seed coats combine adequate compressibility with tunable drug-release kinetics. An in vitro investigation purified these polymeric constituents and employed them in paracetamol tablet formulations, with binders and disintegrants delineated through a suite of quality-control metrics. Evaluative parameters included mass uniformity, tablet hardness, friability resistance and disintegration kinetics, thus providing a comprehensive profile of their technological viability (Ghosh et al., 2025).

18.3.1 Binding, Disintegrant and Film-Forming Properties

Binders, disintegrants and film-forming agents are foundational excipients in solid pharmaceutical dosage forms, influencing overall drug release and bioavailability. Plant-derived swollen gums and mucilages continue to receive attention for their compatibility and functionality in such systems (Badwaik et al., 2020). *M. oleifera* seeds, in particular, have established multipurpose utility; their proteinaceous extract serves as a coagulant for turbidity removal and as an antimicrobial against *Escherichia coli* in aqueous matrices. Isolated lectins from moringa seeds have been highlighted for their potential in achieving targeted drug delivery to mucosal surfaces. Moreover, moringa seed extract demonstrates anticancer activity in vitro and functions as a water-soluble, mucoadhesive polymer suitable for mucoadhesive dosage forms. The seeds, pods and leaves exhibit documented activity in promoting wound healing, thus reinforcing their status in biopharmaceutical applications. Plant gums and mucilages have been validated in academic and industrial settings as both binders and film-forming agents. For instance, okra gum has been characterised as an effective excipient that combines binding and release-modulating capabilities, leading to successful sustained-release formulations of theophylline. When used as a film-forming polymer, okra gum afforded controlled-release profiles for diltiazem hydrochloride tablets. The mucilage from *Sterculia urens* bark has been similarly employed; its viscosity-related properties have enabled the design of orally bioavailable, controlled-release diclofenac sodium tablets. Mucilages derived from *Plantago ovata*, *Lepidium sativum* and fenugreek have been similarly applied as release retardants within the formulation of controlled-release tablets (Kumar & Renuka, 2019).

18.3.2 Comparison with Synthetic Excipients

Synthetic excipients present distinct advantages relative to their natural counterparts, particularly in terms of reproducible performance, reduced ecological footprint, economic viability, ubiquitous regional supply and precise modulation of critical formulation parameters, including release kinetics, stability and particle-size distribution. Within the category of natural excipients, tree-derived gums are ubiquitous; however, their inter-batch heterogeneity necessitates rigorous characterisation. One example is *M. oleifera* gum isolated from bark wounds in southern India, purified to yield a pale cream powder and evaluated against pharmacopeial benchmarks (Kumar & Renuka, 2019). This natural polysaccharide exhibits elevated bulk density, superior flow kinetics, a compressibility index indicative of robust particle interaction and a moisture content within pharmaceutically acceptable limits. It performs admirably as a tablet binder and as a release-retarding agent within matrix architectures, achieving sustained drug liberation for periods exceeding 24 hours. Complementary investigations of *M. oleifera* seed fractions reveal additional excipient potential: purified starch and a protein–polysaccharide complex isolated from seed coats serve respectively as effective binders and disintegrants in paracetamol tablets (Sailaja et al., 2016). The resulting dosage forms meet accepted criteria for weight consistency, mechanical strength, friability and disintegration, supporting the functional viability of moringa seed-derived components in pharmaceutical formulation design (Gopalakrishnan et al., 2016).

18.4 DRUG DELIVERY SYSTEMS USING *M. OLEIFERA*

Controlled drug delivery encompasses the mechanistic regulation of bioactive agent liberation, optimising spatiotemporal exposure to target tissues. The engineering of sustained-release formulations stands as a frontier of contemporary pharmaceutical technology, blending material science, physiological transport phenomena and therapeutics in the pursuit of precise therapeutic outcomes. Oral administration is most preferred because of its ease of ingestion, pain avoidance and being the most cost-effective. However, oral ingestion can also have adverse side effects, such as irritation of the gastric mucosa, nausea and possible destruction of the active ingredient during the prolonged and variable transit through the stomach and intestine. Sustained-release preparations also extend the period of drug effectiveness, frequently reducing the number of doses administered per day, thus improving patient compliance (Kumar & Renuka, 2019).

A drug delivery system based on a natural plant-derived material would meet two essential requirements: biocompatibility and economy. Natural excipients may reduce toxicity, side effects, increase bioavailability and stability and enhance therapeutic activity. *M. oleifera*, belonging to the monogeneric family Moringaceae, is utilised in the treatment of a range of ailments, including cardiovascular, haematological, gastrointestinal, hepatic, skin and joint disorders. Several studies also showed that the gum of *M. oleifera* can be applied in adhesives, cosmetics, pharmaceuticals and foodstuffs, and acts as a binder and release retardant. When compared with the various commercial synthetic sustained-release polymers, *M. oleifera* gum offers excellent sustained-release properties suitable for oral controlled-release drug formulations (Sharma et al., 2013).

18.4.1 Conventional Dosage Forms

Conventional drug delivery systems refer to oral formulations such as tablets and capsules that contain a suitable quantity of the drug that dissolves immediately upon contact with gastrointestinal fluids, becoming available for absorption (Hannan et al., 2016).

The extent and rate of drug absorption from such conventional dosage forms depend on various factors. These include the physicochemical properties of the drug, the nature of the excipients used and physiological factors such as the presence or absence of food. Several issues have been associated with conventional dosage forms. Basic requirements include the need for the drug to be administered several times a day, depending on its biological half-life. Multiple dosing results in increased chances of drug wastage and non-compliance of patients. There is therefore significant interest in the development of drug delivery systems capable of delivering the drug at a rate dictated by the needs of the body throughout the period of application, ensuring that the drug reaches its site of action. Recognition that the rate of drug absorption can be decreased by reducing the drug release rate has led to the development of systems that release medication slowly instead of releasing the entire dose immediately, thus prolonging drug release and sustaining its action (Lakshmi & Swain, 2018).

About two-thirds of the global medication market consists of pharmaceutical tablets, which are the most commonly used oral dosage form. Typically, tablets are formulated by compressing dry powder blends of ingredients with different functionalities. Newer and more efficient oral tablet drug delivery systems are rapidly being developed by the pharmaceutical industry to meet requirements for improved patient acceptability and enhanced bioavailability. Compressed tablets, repeat-action tablets, multiple-compressed tablets, delayed-release tablets, film-coated tablets, sugar-coated tablets and chewable tablets are among the various kinds of tablets that are administered orally. Tablets used in the oral cavity comprise buccal and sublingual tablets (Rodenhuis et al., 2004).

18.4.2 Novel Drug Delivery Systems

Sustained-release medications are formulations that continuously release an active ingredient over extended periods, allowing a single dose to extend therapeutic action beyond that of conventional medications. These medications are designed to release the active ingredient slowly or in specific locations to ensure steady delivery.

The use of natural polymers in formulating sustained-release dosage forms has gained popularity due to their availability, non-toxicity, low cost and ease of modification. Natural gums represent one class of such polymers. Consequently, a study aimed to isolate and characterise *M. oleifera* gum as a novel sustained-release polymer (Kumar & Renuka, 2019).

18.4.3 Nanoparticles and Nanocarriers

Nanotechnology delivers the promise of novel physical, chemical and surface properties that can be tuned through preparation parameters. Nanoparticles and nanocarriers are therefore promising for drug delivery in general, and in particular for *M. oleifera.* Nanomaterials are an emerging branch in current research due to their inherent ability to generate novel shapes and structures. Magnetic nanoparticles (MNPs) represent one of the most prospective platforms regarding drug delivery and are under extensive study because of their high magnetic responsivity, biodegradability, high delivery efficiency and possible targeting function. The magnetic responsivity of MNPs allows one to inject drug-loaded MNPs directly into the tumour and hold them in place using an external magnetic field, so that the supported drug is released in a highly controlled way. Interestingly, starch is regarded as one of the few materials that can combine biodegradability and biocompatibility. Therefore, the polysaccharide is used in a variety of drug delivery systems, for example, tumour targeting, transdermal delivery and brain tumours. Starch microspheres are also exploited in tissue engineering and targeted drug delivery (Vijayakumar et al., 2023).

Among the many metallic nanoparticles investigated for drug and gene delivery, gold nanoparticles (AuNPs) have received considerable attention. This attention stems from the fact that AuNPs are highly functionalised due to their strong affinity to form covalent and non-covalent bonds with various organic and inorganic molecules. Metallic nanoparticles are usually synthesised using several approaches, the most common being the colloidal route. The average size of the produced AuNPs is in the range of 3–200 nm. Added to an extremely high surface area to volume ratio spread over such a large size range, the nanoparticle surface allows loading a large number of molecules. The latter can be drugs, enzymes, plant extracts and so on; gold nanoparticles are generally coated with different types of molecules and can contact the above 'active' substances. Thus, gold nanoparticles can be constructed directly in the direction of drug/gene delivery pressure and function highly efficiently, thereby alleviating the possibility of side effects. PEG and transferrin were employed as capping agents, which are beneficial for drug/gene delivery (Das et al., 2026; Majee et al., 2025).

18.4.4 Liposomes

Liposomes, or bilayer spherical vesicles, provide a versatile system for the encapsulation and delivery of multiple therapeutic agents (Sercombe et al., 2015). This is especially appropriate for enhancing the drug delivery of *M. oleifera* extracts, which are rich in many bioactive compounds that are water-insoluble with low bioavailability. Encapsulation of such molecules in liposomes will shield them from degradation and allow them to be delivered more effectively to their target locations (Shi et al., 2024). One of the most significant benefits of employing liposomes in *M. oleifera* is the increase in therapeutic efficacy. The bilayer of the liposomes can encapsulate aqueous-soluble constituents, some alkaloids and proteins and lipid-soluble constituents, for example, flavonoids and polyphenols, within a single formulation. This property of dual encapsulation is important in the context of moringa because its therapeutic action is due to a wide range of phytochemicals (Tungadi et al., 2023). For instance, liposomal encapsulation of the bioflavonoid quercetin, a highly concentrated antioxidant in moringa leaves, was found to dramatically enhance its stability, solubility and cellular incorporation, with consequences that materialise as enhanced anti-inflammatory and anticancer activity. The amphiphilic character of phospholipids in liposomes allows for the entrapment of additional moringa extract components. Liposomes increase cell uptake, extend circulation time and allow for oral and topical delivery. Entrapment avoids premature metabolism and excretion, extending the drug's therapeutic window for moringa bioactives. Vesicle composition can be engineered to achieve specific release kinetics and targeting needs (Wang et al., 2025).

18.4.5 Hydrogels

Hydrogels, consisting of up to 90% water and at least 5–10% polymeric material, have distinct physical properties suitable for targeted drug delivery and are typically formed by covalent cross-linking of polymers (Narayanaswamy & Torchilin, 2020). *M. oleifera* gum biopolymer, a complex polysaccharide, has been isolated and characterised as a novel sustained-release polymer (Kumar & Renuka, 2019). Nanoparticle-hydrogel hybrid systems can be created by entrapping hydrogels within nanoparticles, cross-linking nanoparticles via hydrophobic interactions or opposite charges, or covalently coupling nanoparticles with hydrogels. Stimuli-responsive liposome-based hydrogels have been developed in which glutathione triggers drug release by degrading cross-links, facilitating delivery of compounds such as doxorubicin and cytochrome-C; the presence of glutathione in tumour microenvironments prompts release. The high adaptability of hydrogels in shape and size enables design tailored to the delivery route. In ophthalmic applications, hydrogels help reduce drug wastage from conventional eye drops, as demonstrated by the FDA-approved Dextenza ocular hydrogel (Ranote et al., 2022).

18.4.6 Buccal and Transdermal Films

Drug delivery through the buccal mucosa offers numerous advantages over oral administration, including easy and convenient administration, bypassing the first-pass effect and a rapid onset of action. Transport across the buccal mucosa predominantly occurs by passive diffusion along a concentration gradient. Lipophilic molecules mainly permeate transcellularly, passing through the epithelial cells, whereas hydrophilic molecules use the intercellular pathway. The molecular weight cut-off of the buccal epithelium is between 500 and 1000 Da, thereby hindering systemic absorption of larger molecules. Saliva turnover generates a washout effect, leading to premature swallowing of the carrier and reduced drug residence time at the absorption site (Badwaik et al., 2020).

In this context, oral films are emerging as a promising alternative to traditional solid dosage forms for systemic or local delivery via the oral route. Unlike tablets and capsules, oral films are mucoadhesive, disintegrate rapidly and require no water for administration. They are particularly advantageous for paediatric, geriatric, bedridden and dysphagic patients (Bala et al., 2013). Incorporating micro or nanocarriers, such as beads, into oral films may prolong drug release by reducing the exposed surface area compared with free drug-loaded films. Biocompatible polymers like alginate provide additional benefits such as mucoadhesion and sustained release. Alginate beads loaded with caffeine can be prepared cost-effectively using the emulsification-internal gelation technique, which offers mild conditions that avoid exposure of the entrapped drug to cross-linking agents. Although alginate beads have been used for local buccal activity, their incorporation into film matrices is an innovative approach that combines the convenience of oral films with the controlled release achievable with alginate beads. Buccal mucoadhesive films are intended for absorption through the oral mucosa to increase systemic bioavailability and reduce the hepatic first-pass effect. Providing sustained drug release further enhances therapeutic outcomes and patient compliance. Films prepared with Eudragit RL100 exhibit transparency, uniform thickness, flexibility and satisfactory dissolution characteristics. Drug release from such films follows Higuchi's model, governed by non-Fickian diffusion. Compared with other polymers like PVP or cellulose derivatives, Eudragit RL100 formulations demonstrate superior uniformity, flexibility and release profiles, making them well-suited for buccal film administration (Bahri-Najafi et al., 2014).

18.4.7 Self-Emulsifying and Nanoemulsifying Systems

Self-microemulsifying drug delivery systems (SMEDDS) and self-nanoemulsifying drug delivery systems (SNEDDS) are lipoid systems that spontaneously create fine emulsions or nano-emulsions in contact with aqueous media, typically stomach contents. They enhance the solubility and bioavailability of lipophilic phytochemicals with the aid of thermodynamically stable dispersions, requiring less mechanical mixing (Barakoti et al., 2019; Zubair et al., 2025). A notable formulation investigation optimised an SMEDDS of *M. oleifera* leaf extract for kaempferol and quercetin solubilisation. Optimal composition consisted of propylene glycol dicaprylocaprate (20 %), polysorbate 80 (60 %) and PEG-400 (20 %). SMEDDS have been widely applied for overcoming the issue of low oral bioavailability in biopharmaceutical classification system (BCS) class II and class IV drugs. SMEDDS are clear, isotropic oil, surfactant and co-solvent or co-surfactant mixtures that can self-emulsify spontaneously to yield oil/water (o/w) emulsions upon contact with an aqueous phase, such as the GI fluid, following exposure to mild agitation by the GI motility. The major benefit of SMEDDS is that they increase the oral bioavailability of drugs without requiring dissolution, which is the rate-determining step of drug absorption. The emulsion droplet size of SMEDDS ranges from 100 to 300 nm, while the diameter of the clear microemulsion formed from SMEDDS is below 50 nm (Sermkaew & Plyduang, 2020).

SNEDDS is among the oil or fatty nanoparticle-based formulations that are a physical mixture of an isotropic mixture between co-surfactant and surfactant and oil phase. Upon mixing with water, it will spontaneously develop an o/w nano-emulsion. SNEDDS have the potential to form a nano-emulsion of oil in water when subjected to mild agitation followed by dilution in a water medium, such that the o/w interface area increases, followed by an increase in solubility, which will further increase bioavailability. SNEDDS of plant extract (moringa leaf extract and *Ketepeng cina* leaf extract) could be prepared with corn oil as the oil phase, Tween 80 as the surfactant and propylene glycol as the co-surfactant (Suryani et al., 2019; Zubair et al., 2025).

18.4.8 pH-Responsive Nanogel

pH-sensitive nanogels of *M. oleifera* are an intelligent, smart drug delivery system utilising the natural biopolymer gum of *M. oleifera* for the targeted and controlled release of drugs, particularly anti-cancer drugs such as doxorubicin (DOX). These nanogels are pH-sensitive hydrophilic nanoscale polymeric networks that contract or expand as a function of environmental pH alterations, such that drug release occurs at desired sites like tumorous tissues, which have acidic microenvironments (pH ranges of 5.5–6.8), but the polymeric network is stable under normal physiological pH (around 7.4). Functionalised *M. oleifera* gum is chemically grafted to prepare a nanogel matrix that can entrap drugs with enhanced entrapment efficiency. When in contact with the acidic microenvironment of the tumour cell, the nanogel swells because of the ionisation change of the polymer, leading to localised drug release at the cancer site. In contrast, the nanogel collapses under the physiological pH of the body, minimising early drug release and possibly reducing toxic side effects on normal tissues (Ranote et al., 2022).

18.4.9 Phytosomes and Phyto-Phospholipid Complexes

Phytosomes represent novel lipid-based drug delivery systems comprising complexes of natural active phytoconstituents and phospholipids. The resulting structures lack a vesicular architecture but nonetheless encapsulate a hydrophilic complex within a lipophilic milieu. The primary chemical interactions that encourage complex formation are hydrogen bonding and electrostatic attractions between the polar head of phospholipids and the polar functionalities of the phytoconstituents. Due to its beneficial properties, phosphatidylcholine is used in the formation of phytosomes in an equimolar ratio with phytoconstituents (Mahapatra et al., 2020).

Phytosomes have broad applications in several therapeutic areas, including wound healing, neurological disorders, hepatitis and inflammatory diseases. Various studies on phytosome characterisation use techniques such as microscopic examination, particle size and zeta potential measurement, thin-layer chromatography, antioxidant activity measurement and hepatoprotective activity evaluation. They exhibit favourable characteristics for topical delivery, enabling drug administration via dermal and transdermal routes. Preparation procedures for phytosomes are non-conventional but uncomplicated and reproducible, which represents a bridge between traditional phytoconstituent delivery and innovative drug delivery systems (Tripathy et al., 2013).

Despite the availability of numerous phytosome products for topical delivery, many phytoconstituents with therapeutic potential for critical pathologies – such as cancer, diabetes, obesity, neurodegeneration and bronchitis – remain to be incorporated into phytosome systems. Ongoing research endeavours to develop highly target-specific phytosomes as phytomedicine-oriented interventions (Chivte et al., 2017).

18.5 MECHANISMS OF DRUG RELEASE AND TARGETING

Drug delivery systems have strict requirements regarding the temporal and spatial pattern of drug release. Such systems should typically allow single-dose, controlled, prolonged-release and rationed administration of medication; furthermore, sustained-release formulations should also direct medicaments to their desired sites of action (Kumar & Renuka, 2019). *M. oleifera* gum is a candidate polymer for prolonged-release formulations, and its utility would be improved by delivery systems incorporating it into porous inorganic matrices. The pore-size distribution, spacing and accessibility determine the ability of mesoporous materials to localise and sustain the release of drugs and thus make these materials particularly appropriate for drug delivery applications. Computational studies of prodrug retention and adsorbent equivalence reveal that bimodal pore-size distributions can provide better delivery performance than their monomodal counterparts, with the best configurations consisting of similar amounts of mesopores and macropores, with sizes of ca. 27 nm and 130 nm, respectively (Ranote et al., 2022).

18.5.1 Sustained and Controlled-Release Profiles

Sustained- and controlled-release drug delivery systems maintain more-constant plasma levels and reduce dosing frequency, so that drug safety, efficacy and patient compliance can all be improved. For oral solid dosage forms, release is usually controlled via dissolution, diffusion or a combination thereof. These are achieved either by incorporating the drug into an inert polymeric carrier or by modifying the drug's molecular structure. Drug release may take place by a variety of mechanisms. These include diffusion through a gel matrix or membrane, slow dissolution of a matrix to liberate the drug, osmotic pressure, ion exchange, chemical reactions and enzyme reactions. The product may be defined as sustained-action, sustained-release, prolonged, controlled-release, extended-action, depot, respiratory, retarded-release, timed-release and targeted delivery systems (Kadam & Shendarkar, 2017).

Benefits include reduced administration frequency, improved patient compliance, reduced blood level oscillations, possible dose reduction and improved safety margins. Limitations occur due to a lack of flexibility in adjusting therapy, fixed and sometimes suboptimal dosing schedules and usually increased manufacturing costs. Important system parameters for design and optimisation of an oral controlled-release dosage form include the elimination rate constant, the absorption rate constant, volume of distribution, loading dose, maintenance dose, dosing intervals, lag time and release rate. Ideal criteria of a sustained-release formulation include a single dose of no more than 500 mg for oral administration (Kumar & Renuka, 2019).

The object of controlled- and sustained-release drug therapy is thus to enable maintaining optimum levels of a drug over as long a period as possible through single-dose formulations. In ideal practice, the active entity should be delivered directly to the site of action; then, the therapeutic effect can be maintained, while systemic toxicity and other side effects are minimised or eliminated completely (Hannan et al., 2016).

18.5.2 Site-Specific Delivery

A colon-targeted drug delivery system (OCTDDS) has been developed to overcome the drawbacks of oral drug delivery for intestinal diseases by preventing premature absorption in the upper gastrointestinal tract and delivering therapeutic agents directly to the colon (Philip & Philip, 2010). This system protects protein drugs from enzymatic degradation in the stomach and small intestine, increases the drug concentration at the pathological site and reduces side effects on healthy tissues. Targeting issues include mechanisms and sites, as well as the selection of appropriate drug carriers and strategies for their modification. OCTDDS exploits physiological conditions such as colonic pH, microbial enzyme activity, slow peristalsis and significant water absorption to achieve localised delivery. Conventional methods – pH-dependent, time-dependent and pressure-controlled systems – as well as approaches triggered by microbial flora, enzyme activity and prodrug formation, are utilised. Current research emphasises chemical modification of carriers to enhance sensitivity, specificity, release stability and biocompatibility (Singhal et al., 2012).

18.6 BIOMEDICAL APPLICATIONS

Moringa oleifera (MO) parts have long been used in traditional medicine owing to the wide range of therapeutic implications, including antimicrobial, anticancer and anti-inflammatory effects. MO gum exuded from the stem is an important exudate polysaccharide that possesses several beneficial pharmaceutical properties (U. Singh et al., 2017). The gum's adequate tensile strength and low vapour transmission rate make it suitable for use as a film coating agent in tablets and for the manufacture of polymeric drug delivery systems. Plant phenolics, such as those from MO and *Morus indica*, have shown stabilising activity and can extend the shelf life of products like pineapple juice stored at 4°C. MO parts are also valuable in cosmetics; seed oil, known as Behen oil, is used as a carrier in cosmetic preparations and is appreciated for its oxidative stability, high oleic acid content, light texture and skin-spreading properties, making it suitable for massage and aromatherapy (Badwaik et al., 2020).

18.6.1 Anticancer Drug Delivery

Plants and aquatic organisms – alongside microbial life – constitute major reservoirs for anticancer therapeutics. Presently, over 60% of oncology drugs have natural origins. The plant *Camptotheca acuminata* provided camptothecin, progenitor of the clinically employed derivatives topotecan and irinotecan. Botanically sourced phytochemicals typically exhibit reduced toxicity and fewer adverse reactions, thereby incentivising intensive investigations of medicinal flora. MO, characterised by low toxicity and a broad repertoire of secondary metabolites, has emerged as a promising candidate. Considerable inquiry has focused on extracts of MO foliage, which are enriched in polyphenols, flavonoids and other redox-active compounds with potential cytostatic effects (Reda et al., 2017). These investigations commonly initiate with assays measuring antioxidant and anti-inflammatory capacities, hypothesising that diminished oxidative stress may interrupt tumour initiation. Subsequent in vitro experiments with neoplastic cell cultures have shown alterations in proliferation, growth kinetics and morphological phenotypes. Favourable data have prompted the dissection of specific intracellular signalling cascades perturbed by the extracts. Breast cancer, marked by elevated incidence and mortality rates globally, poses a persistent therapeutic obstacle. Malignant cells evolve diverse chemoresistant phenotypes, undermining the efficacy of standard cytotoxic regimens (Al-Asmari et al., 2015).

Natural products, particularly secondary metabolites derived from higher plants, have emerged as promising candidates for attenuating multidrug resistance while augmenting the efficacy of standard chemotherapeutic agents. MO, popularly referred to as the 'miracle tree', is used for both dietary and therapeutic purposes across a broad geographic expanse. Cumulative investigations indicate that the species exerts inhibitory effects against a spectrum of neoplasms. Leaf extracts have consistently produced antiproliferative and pro-apoptotic responses in colon, lung, prostate and breast cancer cell lines. Principal active constituents, namely glucosinolates and their derived isothiocyanates, are posited to underlie the observed cytotoxicity. Given the intrinsic limitations of classical chemotherapeutic regimens, interest in MO as a complementary modality is increasing. Current scholarship has concentrated almost exclusively on in vitro experimental paradigms, underscoring the imperative for subsequent in vivo validation to fully delineate the plant's putative anticancer efficacy (Mohammed et al., 2024).

18.6.2 Antimicrobial and Wound-Healing Applications

Wounds are an inherent consequence of trauma, affecting both humans and animals. They initiate a multistage healing process involving inflammation, coagulation, migration, proliferation, epithelialisation, remodelling and scar formation phases. The progression and outcome of healing are influenced by numerous factors, which can impede or enhance the process. Infections pose a significant obstacle to effective wound and burn healing. Natural plant materials offer a potential solution, given their phytochemical content, therapeutic properties and non-toxic nature (Ali et al., 2020). A wide variety of medicinal plants have been employed globally in traditional wound care formulations. MO, known for its anti-inflammatory, antimicrobial, antioxidant and wound-healing activities, stands out as a promising herbal candidate for wound dressings (Shady et al., 2022). Freeze-dried MO extracts exhibit good antioxidant activity for the treatment of free radical diseases (Gothai et al., 2017).

18.6.3 Anti-inflammatory and Other Therapeutic Uses

MO has long had a position in traditional medicine (P. S. Kumar et al., 2010). MO has demonstrated anti-inflammatory and antinociceptive activity, as well as hepatoprotective and antiulcer activity. Leaf, root and other plant parts' extracts were found to have promise against gastric ulcers, hepatotoxicity and blood pressure modulation. Scientific evidence also suggests that the plant constituents exhibit hypotensive and spasmodic activity, supporting the plant's therapeutic value in several inflammatory and gastrointestinal disorders (U. Singh et al., 2017).

18.7 PHARMACOKINETICS AND BIOAVAILABILITY ENHANCEMENT

Pharmacokinetics and the measurement of drug absorption, distribution, metabolism and elimination are used to guide formulation design, delivering drugs at optimal rates to achieve therapeutic plasma concentrations (Hannan et al., 2016). Traditional solid oral dosage forms are normally in need of multiple daily frequent dosing to keep the drug above the minimum effective concentration, which is inconvenient for the patient and risky for non-compliance. The optimum system reduces the frequency of dosing by maintaining drug release and facilitates targeted delivery to the site of action with reduced side effects (Kumar & Renuka, 2019). Increasing interest in natural polymers has encouraged research into MO gum as an emerging material for use in sustained-release preparations. Nasal delivery of bioadhesive starch microspheres was found in in vivo sheep studies to significantly enhance drug absorption, and this effect was further enhanced when the microspheres were used in combination with other absorption enhancers. Starch microspheres loaded with insulin and gentamicin improved absorption efficiency five- and tenfold, respectively, compared with control solutions. Therefore, intranasal administration of rats with such microspheres induced around 30-fold increases in drug bioavailability. Studies on *Aloe vera* leaf materials and extracts revealed changes in both in vitro drug transport and in vivo drug bioavailability. A clinical trial revealed that *Aloe vera* products provided considerable increases in the bioavailability of vitamins C and E due to their protective action against gastrointestinal degradation. Further, a model study involving *Aloe ferox* and *Aloe vera* gel in combination with atenolol showed increased bioavailability, likely due to changes in intestinal absorption and altered enzyme activity (Peterson et al., 2019).

18.7.1 Impact of Moringa-Based Systems on Drug Absorption and Stability

Medicinal use of MO seeds is used in conditions such as arthritis, rheumatism, gout, sexually transmitted diseases and suppurative ulcers. Roasted seeds have also been used in the control of symptoms of epilepsy and skin conditions, as well as weak diuretics. Nutritional and phytochemical studies demonstrate the existence of proteins, lipids, carbohydrates and applicable bioactive compounds, among which are benzylglucosinolate and moringyne. Mucilage isolated from the floral pods, designated drumstick polysaccharide, is predominantly formulated from galactose, dextrose and xylose, supplemented with mineral salts, and is characterised by notably high ash and protein levels. The seeds have also been harnessed, through grinding and settling, to encourage sedimentation of particulate impurities in potable water (Alsammarraie et al., 2021).

Naturally occurring polysaccharides obtained from MO have gained attention as sustainable, biocompatible and biodegradable excipients in pharmaceutical technology. Acting as both cohesive binders and retardants of release kinetics, they have been substantiated in several studies. Controlled in vitro experimentation with starch and mucilage fractions isolated from the seed coats has illustrated their capacity to function synergistically in paracetamol tablet formulations. The resultant dosage forms conformed to pharmacopeial standards, exhibiting acceptable limits of weight variation, tensile strength, friability and target disintegration periods (Kumar & Renuka, 2019).

18.8 BIOAVAILABILITY ENHANCEMENT

MO natural gum may be covalently cross-linked via glutaraldehyde coupling under mildly acidic pH to yield hydrogels that display therapeutically relevant pH-responsive swelling and solute release characteristics (Kumar & Renuka, 2019). Remarkably, these hydrogels demonstrate full reversible volume transition – contracting sharply at pH 2 and expanding at physiological pH 7.4 – thereby enabling well-defined delivery profiles for water-soluble therapeutic agents. For example, propranolol hydrochloride administered in an acidic environment is released in an initial, pulsatile fashion and then a steady, prolonged release at pH 7.4. This new swelling kinetics provides a fast loading and prolonged retention mechanism that can be formulated to coincide with gastrointestinal transit times or targeted to particular tissue microenvironments. Overall, chemically cross-linked MO gum-derived hydrogels are multifunctional, green matrices with selectivity, biocompatibility and a sustained therapeutic effect for pH-controlled drug delivery (Ranote et al., 2022).

18.8.1 Mechanisms of Action

MO, with its wide range of nutraceutical potential, has a range of pharmacological properties such as anti-inflammatory, antioxidant, anticancer and antimicrobial activities (Liyongo et al., 2018). The plant fractions trigger antiproliferative effects and apoptosis in tumour cell lines and affect lipid profiles and glycaemic regulation. For the bioavailability, stability and water solubility of MO compounds, encapsulation processes are increasingly utilised; these processes protect the bioactives from degradation that may occur during processing under diverse pH, heat and oxidative environment conditions (Pop et al., 2022). Well-established methods, including gelation cross-linking, freeze-drying, microemulsion formulation, spray-drying and nanoscale formulation, result in stable microcapsules. Microencapsulation not only maintains the functional integrity of active ingredients but also makes them easier to handle, typically based on a polysaccharide or protein matrix as the carrier. Spray-drying, however, is used most commonly because of its scalability and cost-effectiveness, producing homogeneous, stable light-coloured powders from ultra-concentrated extracts or bioemulsions with high bioactive molecule content. Process yield, granule particle size, water content and extent of retention of activities such as antioxidant and antimicrobial activities are all pre-defined by the major parameters of inlet temperature, feed flow rate and encapsulant concentration. The blend of traditional hydroalcoholic extraction with spray-drying encapsulation guarantees the preservation of MO's phytochemical composition, thereby raising its bioavailability through controlled release and thus its drug activity (B. Singh & Kumar, 2018).

18.8.2 Clinical Implications

MO has been of interest due to its varied pharmacological actions, nutritional importance and potential clinical application. Systematic studies have reported its ability to modulate serum lipids, promote dermal healing and possess antimicrobial action. Its bioactive compounds, which have been isolated, have shown inhibitory activity against multidrug-resistant bacteria and highly virulent pathogens. Complementary studies have established its antidiabetic, antihyperlipidaemic, anticancer, antipyretic, antifertility and anti-inflammatory activity. Careful compositional analysis has qualitatively and quantitatively analysed trace metals, fractionated protein fractions and identified other pertinent chemical compounds (Liyongo et al., 2018). MO mucilaginous gum was also suggested for utilisation as a matrix polymer for a controlled-release product, a formulation approach that requires severe clinical evaluation (Kumar & Renuka, 2019).

18.9 CHALLENGES AND FUTURE PROSPECTS

The idea of the ideal drug delivery system is founded on the realisation of a one-dose treatment in which the active medicinal ingredient of the drug is delivered to the site of action, thereby preventing systemic side effects. As a result, a lot of work has been directed towards the design of delivery matrices that enable graded, sustained delivery of the drug over clinically acceptable periods. Most promising among these formulations are sustained-release polymers that are not just economical and biocompatible but also allow precise control of the release pattern to encompass pharmacokinetic and pharmacodynamic considerations (Kumar & Renuka, 2019). MO gum has emerged as a competitively viable excipient in this philosophy. Formulations from this biocompatible gum allow the attainment of therapeutic levels at the site of action for prolonged periods, with quasi-zero-order release kinetics that slow the rate of systemic absorption and thus extend the pharmacological effect.

Ongoing investigations aim to isolate the proteins from MO, characterise their biochemical behaviours and clarify their underlying functional pathways. Despite the available foundational data, the complete spectrum of their attributes remains partially delineated. Introducing moringa-sourced fractions into pharmaceutical matrices could yield innovative therapeutics that alleviate the limitations of entirely synthetic agents. Such advances, however, underline the necessity for thorough risk evaluation to validate the safety profiles of moringa-customised formulations. In parallel with these inquiries, harmonised dietary recommendations and regulatory protocols could crystallise safe consumption strategies (Patil et al., 2022).

Confronting the dual pressures of antibiotic resistance and the clinical failure of therapeutics attributable to systemic toxicity or insufficient specificity, precision-oriented drug delivery vehicles now command urgent experimental attention. Among the candidates, lipid-core nanoparticles, polymeric nanospheres, dendritic macromolecules and inorganic nanocrystals have been synthesised. Among these, polymeric matrices and liposomal constructs command favour owing to their demonstrated biocompatibility and their proclivity for customisable surface chemistries. Yet, mitigating rapid systemic clearance while forestalling immune-mediated inflammatory cascades that jeopardise normal tissue integrity poses persistent and formidable technical challenges (Obeid et al., 2017).

18.9.1 Limitations in Current Research

Despite the extensive investigation of MO for diverse applications, major challenges remain to restrict its more widespread use. Current research focuses on polymeric drug delivery systems, water vapour transmission and the tensile strengths of films from Moringa polymers; phenolic extracts from Moringa leaves exert a stabilising action on juice and can extend the shelf life of pineapple juice stored at 4°C; Moringa parts have cosmetic value as well – seed oil ('Behen oil') is an ideal carrier oil for aromatic compounds, owing to its stability and similarity to olive oil, and is also an ingredient in massage and aromatherapy treatments (U. Singh et al., 2017). Several sustained-release polymers exist, but the quest for safe and inexpensive alternatives continues (Kumar & Renuka, 2019). Applications consequently extend from food preservation and healing to prophylactic and cosmetic products.

18.9.2 Regulatory and Scalability Considerations

Delivering herbal drug components remains a major challenge in pharmaceutical applications. Herbal extracts have been in existence for quite some time, but do not find wide applications due to the complexity of the mixtures and the need to administer them in large quantities to induce therapeutic action. At the same time, these materials are relatively less toxic and available at lower costs. The current research in the delivery of herbal drug components has been focused on developing controlled-release formulations that will reduce the dosage and minimise toxicity effects. Polymeric materials are a key component in drug delivery (Kumar & Renuka, 2019). The sudden release of the total drug content into the system is the major drawback in its application. All drugs require controlled release into the system to ensure appropriate action at the target site (Patil et al., 2022). Regulation of drug release also controls fluctuations in drug concentrations in the blood and tissues. In therapeutic systems, all drugs are assigned a therapeutic range between the minimum toxic concentration and the minimum effective concentration. Dosages of the drug must be provided so that blood concentrations are maintained within this range. The amount of drug given depends on the flow of the drug from the unit and the rate at which the drug degrades or is eliminated from the system. Hence, controlled-release technology is in great demand because it increases the efficiency of drug delivery and avoids the side effects associated with the concentration fluctuation of the drug during conventional therapy. By adjusting the release rate, it should be possible to achieve a therapy that is more closely time-programmed to the needs of the patient or the circadian rhythm of the disease.

18.10 CONCLUSION

Building on a time-honoured record of benign medicinal use, various studies have indicated that *M. oleifera* offers a wide spectrum of pharmacological and nutritional advantages (Liyongo et al., 2018). Among these are antioxidant, anticancer, anti-inflammatory, antidiabetic, anti-hypertensive and antimicrobial properties. Consequently, the prospect of integrating *M. oleifera* extract into both drug delivery platforms and drug formulations has emerged as a focal point of contemporary research.

The feasibility of *M. oleifera* seed gum as a sustained-release excipient has been systematically investigated, revealing its capacity to modulate the liberation profile of therapeutic agents (Kumar & Renuka, 2019). Experimental outcomes indicate that the gum possesses physicochemical attributes compatible with the requirements for controlled-release delivery systems. Parallel investigations indicate that standardised preparations of *M. oleifera* seed extract, administered either as a component of the formulation or as a concomitant to the drug, enhance analgesic, antipyretic and anti-inflammatory activities. However, the existing literature lacks sufficient data on dose–vehicle interplay, suggesting a need for further exploration to delineate the most efficacious combinations for clinical translation. A thorough pharmacological dissection of the active constituents in the gum and seed extract would not only refine formulation science but also inform dietary supplementation regimens, potentially accelerate the discovery of botanically derived therapeutics and reduce dependency on fully synthetic agents. Nevertheless, in line with established pharmacovigilance protocols, comprehensive risk characterisation, inclusive of stability, purity and long-term safety data, remains imperative prior to the commercial approval of *M. oleifera* for pharmaceutical applications (Patil et al., 2022).

REFERENCES

Abdel-Latif, H. M. R., Abdel-Daim, M. M., Shukry, M., Nowosad, J., & Kucharczyk, D. (2022). Benefits and applications of Moringa oleifera as a plant protein source in Aquafeed: A review. Aquaculture, 547, 737369. https://doi.org/10.1016/j.aquaculture.2021.737369

Al-Asmari, A. K., Albalawi, S. M., Athar, M. T., Khan, A. Q., Al-Shahrani, H., & Islam, M. (2015). Moringa oleifera as an anti-cancer agent against breast and colorectal cancer cell lines. *PloS One*, *10*(8), e0135814.

Al-Ghanayem, A. A., Alhussaini, M. S., Asad, M., & Joseph, B. (2022). Moringa oleifera Leaf Extract Promotes Healing of Infected Wounds in Diabetic Rats: Evidence of Antimicrobial, Antioxidant and Proliferative Properties. Pharmaceuticals, 15(5), 528. https://doi.org/10.3390/ph15050528

Ali, A., Garg, P., Goyal, R., Kaur, G., Li, X., Negi, P., Valis, M., Kuca, K., & Kulshrestha, S. (2020). RETRACTED: A novel herbal hydrogel formulation of Moringa oleifera for wound healing. *Plants*, *10*(1), 25.

Alsammarraie, H. J., Khan, N. A., & Mahmud, R. (2021). Amphiphilic polymer for formulation of ethanol extract of *Moringa oleifera* leaves as solid dispersion: Formulation, evaluation, and stability studies. *International Journal of Drug Delivery Technology*, *11*(3), 926–936.

Álvarez-Román, R., Silva-Flores, P. G., Galindo-Rodríguez, S. A., Huerta-Heredia, A. A., Vilegas, W., & Paniagua-Vega, D. (2020). Moisturizing and antioxidant evaluation of Moringa oleifera leaf extract in topical formulations by biophysical techniques. South African Journal of Botany, Special Issue on Moringa Research, 129, 404–411. https://doi.org/10.1016/j.sajb.2019.10.011

Badwaik, H. R., Hoque, A. A., Kumari, L., Sakure, K., Baghel, M., & Giri, T. K. (2020). Moringa gum and its modified form as a potential green polymer used in biomedical field. *Carbohydrate Polymers*, *249*, 116893. https://doi.org/10.1016/j.carbpol.2020.116893

Bahri-Najafi, R., Tavakoli, N., Senemar, M., & Peikanpour, M. (2014). Preparation and pharmaceutical evaluation of glibenclamide slow release mucoadhesive buccal film. *Research in Pharmaceutical Sciences*, *9*(3), 213–223.

Bala, R., Pawar, P., Khanna, S., & Arora, S. (2013). Orally dissolving strips: A new approach to oral drug delivery system. *International Journal of Pharmaceutical Investigation*, *3*(2), 67.

Barakoti, H., Choudhury, A., & Dey, B. K. (2019). An outlook for a novel approach: Self-Micro Emulsifying Drug Delivery System (SMEDDS). *Research Journal of Pharmacy and Technology*, *12*(4), 2055. https://doi.org/10.5958/0974-360X.2019.00340.8

Brandão, R., Borges, L. P., & Nogueira, C. W. (2009). Concomitant administration of sodium 2, 3-dimercapto-1-propanesulphonate (DMPS) and diphenyl diselenide reduces effectiveness of DMPS in restoring damage induced by mercuric chloride in mice. *Food and Chemical Toxicology*, *47*(8), 1771–1778.

Chivte, P. S., Pardhi, V. S., Joshi, V. A., & Ajitha Rani, A. R. (2017). *A review on therapeutic applications of phytosomes.* https://www.cabidigitallibrary.org/doi/full/10.5555/20193045902

Das, U., Sarkar, A., Bachhar, A., Barman, S., Roy, D., & Mandal, S. (2026). Lipid nanoparticles formulation strategies. In Lipid Nanoparticles-Based mRNA Delivery System (pp. 63–90). Elsevier. https://doi.org/10.1016/B978-0-443-45006-8.00002-1

Ferreira, P. M. P., Farias, D. F., Oliveira, J. T. de A., & Carvalho, A. de F. U. (2008). Moringa oleifera: Bioactive compounds and nutritional potential. *Revista de Nutrição*, *21*, 431–437.

Ghosh, S., Solanki, R., Bhatia, D., & Sankaranarayanan, S. (2025). Nanomaterials for delivery of medicinal plant extracts and phytochemicals: Potential applications and future perspectives. *Plant Nano Biology*, *12*, 100161. https://doi.org/10.1016/j.plana.2025.100161

Gopalakrishnan, L., Doriya, K., & Kumar, D. S. (2016). Moringa oleifera: A review on nutritive importance and its medicinal application. *Food Science and Human Wellness*, *5*(2), 49–56.

Gothai, S., Muniandy, K., Zarin, M. A., Sean, T. W., Kumar, S. S., Munusamy, M. A., Fakurazi, S., & Arulselvan, P. (2017). Chemical composition of Moringa oleifera ethyl acetate fraction and its biological activity in diabetic human dermal fibroblasts. *Pharmacognosy Magazine*, *13*(Suppl 3), S462.

Hannan, P. A., Khan, J. A., Khan, A., & Safiullah, S. (2016). Oral dispersible system: A new approach in drug delivery system. *Indian Journal of Pharmaceutical Sciences*, *78*(1), 2.

Kadam, V. S., & Shendarkar, G. (2017). Role of natural polymer in sustained and controlled release. *American Journal of Pharmaceutical Research*, *7*, 7390.

Kumar, P. S., Mishra, D., Ghosh, G., & Panda, C. S. (2010). Medicinal uses and pharmacological properties of Moringa oleifera. *International Journal of Phytomedicine*, *2*(3). https://core.ac.uk/download/pdf/270171818.pdf

Kumar, R. S., & Renuka, R. (2019). Isolation and characterization of Moringa olifera gum: A novel sustained release polymer. *Journal of Drug Delivery and Therapeutics*, *9*, 484–486.

Lakshmi, S. V., & Swain, R. P. (2018). Formulation and characterization of sustained release matrix tablets of indomethacin using Moringa oleifera gum. *IOSR Journal of Pharmacy*, *8*(4), 57–65.

Liyongo, C. I., Bongo, G. N., Ashande, C. M., Lufualabo, L. G., Gbolo, B. Z., Djoza, D. R., Kwembe, J. T. K., Onautshu, O., & Mpiana, P. T. (2018). An updated review on the bioactivities and phytochemistry of the nutraceutical plant Moringa oleifera Lam (Moringaceae) as valuable phytomedicine of multi-purpose. *Discovery Phytomedicine*, *5*(4), 52–63.

Mahajan, S. G., & Mehta, A. A. (2008). Effect of *Moringa oleifera* Lam. Seed extract on ovalbumin-induced airway inflammation in guinea pigs. *Inhalation Toxicology*, *20*(10), 897–909. https://doi.org/10.1080/08958370802027443

Mahapatra, D. K., Patil, S., & Patil, A. G. (2020). The progressive journey of phytosomes in herbal based pharmacotherapeutics. *Current Bioactive Compounds*, *16*(6), 853–886.

Majee, S. B., Bera, R., Avlani, D., Das, U., & Singh, S. (2025). Safety Profiles and Potential Toxicological Concerns of Herbal Nanomedicine. In A. Farid, Herbal Pharmacopeia (1st ed., pp. 412–439). CRC Press. https://doi.org/10.1201/9781003544418-20

Meireles, D., Gomes, J., Lopes, L., Hinzmann, M., & Machado, J. (2020). A review of properties, nutritional and pharmaceutical applications of Moringa oleifera: Integrative approach on conventional and traditional Asian medicine. Advances in Traditional Medicine, 20(4), 495–515. https://doi.org/10.1007/s13596-020-00468-0

Mohammed, H., Karhib, M. M., Al-Fahad, K. S. J., Atef, A. M., Eskandrani, A., Darwish, A. A., Sary, A. A., Elwakil, B. H., Bakr, B. A., & Eldrieny, A. M. (2024). Newly synthesized chitosan nanoparticles loaded with caffeine/moringa leaf extracts Halt Her2, BRCA1, and BRCA2 expressions. *Scientific Reports*, *14*(1), 18118.

Narayanaswamy, R., & Torchilin, V. P. (2020). Hydrogels and their applications in targeted drug delivery. In *The road from nanomedicine to precision medicine* (pp. 1117–1150). Springer.

Obeid, M. A., Al Qaraghuli, M. M., Alsaadi, M., Alzahrani, A. R., Niwasabutra, K., & Ferro, V. A. (2017). Delivering natural products and biotherapeutics to improve drug efficacy. *Therapeutic Delivery*, *8*(11), 947–956. https://doi.org/10.4155/tde-2017-0060

Pareek, A., Pant, M., Gupta, M. M., Kashania, P., Ratan, Y., Jain, V., Pareek, A., & Chuturgoon, A. A. (2023). Moringa oleifera: An Updated Comprehensive Review of Its Pharmacological Activities, Ethnomedicinal, Phytopharmaceutical Formulation, Clinical, Phytochemical, and Toxicological Aspects. International Journal of Molecular Sciences, 24(3), 2098. https://doi.org/10.3390/ijms24032098

Patil, S. V., Mohite, B. V., Marathe, K. R., Salunkhe, N. S., Marathe, V., & Patil, V. S. (2022). Moringa tree, gift of nature: A review on nutritional and industrial potential. *Current Pharmacology Reports*, *8*(4), 262–280. https://doi.org/10.1007/s40495-022-00288-7

Peterson, B., Weyers, M., Steenekamp, J. H., Steyn, J. D., Gouws, C., & Hamman, J. H. (2019). Drug bioavailability enhancing agents of natural origin (bioenhancers) that modulate drug membrane permeation and pre-systemic metabolism. *Pharmaceutics*, *11*(1), 33.

Philip, A. K., & Philip, B. (2010). Colon targeted drug delivery systems: A review on primary and novel approaches. *Oman Medical Journal*, *25*(2), 79.

Pop, O. L., Kerezsi, A. D., & Ciont (Nagy), C. (2022). A comprehensive review of Moringa oleifera bioactive compounds—Cytotoxicity evaluation and their encapsulation. *Foods*, *11*(23), 3787. https://doi.org/10.3390/foods11233787

Putro, J. N., Lunardi, V. B., Soetaredjo, F. E., Yuliana, M., Santoso, S. P., Wenten, I. G., & Ismadji, S. (2021). A review of gum hydrocolloid polyelectrolyte complexes (Pec) for biomedical applications: Their properties and drug delivery studies. *Processes*, *9*(10), 1796.

Ranote, S., Musioł, M., Kowalczuk, M., Joshi, V., Chauhan, G. S., Kumar, R., Chauhan, S., & Kumar, K. (2022). Functionalized Moringa oleifera gum as pH-responsive nanogel for doxorubicin delivery: Synthesis, kinetic modelling and in vitro cytotoxicity study. *Polymers*, *14*(21), 4697. https://doi.org/10.3390/polym14214697

Reda, F., Borjac, J., Fakhouri, R., & Usta, J. (2017). Cytotoxic effect of Moringa oleiferaon colon cancer cell lines. *Acta Horticulturae*, *1158*, 45–54.

Rodenhuis, N., De Smet, P. A., & Barends, D. M. (2004). The rationale of scored tablets as dosage form. *European Journal of Pharmaceutical Sciences*, *21*(2–3), 305–308.

Roy, D., Sarkar, A., Bachhar, A., Das, U., & Mandal, S. (2025). Gum Odina in focus: Modification strategies, extensive characterization multifunctional applications, and comparative assessment with conventional gums. Journal of Biomaterials Science, Polymer Edition, 1–42. https://doi.org/10.1080/09205063.2025.2558696

Roy, D., Das, U., Bachhar, A., Barman, S., & Mandal, S. (2026). Modified gum odina as dynamic platform: From source to advanced drug delivery and tissue regeneration. Polymer Bulletin, 83(7), 395. https://doi.org/10.1007/s00289-026-06462-x

Sailaja, B., Srilakshmi, S., Vardhani, C. S., & Sravya, K. T. S. (2016). *Isolation and evaluation of seed coat constituents of Moringa oleifera.* https://www.cabidigitallibrary.org/doi/full/10.5555/20163065260

Sánchez-Machado, D. I., Núñez-Gastélum, J. A., Reyes-Moreno, C., Ramírez-Wong, B., & López-Cervantes, J. (2010). Nutritional Quality of Edible Parts of Moringa oleifera. Food Analytical Methods, 3(3), 175–180. https://doi.org/10.1007/s12161-009-9106-z

Sarangi, M. K., & Padhi, S. (2018). Novel Herbal Drug Delivery System: An Overview. Archives of Medicine and Health Sciences, 6(1), 171. https://doi.org/10.4103/amhs.amhs_88_17

Sercombe, L., Veerati, T., Moheimani, F., Wu, S. Y., Sood, A. K., & Hua, S. (2015). Advances and challenges of liposome assisted drug delivery. *Frontiers in Pharmacology*, *6*. https://doi.org/10.3389/fphar.2015.00286

Sermkaew, N., & Plyduang, T. (2020). Self-microemulsifying drug delivery systems of *Moringa oleifera* extract for enhanced dissolution of kaempferol and quercetin. *Acta Pharmaceutica*, *70*(1), 77–88. https://doi.org/10.2478/acph-2020-0012

Shady, N. H., Mostafa, N. M., Fayez, S., Abdel-Rahman, I. M., Maher, S. A., Zayed, A., Saber, E. A., Khowdiary, M. M., Elrehany, M. A., & Alzubaidi, M. A. (2022). Mechanistic wound healing and antioxidant potential of Moringa oleifera seeds extract supported by metabolic profiling, in silico network design, molecular docking, and in vivo studies. *Antioxidants*, *11*(9), 1743.

Sharma, J., Kaur, L., Kanuja, N., Nagpal, M., & Bala, R. (2013). Natural polymers-promising potential in drug delivery. *International Journal of Pharmaceutical Technology Research*, *5*, 684–699.

Shi, F., Gong, M., Adu-Frimpong, M., Jiang, X., Wang, X., Hua, Q., Li, T., Li, J., Yu, J., Toreniyazov, E., Cao, X., Wang, Q., & Xu, X. (2024). Isolation, purification of phenolic glycoside 1 from *Moringa oleifera* seeds and formulation of its liposome delivery system. *AAPS PharmSciTech*, *25*(7), 196. https://doi.org/10.1208/s12249-024-02911-w

Singh, B., & Kumar, A. (2018). Network formation of Moringa oleifera gum by radiation induced crosslinking: Evaluation of drug delivery, network parameters and biomedical properties. *International Journal of Biological Macromolecules*, *108*, 477–488.

Singh, U., Dwivedi, C., Kulsum, U., Verma, N., Saraf, S., & Pradhan, D. K. (2017). Phytochemistry and medicinal uses of moringa oleifera: An overview. *Journal of Drug Delivery and Therapeutics*, *7*(6), 104–116.

Singhal, A. K., Jarald, E. E., Showkat, A., & Daud, A. (2012). In vitro evaluation of Moringa oleifera gum for colon-specific drug delivery. *International Journal of Pharmaceutical Investigation*, *2*(1), 48.

Suryani, Zubaydah, W. O. S., Sahumena, M. H., Adawia, S., Wahyuni, R., Adjeng, A. N. T., Nisa, M., Kasmawati, H., Ihsan, S., Ruslin, & Aswan, M. (2019). *Preparation and characterization of self-nanoemulsifying drug delivery system (SNEDDS) from Moringa oleifera L.and Cassia alata L. leaves extracts. AIP Conference Proceedings*, *2175*, 070011. https://doi.org/10.1063/1.5141325

Tripathy, S., Patel, D. K., Barob, L., & Naira, S. K. (2013). A review on phytosomes, their characterization, advancement & potential for transdermal application. *Journal of Drug Delivery and Therapeutics*, *3*(3), 147–152.

Tungadi, R., Tuloli, T. S., & Pateda, S. M. (2023). Stability and antioxidant tests of ethanol extract liposome of moringa leaves (Moringa oleifera). In J. Sukweenadhi & F. Setiawan (Eds.), *Proceedings of the Conference on Natural Resources and Life Sciences 2022 (NRLS-BIO 2022)* (Vol. 38, pp. 165–169). Atlantis Press International BV. https://doi.org/10.2991/978-94-6463-322-1_22

Vergara-Jimenez, M., Almatrafi, M. M., & Fernandez, M. L. (2017). Bioactive components in Moringa oleifera leaves protect against chronic disease. *Antioxidants*, *6*(4), 91.

Vijayakumar, S., Chen, J., Sanchez, Z. I. G., Tungare, K., Bhori, M., Durán-Lara, E. F., & Anbu, P. (2023). Moringa oleifera gum capped MgO nanoparticles: Synthesis, characterization, cyto-and ecotoxicity assessment. *International Journal of Biological Macromolecules*, *233*, 123514.

Wang, X., Jiang, X., Yang, J., Adu-Frimpong, M., Gong, M., Hua, Q., Li, T., Li, J., Pan, P., Toreniyazov, E., Cao, X., Yu, J., Wang, Q., & Xu, X. (2025). Preparation of chitosan-modified Moringa A liposomes and its protective effect on acute alcoholic liver injury. *Journal of Liposome Research*, 1–17. https://doi.org/10.1080/08982104.2025.2516033

Zubair, M., Poluan, J., Hayati, F., Ramadani, A., Sulastri, E., & Ihwan. (2025). Self-nanoemulsifying drug delivery system (SNEDDS) prepared from Begonia medicinalis and Moringa oleifera ethanol extracts. *Tropical Journal of Pharmaceutical Research*, *24*(1), 1–8. https://doi.org/10.4314/tjpr.v24i1.1

19 *Moringa oleifera* Gum in Pharmaceutics

Muhammad Irfan, Farwa Nasir, Sofia Javed, and Zainab Yousaf

19.1 INTRODUCTION

Moringa oleifera gum (MG), locally known as *Gond Sohanjna* and often referred to as 'miracle tree gum', oozes from the stem and bark of the *Moringa oleifera* (MO) tree (Figure 19.1). MO has gained global attention due to the variety of phytochemicals present in its various parts, including fruiting bodies, leaves, seeds, flowers and gum. Among these, MG stands out as a promising natural biopolymer with significant potential in pharmaceutics.

Plant-derived gums, such as MG, are natural polymers composed mainly of complex polysaccharides, with minor components of proteins and minerals (Saha, Tyagi, Gupta & Tyagi, 2017). Plant gums (PGs) are usually water-soluble, insoluble in most organic solvents and translucent. When mixed with water, they absorb it and swell, producing a dense, viscous, gel-like product. These PGs are non-toxic, biodegradable and biocompatible, and such properties make them ideal for use in drug formulation and delivery (Iravani, 2020). Among PGs, MG exhibits unique physicochemical properties, including high viscosity, emulsifying ability, mucoadhesive potential and film-forming characteristics. These properties allow MG to serve as a suitable excipient (base material) in different pharmaceutical formulations.

Excipients are base materials and are inactive ingredients usually used in drug formulations to enhance the physical and chemical stability, bioavailability and acceptability of active pharmaceutical ingredients (Rayaprolu, Strawser & Anyarambhatla, 2018). MG has shown substantial utility in this regard. It has been effectively used as a disintegrant in solid dosage forms, a binder in tablet formulations and a stabiliser in liquid formulations. Additionally, MG's gel-forming potential and swelling abilities make it suitable for controlled and sustained drug-release systems. These properties enable drugs to be delivered at a desired rate, improving their therapeutic outcomes (Singh et al., 2023).

In recent years, advances in green nanotechnology have provided new opportunities for utilising MG in pharmaceuticals. The gum has been successfully used in the synthesis of green nanoparticles (NPs), where its phytochemicals act as reducing, stabilising and capping agents for metallic NP synthesis (Irfan, Munir & Ismail, 2021). Various metallic NPs, including gold, silver, copper and zinc oxide NPs, have been effectively prepared using MG. Such NPs have shown remarkable results in treating chronic diseases, such as cancer, diabetes and inflammatory disorders (Perumalsamy et al., 2024). Furthermore, phytochemicals present in MG, like other parts of this amazing tree, also reveal antioxidant, antimicrobial and anti-inflammatory properties (Badwaik et al., 2020), which enhance the therapeutic efficacy of NPs created using MG.

The pharmaceutical industry is increasingly focused on sustainable alternatives to synthetic polymers. MG offers an alternative solution that combines environmental sustainability with superior functional properties. Unlike traditional synthetic excipients used in the pharmaceutical industry, MG is completely non-toxic, biodegradable and biocompatible, aligning with the concept of green chemistry and sustainable development. Its abundant availability, ease of

DOI: 10.1201/9781003617860-22

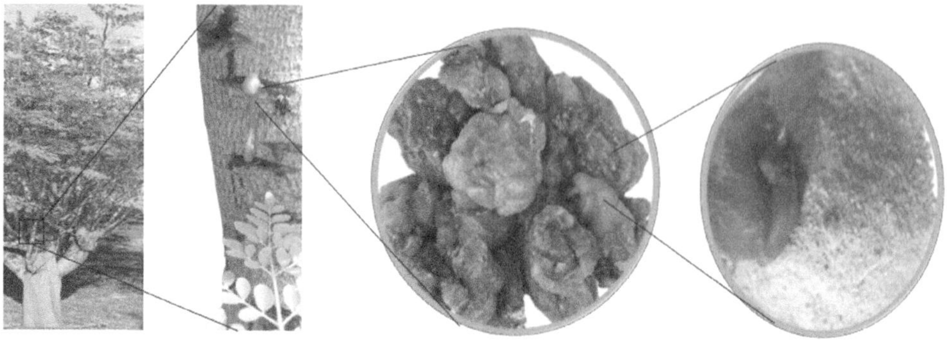

FIGURE 19.1 *Moringa oleifera* Gum. From oozing exudate on the tree to its dried powdered form.

extraction and convenient storage further increase its attractiveness to the pharmaceutical industry (Badwaik et al., 2020).

While MG has shown significant potential, challenges remain. Issues such as batch-to-batch variability in quality, stability concerns and limited scalability of extraction processes need to be addressed to ensure its widespread adoption in the pharmaceutical industry. Despite these challenges, ongoing research and development efforts are paving the way to overcome them and introduce innovative applications of this natural gum.

19.2 EXTRACTION, PROCESSING AND CHARACTERISATION

19.2.1 Methods of Extraction and Purification

MG is produced as a natural exudate by the *M. oleifera* plant and is mainly composed of glycoproteins and polysaccharides. The extraction of MG involves highly systematic approaches that maintain a high-quality yield, enabling its utilisation for various commercial purposes (Ekrami, Shakouri, Nikkhou & Emam-Djomeh, 2023). Various extraction methods are significant because they not only affect yield but also influence the structure and bioavailability of polysaccharides. Various extraction methods, involving conventional and non-conventional, have been utilised for the extraction of MG according to their end-point use and applications (K. Sharma et al., 2022).

The following sections are various conventional and non-conventional methods for the extraction of gum.

19.2.1.1 Tapping Method

This method involves extracting gum by making incisions in the tree's bark. The exudate seeps out and is collected (S. C. Sharma, Pandey, Prasad & Raj, 2024).

19.2.1.2 Mechanical Extraction

This method involves the extraction of gum by using different mechanical equipment that makes cuts in the bark. Gum is released through these ruptured parts.

19.2.1.2.1 Solvent Extraction

This method involves the use of various solvents, including alcohol or acetone. During this process, solvents cause the breakdown of complex structures. The mixture is evaporated and dried to obtain gum (Tiwari, 2021).

19.2.1.2.2 Heat Extraction

This method involves the use of heat to exude gum. During this process, a controlled temperature is essential to avoid the degradation of the gum.

19.2.1.3 Ultrasound-Assisted Extraction

This method uses ultrasound waves that cause the breakdown of plant tissues and assist in the release of gum (Chen, Zhang, Huang, Fu & Liu, 2017).

After extraction, various purification methods are used to obtain a high-quality product (K. Sharma et al., 2022). The following sections are various methods of purification.

19.2.1.4 Filtration

After the extraction of MG, the filtration process is carried out to remove various impurities such as plant debris. Various types of filters can be used to carry out the filtration process.

19.2.1.5 Decolourisation

Various bleaching agents are used to remove any colour present in the gum. These bleaching agents may include clay or activated charcoal, as these agents absorb colour compounds.

19.2.1.6 Precipitation

In order to remove water-soluble compounds from MG, various precipitating agents, such as acetone or ethanol, are added. These agents cause the precipitation of purified gum, leaving behind water-soluble impurities.

19.2.1.7 Centrifugation

In order to remove any suspended particles or impurities present in the gum, centrifugation is carried out, in which the gum is centrifuged at a very high speed. It purifies gum from other impurities using a density gradient.

19.2.1.8 Dialysis

This method involves the use of a semipermeable membrane to purify gum. It causes the separation of low molecular weight contaminants present in the gum solution.

19.2.2 Quality Control and Characterisation Techniques

In order to access various chemical, physical and mechanical properties of MG, several characterisation techniques are carried out (Dong et al., 2018). These techniques may include Fourier transform infrared (FTIR) spectroscopy, chromatography, XRD and UV-vis spectroscopy and so on, for physicochemical, molecular, structural, chemical, mechanical and biological characterisation (Cortez-Trejo, Manríquez & Mendoza, 2023). Knowledge of gum composition, including bonding patterns and various functional groups such as amino and carboxylic acid groups, is important to determine different functional properties associated with the respective gum. The following sections show the techniques that are mostly used for the characterisation of gum.

19.2.2.1 FTIR Spectroscopy

This technique uses absorption spectra to identify various functional groups present in the gum (Deshmukh & Parate, 2024). The molecular structure of the gum, along with the presence of different compounds, is also provided by FTIR spectroscopy.

19.2.2.2 Chromatography

High-performance liquid chromatography (HPLC) or gas chromatography (GC) techniques are used to identify different sugars and chemical components present in the gum, such as amino acids and polyphenols (Gaafar, Ibrahim, Asker, Moustafa & Salama, 2016).

19.2.2.3 UV-Vis Spectroscopy

This technique can be used to analyse the absorption characteristics of the gum (Vettumperumal, Dhineshbabu & Karthikeyan, 2022). It involves the identification of bioactive compounds present in the gum, such as polysaccharides, alkaloids, flavonoids and polyphenols.

19.2.2.4 Elemental Analysis

Elemental analysis (EDX) is carried out to determine the carbon, hydrogen and nitrogen content present in gum. It provides information about the elemental composition of gums.

19.2.2.5 X-Ray Diffraction

X-ray diffraction (XRD) is carried out for crude, soluble and non-soluble fractions of gum that provide information about the crystalline or amorphous properties of the gum that actually affect functional properties (Gupta, Kachhwaha, Kothari, Bohra & Jain, 2020).

19.2.2.6 Atomic Force Microscopy

At the nanoscale, the mechanical properties of gum are determined by using atomic force microscopy (AFM). It provides information about surface topography and the interaction of gum with other materials.

19.2.3 Scalability and Commercial Feasibility

MG is gaining considerable attention due to its wide range of applications in various fields, including cosmetics, food, pharmaceuticals and water purification (Sonika, Singh, Arora & Arora, 2020). For the development of a viable market, the scalability and commercial feasibility of the extraction and processing of MG are essential.

In view of scalability, MO is highly adaptable and is capable of thriving in dry regions; this makes large-scale cultivation of these plants easy across diverse geographical regions. These plants also grow at a very fast rate, along with low maintenance requirements. These characteristics enable these plants to grow on a large scale in regions with very limited resources.

In terms of commercial feasibility, the growing population's demand for organic, biocompatible and sustainable products aligns well with the potential of MG, as consumers opt for natural additives over hazardous synthetic ones. As MG holds significant promise as a commercially feasible and scalable product in various industries, extraction methods should be standardised to ensure better-quality gum production to cope with market demands.

19.3 MG: COMPOSITION AND PROPERTIES

19.3.1 Chemical and Functional Properties

MG is composed of complex carbohydrates, predominantly D-galactose, D-glucuronic acid, D-mannose, D-xylose, L-rhamnose and L-arabinose, which are present in specific ratios. These components contribute significantly to the gum's emulsifying and thickening properties (Figure 19.2). B-type linkages (1→3) and the highly branched structure of the polysaccharide of MG have been reported by Bhattacharya et al. (1982). GC-MS analysis of MG revealed the presence of protein content (7% w/w), asparagine/aspartic acid (17.5%), glycine (9.2%), lysine (9.5%) and glutamine/glutamic acid (8.9%) (Badwaik et al., 2020). The linkage pattern of the monosaccharide units of MG showed a highly branched structure of arabinogalactan with non-reducing units of arabinopyranose, galactopyranose, arabinofuranose, xylopyranose and rhamnopyranose. MG is non-toxic and environmentally friendly, contains biodegradable polymers and has the capability of water retention and hydrogel formation.

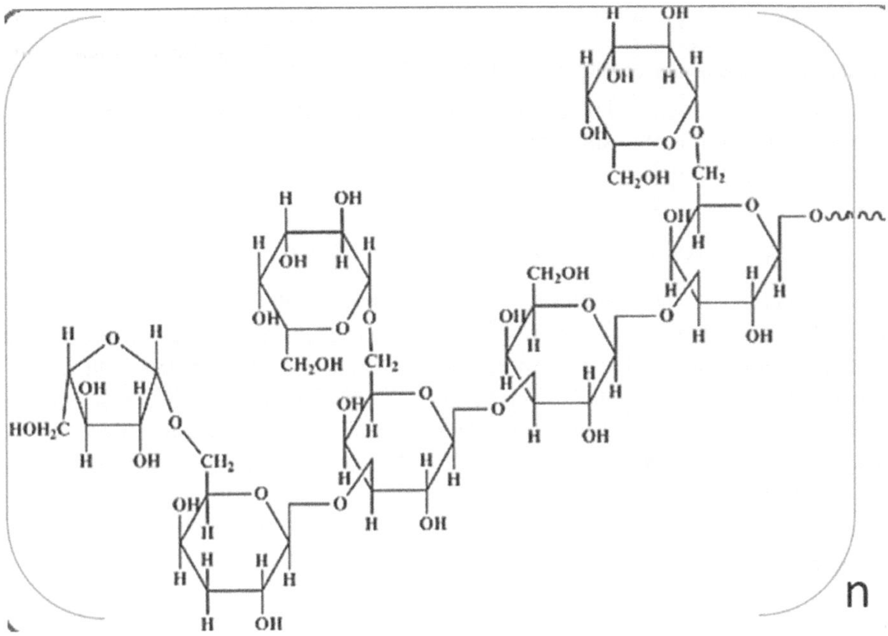

FIGURE 19.2 Chemical Structures of the Carbohydrate Constituents of *M. oleifera* Gum.

MG possesses antimicrobial properties against *Streptococcus pyogenes*, *Proteus vulgaris*, *Micrococcus kristinae*, *Staphylococcus aureus* and *S. enteritidis*, and anti-inflammatory properties due to the presence of quercetin, 4-[(α-L-rhamnosyloxy)benzyl] isothiocyanate, aurantiamide acetate, 4-[(3-O-acetyl-α-rhamnosyloxy)benzyl] isothiocyanate and 1,3-dibenzyl urea (Farooq, Rai, Tiwari, Khan & Farooq, 2012; Lee et al., 2013; Muangnoi, Chingsuwanrote, Praengamthanachoti, Svasti & Tuntipopipat, 2012; Sulaiman et al., 2008). Binding and disintegrating properties have been investigated for the preparation of paracetamol & aceclofenac tablets, respectively. Different compounds in MG, such as β-carotene, vitamin C, quercetin and phenolic acids (Satish, Kumar, Rakshith, Satish & Ahmed, 2013; Sonika et al., 2020; Sudha, Asdaq, Dhamingi & Chandrakala, 2010), contribute to antioxidant properties. Antitumour and cancer activities have also been shown by MG. Water-soluble proteins in MG make it suitable for water purification (Ramachandran, Peter & Gopalakrishnan, 1980). It has high nutritional values for magnesium, vitamins (B2, B6 and C), calcium, iron, provitamin A, proteins and carotenoids (Mehta, Shukla, Bukhariya & Charde, 2011).

19.3.2 Physical and Rheological Properties

MG is initially white in colour, and after exposure to light, it is reddish-brown and finally black in colour, while having a characteristic odour and mucilaginous taste. It is minimally soluble in cold water, and its solubility can be increased through the deacetylation of aqueous gum extract. It readily swells up and makes an adhesive solution when dissolved in water. It can behave as a liquid and a solid at the same time because of its viscoelastic property, although the pH of the solution is affected by the solubility and viscosity of the gum (high viscosity at lower shear rates).

MG can be used in tablet formulations as a release retardant due to its high swelling index (Korsmeyer et al., 1983). MG is soluble in water, making a viscous solution, while insoluble in ether, acetone and alcohol. It exhibits high stability to heat and light. The hygroscopic nature of MG affects its behaviour during storage (Deshmukh & Parate, 2024), including different physical parameters such as swelling index (3.6%), ash content (3.4%), pH (6.21±0.02), viscosity (3.6%) and loss on drying (0.47%).

Panda et al. (2013) (D. S. Panda & Ansari, 2013) examined different parameters affecting rheological properties of MG, such as temperature, sugar type and blend concentration. Solutions of MG at different concentrations show pseudoplastic and non-Newtonian rheological behaviour at all concentrations, with pseudoplastic behaviour decreasing at low concentrations. No significant thixotropy was observed in gum solutions, and increased temperature reduced the viscosity of different concentrations of gum.

19.3.3 Comparative Analysis with Other Natural Gums

MG exhibits superior emulsifying characteristics compared with gum acacia (D. S. Panda, 2014) because the MG emulsion is less sensitive to pH, temperature alterations, electrolytes and centrifugation. The monosaccharide composition of MG is related to gum acacia, and its gelling and thickening properties are similar. On the other hand, the swelling index of MG (3.4%) is lower than that of acacia, which is 5.6% (Bnuyan, Hindi, Jebur & Mahdi, 2015; Sravani et al., 2014). The solubility index of MG is comparatively lower than that of karaya gum, tragacanth gum and ghatti gum (Sarkar et al., 2018).

19.4 PHARMACEUTICAL APPLICATIONS OF MG

MG has a wide range of applications in pharmaceutics due to its unique physicochemical properties. The following sections describe some of its roles.

19.4.1 Drug Formulation

Due to the several beneficial properties of MG discussed previously, it is suitable to use as a natural excipient in pharmaceutical formulations. It is able to act as an excellent binder, stabiliser, emulsifier and so on.

19.4.1.1 As a Binder and Disintegrant

A binding agent helps to hold the drug's active pharmaceutical ingredients together. MG is a plant-derived gum and is non-toxic, biocompatible and biodegradable in nature. These properties make it safe for pharmaceutical use. The polysaccharide nature (especially D-galactose and D-glucuronic acid) of MG makes it more suitable to be used as a binder (Bhattacharya et al., 1982). The gum was used as a binding agent in the formulation of paracetamol tablets. The results show that the gum slows down the release of the drug in the gastrointestinal tract (GIT) (RS Kumar, Sundari, Kasturi & Parkavi, 2017). In another study, it was observed that the binding property of MG was as good as that of potato starch in the formulation of chloroquine phosphate tablets. This study revealed that increasing gum concentration can increase tablet hardness and disintegration time (Basawaraj, Srinivas, Kulkarni Upendra & Prakassh, 2010).

In addition to being a binder, MG can also act as a disintegrant. A disintegrant ensures that the drug breaks down efficiently after ingestion and allows the drug to be absorbed in the body. The high swelling index (19.7 mL/g) and hygroscopicity (17%) (D. S. Panda & Ansari, 2013) of MG can make it a good candidate for a disintegrant (Korsmeyer, Gurny, Doelker, Buri & Peppas, 1983). Patel and Chobey (2012) observed that MG can be effectively used as a disintegrant in drug formulation. They prepared an aceclofenac tablet with MG as a disintegrant. The disintegration time was decreased with increased gum concentration. In their study, maximum drug release was observed in the in vitro dissolution profile (V. Patel & Chobey, 2012; M. T. Patel, Patel & Upadhyay, 2012) and they concluded that MG can be a good disintegrant and binder in drug formulation.

19.4.1.2 As a Stabiliser, Thickener and Emulsifier

A stabiliser ensures the integrity of the drug over time and prevents any change in the drug's chemical composition. The viscosity (D. S. Panda & Ansari, 2013), hydrophilicity and gelling property (D. Panda, Si, Swain, Kanungo & Gupta, 2006) of MG make it a strong candidate as a stabilising agent. The hydroxyl groups of the polysaccharide chain make it water-attracting, and when dissolved in water, it makes a viscous solution (Gupta, Kachhwaha, Kothari, Bohra & Jain, 2020). Stability studies of aceclofenac tablets prepared with MG revealed that the tablets were stable and unaffected by environmental changes (Shah et al., 2011). MG is also valuable in liquid formulations. It can act as a thickening agent. It absorbs water and forms a viscous gel-like structure, which helps in improving the texture of the desired product. The viscosity of the product can be controlled by adjusting the gum concentration (Shah et al., 2011).

MG can also act as an emulsifying agent, especially in drug formulation, where APIs are insoluble in water. By stabilising the emulsion, MG improved the bioavailability of such APIs. MG possesses better emulsifying properties than acacia gum and is less sensitive to pH, centrifugation, electrolytes and temperature (D. S. Panda, 2014).

19.4.2 Controlled and Sustained Drug Release

MG is gaining recognition for its role in the formation of controlled and sustained-release formulations. The sustained release of drugs over a specific time period to maintain therapeutic efficiency while minimising the side effects is still a challenge. The swelling and gelling properties of MG make it a candidate for use in tablets, where MG can control drug release over an extended time period. A matrix tablet of indomethacin, a non-steroidal anti-inflammatory drug, was prepared by using MG as a polymer. The gum forms a gel layer around the tablet upon contact with body fluids, thereby retarding the release and diffusion of the drug. It also increases the effectiveness of the drug by localising the drug at the action site (Lakshmi & Swain, 2018). MG has also been successfully reported for the controlled delivery of amoxicillin. Amoxicillin loaded into the hydrogel matrix of MG and carboxymethyl cellulose showed controlled delivery in 24 hours compared with the amoxicillin directly loaded into a polymer matrix (Kaur, Kaur, Sharma & Jindal, 2024). As research on MG evolves, it is likely to become an important excipient for effective drug delivery systems.

19.4.3 Topical Drug Delivery and Wound Healing

MG has been investigated for its potential in topical drug delivery and wound healing. The biocompatibility, anti-inflammatory and antioxidant nature of MG make it suitable for use in topical creams and gels. The MG also possesses antimicrobial properties, which can help to prevent infection in wounds. MG has been used in the formation of wound-healing products. It provides a protective barrier over the wound and also helps in moisture retention, which is critical for the healing process (Singh & Sharma, 2021). The gum can act as a hydrophilic matrix, which controls the release of APIs of the drug at the wound site. This ensures the consistent supply of therapeutic agents to the wound. Studies also show that MG can promote tissue proliferation, which is important for wound healing (Rai, Bal & Swain, 2020). Due to these satisfactory effects, MG can be explored for use in various formulations to promote wound healing. Its chemical composition and non-toxic nature also make it suitable for use in wound and skin treatment.

19.5 MG-BASED NANOPARTICLES

Nanoparticles (NPs) can enhance drug penetration due to their small size, facilitating diffusion across biological barriers (Ihsan et al., 2024). A boost in drug solubility and protection against degradation is a result of nanoscience (Khan et al., 2022). NPs lead to the targeted delivery of drugs that reduce the chances of damaging healthy tissues in the body (Mitchell et al., 2021). Surface properties, such as charge and hydrophilicity of NPs, can be customised to maximise drug administration (Masud, Munir, Irfan & Tayyab, 2023).

Multiple green routes are opted for the synthesis of desired NPs. It is considered an eco-friendly approach that uses different plant parts, such as leaves, roots, stems, fruits, resin and seeds. $AgNO_3$ and gum powder were mixed and autoclaved at 120°C for two minutes. Centrifugation at 10,000 g for 120 min led to the formation of an NPs pellet. ZnO NPs were obtained by mixing salt and MG powder, followed by heating at 70°C and stirring for four hours at 100 rpm. The pellet was calcined after washing (Irfan et al., 2021). AuNPs can be synthesised using MG extract and an $HAuCl_4$ solution. The rapid formation of a red-brown colour within seconds indicates the occurrence of surface plasmon resonance (Tiloke, Phulukdaree, Anand, Gengan & Chuturgoon, 2016). Titanium tetra isopropoxide (TTIP) reacts with MG extract, along with uninterrupted stirring at 50°C for four hours. This led to the formation of tetragonal TiO_2 NPs having an average size of 12 nm (Perumalsamy et al., 2024). Iron oxide nanorods (FeO-NRs) were obtained from $FeCl_3$ and MG, with an average particle size of 15.01±6.03 nm. Visible colour changes were seen when NPs formed (Aisida et al., 2020). Biochemical reactions between $FeCl_3$ and MG can lead to the emergence of iron oxide nanorods (FeO-NRs) that can be observed via a visible colour change. X-ray diffraction depicts a spinel cubic lattice in the tetrahedral hematite structure of these NRs. They have an average particle size of 15.01±6.03 nm (Aisida et al., 2020).

To understand the biological, physical and chemical properties of MG-based NPs, it is essential to characterise them (Irfan, Munir & Ismail, 2022). Size and morphology of the NPs can be examined using scanning electron microscopy (SEM), which provides information by scanning the surface with the help of a focused beam of electrons, and transmission electron microscopy (TEM), which examines the internal structures using electron beams (Irfan et al., 2023). FTIR is used to study the surface properties of NPs, such as the functional group present (Figure 19.3).

NPs bring revolutionary changes to pharmaceutics, especially in drug delivery systems, due to their smart characteristics such as minute size, large surface area and the ability to be functionalised for targeted therapy. MG-based green NP minimises the risk of off-target drug delivery. This

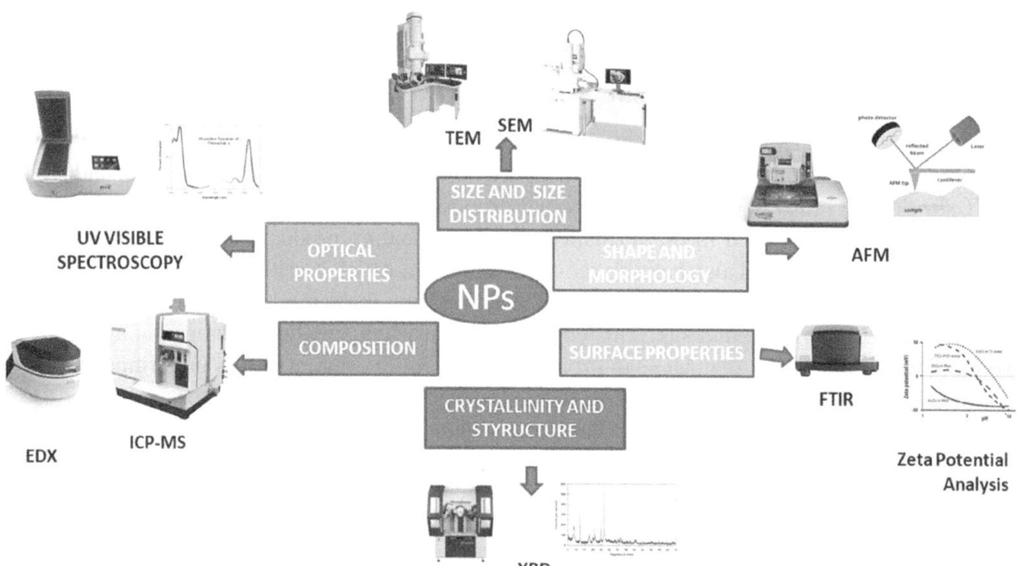

FIGURE 19.3 The Characterisation of MG-Based NPs Highlighting Key Techniques such as Size and Morphology Analysis (TEM, SEM, DLS), Surface Chemistry (FTIR, XPS), Crystallinity (XRD) and Stability (Zeta Potential).

enhances treatment efficacy and minimises the side effects. There are several ways that help us in this mechanism, such as active targeting, in which NPs work along with ligands like antibodies, peptides and aptamers. Passive targeting relies on the enhanced permeability and retention (EPR) effect, and lastly, NPs are engineered to liberate drugs in response to specific triggers, including both internal and external factors (Ashique, Sandhu, Chawla & Chawla, 2021). NPs are used in different modes like liposomes, polymeric NPs, AuNPs, magnetic NPs, dendrimers and silica nanoparticles.

19.6 THE PHARMACOLOGICAL AND THERAPEUTIC POTENTIAL OF MG-BASED NPS

19.6.1 Antimicrobial and Antioxidant Properties

MG also contains some antioxidants, flavonoids, tannins, phenolic acids and steroids. These compounds exhibit antimicrobial and antioxidant potential. MG-based zinc oxide and silver NPs show good antibacterial potential against Gram-negative (*Escherichia coli*), Gram-positive (*S. aureus*) and methicillin-resistant *S. aureus* (Irfan et al., 2021). Immobilised AgNPs synthesised from MG showed suitable antibacterial potential against different bacteria like *Bacillus subtilis*, *E. coli*, *Klebsiella pneumoniae*, *Proteus mirabilis*, *Pseudomonas aeruginosa* and *Salmonella typhi*, as well as exhibiting appropriate antioxidant activity (Ali et al., 2024). MG-based binary nanostructured TiO_2-ZrO_2 composites exhibit antibacterial activity against *E. coli* and *S. aureus* (Korde, Thombre, Sangshetti, Rajbhoj & Gaikwad, 2025).

Antifungal activity was observed by MG-based AgNPs on a fungal pathogen from coconut palm leaves (Jenish, Ranjani & Hemalatha, 2022). *Candida albicans* biofilm growth was inhibited through AgNPs prepared using MG (Imelda et al., 2024). A functional copolymeric hydrogel network derived from MG contributes to its antioxidant properties. When this gel is mixed with tetracycline, antibacterial activity against *S. aureus*, *E. coli* and *P. aeruginosa* is observed (Rajesh Kumar & Singh, 2024). CoAgxFe2–xO4 nanohybrids synthesised using MG were evaluated for their noticeable antibacterial activity against Gram-positive (*B. subtilis* and *S. aureus*) and Gram-negative (*P. aeruginosa* and *E. coli*) bacteria, as well as for their antioxidant (DPPH) and antibiofilm (*E. coli*) activities (Tamboli, Bushnak, Patange, Zakde & Jadhav, 2025).

19.6.2 Anti-Inflammatory and Wound-Healing Properties

MG-based hydrogel dressings have been designed and evaluated for cutaneous wound healing (Iqbal, Chatha, Chauhdary, Ijaz Hussain & Khan, 2023). Using 'green' methods, MG-based NPs have been synthesised and exhibit potential anti-inflammatory properties. However, these NPs shrink the expression of Pc, IκBα and p-IκBα by regulating the NFkB pathway. The decline of these proteins leads to the repression of pro-inflammatory cytokines (Perumalsamy et al., 2024). Due to the high swelling index of the gum, hydrogels and gum-based wound dressings can be designed for better results.

19.6.3 Potential in Cancer Therapy and Other Diseases

Green (MG-based) NPs are being explored to silence oncogenes or restore tumour suppressor genes by delivering genetic material. NPs have the potential to deliver antigens that could boost the immune system. AuNPs obtained from MG have proven effective and lifesaving in cancer treatment due to their antiproliferative potential (Tiloke, Anand, Gengan, Chuturgoon & Pharmacotherapy, 2018). AgNPs synthesised from MG have also shown anticolon cancer activities (Althomali et al., 2022).

19.7 CHALLENGES AND LIMITATIONS

19.7.1 Stability and Compatibility Issues in Formulations

The lower solubility of MG in certain solvents, the need for extensive purification and the loss of functional properties over time due to environmental factors or microbial activity are some of the limiting factors for its use in pharmaceuticals. Additionally, the gum may not be compatible with certain APIs, further restricting its applicability. Scaling up gum-based pharmaceuticals and maintaining consistency in batch-to-batch purification of the gum can pose additional challenges. Regarding MG-based NPs, over time, they may aggregate, leading to various changes in their shape, size and surface properties. The stability of MG-based NPs is also affected by changes in pH and temperature. The stability of such NPs is also affected by oxidation as the polymeric structure of the gum undergoes oxidation. The interactions of NPs with other excipients, such as co-solvents and surfactants, can also affect their stability.

19.7.2 Regulatory and Safety Considerations

Evolution in regulatory and safety considerations has been observed along with advancements in pharmaceuticals and nanobiotechnology. The gum must achieve generally recognised as safe (GRAS) status or equivalent recognition to be used in pharmaceuticals. Allergenic potential, toxicity, stability and microbial contamination can be limiting factors when selecting MG for pharmaceutical applications. Detailed characterisation of gums and gum-based NPs, including size, shape, composition and surface charge, should be known by the regulatory bodies. Quality and reproducibility of NPs should be considered. Preclinical trials can help us to understand the cytotoxicity, genotoxicity and pharmacokinetics. There is room to know about the possible environmental implications of NP disposal. Continuous checks on NPs are needed to understand the long-lasting and rare side effects.

19.8 CONCLUSION

MG is a natural biopolymer and a potential candidate in the pharmaceutical industry due to its diverse phytochemical profile, biocompatibility, physicochemical properties and biodegradability. From serving as a binder, disintegrant and stabiliser in drug formulations to facilitating controlled and sustained drug release, MG demonstrates versatility and environmental sustainability. Its role in green nanobiotechnology, particularly in synthesising metallic NPs, has opened new horizons for its application in targeted drug delivery and antimicrobial therapies. Additionally, MG-based hydrogels and wound dressings have shown promising results in wound healing and topical drug delivery.

Despite these advances, challenges such as quality consistency, large-scale production and regulatory issues must be addressed for the broader application of MG in pharmaceutics. Optimised extraction and purification processes, along with comprehensive safety evaluations, will be critical to unlocking the full potential of MG. In conclusion, MG represents a promising, sustainable alternative to synthetic excipients and polymers, aligning with the principles of green chemistry and sustainable development.

REFERENCES

Aisida, S. O., Madubuonu, N., Alnasir, M. H., Ahmad, I., Botha, S., Maaza, M., & Ezema, F. I. J. A. N. (2020). Biogenic synthesis of iron oxide nanorods using Moringa oleifera leaf extract for antibacterial applications. *Applied Nanoscience*, *10*, 305–315.

Ali, L., Ahmad, N., Uddin, M. N., Abdel-Maksoud, M. A., Fazal, H., Fatima, S., . . . Rahat, M. A. (2024). Immobilization of silver nanoparticles with defensive gum of Moringa oleifera for antibacterial efficacy against resistant bacterial species from human infections. *Pharmaceuticals*, *17*(11), 1546.

Althomali, A., Daghestani, M. H., Basil Almukaynizi, F., Al-Zahrani, S. A., Awad, M. A., Merghani, N. M., . . . Synthesis. (2022). Anti-colon cancer activities of green-synthesized Moringa oleifera–AgNPs against human colon cancer cells. *Journal of Nanostructure in Chemistry, 11*(1), 545–554.

Ashique, S., Sandhu, N. K., Chawla, V., & Chawla, P. A. J. C. d. d. (2021). Targeted drug delivery: trends and perspectives. *Current Drug Delivery, 18*(10), 1435–1455.

Badwaik, H. R., Al Hoque, A., Kumari, L., Sakure, K., Baghel, M., & Giri, T. K. (2020). Moringa gum and its modified form as a potential green polymer used in biomedical field. *Carbohydrate Polymers, 249*, 116893.

Basawaraj, S., Srinivas, R., Kulkarni Upendra, K. U., & Prakassh, G. (2010). Evaluation of Moringa oleifera gum as a binder in tablet formulation. *International Journal of Pharmacy and Pharmaceutical Sciences, 2*(3), 123–127.

Bhattacharya, S. B., Das, A. K., & Banerji, N. (1982). Chemical investigations on the gum exudate from sajna (Moringa oleifera). *Carbohydrate Research, 102*(1), 253–262.

Bnuyan, I., Hindi, N., Jebur, M., & Mahdi, M. (2015). In vitro antimicrobial activity of gum Arabic (Al Manna and Tayebat) prebiotics against infectious pathogens. *IJPPR Human, 3*(3), 77–85.

Chen, C., Zhang, B., Huang, Q., Fu, X., & Liu, R. H. (2017). Microwave-assisted extraction of polysaccharides from Moringa oleifera Lam. leaves: Characterization and hypoglycemic activity. *Industrial Crops and Products, 100*, 1–11.

Cortez-Trejo, M., Manríquez, J., & Mendoza, S. (2023). Characterization of natural gums: Emphasizing distinctive spectroscopic techniques. In *Natural gums* (pp. 123–161). Elsevier.

Deshmukh, S. H., & Parate, V. R. (2024). Comparative physicochemical and FTIR spectroscopic studies of exudate and purified forms of Moringa oleifera gum and Jhingan gum. *Chemical Papers*, **78**, 4857–4865 .

Dong, Z., Li, C., Huang, Q., Zhang, B., Fu, X., & Liu, R. H. (2018). Characterization of a novel polysaccharide from the leaves of Moringa oleifera and its immunostimulatory activity. *Journal of Functional Foods, 49*, 391–400.

Ekrami, M., Shakouri, M., Nikkhou, S., & Emam-Djomeh, Z. (2023). Extraction and physicochemical characterization of gum. In *Handbook of Natural polymers* (Vol. 1, pp. 597–630): Elsevier.

Farooq, F., Rai, M., Tiwari, A., Khan, A. A., & Farooq, S. (2012). Medicinal properties of Moringa oleifera: An overview of promising healer. *Journal of Medicinal Plants Research, 6*(27), 4368–4374.

Gaafar, A. A., Ibrahim, E. A., Asker, M. S., Moustafa, A. F., & Salama, Z. A. (2016). Characterization of polyphenols, polysaccharides by HPLC and their antioxidant, antimicrobial and antiinflammatory activities of defatted Moringa (Moringa oleifera L.) meal extract. *International Journal of Pharmaceutical and Clinical Research, 8*(6), 565–573.

Gupta, S., Kachhwaha, S., Kothari, S., Bohra, M. K., & Jain, R. (2020). Surface morphology and physicochemical characterization of thermostable moringa gum: a potential pharmaceutical excipient. *ACS Omega, 5*(45), 29189–29198.

Ihsan, S., Munir, H., Meng, Z., Tayyab, M., Zeeshan, N., Rehman, A., . . . Irfan, M. (2024). Tragacanth gum-based copper oxide nanoparticles: Comprehensive characterization, antibiofilm, antimicrobial and pho tocatalytic potentials. *International Journal of Biological Macromolecules, 268*, 131600.

Imelda, I., Astuty, S. D., Gareso, P. L., Dwyana, Z., Arifin, N. F., & Astuti, S. D. (2024). Green synthesis of silver nanoparticles using Moringa oleifera: Implementation to photoantimicrobial of Candida albicans with LED Light. *Trends in Sciences, 21*(9), 8032–8032.

Iravani, S. (2020). Plant gums for sustainable and eco-friendly synthesis of nanoparticles: Recent advances. *Inorganic and Nano-Metal Chemistry, 50*(6), 469–488.

Irfan, M., Bagherpour, S., Munir, H., Perez-Garcia, L., Abelha, T. F., Afroz, A., . . . Rashid, U. (2023). GC–MS metabolomics profile of methanol extract of Acacia modesta gum and gum-assisted fabrication and characterization of gold nanoparticles through green synthesis approach. *International Journal of Biological Macromolecules, 252*, 126215.

Irfan, M., Munir, H., & Ismail, H. (2021). Moringa oleifera gum based silver and zinc oxide nanoparticles: Green synthesis, characterization and their antibacterial potential against MRSA. *Biomaterials Research, 25*(1), 17.

Irfan, M., Munir, H., & Ismail, H. (2022). Characterization and fabrication of zinc oxide nanoparticles by gum Acacia modesta through green chemistry and impregnation on surgical sutures to boost up the wound healing process. *International Journal of Biological Macromolecules, 204*, 466–475.

Iqbal, Y., Chatha, S. A. S., Chauhdary, Z., Ijaz Hussain, A., & Khan, I. U. (2023). Design and evaluation of moringa gum-based hydrogel dressings for cutaneous wound healing. *Journal of Bioactive and Compatible Polymers, 38*(6), 417–436.

Jenish, A., Ranjani, S., & Hemalatha, S. (2022). Moringa oleifera nanoparticles demonstrate antifungal activity against plant pathogenic fungi. *Applied Biochemistry and Biotechnology, 194*(10), 4959–4970.

Kaur, K., Kaur, L., Sharma, M., & Jindal, R. (2024). A comparison of controlled release from cyclodextrin mediated and non-mediated hydrogel matrices of Moringa Oleifera gum and carboxymethyl cellulose. *Polymer Bulletin*, 81, 12841–12865 .

Khan, K. U., Minhas, M. U., Badshah, S. F., Suhail, M., Ahmad, A., & Ijaz, S. J. L. S. (2022). Overview of nanoparticulate strategies for solubility enhancement of poorly soluble drugs. *Life Sciences, 291*, 120301.

Korde, S. A., Thombre, P. B., Sangshetti, J. N., Rajbhoj, A. S., & Gaikwad, S. T. (2025). Green synthesis of binary nanostructured TiO2-ZrO2 composites using moringa oleifera gum: Photocatalytic and antimicrobial applications. *Journal of Molecular Structure, 1323*, 140725.

Korsmeyer, R. W., Gurny, R., Doelker, E., Buri, P., & Peppas, N. A. (1983). Mechanisms of solute release from porous hydrophilic polymers. *International Journal of Pharmaceutics, 15*(1), 25–35.

Kumar, R., & Singh, B. (2024). Functional network copolymeric hydrogels derived from moringa gum: Physiochemical, drug delivery and biomedical properties. *International Journal of Biological Macromolecules, 275*, 133352.

Kumar, R., Sundari, A. S., Kasturi, M., & Parkavi, P. (2017). Application of Moringa oleifera and Terminalia Catappa gum as drug binder. *World Journal of Pharmaceutical Research, 6*, 546–549.

Lakshmi, S. V., & Swain, R. (2018). Formulation and characterization of sustained release matrix tablets of indomethacin using Moringa oleifera gum. *IOSR Journal of Pharmacy, 8*(4), 57–65.

Lee, H.-J., Jeong, Y.-J., Lee, T.-S., Park, Y.-Y., Chae, W.-G., Chung, I.-K., . . . Kim, W.-J. (2013). Moringa fruit inhibits LPS-induced NO/iNOS expression through suppressing the NF-κ B activation in RAW264. 7 cells. *The American Journal of Chinese Medicine, 41*(05), 1109–1123.

Masud, S., Munir, H., Irfan, M., & Tayyab, M. (2023). Allium cepa-based zinc oxide nanoparticles: Characterization and biochemical potentials. *Bioinspired, Biomimetic and Nanobiomaterials, 12*(2), 82–93.

Mehta, J., Shukla, A., Bukhariya, V., & Charde, R. (2011). The magic remedy of Moringa oleifera: An overview. *International Journal of Biomedical and Advance Research, 2*(6), 215–227.

Mitchell, M. J., Billingsley, M. M., Haley, R. M., Wechsler, M. E., Peppas, N. A., & Langer, R. (2021). Engineering precision nanoparticles for drug delivery. *Nature Reviews Drug Discovery, 20*(2), 101–124.

Muangnoi, C., Chingsuwanrote, P., Praengamthanachoti, P., Svasti, S., & Tuntipopipat, S. (2012). Moringa oleifera pod inhibits inflammatory mediator production by lipopolysaccharide-stimulated RAW 264.7 murine macrophage cell lines. *Inflammation, 35*, 445–455.

Panda, D., Si, S., Swain, S., Kanungo, S., & Gupta, R. (2006). Preparation and evaluation of gels from gum of Moringa oleifera. *Indian Journal of Pharmaceutical Sciences, 68*(6), 777.

Panda, D. S. (2014). Studies on gum of Moringa oleifera for its emulsifying properties. *Journal of Pharmacy and Bioallied Sciences, 6*(2), 92–96.

Panda, D. S., & Ansari, S. A. (2013). Preformulation study on the gum of Moringa oleifera. *Malaysian Journal of Pharmaceutical Sciences, 11*(2), 41.

Patel, M. T., Patel, J. K., & Upadhyay, U. M. (2012). Assessment of various pharmaceutical excipient properties of natural Moringa oleifera gum [Mucoadhesion, disintegration, binder]. *International Journal of Pharmacy & Life Sciences, 3*(7), 2045–2052.

Patel, V., & Chobey, N. (2012). Evaluation of MoringaOleifera gum as tablet disintegrant. *International Journal of Pharmacy and Pharmaceutical Sciences, 4*(1), 210–214.

Perumalsamy, H., Balusamy, S. R., Sukweenadhi, J., Nag, S., MubarakAli, D., El-Agamy Farh, M., . . . Rahimi, S. (2024). A comprehensive review on Moringa oleifera nanoparticles: importance of polyphenols in nanoparticle synthesis, nanoparticle efficacy and their applications. *Journal of Nanobiotechnology, 22*(1), 71.

Rai, N., Bal, T., & Swain, S. (2020). In vitro evaluations of biodegradable polyacrylamide grafted moringa bark gum graft copolymer (MOG-g-PAAM) as biomedical and controlled drug delivery device synthesized by microwave accelerated free radical synthesis. *Indian Journal of Pharmaceutical Education and Research, 54*(2), 385–396.

Ramachandran, C., Peter, K., & Gopalakrishnan, P. (1980). Drumstick (Moringa oleifera): A multipurpose Indian vegetable. *Economic Botany, 34*(3), 276–283.

Rayaprolu, B. M., Strawser, J. J., & Anyarambhatla, G. (2018). Excipients in parenteral formulations: selection considerations and effective utilization with small molecules and biologics. *Drug Development and Industrial Pharmacy, 44*(10), 1565–1571.

Saha, A., Tyagi, S., Gupta, R. K., & Tyagi, Y. K. (2017). Natural gums of plant origin as edible coatings for food industry applications. *Critical Reviews in Biotechnology, 37*(8), 959–973.
Sarkar, P., Sahu, U., Binsi, P., Nayak, N., Ninan, G., & Ravishanker, C. (2018). Studies on physico-chemical and functional properties of some natural Indian gums. *Asian Journal of Dairy and Food Research, 37*(2), 126–131.
Satish, A., Kumar, R. P., Rakshith, D., Satish, S., & Ahmed, F. (2013). Antimutagenic and antioxidant activity of Ficus benghalensis stem bark and Moringa oleifera root extract. *International Journal of Chemical and Analytical Science, 4*(2), 45–48.
Shah, D. P., Jain, V. C., Daldavi, H. P., Ramani, V. D., Patel, K. G., Saralai, M. G., & Jani, G. K. (2011). A preliminary investigation of Moringa oleifera Lam gum as a pharmaceutical excipient. *International Journal of Pharmacy Research & Technology (IJPRT), 1*(1), 12–16.
Sharma, K., Kumar, M., Waghmare, R., Suhag, R., Gupta, O. P., Lorenzo, J. M., . . . Thappa, C. (2022). Moringa (Moringa oleifera Lam.) polysaccharides: Extraction, characterization, bioactivities, and industrial application. *International Journal of Biological Macromolecules, 209*, 763–778.
Sharma, S. C., Pandey, S. K., Prasad, N., & Raj, H. (2024). Unveiling the potential of Moringa oleifera Lam. gums: An investigation into optimal tapping strategies. *Industrial Crops and Products, 214*, 118511.
Singh, B., & Sharma, V. (2021). Designing moringa gum-sterculia gum-polyacrylamide hydrogel wound dressings for drug delivery applications. *Carbohydrate Polymer Technologies and Applications, 2*, 100062.
Singh, B., Singh, J., Sharma, V., Sharma, P., & Kumar, R. (2023). Functionalization of bioactive moringa gum for designing hydrogel wound dressings. *Hybrid Advances, 4*, 100096.
Sonika, S. D., Singh, T. G., Arora, G., & Arora, S. (2020). Moringa gum: A comprehensive review on its physicochemical and functional properties. *Plant Archives, 20*, 3794–3805.
Sravani, P., Kiranmayee, Y., Narasimha, M., Reddy, V., Asha, S., & Kumar, R. B. (2014). In-vitro experimental studies on selected natural gums and resins for their antimicrobial activity. *International Journal of Pharmaceutical Research & Allied Sciences, 3*(1), 100–107.
Sudha, P., Asdaq, S., Dhamingi, S. S., & Chandrakala, G. K. (2010). Immunomodulatory activity of methanolic leaf extract of Moringa oleifera in animals. *Indian Journal of Physiology and Pharmacology, 54*(2), 133–140.
Sulaiman, M. R., Zakaria, Z., Bujarimin, A., Somchit, M., Israf, D., & Moin, S. (2008). Evaluation of Moringa oleifera aqueous extract for antinociceptive and anti-inflammatory activities in animal models. *Pharmaceutical Biology, 46*(12), 838–845.
Tamboli, Q. Y., Bushnak, I., Patange, S. M., Zakde, K. R., & Jadhav, K. (2025). Effect of Ag on dielectric, elastic, electrical impedance, and Raman spectroscopic properties of CoFe2O4. *Journal of Sol-Gel Science and Technology*, 1–10.
Tiloke, C., Anand, K., Gengan, R. M., Chuturgoon, A. A. J. B., & Pharmacotherapy. (2018). Moringa oleifera and their phytonanoparticles: Potential antiproliferative agents against cancer. *Biomedicine & Pharmacotherapy, 108*, 457–466.
Tiloke, C., Phulukdaree, A., Anand, K., Gengan, R. M., & Chuturgoon, A. A. (2016). Moringa oleifera gold nanoparticles modulate oncogenes, tumor suppressor genes, and caspase-9 splice variants in a549 cells. *Journal of Cellular Biochemistry, 117*(10), 2302–2314.
Tiwari, P. (2021). *Extraction and characterization of gum from moringa tree (moringa oleifera).* Indira gandhi krishi vishwavidyalaya raipur (chhattisgarh).
Vettumperumal, R., Dhineshbabu, N., & Karthikeyan, B. (2022). Material characterizations of Moringa oleifera gum (MOG). *Phosphorus, Sulfur, and Silicon and the Related Elements, 197*(3), 231–236.

Part 4

Crop Production and Agriculture Practices

20 *Moringa oleifera* Harvesting, Handling and Storage Practices

Danilo L. Dantas, Alécia R. A. Silva, Ana R. N. Campos, Renato A. C. Santana, and Juliano C. R. Freitas

20.1 INTRODUCTION

Moringa oleifera Lam. (*M. oleifera*) is one of the best-known and most widely used non-conventional food plants (UFPs) worldwide due to its versatility in various applications, including medicine, human and animal supplementation, technology and personal hygiene (such as water clarification and biodiesel production) (Milião et al., 2022). Morphologically, *M. oleifera* is classified as a perennial, arboreal plant with a slender trunk, compound leaves, numerous flowers (which appear in inflorescences) and long, pod-shaped fruits containing several seeds (Rodrigues et al., 2016; Pareek et al., 2023).

This plant is native to the higher regions of the Himalayas, in northwestern India (Gandji et al., 2020) and belongs to the Moringaceae family, which is classified within the order Brassicales (Ravi, Nair and Siril, 2021; Liang et al., 2025). Among the 13 species belonging to the Moringaceae family, *M. oleifera* is the most studied. Its first botanical description was made in 1785 by the French naturalist Jean-Baptiste Lamarck, who described it as a tree with thin trunks and dense foliage (Ghimire et al., 2021).

M. oleifera exhibits rapid growth, reaching a height of 5–12 m. Its trunk is cylindrical, smooth, greyish in colour and has a high regenerative capacity when cut. The leaves are compound, pinnate and dark green. The small flowers are white or beige, with a slightly sweet aroma and, depending on climatic conditions, they can bloom throughout the year (Padayachee and Baijnath, 2020; Queiroga and Figueiredo, 2022).

M. oleifera has long, pod-shaped fruits with a triangular cross-section, measuring between 20 and 60 cm in length. Inside the pods, there are normally between 12 and 21 seeds, wrapped in a thin and translucent film known as 'wings', which aid in wind dispersal. The seeds contain an oil endosperm, which is used in high-quality vegetable oil extraction. In addition, the plant exhibits a deep root system, giving it resistance to drought (with an annual rainfall of 250 mm) and high adaptability to distinct soil types, including low-fertility soils. These characteristics give *M. oleifera* both scientific and industrial significance (Kumar et al., 2017).

Regarding planting, the same can happen with seeds, cuttings and/or grafts (Ravi, Nair and Siril, 2021), with an average growth rate of 1–3 m per year (Pina et al., 2018). Abdelwanis, Abdelaty and Saleh (2024) reported that sowing can achieve an 80–90% germination rate, with an average germination time of nine to ten days.

M. oleifera exhibits a high concentration of macro- and micronutrients in its leaves, seeds, flowers and stem. The seeds and leaves, in particular, are considered the most nutritious parts. Notably, the leaves are especially rich in vitamins, minerals, amino acids and other essential compounds necessary for human nutrition (Saucedo-Pompa et al., 2018; Hodas, Rosa and Milani, 2021), being commonly used for food fortification (Khan et al., 2023; Sibanda, Jideani and Obilana, 2024) or consumed in powdered form.

DOI: 10.1201/9781003617860-24

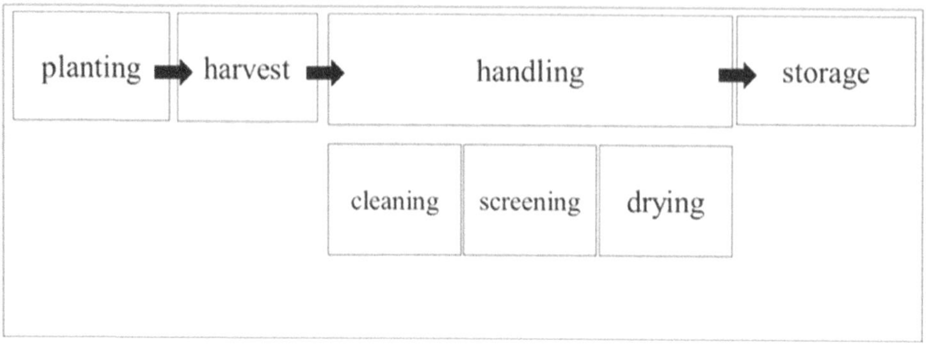

FIGURE 20.1 *M. oleifera* Processing Steps.

According to Prajapati et al. (2022), *M. oleifera* has been studied for its antitumour, anti-inflammatory, antiulcer, antipyretic, antiepileptic, antispasmodic, diuretic, antihypertensive, antidiabetic, cholesterol-lowering, cytoprotective and hepatoprotective properties.

M. oleifera can be used in various ways, depending on its purpose. The leaves *in natura* can be used for infusion, producing teas with diverse biologically active compounds or added to salads (Padayachee and Baijnath, 2020). However, dried leaves are preferred for use due to the benefits of the drying process, such as increasing the shelf life of the biological material and the concentration of essential nutrients, ensuring medicinal, food and industrial use with high efficiency.

In view of *M. oleifera*'s importance in the scientific and industrial contexts, this chapter describes a review of practices related to its planting, harvesting, handling, drying and storage, as shown in Figure 20.1.

20.2 PLANTING AND HARVESTING *M. OLEIFERA*

M. oleifera is a multifunctional plant, cultivated in different parts of the world, especially in tropical and subtropical regions. However, its adaptability is greatest in semi-arid areas. Proper planting management is essential for maximising production and preserving its properties, with different planting techniques being explored, such as direct sowing, propagation by cuttings and grafting (Noguera-Talavera, Reyes-Sánchez and Mendieta-Araica, 2017; Núñez-Gastélum et al., 2023).

Direct sowing is the most common planting technique, especially in extensive systems, due to its simplicity and low cost. Seeds should be sown in cavities 2–3 cm deep, in well-drained soils rich in organic material, which allows the development of a deep root system, giving it resistance to drought conditions (Abdelwanis, Abdelaty and Saleh, 2024). In addition, it is recommended to pre-soak the seeds for 24 hours, as this facilitates initial hydration, accelerates germination and increases the germination rate (Viviana et al., 2022).

It is important to note that *M. oleifera* grows in distinct soil types; nevertheless, in marginal soils, it is necessary to apply macronutrients, such as nitrogen and potassium, as they are essential for development, particularly during the early growth stages (Abdelwanis, Abdelaty and Saleh, 2024). According to Abdelwanis et al. (2022), the development of *M. oleifera* cultivated by sowing is affected when grown under high water demand, leading to reduced growth and compromising fruit quality.

Rodrigues et al. (2016) evaluated the development of *M. oleifera* seedlings, cultivated from seeds in different substrates (green coconut fibre, organic matter and vermicomposts). The results of this research showed that in all tested conditions, the seedlings showed satisfactory development, which was attributed to the species' high adaptive capacity, even under adverse conditions. However, Rodrigues et al. (2016) found better development for the *M. oleifera* seedlings using a substrate composed of 100% organic material, likely because this substrate offers greater protection to the plant's root system.

Propagation by cuttings is widely used in agroforestry systems and in intensive plantings for leaf production. This method utilises segments of woody branches, 1–1.5 m long, with a diameter of 4–5 cm. The cuttings should be inserted directly into the soil or in appropriate substrates, and they must be kept under appropriate humidity conditions to stimulate rooting (Queiroga and Figueiredo, 2022).

Planting using this method ensures greater uniformity in plant development and anticipated harvesting compared with planting by seed. In this context, Prajapati et al. (2022) highlighted in their study that *M. oleifera* cuttings should be obtained from the stem, preferably during spring and summer, as these warmer periods favour better root establishment, while mature tree cuttings should be planted in rainy seasons.

Grafting, although less common, has been gaining popularity, as it acts as an alternative to the two prior techniques and improves *M. oleifera*'s quality and productivity, especially in commercial systems. This technique combines resistant rootstocks to adverse conditions with crown varieties that present a higher productive potential or nutritional quality of leaves and fruits. Additionally, grafting simplifies pruning and disease control, as the rootstocks can provide greater plant vigour and resistance (Chilukamarri, Ashrafzadeh and Leung, 2021; Viviana et al., 2022).

M. oleifera harvesting involves a set of techniques aimed at the precise removal of its useful parts, such as leaves, seeds and pods, considering the optimal time to preserve the integrity of the plant and maximise yield. For this reason, harvesting requires strategies that take into consideration the plant's development stage and the final purpose of cultivation. Leaf harvesting is carried out in a recurring and selective manner, ensuring the sustainability of the reproductive system and the plant's regenerative potential. For seeds and pods, the ideal moment is defined by physiological maturity, when biologically active compounds and the viability of genetic material are at their peak. Efficiency in harvesting not only guarantees a high-quality product but also reduces planting costs (Krah et al., 2021).

Specifically, leaf harvesting can be carried out throughout the year, at intervals of 30 to 45 days, depending on the climatic conditions and nutritional management. Harvesting should preferably be done manually or using simple tools, cutting the leaves 30 to 50 cm from the base, to promote plant regeneration. This type of selective harvesting ensures the sustainability of cultivation (Queiroga and Figueiredo, 2022).

Seed harvesting occurs when the pods reach physiological maturity, identified by the colour change from green to light brown. In contrast to leaf harvesting, the pods can be manually collected or collected using mechanical techniques, depending on the scale of cultivation. After collection, the seeds should be separated and dried in a shaded, well-ventilated area to maintain viability and preserve quality (Abdoun et al., 2022).

Intercropping *M. oleifera* with other crops has been studied in order to increase land use efficiency, reduce the risks associated with monoculture and minimise pest incidence. This practice allows for maximising the agricultural area and production diversification, while also improving soil fertility due to moringa's ability to fix nitrogen when associated with leguminous planting (Motis et al., 2017).

The introduction of agroecological practices in *M. oleifera* planting, such as the use of biofertilisers and integrated pest management, contributes positively to sustainability and production maximisation. The application of mulch helps keep the soil moist, reduces erosion and promotes microbial activity, which inevitably benefits plant growth. Combined, these agroecological practices enhance the efficiency of natural resource use and have a low environmental impact, reinforcing *M. oleifera*'s potential as a sustainable and highly productive crop (Chanthanousone et al., 2023). Haouvang et al. (2017) emphasised the benefits of using cow manure in association with soil for *M. oleifera* cultivation. In their study, Haouvang et al. (2017) observed significant improvements in plant growth, including increased stem diameter and enhanced leaf development when compared with non-fertilised plants.

20.3 HANDLING *M. OLEIFERA*

Handling vegetable species consists of all practices performed after harvesting to preserve their quality, nutritional value and functional properties in useful parts, such as leaves, seeds and pods. These practices involve initial cleaning and storage, which help ensure the integrity of the plant's biologically active compounds, such as vitamins, proteins and antioxidants, and also influence its commercial viability. The efficiency of handling depends on strategies that minimise losses during the process and ensure stability of plant material (EMBRAPA, 2015).

The initial stage of handling involves cleaning, which aims to remove dirt, residues and contaminants from the freshly harvested material. The fresh leaves should be washed with clean or sanitised water (such as sodium hypochlorite solution), followed by drainage on sterile surfaces to avoid microbiological contamination. Seeds should be cleaned manually or using specific sieves, removing the residues before the processing or storage (UNIRIO, 2021).

Sorting is the next stage, aimed at separating damaged parts or those that are improper for the intended use. For example, in the case of leaves, it is recommended to remove yellowed, wilted samples or those showing signs of pests and diseases, which helps improve the final product quality. Seed sorting is based on size, shape and integrity, characteristics that directly impact oil extraction efficiency or germination viability (EMBRAPA, 2021).

Drying is the third stage and one of the most critical in handling, as it directly influences the stability and durability of *M. oleifera*-derived products. Research on moringa leaves shows that drying in shaded, ventilated areas is the best practice to preserve heat-sensitive compounds, such as vitamin C and polyphenols. In the case of seeds, it is recommended to dry them in the sun or specific dryers, as long as the temperature is controlled to avoid thermal degradation or losses in essential oil composition (Alakali, Kucha and Rabiu, 2015; Queiroga and Figueiredo, 2022).

Drying is an ancient technique used to preserve food, dating back to the first hominids (Igual, Purificación and Javier, 2020). It is considered a complex process that involves the simultaneous transfer of heat and mass, aiming to reduce the biological volume of the material (Kilic et al., 2024). Drying removes water that is distributed in a heterogeneous way on the plant material, impacting the final product's durability (Arslan and Alibaş, 2024). According to Kilic et al. (2024), the drying process, especially for fruits and vegetables, increases the product's durability.

Radojčin et al. (2021) pointed out numerous benefits, such as preserving the nutritional value of food, increasing the relative concentration of nutrients, extending shelf life and minimising packaging, handling and transportation costs.

Generally speaking, food drying methods can be divided into two types: natural and artificial methods. The natural drying method, specifically shade drying, is highlighted for its simplicity and low cost. Shade drying preserves most heat-sensitive biologically active compounds (e.g. vitamin C and polyphenols), provided that the material is laid out in thin layers in a ventilated area. However, its efficiency is limited, especially on a large scale, due to its dependence on climatic conditions and vulnerability to contamination by microorganisms and foreign materials (Silva et al., 2021).

The artificial drying method involves the use of equipment that controls operational conditions, such as temperature, humidity, airflow and equipment power. These methods are based on conduction, convection, irradiation, vacuum drying and freeze-drying (Cobos and Díaz, 2015; Mediani et al., 2022). Among the main benefits of artificial drying methods are greater control over drying conditions, preservation of flavour and nutrients, a decrease in drying time and an increase in food products' shelf life (Calín-Sánchez et al., 2020).

Vega-Mercado, Góngora-Nieto and Barbosa-Cánovas (2001) classified artificial drying technologies into four generations, as shown in Figure 20.2.

First-generation drying technology consists of a cabinet-type system (e.g. a forced-air circulation stove), in which air circulation removes moisture from food. The second generation is based on spraying the plant in droplets, which then form dry particles; this process is extremely fast due to an increase in the surface area to be dried (Cobos and Díaz, 2015). The drying technologies of the

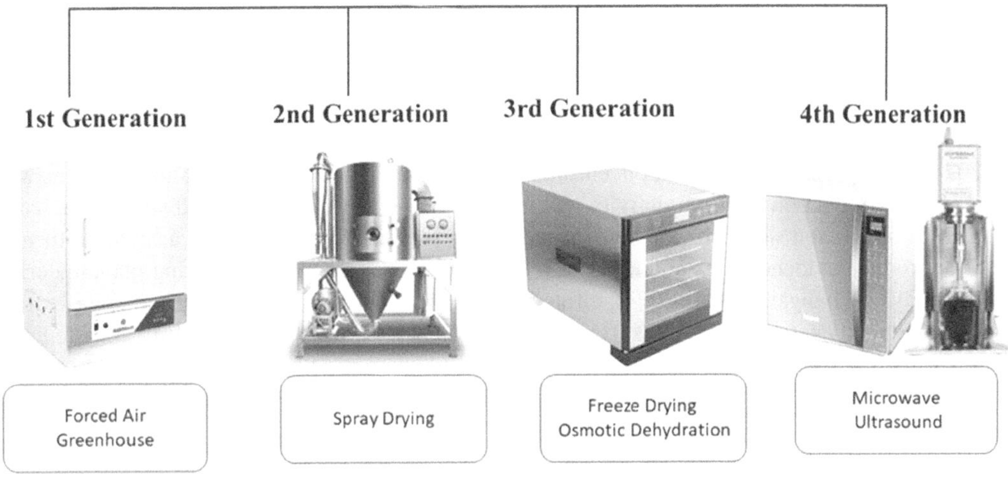

FIGURE 20.2 Artificial Drying Technologies.

third generation, according to Vega-Mercado (2001), are subdivided into two groups: lyophilisation and osmotic dehydration.

Lyophilisation involves simultaneous heating and cooling of the plant, resulting in better durability against microorganisms (Mediani et al., 2022), while osmotic dehydration is based on immersing plant material in a hypertonic solution to remove part of the water from plant tissues. Osmotic dehydration has the advantage of maintaining food organoleptic properties due to the low-temperature process, preventing enzymatic browning reactions and reducing colour loss (Wang et al., 2025).

The fourth generation includes the use of microwave and ultrasound irradiation. The drying process using microwave irradiation directs almost all the equipment's energy towards food drying, uniform drying, less drying time, better final product quality, colour conservation and better control of drying parameters (Vega-Mercado, Góngora-Nieto and Barbosa-Cánovas, 2001; Radojčin et al., 2021; Teleken et al., 2021). On the other hand, drying by ultrasound irradiation enhances food quality and increases shelf life, as this process results in the destruction of microorganisms, enzyme inactivation and the facilitation of the extraction of bioactive compounds (Alvarenga et al., 2021).

However, the selection of the proper drying method depends on numerous factors, such as the nature and freshness of the food, and appropriate conditions (temperature, drying time) (Arslan and Alibas, 2024).

Regarding the drying of *M. oleifera*, the literature describes the drying process of this plant species using all of these techniques. Khaled et al. (2024) investigated the drying process of *M. oleifera* leaves using shaded drying on a forced-air stove (at temperatures of 50, 60 and 70°C). Khaled et al. (2024) observed that both drying methods significantly affected the quality of the flours obtained. Shaded drying resulted in a greater loss of biologically active compounds, as indicated by reduced levels of vitamin C and phenolic compounds. In contrast, stove drying showed more efficiency, preserving higher amounts of protein, fibre and ash. Additionally, they highlighted that the stove temperature directly influenced the product's characteristics, with higher temperatures leading to a higher concentration of nutrients and better compaction of the flour.

Macambira et al. (2018) verified the production of *M. oleifera* leaf flour for broiler chicken feed and found that using a forced-air circulation stove (at 65°C for 72 hours) provided significant benefits, such as a higher concentration of macro- and micronutrients, facilitating its use for animal feed.

Pandidurai et al. (2022) researched the drying process of *M. oleifera* flowers using lyophilisation. The authors emphasise that the product obtained at –55°C for 18 hours presented an excellent concentration of vitamin C, preserving the flower's original characteristics, low hygroscopicity and high antioxidant activity, thereby enhancing and making this method viable.

Dantas et al. (2024) conducted a study testing *M. oleifera* leaves and seeds in a drying process using microwave irradiation. The results indicated that this drying method provided greater preservation of the physical and chemical characteristics of the biological materials analysed. Dantas et al. (2024) highlighted a significant concentration of nutrients, especially proteins and fibres, as well as an efficient reduction in water activity and moisture content. Additionally, advantages in using microwave irradiation were proven, since it requires a shorter drying time and a higher return of the final product, evidencing its efficiency as a technological alternative for treating plant material.

In a study by Thangaiah et al. (2024), ultrasound irradiation was used in drying and in the extraction of biologically active compounds from *M. oleifera* leaves. This technique significantly reduced the drying time, preserved heat-sensitive compounds and increased the extraction yield. Other benefits associated with ultrasounds included the lowest solvent and energy costs, as well as the quality and stability of biologically active compounds.

20.4 STORAGE OF *M. OLEIFERA*

In general, the storage process involves a set of practices applied to plant material after harvesting and drying, with the objective of maintaining its physicochemical stability and protecting it from factors that could cause deterioration, oxidation, humidity and the proliferation of microorganisms. Storage efficiency relies on the parameters of environmental control, the choice of suitable packaging and continuous quality monitoring (Ariwaodo, 2022).

Proper storage of plant material depends on the fact that *M. oleifera* presents a high concentration of biologically active compounds (e.g. vitamins, proteins and phenolics), which makes its products sensitive to adverse conditions. Studies have shown that factors such as exposure to light, oxygen and humidity can lead to the degradation of these biologically active compounds, significantly reducing the nutritional value and *M. oleifera* functionality. For example, *M. oleifera* leaf powder requires strict storage conditions to preserve its quality and extend its shelf life (Bezerra et al., 2004; Saeed, Ali and Ramzan, 2021).

One important factor related to efficient storage is the use of appropriate packaging that protects the material from light, humidity and oxygen. Opaque and hermetic packaging, such as metallised bags or high-density plastic containers, is recommended for storing chalky products. These packages create effective physical barriers, reducing oxidation and microorganism contamination. However, it is recommended that seeds be stored in fabric bags or ventilated containers to prevent moisture accumulation, which can compromise their viability for planting or oil extraction (Ahmad et al., 2021).

Environmental control plays an important role in the storage of *M. oleifera*-derived products, in which the ideal temperature should range between 10°C and 25°C, while relative air humidity must not exceed 60%. Environments with climate control, such as cold rooms or dehumidified rooms, prevent changes in the physicochemical compositions of the plant material. In tropical regions, this control helps to maintain stable temperature and moisture levels, which are naturally high and conducive to microorganism proliferation (Cruz et al., 2012).

The choice of storage method should also consider the final use of the plant material. For example, if leaves are intended for tea or supplements production, long-term storage requires greater care of temperature and moisture levels, while seeds meant for propagation or oil extraction can tolerate slightly less rigorous conditions. Additionally, products with a higher added value, such as extracts or capsules, require more controlled storage conditions due to the sensitivity of their biologically active compounds (Khan, Sajad and Narayan, 2017; Wang et al., 2025).

Regular monitoring of storage conditions is essential to ensure the quality of the plant material. Periodic inspections help detect mould formation, the presence of pests or changes in colour and material odour. Furthermore, the adoption of temperature and moisture regulators ensures that products of plant origin are stored within ideal limits, minimising losses and maximising economic utilisation (Wiethölter and Fassina, 2017; Simensato and Silvia, 2019).

In addition to preserving quality, proper storage of *M. oleifera* has direct implications for food safety and market acceptance. Products that meet high international quality standards are more competitive, especially in strict markets, such as those from North America and Europe.

During the last few years, trade involving *M. oleifera* products has been growing considerably. In 2018, the global moringa market was valued at around US$5.5 billion, with steady growth expected in the following years (Grosshagauer et al., 2021; Islam et al., 2021). Grosshagauer et al. (2021) emphasised that the growing development of new products and the popularity of this plant species on a global scale highlight the need for efficient and safe control at all stages of the *M. oleifera* production process.

Moreover, on a global scene, the demand for moringa powder leaves is rising, driven by the expanding market for diuretics and food products (Islam et al., 2021). This trend reinforces the necessity of product traceability, a requirement demanded by consumers and regulators, highlighting the importance of this aspect in the production chain.

Peraza-Juan et al. (2022) observed the nutraceutical properties of protein-derived peptides contained in *M. oleifera* seeds. Although the primary focus of that study was to assess the chemical properties of proteins, the authors highlighted that proper storage plays a crucial role in this process, as proteins in inappropriate storage conditions can easily degrade.

Thus, all researchers agreed that efficient storage is achieved when the humidity and water activity of plant-derived products are controlled. The lack of control over these parameters can accelerate microorganism growth in chalky products, thereby increasing enzymatic activity (Lancelot et al., 2021). Optimal storage conditions help reduce the activities of microorganisms over time (Pacheco and Mendes, 2021).

It was noted that there are practically no studies investigating the stability of *M. oleifera*. The few available works focus on the stability of isolated compounds extracted from the plant or the shelf life of products supplemented or fortified with *M. oleifera* (Saeed, Ali and Ramzan, 2021; Peraza-Juan et al., 2022). This limitation highlights a meaningful gap in scientific knowledge about *M. oleifera*, emphasising the need for more studies on this topic.

REFERENCES

Abdelwanis, F. M., Abdelaty, H. S., & Saleh, S. A. 2024. Exploring the multifaceted uses of *Moringa oleifera*: Nutritional, industrial and agricultural innovations in Egypt. *Discover Food* 4(1): 1–19.

Abdelwanis, F. M., Hosni, A. M., Abdelhamid, A. N., Sulman, A. A., Ezoo, M. I., & Saleh, S. A. 2022. Effect of Zinc and Boron foliar application on leaf chemical composition of *Moringa oleifera* and on yield and characters of its seed oil. *Egyptian Journal of Chemistry* 65(12): 87–93.

Abdoun, K., Alsagan, A., Altahir, O., Suliman, G., Al-Haidary, A., & Alsaiady, M. 2022. Cultivation and uses of *Moringa oleifera* as non-conventional feed stuff in livestock production: A review. *Life* 13(1): 63.

Ahmad, S. S., Yousuf, O., Islam, R. U., & Younis, K. 2021. Silver nanoparticles as an active packaging ingredient and its toxicity. *Packaging Technology and Science* 34(11–12): 653–663.

Alakali, J. S., Kucha, C. T., & Rabiu, I. A. 2015. Effect of drying temperature on the nutritional quality of *Moringa oleifera* leaves. *African Journal of Food Science* 9(7): 395–399.

Alvarenga, P. D. L., Cavatti, L. S., Valiati, B. S. et al. 2021. Aplicação do ultrassom no processamento de frutas e hortaliças. *Brazilian Journal of Food Technology* 24: e2020274.

Ariwaodo, C. A. 2022. Handling strategies and facilities for horticultural crops. *Open Access Library Journal* 9(5): 1–29.

Arslan, A., & Alibaş, İ. 2024. Assessing the effects of different drying methods and minimal processing on the sustainability of the organic food quality. *Innovative Food Science & Emerging Technologies* 94: 103681.

Atta-Krah, K., Chotte, J.L., Gascuel, C., Gitz, V., Hainzelin, E. et al. 2021. Agroecological transformation for sustainable food systems: insight on France-CGIAR research Rep. 26 Agropolis Int. Montpellier, France.

Bezerra, A. M. E., Medeiros Filho, S., Freitas, J. B. S., & Teófilo, E. M. 2004. Evaluation of quality of the drumstick seeds during the storage. *Ciencia e Agrotecnologia* 28: 1240–1246.

Calín-Sánchez, Á., Lipan, L., Cano-Lamadrid, M. et al. 2020. Comparison of traditional and novel drying techniques and its effect on quality of fruits, vegetables and aromatic herbs. *Foods* 9(9): 1261.

Chanthanousone, H., Phan, T. T., Nguyen, C. Q., Nguyen, T. D. T., Pham, H. T. T., & Truong, H. T. H. 2023. Influence of biofertilizer produced using drumstick (*Moringa oleifera* L.) unused parts on the growth performance of two leafy vegetables. *Journal of Experimental Biology and Agricultural Sciences* 11(2): 280–289.

Chilukamarri, L., Ashrafzadeh, S., & Leung, D. W. 2021. In-vitro grafting–Current applications and future prospects. *Scientia Horticulturae* 280: 109899.

Cobos, A., & Díaz, O. 2015. Chemical composition of meat and meat products. *Handbook of Food Chemistry* 1: 471–510.

Cruz Silva, P. C., Andrade, L. A., Souza, V. C., Fabricante, J. R., & Silva, M. L. M. 2012. Comportamento germinativo de sementes de *moringa oleifera* L. em diferentes ambientes e tempos de armazenamento. *Agropecuária Científica no Semiarido* 8(1): 1–6.

Dantas, D. L., da Costa Alves, M., Dantas, G. M. S. et al. 2024. Supplementation with *Moringa oleifera* Lam leaf and seed flour during the pregnancy and lactation period of Wistar rats: Maternal evaluation of initial and adult neurobehavioral development of the rat progeny. *Journal of Ethnopharmacology* 325: 117904.

Drisya Ravi, R. S., Nair, B. R., & Siril, E. A. 2021. Morphological diversity, phenotypic and genotypic variance and heritability estimates in *Moringa oleifera* Lam: A less used vegetable with substantial nutritional value. *Genetic Resources and Crop Evolution* 68(8): 3241–3256.

EMBRAPA. 2021. *Boas Práticas na Colheita e Pós-Colheita de Hortaliças Folhosas*. Brasília: Embrapa. Available at: https://www.infoteca.cnptia.embrapa.br/bitstream/doc/1133593/1/MMLana-Doc-190-BPA-Folhosas.pdf. Accessed January 27, 2025.

Empresa Brasileira de Pesquisa Agropecuária (EMBRAPA). 2015. *Postharvest Pathology: Challenges and Solutions*. Brasília: Embrapa. Available at: https://www.infoteca.cnptia.embrapa.br.

Gandji, K., Tovissodé, F. C., Azihou, A. F. et al. 2020. Morphological diversity of the agroforestry species *Moringa oleifera* Lam. as related to ecological conditions and farmers management practices in Benin (West Africa). *South African Journal of Botany* 129: 412–422.

Ghimire, S., Subedi, L., Acharya, N., & Gaire, B. P. 2021. *Moringa oleifera*: A tree of life as a promising medicinal plant for neurodegenerative diseases. *Journal of Agricultural and Food Chemistry* 69(48): 14358–14371.

Grosshagauer, S., Pirkwieser, P., Kraemer, K., & Somoza, V. 2021. The future of moringa foods: A food chemistry perspective. *Frontiers in Nutrition* 8: 751076.

Haouvang, L. C., Albert, N., Martin, Y., & Mbaiguinam, M. 2017. Growth response of *Moringa oleifera* Lam. as affected by various amounts of compost under greenhouse conditions. *Annals of Agricultural Sciences* 62(2): 221–226.

Hodas, F., Rosa, M. T. Z., & Milani, P. G. 2021. *Moringa oleifera* potential as a functional food and a natural food additive: A biochemical approach, *Anais da Academia Brasileira de Ciências* 93: 1–15.

Igual, M., Purificación, G. S., & Javier, M. M. 2020. Effect of Acheta domesticus (house cricket) addition on protein content, colour, texture, and extrusion parameters of extruded products. *Journal of Food Engineering* 282: 110032.

Islam, Z., Islam, S. R., Hossen, F., Mahtab-ul-Islam, K., Hasan, M. R., & Karim, R. 2021. *Moringa oleifera* is a prominent source of nutrients with potential health benefits. *International Journal of Food Science* 1: 6627265.

Khaled, B. M., Das, A. K., Alam, S. S. et al. 2024. Effect of different drying techniques on the physicochemical and nutritional properties of *Moringa oleifera* leaves powder and their application in bakery product. *Applied Food Research* 4(2): 100599.

Khan, F. A., Sajad, A. B., & Narayan, S. 2017. Storage methods for fruits and vegetables. *Sher-e-Kashmir University of Agricultural Sciences and Technology of Kashmir* 1: 1–9.

Khan, M. A., Shakoor, S., & Ameer, K. et al. 2023. Effects of dehydrated moringa (*Moringa oleifera*) leaf powder supplementation on physicochemical, antioxidant, mineral, and sensory properties of whole wheat flour leavened bread. *Journal of Food Quality* 1: 1–13.

Kilic, M., Sahin, M., Hassan, A., & Ullah, A. 2024. Preservation of fruits through drying - A comprehensive review of experiments and modeling approaches. *Journal of Food Process Engineering* 47(3): e14568.

Krah, C. Y., & Krisnadi, D. 2022. Effect of harvest and postharvest handling on quality of moringa leaf powder. *IOP Conference Series: Earth and Environmental Science* 1038(1): 012074.

Kumar, Y., Thakur, T. K., Sahu, M. L., & Thakur, A. 2017. A multifunctional wonder tree: *Moringa oleifera* Lam open new dimensions in field of agroforestry in India. *International Journal of Current Microbiology and Applied Sciences* 6(8): 229–235.

Lancelot, E., Fontaine, J., Grua-Priol, J., & Le-Bail, A. 2021. Effect of long-term storage conditions on wheat flour and bread baking properties. *Food Chemistry* 346: 128902.

Liang, K., Cao, J., Wang, J., Cao, F., & Su, E. 2025. A comprehensive review of *Moringa oleifera* leaves and seeds with a focus on their food-related aspects. *Journal of Food Composition and Analysis* 140: 107247.

Macambira, G. M., Rabello, C. B. V., Navarro, M. I. V. et al. 2018. Caracterização nutricional das folhas de *Moringa oleifera* (MOL) para frangos de corte. *Arquivo Brasileiro de Medicina Veterinária e Zootecnia* 70: 570–578.

Mediani, A., Hamezah, H. S., Jam, F. A. et al. 2022. A comprehensive review of drying meat products and the associated effects and changes. *Frontiers in Nutrition* 9: 1057366.

Milião, G. L., de Oliveira, A. P., de Souza Soares, L. et al. 2022. Unconventional food plants: Nutritional aspects and perspectives for industrial applications. *Future Foods* 5: 100124

Motis, T. N., Longfellow, J. M., Jani, A. D., Lingbeek, B. J., D'Aiuto, C. J., & Bergen, J. C. J. 2017. Productivity of *Moringa oleifera* augmented with intercropped tropical legumes. *Acta Hortic* 1158: 85–96.

Noguera-Talavera, A., Reyes-Sánchez, N., & Mendieta-Araica, B. 2017. Diversidad y distribución de la macrofauna edáfica en dos sistemas de manejo de *Moringa oleífera* (Lam.): Relación con las propiedades del suelo. *La Calera* 17(29): 78–86.

Núñez-Gastélum, J. A., Arguijo-Sustaita, A. A., López-Díaz, J. A., Díaz-Sánchez, Á. G., Hernández-Peña, C. C., & Cota-Ruiz, K. 2023. Seed germination and sprouts production of *Moringa oleifera*: A potential functional food?. *Journal of the Saudi Society of Agricultural Sciences* 22(4): 223–230.

Pacheco, T. F., & Mendes, T. D. 2021. *Guia Prático para Caracterização de Enzimas*. Brasília, DF: Embrapa Agroenergia. https://www.infoteca.cnptia.embrapa.br/infoteca/bitstream/doc/1131161/1/Guia-pra769tico-para-a-caracterizac807a771o-de-enzimas.pdf.

Padayachee, B., & Baijnath, H. J. S. A. J. O. B. 2020. An updated comprehensive review of the medicinal, phytochemical and pharmacological properties of *Moringa oleifera*. *South African Journal of Botany* 129: 304–316.

Pandidurai, G., Amutha, S., Kanchana, S., Vellaikumar, S., & Prabhakaran, K. 2022. Optimization of freeze drying parameters for moringa (*Moringa oleifera*) flower powder by using response surface methodology and principal component analysis. *Journal of Horticultural Sciences* 17(2): 388–396.

Pareek, A., Pant, M., Gupta, M. M. et al. 2023. *Moringa oleifera*: An updated comprehensive review of its pharmacological activities, ethnomedicinal, phytopharmaceutical formulation, clinical, phytochemical, and toxicological aspects. *International Journal of Molecular Sciences* 24(3): 2098.

Peraza-Juan, M., Espinoza-Sánchez, E. A., Lara-Reyes, J. A. et al. 2022. Characterization of seed storage protein of *Moringa oleifera* and analyses of their bioactive peptide's nutraceutical properties. *Acta Scientiarum Polonorum Technologia Alimentaria* 21(3): 311–320.

Pina, J. C., Oliveira, A. K. M. D., Matias, R., & Silva, F. D. 2018. Influência de diferentes substratos na produção de fitoconstituintes de *Moringa oleifera* Lam. cultivada a pleno sol. *Ciência Florestal* 28(3): 1076–1087.

Prajapati, C., Ankola, M., Upadhyay, T. K. et al. 2022. *Moringa oleifera*: Miracle plant with a plethora of medicinal, therapeutic, and economic importance. *Horticulturae* 8(6): 492

Queiroga, V., & Figueiredo, N. A. A. 2022. *Moringa oleifera* Lam*: Sistemas de produção e utilização*. 1ª ed. Brasília. Embrapa.

Radojčin, M., Pavkov, I., Bursać Kovačević, D. et al. 2021. Effect of selected drying methods and emerging drying intensification technologies on the quality of dried fruit: A review. *Processes* 9(1): 132.

Rodrigues, L. A., Muniz, T. A., Samarão, S. S., & Cyrino, A. E. 2016. Qualidade de mudas de *Moringa oleifera* Lam. cultivadas em substratos com fibra de coco verde e compostos orgânicos. *Revista Ceres* 63(4): 545–552.

Saeed, M., Ali, S. W., & Ramzan, S. 2021. Physicochemical analysis of mango flavored yogurt supplemented with *moringa oleifera* leaf powder. *Journal of Food science and Technology* 58(12): 4805–4814.

Saucedo-Pompa, S., Torres-Castillo, J. A., Castro-López, C. et al. 2018. Moringa plants: Bioactive compounds and promising applications in food products. *Food Research International* 111: 438–450.

Sibanda, F., Jideani, V. A., & Obilana, A. O. 2024. Nutritional, biochemical, and functional properties of pearl millet and *moringa oleifera* leaf powder composite meal powders. *Foods* 13(5): 743.

Silva, A., Silva, E., Rigo, L. et al. 2021. Técnicas De Secagem De Frutas: Uma revisão. *Scientific Electronic Archives* 14(10).

Simensato, L. A. & Silvia, M. B. 2019. Importância da Qualidade da Água na Indústria de Alimentos. *Revista do Curso de Engenharia de Alimentos - UNILAGO* 1(1): 1–9.
Teleken, J., Porciúncula, B. D. A., Teleken, J. G., & Carciofi, B. A. M. 2021. Drying of foods under intermittent supply of microwave energy: proposal for a mathematical model. *Acta Scientiarum - Technology* 43: 1–12.
Thangaiah, A., Gunalan, S., Kulandaivelu Rathnasamy, V. et al. 2024. Optimization of ultrasound-assisted phytomolecules extraction from moringa leaves (*Moringa oleifera* Lam) using response surface methodology. *Cogent Food & Agriculture* 10(1): 2309834.
Universidade Federal do Estado do Rio de Janeiro (UNIRIO). *Higienização de frutas, legumes e verduras.* Boletim nº 14, junho de 2021. Available at: https://www.unirio.br/prae/nutricao-prae-1/quarentena/carregamento-boletins-setan-2021/boletim-no-14-2021#:~:text=Assuntos%20Estudantis%2FUNIRIO.-,Boletim%20n%C2%BA,Junho%2F2021.&text=Manter%20sob%20refrigera%C3%A7%C3%A3o%20at%C3%A9%20a%20hora%20de%20servir.&text=FOLHAS%3A%20d%C3%AA%20aten%C3%A7%C3%A3o%20especial%20as,terra%20e%20pequenos%20in%2D%20setos.
Vega-Mercado, H., Góngora-Nieto, M. M., & Barbosa-Cánovas, G. V. 2001. Advances in dehydration of foods. *Journal of Food Engineering* 49(4): 271–289.
Viviana, P., Sumathi, T., Pugalendhi, L., & Djanaguiraman, M. 2022. Standardization of optimum stage for grafting in Moringa [*Moringa oleifera* Lam. (Moringaceae)]. *International Journal of Environment and Climate Change* 12(11): 1705–1710.
Wang, Y., Wu, Y., Chen, Z., Zhong, B., & Liu, B. 2025. Intelligent food packaging materials: Principles, types, applications, and hydrophobization. *Food Control* 171: 111138.
Wiethölter, M. J., & Fassina, P. 2017. Temperaturas de armazenamento e distribuição dos alimentos. *Segurança Alimentar e Nutricional* 24(1): 17–25.

21 *Moringa oleifera*
Bridging Nutrition and Agriculture

Shraddha R. Parmar and Shraddha B. Bhatt

21.1 INTRODUCTION

Native to the Indian subcontinent, *Moringa oleifera* Lam. is an adaptable tree that belongs to the Moringaceae family. According to Razis et al. (2014), it is frequently referred to by a number of names in different places, such as horseradish tree, mulangay, mlonge, benzolive, drumstick tree, sajna, kelor, saijihan and marango. A member of the Moringaceae family, *M. oleifera* is a fast-growing, drought-resistant tree. Because its immature seed pods and green leaves can be used as both food and medicine, it is commonly grown. It is regarded as an excellent supplement due to its high protein content.

India produces and exports the majority of *M. oleifera* in the form of fresh fruits, seeds, oil, canned vegetables and leaf powder. India produces between 1.1 and 1.3 million tonnes of tender pods annually.

21.2 WHY IS MORINGA KNOWN AS THE MIRACLE TREE?

The moringa tree is well known for its rapid growth and ease of cultivation from cuttings or seeds. Surprisingly, it can blossom quickly and can grow even in poor soil. It is inherently drought-resistant and extremely sustainable, particularly in hot, arid conditions. Its edible leaves, fruits, blossoms and immature pods – all of which have long been a staple of many tropical and subtropical regions' diets – are what really set moringa apart. Because its leaves, pods and seeds contain a variety of vital compounds, moringa is a nutritional powerhouse. Moringa has outstanding nutritional value: it has seven times the vitamin C of oranges, ten times the vitamin A of carrots, 17 times the calcium of milk, nine times the protein of yoghurt, 15 times the potassium of bananas and 25 times the iron of spinach. Its tiny leaves are particularly nutrient-dense; they provide more calcium, iron and vitamin A than spinach, carrots and milk, respectively, and more protein than eggs.

Moringa has potential use in cosmetics and pharmaceuticals in addition to its nutritional benefits. Its seeds yield an oil that is good for skin and hair care. The seeds themselves have antimicrobial properties and are rich in vitamins and minerals. They serve as natural water filters. Moringa seeds may help lower blood pressure, blood sugar, inflammation and oxidative stress, according to research. Moringa is frequently considered a superfood due to its high nutritional content, making it particularly beneficial for those who are poor and suffering from malnutrition.

Pakistan, India, Nepal, Afghanistan, Bangladesh, Sri Lanka, Southeast and West Asia, the Arabian Peninsula, East and West Africa, the West Indies, southern Florida and Central and South America – from Mexico to Peru, including Brazil and Paraguay – are just a few of the places in the world where moringa is widely grown. Usually found in tropical and subtropical areas, this tree can grow up to 1200 m above sea level. Although the ideal temperature range for growth is between 25°C and 35°C, it can withstand high temperatures of up to 48°C (Ebert & Palada, 2017). The fast-growing perennial moringa is known as the 'miracle tree' because of its exceptional nutritional and medicinal qualities (Daba, 2016). It grows well across a range of soil types, except wet soils, since standing water can lead to root rot. According to Nouman et al. (2014), sandy loam or slightly

DOI: 10.1201/9781003617860-25

alkaline clay soils with adequate drainage provide the optimum development. According to Foidl et al. (2001), moringa trees can attain heights of 7–12 m and trunk diameters of 20–40 cm.

Every component of the moringa plant, including the stem, roots, leaves and flowers, has special advantages and applications. Moringa is eaten as a vegetable in several countries because it is thought to help cool the body in hot, dry weather. With more protein than milk and eggs, more calcium than milk, more iron than spinach, more vitamin C than oranges and more vitamin A than carrots, the leaves are very nutrient-dense (Awad et al., 2014). In addition to being used to make carpets and ropes, the trunk's bark is useful for adsorbing heavy metals (Orestes et al., 2013). According to scientific research, extracts from moringa leaves contain several advantageous qualities, such as antihypertensive, antifungal and anticancer activities. Because of their high calorific content, moringa seeds are prized for their potential as a source of biodiesel and other renewable energy (Granella et al., 2021). The seeds can also be used to make curry powders, culinary oils and wastewater treatment because of their potent coagulating and antibacterial qualities. Ben oil, a transparent, odourless, non-drying oil that makes up roughly 38–40% of the seed content, is one of the main products made from moringa seeds. This oil is highly valued for lubricating delicate objects, such as timepieces, and it never becomes rancid. Ben oil is edible, appetising and occasionally used as a salad dressing or as an alternative to more costly olive oil because it tastes and is chemically similar to olive oil. Additionally, it is highly valued in the fragrance and cosmetics sectors (Seifu & Teketay, 2020). It has long been thought that the roots of the moringa plant have antispasmodic and antibacterial qualities. They are used to treat ailments including gastritis and asthma, and in certain impoverished nations, women also use them as a natural form of birth control.

21.3 PROPAGATION

Cuttings or seeds can be used to propagate *M. oleifera*. It can be directly seeded in the field because of its high germination rate. If you have well-draining soil, you can plant moringa seeds year-round. Cuttings that are at least 4 cm in diameter and 1 m long are advised for vegetative propagation.

Plants are usually spaced 15×15 cm or 20×10 cm apart when growing moringa for intensive leaf output. Wider alleys, every 4 m, are used to make management and harvesting easier. High planting densities, however, may make disease and weed control more difficult.

A 0.5×1 m spacing is frequently used for semi-intensive growing because it provides a fair mix between production and ease of care.

Additionally, moringa trees make excellent natural fences. They can be interplanted with other species or mixed alongside alleys. Moringa rows in agroforestry systems are often spaced 2–4 m to make room for other crops and techniques.

21.4 HARVESTING

Particularly in high-density agriculture, moringa trees can be harvested when they are 1.5–2 m tall. The leaves can be collected by gently snapping them off by hand or by cutting the stems with a sharp knife. Harvesting should be done between 20 and 45 cm above the ground to promote continuous development and encourage the production of new shoots (Fredalette, 2018).

When grown for fodder, moringa should be harvested every 75 days, but under normal circumstances, it can be harvested every 35 to 40 days. The leaves should not be piled up after harvesting, as this can cause spoilage, and the early morning is the best time to harvest in order to reduce moisture loss.

If the leaves are intended for fresh sale, they should be sold the same day, as they tend to lose moisture quickly after being harvested. Additionally, moringa can be harvested at 50 cm above ground level, which makes it easier to use mechanical harvesting methods (Mabapa et al. 2017).

21.5 CULTIVATION CONDITIONS

M. oleifera is a small to medium-sized deciduous or evergreen tree commonly found in tropical regions, foothills and some mid-hill areas of Nepal. It can grow up to 25–30 feet (7.5–9 m), with the Siwalik, Terai and mid-hill regions being most suitable for its cultivation (Pokhrel et al. 2016).

The tree grows well at heights between 0 and 1000 m above sea level and does best in direct sunlight. Although moringa may grow in a variety of soil types, it likes sandy or loamy soils that drain well and have a pH of 6.3 to 7.0, which is neutral to slightly acidic. It can withstand up to 3000 mm of rainfall every year, but it needs at least 250 mm. However, because its roots are prone to rot, the plant is susceptible to wet circumstances. It is advised to plant on small hills or mounds in regions with high rainfall to enhance water runoff (Singh, 2017).

Because of its deep taproot, moringa is incredibly drought-resistant and can withstand prolonged dry spells. Although it can withstand temperatures as high as 48°C in shaded regions, the optimal temperature range for growth is 12°C to 40°C. Although freezing temperatures may kill the plant's above-ground components, it may also tolerate light frost. Fortunately, when cut or frozen, it can swiftly regrow from the base or trunk. Stream banks and savanna regions with well-drained soils and a year-round steady water table are common habitats for this species. Even though the plant can withstand drought, persistent water stress causes it to produce noticeably fewer leaves (Leone et al., 2015).

21.6 METHODOLOGY

Secondary data sources were used to gather information for this investigation. A review was conducted to carefully assess and evaluate the body of available literature. Journal articles received extra consideration to guarantee the review's calibre and legitimacy. The main resource for finding pertinent research papers and scholarly materials was Google Scholar.

21.7 RESULTS AND DISCUSSION

Moringa, sometimes known as the 'miracle tree', is prized for its many uses in a variety of industries. By raising stomach serotonin levels, it has been utilised in medicine to treat conditions including gastritis (Posmontier, 2011). Moringa seeds are used in environmental research as a natural water purification coagulant. The turbidity of untreated water can be considerably reduced by using these seeds. However, a fresh aqueous extract usually needs to be prepared daily to maintain efficacy in clarifying water (Garcia-Fayos et al., 2015).

Moringa is becoming increasingly acknowledged in the agricultural industry as a sustainable substitute for synthetic chemicals, which, since the start of the Green Revolution, have seriously harmed human health and soil health. Using moringa as an organic foliar fertiliser is one of its agricultural applications.

21.8 ORGANIC FERTILISER (FOLIAR FERTILISER APPLICATION)

Compared with conventional soil treatment, foliar fertilisation enables faster nutrient uptake through the leaves and can address nutrient deficiencies more quickly (Fageria et al., 2009). Direct application of chelated nutrients to leaves has been demonstrated to increase fertiliser efficiency while reducing the overall amount required (Niu et al., 2021).

Applying 2% water extracts of brassica and moringa as foliar sprays every 15 days will increase canola production by 10%, according to a field trial conducted by Arshad (2014). Considering the growing expense of inorganic fertilisers, this approach proved both economical and successful. In a similar vein, Yasmeen et al. (2012) demonstrated that using moringa leaf extract as a foliar spray to late-planted wheat during booting, heading, jointing and tillering prolonged the leaf area (Seasonal

LAD), slowed down grain filling and postponed maturity. When compared with untreated crops, these impacts helped to improve output by 10%. Furthermore, under field conditions, a single foliar spray at the heading stage increased production by 6.8%.

According to a study by Zaki and Rady (2015), common bean plants can be shielded from the negative effects of salt stress by externally applying moringa leaf extract (MLE). This treatment lessens the detrimental effects of salt on plant growth and output. The best method for increasing bean plants' ability to withstand salt under moderate soil salinity (EC = 6.23–6.28) was to soak the seeds and then spray the leaves with MLE, prepared by combining one part extract paste with 30 parts water by volume.

21.9 LIVESTOCK FEED

Many trees and shrubs are used by farmers worldwide as supplements or animal feed, and because of its high vitamin content, moringa is an especially suitable option. Moringa trees can provide 4.2 to 8.3 tonnes of dry matter per hectare, depending on the season, ecological zone and type of fertiliser. Iron, potassium, calcium and other vital elements are abundant in moringa leaves and are necessary for boosting animal weight gain and milk production (Nouman et al., 2014).

Essential amino acids such as lysine, leucine, histidine, glutamic acid, valine, isoleucine, alanine, phenylalanine and arginine are substantially more abundant in moringa than in other woody plants. On a dry matter basis, moringa leaves contain 22.05 mg/kg of zinc, 318.81 mg/kg of iron, 190 mg/kg of magnesium, 24,700 mg/kg of calcium and 4400 mg/kg of phosphorus. Farmers are aware of the significance of these minerals – calcium, magnesium, phosphorus, iron and zinc – and their concentrations are comparatively higher than those of many other tree leaves (Su & Chen, 2020; Price, 2007).

Moringa is also recognised for its high concentration of polyphenols and lipophilic antioxidants, including tocopherols and carotenoids, which make it a nutritious animal feed (Cohen-Zinder et al., 2017). According to their findings, dairy cow productivity and health can be enhanced by switching to moringa silage from maize silage.

21.10 PLANT PROTECTION

21.10.1 Biopesticide

Biopesticides are natural compounds made from microorganisms, plants, animals and specific minerals that are used to control pest populations in an eco-friendly manner. *Bacillus thuringiensis* (Bt) is the basis for around 90% of the microbial biopesticides now on the market (Damalas & Koutroubas, 2018). Biopesticides are becoming increasingly popular as a sustainable substitute for chemical pesticides due to their detrimental effects on the environment.

Moringa leaf extract dramatically decreased aphid infestations during all important growth stages – booting, milking and heading – according to field tests conducted on wheat by Ali et al. (2015). Similarly, Rhamnosyloxy-benzyl isothiocyanate, a component in moringa, showed harmful effects against adult female mites and mealybugs, according to research by El-Masry et al. (2017).

Subsequent experiments revealed that, in comparison with control groups, higher concentrations (300–400 mg) of *M. oleifera* seed methanol extract (MOSME) and leaf methanol extract (MOLME) significantly decreased the ability of stored grain pests, such as the rice weevil (*Sitophilus oryzae*) and the cowpea weevil (*Callosobruchus maculatus*), to lay eggs. Outperforming MOLME, stem bark extract (MOSBME), and their corresponding controls, MOSME had the strongest impact on lowering these pests' foraging behaviour among the studied extracts. El-Masry et al. (2017) suggested MOSME for controlling *S. oryzae* and *C. maculatus* in stored grains in light of these findings.

21.10.2 Crop Disease Management

Moringa is well known for its ability to control a variety of crop diseases. Its antifungal and antibacterial qualities have been shown in numerous studies, underscoring its usefulness in the management of crop diseases. Dania and Thomas (2019) assessed how well moringa extracts shielded sweet potatoes from fungi that cause rot in both in vitro and in vivo settings. Applying a 75% w/v moringa extract decreased the severity of rot in infected tubers by 41.4% to 52.0%, according to the in vivo results. According to the study's findings, moringa extracts have a strong potential to prevent post-harvest rot caused by fungal pathogens in sweet potatoes. According to Adekanmi et al.'s (2020) phytochemical investigation, moringa leaves have high concentrations of tannins, alkaloids and phenolics, all of which support the plant's antibacterial properties.

In a different study, El-Mohamedy and Abdalla (2014) examined the antifungal qualities of extracts from moringa roots, leaves and pod husks against seven common phytopathogenic fungi, including *Macrophomina phaseolina*, *Rhizoctonia solani*, *Alternaria alternata*, *Fusarium oxysporum* and *Fusarium solani*. Strong inhibitory effects were demonstrated by all extracts, which markedly decreased dry mycelial mass, spore germination and fungal radial growth.

Flavonoids, alkaloids, tannins and saponins are among the bioactive substances that give moringa its antibacterial properties. Because of these chemicals, moringa is a promising natural antibacterial agent that may be used in the pharmaceutical industry to combat harmful microorganisms (Abalaka et al., 2012).

21.10.3 Plant Growth Hormones

MLE has been demonstrated to dramatically increase crop yields and plant growth. A study by Mvumi et al. (2013) found that using MLE increased the height and root development of maize and common bean plants. Zeatin, a cytokinin hormone found in the extract, is primarily responsible for these growth advantages. According to the study, MLE should be applied every two weeks until the crops reach maturity, beginning two weeks after germination.

Iqbal et al. (2020) conducted additional research on sunflowers and discovered that moringa extract derived from leaves is superior to extracts derived from roots in terms of promoting growth. Zeatin and other growth-promoting hormones found in moringa leaf extract can increase crop yields by as much as 45%. Additionally, it has well-balanced micronutrients that improve productivity, growth and yield components.

Maishanu et al. (2017), for example, found that regular MLE application in cowpea increased dry matter, plant height and yield. Fresh moringa leaf juice has been shown to boost yields by 25–30% for a variety of crops, including melons, peppers, tomatoes, soybeans, onions, sorghum, tea and coffee (Kutawa et al., 2016).

21.10.4 Green Manure

Enhancing the soil and supplying nutrients for future crops is the primary goal of green manure (GM). Green manures made on-site do not require the handling and transportation costs that other organic inputs often do. The delayed release of nitrogen from decomposing GM residues may be more in line with plant absorption than from sources of inorganic nitrogen, increasing crop yield and N-uptake efficiency while reducing N leaching losses (Cherr et al., 2006).

Moringa plants, which grow quickly, offer a great opportunity for use as a green manure crop, especially for plantation crops such as olive and date palm. The green leaves are suitable for mulching, while the dry, shaded leaves can be used as organic manure (Mridha, 2015).

21.10.5 Agroforestry

Agroforestry based on moringa can raise agricultural output and rural populations' standard of living (Leakey, 2017). Traditional agroforestry has long employed moringa, but its entire range of advantages has not always been recognised. Moringa does not compete with neighbouring crops for water or nutrients because of its deep root structure. It is ideal for intercropping in residential gardens because of its canopy, which also offers filtered shade. Because of this, it is a perfect partner plant for crops that can withstand shade, such as aerial yam, coffee and wild bananas (Tenaye et al., 2009).

There are financial and environmental benefits to incorporating moringa into agroforestry systems. Increasing revenue and fostering more resilient agricultural landscapes help smallholder farmers. Monoculture farming is less economical than agroforestry systems that combine moringa with other crops, according to studies (Devkota et al., 2020). Furthermore, studies by Ndaginna and Ozobia (2017) showed that using moringa as an alley crop produced favourable results, and they strongly advised its inclusion in farming systems.

21.10.6 Wind Breaks and Live Fences

Barriers called windbreaks are made to slow and reroute wind. In addition to fences, grasses and different annual or perennial crops, these can be constructed from trees and plants. Windbreaks alter the local microclimate in the protected area by reducing wind speeds (Brandle & Finch, 1991). *Moringa stenopetala* is a bushy, thick plant that can be used as a windbreak (Price, 2007).

Moringa trees are frequently utilised as living fence posts around gardens throughout Africa (Bancessi et al., 2019). When used in a fencing system, moringa trees' wood is flexible, making it simple to attach wires or other materials (Price, 2007).

21.11 BENEFITS OF *M. OLEIFERA*

21.11.1 Nutritional Properties

Because of its many nutritional advantages and therapeutic qualities, *M. oleifera* is frequently referred to as the 'miracle tree'. The plant is rich in important nutrients in almost every part. Essential minerals like calcium, potassium, zinc, magnesium, iron and copper are abundant in the leaves. Additionally, they are low in calories, which makes them appropriate for diets for obese people and for managing weight. Moringa leaves are also high in protein and important minerals, and contain all the essential amino acids (Janick & Paull, 2018).

With about 46.78% fibre and 20.66% protein, the pods are also quite nutrient-dense. They are particularly prized for their contribution to colon cancer prevention and digestive health support (Gopalkrishan et al., 2016).

Vitamin A, vitamin B complex, β-carotene, pyridoxine (B6), nicotinic acid (niacin), vitamin C, vitamin D and vitamin E are among the many vitamins that are abundant in moringa, as well as numerous significant phytochemicals, such as tannins, sterols, terpenoids, flavonoids, saponins, anthraquinones, alkaloids and reducing sugars. It also contains a number of anticancer substances, including glycerol-1-9-octadecanoate, glycosides, isothiocyanates and glucosinolates (Freiberger et al. 1998).

However, moringa leaves also contain certain antinutritional components, such as oligosaccharides and oxalates. Dried moringa leaves can keep up to 87.5% of β-carotene after four months and 50% after three months of storage, according to studies. This suggests that they are suitable for processing and long-term application in a variety of food products (Glover-Amengor & Mensah, 2012).

The following describes how various portions of Moringa oleifera are used:

- Leaves: *M. oleifera* has several uses for its different parts, but its leaves are particularly useful. Rich in vitamins and minerals, they are also well known for their potent antioxidant properties, largely due to the presence of phenolic compounds such as kaempferol and quercetin. Moringa leaves are used to make tea, powder and other medicinal items, or they can be eaten as a vegetable. Furthermore, fresh leaf juice has the potential to function as a natural growth hormone, boosting crop yields by 25–35%. According to a 2009 study from Japan, the rate at which carbon dioxide is absorbed and used by moringa leaves can be up to 20 times that of other types of plants.

- Roots: *M. oleifera* root bark has shown potential in the treatment of gastric ulcers and gastric mucosal lesions. It helps by reducing stomach acidity and increasing the pH of gastric juice, which contributes to its antiulcer and antisecretory properties. These findings suggest that moringa could be a promising natural source for the development of future antiulcer medications.
- Pod husks: *M. oleifera* pod husks contain a variety of bioactive compounds, including alkaloids, flavonoids, tannins, triterpenoids, diterpenoids and cardiac glycosides. Extracts from the pod husks have demonstrated antimicrobial activity, showing effectiveness against Gram-positive bacteria such as *Staphylococcus epidermidis* and *Enterococcus faecalis*, as well as Gram-negative bacteria such as *Klebsiella pneumoniae* and *Salmonella typhimurium*.
- Seeds: strong antioxidants such as vitamin C, β-carotene, α- and γ-tocopherol, β-sitosterol and vitamin A are among the many phytochemicals found in *M. oleifera* seeds. Along with flavonoids and anthocyanins, they also include noteworthy phenolic components, including kaempferol and quercetin. Furthermore, the seeds include a few uncommon bioactive substances, such as isothiocyanates, glucosinolates and alkaloids (Ghazali & Mohammed, 2011). The oil content of mature moringa seeds is especially high; roughly 22–40% of their makeup is crude fat. Monounsaturated fatty acids, particularly oleic acid, make up the majority of this oil. Reduced risk of coronary heart disease has been associated with increased consumption of monounsaturated fats, mainly oleic acid. Furthermore, extracts from *M. oleifera* seeds have demonstrated promise as natural mosquito control agents, especially against *Anopheles stephensi*, a vector of malaria. These seed extracts' phytochemicals show remarkable efficacy in reducing mosquito populations, and they may prove useful in integrated pest management initiatives.

21.12 MEDICINAL PROPERTIES

M. oleifera is widely recognised for its numerous medicinal advantages and its capacity to treat a wide range of ailments. It has been used to treat numerous conditions, including diabetes, heart disease, anaemia, arthritis, respiratory diseases, skin and liver disorders, paralysis, infertility, rheumatism, digestive problems and many more. The National Institute of Health and Family Care in India named it the 'Plant of the Year' in 2008. In several African countries, moringa is also used to treat ascites, pneumonia and toxic stings.

Moringa powder can be used as a natural alternative to iron supplementation, making it a helpful therapy for anaemia. Moringa leaves have also been discovered to have antifungal, antiviral and anti-abortifacient properties. They also function as flocculating agents and stimulants. This incredible tree seems to have countless health advantages.

Apart from the qualities discussed earlier, regular intake of *M. oleifera* helps with key biological activities, for example:

- Antimicrobial and anthelmintic activity
- Antiasthmatic activity

- Anticancer and antitumour activity
- Antidiabetic and wound healing activity
- Cardiac and circulatory stimulant, as well as antidiuretic activity
- Analgesic (pain-relieving) activity
- Antipyretic (fever-reducing) activity
- Hepatoprotective (liver-protecting) activity
- Antispasmodic and antiulcer effects

21.13 CONCLUSION

It is true that the moringa tree is a miracle. It can be used to treat a variety of illnesses and as a nutritional substitute for eating. It offers numerous advantages for raising crop productivity and yield in the agricultural industry. Using moringa can help lessen the negative impacts of chemicals on the environment and human health (ecological balance). Plans and strategies must be developed to fully profit from the moringa tree. Moringa should be promoted more in poor nations like Nepal, as it is cost-effective and should be used exclusively rather than relying on wealthy nations.

Moringa is our chance to combat the damage brought about by the green revolution. Plants that produce biogas and biodiesel should be established to protect the environment and create jobs. The moringa tree and its significance need to be better understood by young people who are interested in agriculture. Moringa should be promoted in a variety of fields. It is necessary to make use of and market this natural gift.

REFERENCES

Abalaka, M. E., Daniyan, S. Y., Oyeleke, S. B., & Adeyemo, S. O. (2012). The antibacterial evaluation of *Moringa oleifera* leaf extracts on selected bacterial pathogens. *Journal of Microbiology Research*, *2*(2), 1–4. https://doi.org/10.5923/j.microbiology.20120202.01

Adekanmi, A. A., Adekanmi, S. A., & Adekanmi, S. O. (2020). Qualitative and quantitative phytochemical constituents of moringa leaf. *International Journal of Engineering and Information Systems*, *4*(5), 10–17. www.ijeais.org

Ali, H., Muhammad, A., & Idrees, A. (2015). Potential of moringa (*Moringa oleifera*: Moringaceae) as plant growthregulator and bio-pesticide against wheat aphids on wheat crop (*Triticum aestivum*; Poaceae). *Article in Journal of Biopesticides*, *8*(2), 120–127. https://www.researchgate.net/publication/289672172 .

Arshad, M. (2014). Screening rhizobacteria for improving the growth, yield, and oil content of canola (Brassica napus L.). *American-Eurasian Journal of Agriculture & Environmental Sciences, 14*(10), 1067–1073. https://doi.org/10.5829/idosi.aejaes.2014.14.10.12429

Awad, S., Ameri, A., Yaaqoub, F., Shaibani, A., Cheruth, A. J., Al-Awad, I., Al-Yafei, M. A. S., Karthishwaran, K., & Kurup, S. S. (2014). Comparative phytochemical analysis of *Moringa oleifera* and *Moringa peregrina*. *Pharmacologyonline*, *3*, 216–221. http:// pharmacologyonline.silae.it

Bancessi, A., Bancessi, Q., Baldé, A., & Catarino, L. (2019). Present and potential uses of *Moringa oleifera* as a multipurpose plant in Guinea-Bissau. *South African Journal of Botany*, *129*, 206–208. http:// doi:10.1016/j.sajb.2019.06.013

Brandle, J. R., & Finch, S. (1991). *How windbreaks work*. https://digitalcommons.unl.edu/ natrespapers

Cherr, C. M., Scholberg, J. M. S., & McSorley, R. (2006). Green manure approaches to crop production: A synthesis. *Agronomy Journal*, *98*(2), 302–319. https://doi.org/10.2134/ AGRONJ2005.0035

Cohen-Zinder, M., Weinberg, Z., Leibovich, H., Chen, Y., Rosen, M., Sagi, G., Orlov, A., Agmon, R., Yishay, M., Miron, J., & Shabtay, A. (2017). Ensiled *Moringa oleifera*: An antioxidant-rich feed that improves dairy cattle performance. *Journal of Agricultural Science*, *155*(7), 1174–1186. https://doi.org/10.1017/ S0021859617000387

Daba, M. (2016). Miracle Tree: A review on multi-purposes of *Moringa oleifera* and its implication for climate change mitigation. *Journal of Earth Science & Climatic Change*, *7*(8). https://doi.org/10.4172/2157-7617 .1000366

Damalas, C. A., & Koutroubas, S. D. (2018). Current status and recent developments in biopesticide use. *Agriculture, 8*(1), 13. https://doi.org/10.3390/ AGRICULTURE8010013
Dania, V. O., & Thomas, A. S. (2019). Bioactivity and phytochemical composition of *Moringa oleifera* in the management of fungal rot disease of orange-fleshed sweet potato. *Nigerian Journal of Mycology, 11*, 26–45.
Devkota, S., & Bhusal, K. K. (2020). *Moringa oleifera*: A miracle multipurpose tree for agroforestry and climate change mitigation from the Himalayas – A review. *Cogent Food and Agriculture, 6*(1) https://doi.org/10.1080/23311932.2020.1805951
Ebert, A. W., & Palada, M. C. (2017). Moringa - A vegetable tree for improved nutrition, health and income of smallholder farmers. *Acta Horticulturae, 1158*, 309–316. https:// doi.org/10.17660/ActaHortic.2017.1158.35
El-Masry, G. N., Saleh, F. M., & Abd El-Mageed, S. A. M. (2017). Moringa plant, *Moringa oleifera* Lam., as a source of biopesticide. *SINAI Journal of Applied Sciences, 6*, 2314– 6079.
El-Mohamedy, R. S. R., & Abdalla, A. M. (2014). Evaluation of antifungal activity of *Moringa oleifera* extracts as natural fungicide against some plant pathogenic fungi in-vitro. *Journal of Agricultural Technology, 10*(4), 963–982. http://www.ijat-aatsea.com
Fageria, N. K., Filho, M. P. B., Moreira, A., & Guimarães, C. M. (2009). Foliar fertilization of crop plants. *Journal of Plant Nutrition, 32*(6), 1044–1064. https://doi. org/10.1080/01904160902872826
Fredalette. (2018). *Moringa production: Production guidelines for the Miracle tree.*
Freiberger, C. E., Vanderjagt, D. J., Pastuszyn, A., Glew, R. S., Mounkaila, G., Millson, M., & Glew, R. H. (1998). Nutrient content of the edible leaves of seven wild plants from Niger. *Plant Foods for Human Nutrition, 53*, 57–69.
Foidl, N., Makkar, H. P. S., & Becker, K. (2001, October 20–November 2). The potential of *Moringa oleifera* for agricultural and industrial uses. In *What development potential for moringa products?*. Dar es Salaam.
Garcia-Fayos, B., Arnal, J. M., Ruiz, V., & Sancho, M. (2015). Use of *Moringa oleifera* in drinking water treatment: Study of storage conditions and performance of the coagulant extract. *Desalination and Water Treatment,* 57(48–49), 23365–23371. http://dx.doi.org /10.1080/19443994.2015.1117820
Ghazali, H. M., & Mohammed, A. S. (2011). Moringa (Moringa oleifera) seed oil: composition, nutritional aspects, and healthattributes. In *Nuts and seeds in health and disease prevention* (pp. 787–793). Academic Press.
Glover-Amengor, M., & Mensah, F. (2012). Nutritional evaluation of Moringa oleifera leaves using three drying methods. *Journal of Research in Biology, 2*(5), 469–473.
Granella, S. J., Bechlin, T. R., Christ, D., Coelho, S. R. M., & Paz, C. H. de O. (2021). An approach to recent applications of *Moringa oleifera* in the agricultural and biofuel industries. *South African Journal of Botany, 137*, 110–116. https://doi.org/10.1016/j. sajb.2020.10.006
Iqbal, J., Irshad, J., Bashir, S., Khan, S., Yousaf, M., & Shah, A. N. (2020). Comparative study of water extracts of Moringa leaves and roots to improve the growth and yield of sunflower. *South African Journal of Botany, 129*, 221–224. https://doi.org/10.1016/J. SAJB.2019.06.032
Janick, J., & Paull, R. E. (2018). *The encyclopedia of fruits and nuts.* Cabi Publishing.
Kutawa, A. B., Musa, S. P. D. D., Bashir, K. A., Waziri, A. F., & Musa, D. D. (2016). *Moringa oleifera*, A potential miracle tree; A review. *Journal of Pharmacy and Biological Sciences, 11*(6), 25–30. https://doi.org/10.9790/3008-1106012530
Leakey, R. R. B. (2017). *Multifunctional agriculture: Achieving sustainable development in Africa.* Academic Press.
Leone, A., Spada, A., Battezzati, A., Schiraldi, A., Aristil, J., & Bertoli, S. (2015). Cultivation, genetic, ethnopharmacology, phytochemistry and pharmacology of Moringa oleifera leaves: An overview. *International Journal of Molecular Sciences, 16*(6), 12791–12835.
Mabapa, M., et al. (2017). Effect of planting density and harvest interval on the leaf yield and quality of *Moringa* under diverse agroecological conditions of northern South Africa. *International Journal of Agronomy, 9,* 103–118..
Maishanu, H. M., Mainasara, M. M., Yahaya, S., & Yunusa, A. (2017). The use of moringa leaves extract as a plant growth hormone on cowpea (*Vigna anguiculata*). *Path of Science, 3*(12). https://doi.org/10.22178/pos.29-4
Mridha, M. A. U. (2015). Prospects of moringa cultivation in Saudi Arabia. *Journal of Applied Environmental Biological Science, 5*(3), 39–46. www.textroad.com

Mvumi, C., Tagwira, F., & Chiteka, A. Z. (2013). Effect of moringa extract on growth and yield of maize and common beans. *Greener Journal of Agricultural Sciences*, *3*(1), 55–62. https://doi.org/10.15580/GJAS.2013.1.111512264

Ndaginna, I., & Ozobia, P. (2017). The performance of soybean using moringa as alley to improve soil productivity in North-central Nigeria. *African Journal of Agricultural Research*, *12*(14), 1182–1189. https://doi.org/10.5897/AJAR2017.12128

Niu, J., Liu, C., Huang, M., Liu, K., & Yan, D. (2021). Effects of foliar fertilization: A review of current status and future perspectives. *Journal of Soil Science and Plant Nutrition*, *21*(1), 104–118. https://doi.org/10.1007/s42729-020-00346-3

Nouman, W., Basra, S. M. A., Siddiqui, M. T., Yasmeen, A., Gull, T., & Alcayde, M. A. C. (2014). Potential of Moringa oleifera L. as livestock fodder crop: A review. *Turkish Journal of Agriculture and Forestry*, *38*(1), 1–14. https://doi.org/10.3906/tar-1211-66

Orestes, C., Medina, M., Martín, C., Martín, G., García, A., Fernández, T., Hernández, E., & Puls, J. (2013). Potential applications of *Moringa oleifera*. A critical review. *Pastos y Forrajes*, *36*(2), 150–158.

Posmontier, B. (2011). The medicinal qualities of Moringa oleifera. *Holistic Nursing Practice, 25*(2), 80–87. https://doi.org/10.1097/HNP.0b013e31820dbb27

Pokhrel, C. P., Timilsina, A., Yadav, R. K. P., & Khanal, R. (2016). *Moringa oleifere: A potential cash crop in Nepal*. International Symposium on Healthy Society & Healthy World. Kathmandu, Nepal, pp. 33–41.

Price, M. L. (2007). The moringa tree. *ECHO Technical Note*, *17391*, 1–19. http://www. ecohonet.org

Razis, A. F. A., Ibrahim, M. D., & Kntayya, S. B. (2014). Health benefits of *Moringa oleifera*. *Asian Pacific Journal of Cancer Prevention*, *15*(20), 8571–8576. https://doi.org/10.7314/APJCP.2014.15.20.8571

Seifu, E., & Teketay, D. (2020). Introduction and expansion of *Moringa oleifera* Lam. in Botswana: Current status and potential for commercialization. *South African Journal of Botany*, *129*, 471–479. https://doi.org/10.1016/j.sajb.2020.01.020

Singh, M. (2017). *The tree of life: Moringa oleifera*. Online Khabar.

Su, B., & Chen, X. (2020). Current status and potential of *Moringa oleifera* leaf as an alternative protein source for animal feeds. *Frontiers in Veterinary Science*, *7*(53). https://doi.org/10.3389/fvets.2020.00053

Tenaye, A., Geta, E., & Hebana, E. (2009). A multipurpose cabbage tree (*Moringa stenopetala*): Production, utilization and marketing in SNNPR, Ethiopia. *Acta Horticulturae*, 115–120. https://doi.org/10.17660/ActaHortic.2009.806.12

Yasmeen, A., Maqsood, S., Basra, A., Ahmad, R., & Wahid, A. (2012). Performance of late sown wheat in response to foliar application of *Moringa oleifera* Lam. leaf extract. *Chilean Journal of Agricultural Research*, *72*(1). https://doi.org/10.4067/S0718-58392012000100015

Zaki, S. N., & Rady, M. (2015). *Moringa oleifera* leaf extract improves growth, physiochemical attributes, antioxidant defense system and yields of salt-stressed *Phaseolus vulgaris* L. plants. *International Journal of ChemTech Research*, *8*(11), 120–134. https://www. researchgate.net/publication/290453500

Index

For Product Safety Concerns and Information please contact our EU representative GPSR@taylorandfrancis.com Taylor & Francis Verlag GmbH, Kaufingerstraße 24, 80331 München, Germany